Statistical Pattern Recognition

Statistical Pattern Recognition

Second Edition

Andrew R. Webb
QinetiQ Ltd., Malvern, UK

JOHN WILEY & SONS, LTD

First edition published by Butterworth Heinemann

Telephone (+44) 1243 779777

Email (for orders and customer service enquiries): cs-books@wiley.co.uk
Visit our Home Page www.wileyeurope.com or www.wiley.com

Reprinted May 2003, September 2004

Other Wiley Editorial Offices

John Wiley & Sons Inc., 111 River Street, Hoboken, NJ 07030, USA

Jossey-Bass, 989 Market Street, San Francisco, CA 94103-1741, USA

Wiley-VCH Verlag GmbH, Boschstr. 12, D-69469 Weinheim, Germany

John Wiley & Sons Australia Ltd, 33 Park Road, Milton, Queensland 4064, Australia

John Wiley & Sons (Asia) Pte Ltd, 2 Clementi Loop #02-01, Jin Xing Distripark, Singapore 129809

John Wiley & Sons (Canada) Ltd, 22 Worcester Road, Etobicoke, Ontario M9W 1L1

Wiley also publishes its books in a variety of electronic formats. Some content that appears in print may not be available in electronic books.

British Library Cataloguing in Publication Data

A catalogue record for this book is available from the British Library

ISBN 0–470–84513–9 (Cloth)
ISBN 0–470–84514–7 (Paper)

Typeset from LaTeX files produced by the author by Laserwords Private Limited, Chennai, India
Printed and bound in Great Britain by Biddles Ltd, King's Lynn, Norfolk
This book is printed on acid-free paper responsibly manufactured from sustainable forestry in which at least two trees are planted for each one used for paper production.

To Rosemary,
Samuel, Miriam, Jacob and Ethan

Contents

Preface

This book provides an introduction to statistical pattern recognition theory and techniques. Most of the material presented is concerned with discrimination and classification and has been drawn from a wide range of literature including that of engineering, statistics, computer science and the social sciences. The book is an attempt to provide a concise volume containing descriptions of many of the most useful of today's pattern processing techniques, including many of the recent advances in nonparametric approaches to discrimination developed in the statistics literature and elsewhere. The techniques are illustrated with examples of real-world applications studies. Pointers are also provided to the diverse literature base where further details on applications, comparative studies and theoretical developments may be obtained.

Statistical pattern recognition is a very active area of research. Many advances over recent years have been due to the increased computational power available, enabling some techniques to have much wider applicability. Most of the chapters in this book have concluding sections that describe, albeit briefly, the wide range of practical applications that have been addressed and further developments of theoretical techniques.

Thus, the book is aimed at practitioners in the 'field' of pattern recognition (if such a multidisciplinary collection of techniques can be termed a field) as well as researchers in the area. Also, some of this material has been presented as part of a graduate course on information technology. A prerequisite is a knowledge of basic probability theory and linear algebra, together with basic knowledge of mathematical methods (the use of Lagrange multipliers to solve problems with equality and inequality constraints, for example). Some basic material is presented as appendices. The exercises at the ends of the chapters vary from 'open book' questions to more lengthy computer projects.

Chapter 1 provides an introduction to statistical pattern recognition, defining some terminology, introducing supervised and unsupervised classification. Two related approaches to supervised classification are presented: one based on the estimation of probability density functions and a second based on the construction of discriminant functions. The chapter concludes with an outline of the pattern recognition cycle, putting the remaining chapters of the book into context. Chapters 2 and 3 pursue the density function approach to discrimination, with Chapter 2 addressing parametric approaches to density estimation and Chapter 3 developing classifiers based on nonparametric schemes.

Chapters 4–7 develop discriminant function approaches to supervised classification. Chapter 4 focuses on linear discriminant functions; much of the methodology of this chapter (including optimisation, regularisation and support vector machines) is used in some of the nonlinear methods. Chapter 5 explores kernel-based methods, in particular, the radial basis function network and the support vector machine, techniques for discrimination and regression that have received widespread study in recent years. Related nonlinear models (projection-based methods) are described in Chapter 6. Chapter 7 considers a decision-tree approach to discrimination, describing the classification and regression tree (CART) methodology and multivariate adaptive regression splines (MARS).

Chapter 8 considers performance: measuring the performance of a classifier and improving the performance by classifier combination.

The techniques of Chapters 9 and 10 may be described as methods of exploratory data analysis or preprocessing (and as such would usually be carried out prior to the supervised classification techniques of Chapters 2–7, although they could, on occasion, be post-processors of supervised techniques). Chapter 9 addresses feature selection and feature extraction – the procedures for obtaining a reduced set of variables characterising the original data. Such procedures are often an integral part of classifier design and it is somewhat artificial to partition the pattern recognition problem into separate processes of feature extraction and classification. However, feature extraction may provide insights into the data structure and the type of classifier to employ; thus, it is of interest in its own right. Chapter 10 considers unsupervised classification or *clustering* – the process of grouping individuals in a population to discover the presence of structure; its engineering application is to vector quantisation for image and speech coding.

Finally, Chapter 11 addresses some important diverse topics including model selection. Appendices largely cover background material and material appropriate if this book is used as a text for a 'conversion course': measures of dissimilarity, estimation, linear algebra, data analysis and basic probability.

The website `www.statistical-pattern-recognition.net` contains references and links to further information on techniques and applications.

In preparing the second edition of this book I have been helped by many people. I am grateful to colleagues and friends who have made comments on various parts of the manuscript. In particular, I would like to thank Mark Briers, Keith Copsey, Stephen Luttrell, John O'Loghlen and Kevin Weekes (with particular thanks to Keith for examples in Chapter 2); Wiley for help in the final production of the manuscript; and especially Rosemary for her support and patience.

Notation

Some of the more commonly used notation is given below. I have used some notational conveniences. For example, I have tended to use the same symbol for a variable as well as a measurement on that variable. The meaning should be obvious from the context. Also, I denote the density function of x as $p(x)$ and y as $p(y)$, even though the functions differ. A vector is denoted by a lower-case quantity in bold face, and a matrix by upper case.

p	number of variables
C	number of classes
n	number of measurements
n_i	number of measurements in class i
ω_i	label for class i
$X_1, \ldots, X_p$	p random variables
$x_1, \ldots, x_p$	measurements on variables $X_1, \ldots, X_p$
$\boldsymbol{x} = (x_1, \ldots, x_p)^T$	measurement vector
$\boldsymbol{X} = [\boldsymbol{x}_1, \ldots, \boldsymbol{x}_n]^T$	$n \times p$ data matrix
$\boldsymbol{X} = \begin{bmatrix} x_{11} & \ldots & x_{1p} \\ \vdots & \ddots & \vdots \\ x_{n1} & \ldots & x_{np} \end{bmatrix}$	
$P(\boldsymbol{x}) = \text{prob}(X_1 \leq x_1, \ldots, X_p \leq x_p)$	
$p(\boldsymbol{x}) = \partial P / \partial \boldsymbol{x}$	
$p(\omega_i)$	prior probability of class i
$\boldsymbol{\mu} = \int \boldsymbol{x} p(\boldsymbol{x}) d\boldsymbol{x}$	population mean
$\boldsymbol{\mu}_i = \int \boldsymbol{x} p(\boldsymbol{x}) d\boldsymbol{x}$	mean of class $i, i = 1, \ldots, C$
$\boldsymbol{m} = (1/n) \sum_{r=1}^{n} \boldsymbol{x}_r$	sample mean
$\boldsymbol{m}_i = (1/n_i) \sum_{r=1}^{n} z_{ir} \boldsymbol{x}_r$	sample mean of class $i, i = 1, \ldots, C$ $z_{ir} = 1$ if $\boldsymbol{x}_r \in \omega_i$, 0 otherwise n_i = number of patterns in $\omega_i = \sum_{r=1}^{n} z_{ir}$
$\hat{\boldsymbol{\Sigma}} = \frac{1}{n} \sum_{r=1}^{n} (\boldsymbol{x}_r - \boldsymbol{m})(\boldsymbol{x}_r - \boldsymbol{m})^T$	sample covariance matrix (maximum likelihood estimate)
$n/(n-1)\hat{\boldsymbol{\Sigma}}$	sample covariance matrix (unbiased estimate)

$\hat{\mathbf{\Sigma}}_i = (1/n_i)\sum_{j=1}^{n} z_{ij}(\mathbf{x}_j - \mathbf{m}_i)(\mathbf{x}_j - \mathbf{m}_i)^T$	sample covariance matrix of class i (maximum likelihood estimate)
$\mathbf{S}_i = \frac{n_i}{n_i - 1}\hat{\mathbf{\Sigma}}_i$	sample covariance matrix of class i (unbiased estimate)
$\mathbf{S}_W = \sum_{i=1}^{C} \frac{n_i}{n}\hat{\mathbf{\Sigma}}_i$	pooled within-class sample covariance matrix
$\mathbf{S} = \frac{n}{n-C}\mathbf{S}_W$	pooled within-class sample covariance matrix (unbiased estimate)
$\mathbf{S}_B = \sum_{i=1}^{C} \frac{n_i}{n}(\mathbf{m}_i - \mathbf{m})(\mathbf{m}_i - \mathbf{m})^T$	sample between-class matrix
$\mathbf{S}_B + \mathbf{S}_W = \hat{\mathbf{\Sigma}}$	
$\|\mathbf{A}\|^2 = \sum_{ij} A_{ij}^2$	
$N(\mathbf{m}, \mathbf{\Sigma})$	normal distribution, mean, $\mathbf{m}$ covariance matrix $\mathbf{\Sigma}$
$\mathrm{E}[Y\|X]$	expectation of Y given X
$\mathrm{I}(\theta)$	=1 if θ = `true` else 0

Notation for specific probability density functions is given in Appendix E.

1

Introduction to statistical pattern recognition

Overview

Statistical pattern recognition is a term used to cover all stages of an investigation from problem formulation and data collection through to discrimination and classification, assessment of results and interpretation. Some of the basic terminology is introduced and two complementary approaches to discrimination described.

1.1 Statistical pattern recognition

1.1.1 Introduction

This book describes basic pattern recognition procedures, together with practical applications of the techniques on real-world problems. A strong emphasis is placed on the statistical theory of discrimination, but clustering also receives some attention. Thus, the subject matter of this book can be summed up in a single word: 'classification', both supervised (using class information to design a classifier – i.e. discrimination) and unsupervised (allocating to groups without class information – i.e. clustering).

Pattern recognition as a field of study developed significantly in the 1960s. It was very much an interdisciplinary subject, covering developments in the areas of statistics, engineering, artificial intelligence, computer science, psychology and physiology, among others. Some people entered the field with a real problem to solve. The large numbers of applications, ranging from the classical ones such as automatic character recognition and medical diagnosis to the more recent ones in *data mining* (such as credit scoring, consumer sales analysis and credit card transaction analysis), have attracted considerable research effort, with many methods developed and advances made. Other researchers were motivated by the development of machines with 'brain-like' performance, that in some way could emulate human performance. There were many over-optimistic and unrealistic claims made, and to some extent there exist strong parallels with the growth of research on knowledge-based systems in the 1970s and neural networks in the 1980s.

Nevertheless, within these areas significant progress has been made, particularly where the domain overlaps with probability and statistics, and within recent years there have been many exciting new developments, both in methodology and applications. These build on the solid foundations of earlier research and take advantage of increased computational resources readily available nowadays. These developments include, for example, kernel-based methods and Bayesian computational methods.

The topics in this book could easily have been described under the term *machine learning* that describes the study of machines that can adapt to their environment and learn from example. The emphasis in machine learning is perhaps more on computationally intensive methods and less on a statistical approach, but there is strong overlap between the research areas of statistical pattern recognition and machine learning.

1.1.2 The basic model

Since many of the techniques we shall describe have been developed over a range of diverse disciplines, there is naturally a variety of sometimes contradictory terminology. We shall use the term 'pattern' to denote the p-dimensional data vector $\boldsymbol{x} = (x_1, \dots, x_p)^T$ of measurements (^{T} denotes vector transpose), whose components x_i are measurements of the features of an object. Thus the features are the variables specified by the investigator and thought to be important for classification. In discrimination, we assume that there exist C groups or *classes*, denoted $\omega_1, \dots, \omega_C$, and associated with each pattern $\boldsymbol{x}$ is a categorical variable z that denotes the class or group membership; that is, if $z = i$, then the pattern belongs to ω_i, $i \in \{1, \dots, C\}$.

Examples of patterns are measurements of an acoustic waveform in a speech recognition problem; measurements on a patient made in order to identify a disease (diagnosis); measurements on patients in order to predict the likely outcome (prognosis); measurements on weather variables (for forecasting or prediction); and a digitised image for character recognition. Therefore, we see that the term 'pattern', in its technical meaning, does not necessarily refer to structure within images.

The main topic in this book may be described by a number of terms such as *pattern classifier design* or *discrimination* or *allocation rule design*. By this we mean specifying the parameters of a pattern classifier, represented schematically in Figure 1.1, so that it yields the optimal (in some sense) response for a given pattern. This response is usually an estimate of the class to which the pattern belongs. We assume that we have a set of patterns of known class $\{(\boldsymbol{x}_i, z_i), i = 1, \dots, n\}$ (the *training* or *design* set) that we use to design the classifier (to set up its internal parameters). Once this has been done, we may estimate class membership for an unknown pattern $\boldsymbol{x}$.

The form derived for the pattern classifier depends on a number of different factors. It depends on the distribution of the training data, and the assumptions made concerning its distribution. Another important factor is the misclassification cost – the cost of making an incorrect decision. In many applications misclassification costs are hard to quantify, being combinations of several contributions such as monetary costs, time and other more subjective costs. For example, in a medical diagnosis problem, each treatment has different costs associated with it. These relate to the expense of different types of drugs,

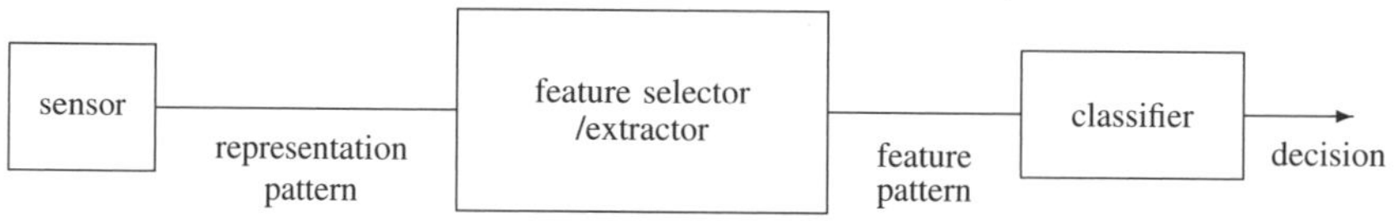

Figure 1.1 Pattern classifier

the suffering the patient is subjected to by each course of action and the risk of further complications.

Figure 1.1 grossly oversimplifies the pattern classification procedure. Data may undergo several separate transformation stages before a final outcome is reached. These transformations (sometimes termed preprocessing, feature selection or feature extraction) operate on the data in a way that usually reduces its dimension (reduces the number of features), removing redundant or irrelevant information, and transforms it to a form more appropriate for subsequent classification. The term *intrinsic dimensionality* refers to the minimum number of variables required to capture the structure within the data. In the speech recognition example mentioned above, a preprocessing stage may be to transform the waveform to a frequency representation. This may be processed further to find formants (peaks in the spectrum). This is a feature extraction process (taking a possible nonlinear combination of the original variables to form new variables). Feature selection is the process of selecting a subset of a given set of variables.

Terminology varies between authors. Sometimes the term 'representation pattern' is used for the vector of measurements made on a sensor (for example, optical imager, radar) with the term 'feature pattern' being reserved for the small set of variables obtained by transformation (by a feature selection or feature extraction process) of the original vector of measurements. In some problems, measurements may be made directly on the feature vector itself. In these situations there is no automatic feature selection stage, with the feature selection being performed by the investigator who 'knows' (through experience, knowledge of previous studies and the problem domain) those variables that are important for classification. In many cases, however, it will be necessary to perform one or more transformations of the measured data.

In some pattern classifiers, each of the above stages may be present and identifiable as separate operations, while in others they may not be. Also, in some classifiers, the preliminary stages will tend to be problem-specific, as in the speech example. In this book, we consider feature selection and extraction transformations that are not application-specific. That is not to say all will be suitable for any given application, however, but application-specific preprocessing must be left to the investigator.

1.2 Stages in a pattern recognition problem

A pattern recognition investigation may consist of several stages, enumerated below. Further details are given in Appendix D. Not all stages may be present; some may be merged together so that the distinction between two operations may not be clear, even if both are carried out; also, there may be some application-specific data processing that may not be regarded as one of the stages listed. However, the points below are fairly typical.

1. Formulation of the problem: gaining a clear understanding of the aims of the investigation and planning the remaining stages.

2. Data collection: making measurements on appropriate variables and recording details of the data collection procedure (ground truth).

3. Initial examination of the data: checking the data, calculating summary statistics and producing plots in order to get a feel for the structure.

4. Feature selection or feature extraction: selecting variables from the measured set that are appropriate for the task. These new variables may be obtained by a linear or nonlinear transformation of the original set (feature extraction). To some extent, the division of feature extraction and classification is artificial.

5. Unsupervised pattern classification or clustering. This may be viewed as exploratory data analysis and it may provide a successful conclusion to a study. On the other hand, it may be a means of preprocessing the data for a supervised classification procedure.

6. Apply discrimination or regression procedures as appropriate. The classifier is designed using a training set of exemplar patterns.

7. Assessment of results. This may involve applying the trained classifier to an independent *test set* of labelled patterns.

8. Interpretation.

The above is necessarily an iterative process: the analysis of the results may pose further hypotheses that require further data collection. Also, the cycle may be terminated at different stages: the questions posed may be answered by an initial examination of the data or it may be discovered that the data cannot answer the initial question and the problem must be reformulated.

The emphasis of this book is on techniques for performing steps 4, 5 and 6.

1.3 Issues

The main topic that we address in this book concerns classifier design: given a training set of patterns of known class, we seek to design a classifier that is optimal for the expected operating conditions (the test conditions).

There are a number of very important points to make about the sentence above, straightforward as it seems. The first is that we are given a *finite* design set. If the classifier is too complex (there are too many free parameters) it may model noise in the design set. This is an example of *over-fitting*. If the classifier is not complex enough, then it may fail to capture structure in the data. An example of this is the fitting of a set of data points by a polynomial curve. If the degree of the polynomial is too high, then, although the curve may pass through or close to the data points, thus achieving a low fitting error, the fitting curve is very variable and models every fluctuation in the data (due to noise). If the degree of the polynomial is too low, the fitting error is large and the underlying variability of the curve is not modelled.

Thus, achieving optimal performance on the design set (in terms of minimising some error criterion perhaps) is not required: it may be possible, in a classification problem, to achieve 100% classification accuracy on the design set but the *generalisation performance* – the expected performance on data representative of the true operating conditions (equivalently, the performance on an infinite test set of which the design set is a sample) – is poorer than could be achieved by careful design. Choosing the 'right' model is an exercise in *model selection*.

In practice we usually do not know what is structure and what is noise in the data. Also, training a classifier (the procedure of determining its parameters) should not be considered as a separate issue from model selection, but it often is.

A second point about the design of optimal classifiers concerns the word 'optimal'. There are several ways of measuring classifier performance, the most common being error rate, although this has severe limitations. Other measures, based on the closeness of the estimates of the probabilities of class membership to the true probabilities, may be more appropriate in many cases. However, many classifier design methods usually optimise alternative criteria since the desired ones are difficult to optimise directly. For example, a classifier may be trained by optimising a squared error measure and assessed using error rate.

Finally, we assume that the training data are representative of the test conditions. If this is not so, perhaps because the test conditions may be subject to noise not present in the training data, or there are changes in the population from which the data are drawn (population drift), then these differences must be taken into account in classifier design.

1.4 Supervised versus unsupervised

There are two main divisions of classification: *supervised classification* (or discrimination) and *unsupervised classification* (sometimes in the statistics literature simply referred to as classification or clustering).

In supervised classification we have a set of data samples (each consisting of measurements on a set of variables) with associated labels, the class types. These are used as exemplars in the classifier design.

Why do we wish to design an automatic means of classifying future data? Cannot the same method that was used to label the design set be used on the test data? In some cases this may be possible. However, even if it were possible, in practice we may wish to develop an automatic method to reduce labour-intensive procedures. In other cases, it may not be possible for a human to be part of the classification process. An example of the former is in industrial inspection. A classifier can be trained using images of components on a production line, each image labelled carefully by an operator. However, in the practical application we would wish to save a human operator from the tedious job, and hopefully make it more reliable. An example of the latter reason for performing a classification automatically is in radar target recognition of objects. For vehicle recognition, the data may be gathered by positioning vehicles on a turntable and making measurements from all aspect angles. In the practical application, a human may not be able to recognise an object reliably from its radar image, or the process may be carried out remotely.

In unsupervised classification, the data are not labelled and we seek to find groups in the data and the features that distinguish one group from another. Clustering techniques, described further in Chapter 10, can also be used as part of a supervised classification scheme by defining prototypes. A clustering scheme may be applied to the data for each class separately and representative samples for each group within the class (the group means, for example) used as the prototypes for that class.

1.5 Approaches to statistical pattern recognition

The problem we are addressing in this book is primarily one of pattern classification. Given a set of measurements obtained through observation and represented as a pattern vector $\boldsymbol{x}$, we wish to assign the pattern to one of C possible classes ω_i, $i = 1, \ldots, C$. A *decision rule* partitions the measurement space into C regions Ω_i, $i = 1, \ldots, C$. If an observation vector is in Ω_i then it is assumed to belong to class ω_i. Each region may be multiply connected – that is, it may be made up of several disjoint regions. The boundaries between the regions Ω_i are the *decision boundaries* or *decision surfaces*. Generally, it is in regions close to these boundaries that the highest proportion of misclassifications occurs. In such situations, we may reject the pattern or withhold a decision until further information is available so that a classification may be made later. This option is known as the *reject option* and therefore we have $C + 1$ outcomes of a decision rule (the reject option being denoted by ω_0) in a C-class problem.

In this section we introduce two approaches to discrimination that will be explored further in later chapters. The first assumes a knowledge of the underlying class-conditional probability density functions (the probability density function of the feature vectors for a given class). Of course, in many applications these will usually be unknown and must be estimated from a set of correctly classified samples termed the *design* or *training* set. Chapters 2 and 3 describe techniques for estimating the probability density functions explicitly.

The second approach introduced in this section develops decision rules that use the data to estimate the decision boundaries directly, without explicit calculation of the probability density functions. This approach is developed in Chapters 4, 5 and 6 where specific techniques are described.

1.5.1 Elementary decision theory

Here we introduce an approach to discrimination based on knowledge of the probability density functions of each class. Familiarity with basic probability theory is assumed. Some basic definitions are given in Appendix E.

Bayes decision rule for minimum error

Consider C classes, $\omega_1, \ldots, \omega_C$, with *a priori* probabilities (the probabilities of each class occurring) $p(\omega_1), \ldots, p(\omega_C)$, assumed known. If we wish to minimise the probability of making an error and we have no information regarding an object other than the class probability distribution then we would assign an object to class ω_j if

$$p(\omega_j) > p(\omega_k) \quad k = 1, \ldots, C; \; k \neq j$$

This classifies all objects as belonging to one class. For classes with equal probabilities, patterns are assigned arbitrarily between those classes.

However, we do have an *observation vector* or *measurement vector* $\boldsymbol{x}$ and we wish to assign $\boldsymbol{x}$ to one of the C classes. A decision rule based on probabilities is to assign $\boldsymbol{x}$ to class ω_j if the probability of class ω_j given the observation $\boldsymbol{x}$, $p(\omega_j|\boldsymbol{x})$, is greatest over all classes $\omega_1, \ldots, \omega_C$. That is, assign $\boldsymbol{x}$ to class ω_j if

$$p(\omega_j|\boldsymbol{x}) > p(\omega_k|\boldsymbol{x}) \quad k = 1, \ldots, C; k \neq j \tag{1.1}$$

This decision rule partitions the measurement space into C regions $\Omega_1, \ldots, \Omega_C$ such that if $\boldsymbol{x} \in \Omega_j$ then $\boldsymbol{x}$ belongs to class ω_j.

The *a posteriori* probabilities $p(\omega_j|\boldsymbol{x})$ may be expressed in terms of the *a priori* probabilities and the class-conditional density functions $p(\boldsymbol{x}|\omega_i)$ using Bayes' theorem (see Appendix E) as

$$p(\omega_i|\boldsymbol{x}) = \frac{p(\boldsymbol{x}|\omega_i)p(\omega_i)}{p(\boldsymbol{x})}$$

and so the decision rule (1.1) may be written: assign $\boldsymbol{x}$ to ω_j if

$$p(\boldsymbol{x}|\omega_j)p(\omega_j) > p(\boldsymbol{x}|\omega_k)p(\omega_k) \quad k = 1, \ldots, C; k \neq j \tag{1.2}$$

This is known as Bayes' rule for *minimum error*.

For two classes, the decision rule (1.2) may be written

$$l_r(\boldsymbol{x}) = \frac{p(\boldsymbol{x}|\omega_1)}{p(\boldsymbol{x}|\omega_2)} > \frac{p(\omega_2)}{p(\omega_1)} \text{ implies } \boldsymbol{x} \in \text{class } \omega_1$$

The function $l_r(\boldsymbol{x})$ is the *likelihood ratio*. Figures 1.2 and 1.3 give a simple illustration for a two-class discrimination problem. Class ω_1 is normally distributed with zero mean and unit variance, $p(\boldsymbol{x}|\omega_1) = N(\boldsymbol{x}|0, 1)$ (see Appendix E). Class ω_2 is a *normal mixture* (a weighted sum of normal densities) $p(\boldsymbol{x}|\omega_2) = 0.6N(\boldsymbol{x}|1, 1) + 0.4N(\boldsymbol{x}|-1, 2)$. Figure 1.2 plots $p(\boldsymbol{x}|\omega_i)p(\omega_i)$, $i = 1, 2$, where the priors are taken to be $p(\omega_1) = 0.5$, $p(\omega_2) = 0.5$. Figure 1.3 plots the likelihood ratio $l_r(\boldsymbol{x})$ and the threshold $p(\omega_2)/p(\omega_1)$. We see from this figure that the decision rule (1.2) leads to a disjoint region for class ω_2.

The fact that the decision rule (1.2) minimises the error may be seen as follows. The probability of making an error, $p(\text{error})$, may be expressed as

$$p(\text{error}) = \sum_{i=1}^{C} p(\text{error}|\omega_i)p(\omega_i) \tag{1.3}$$

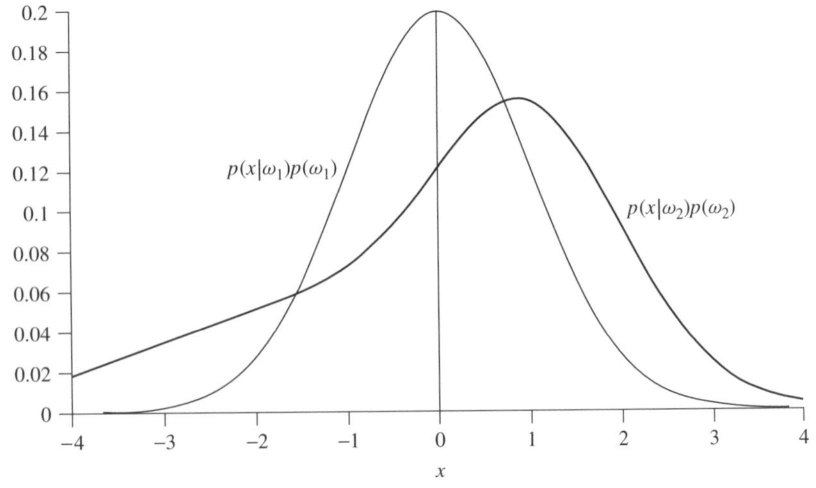

Figure 1.2 $p(x|\omega_i)p(\omega_i)$, for classes ω_1 and ω_2

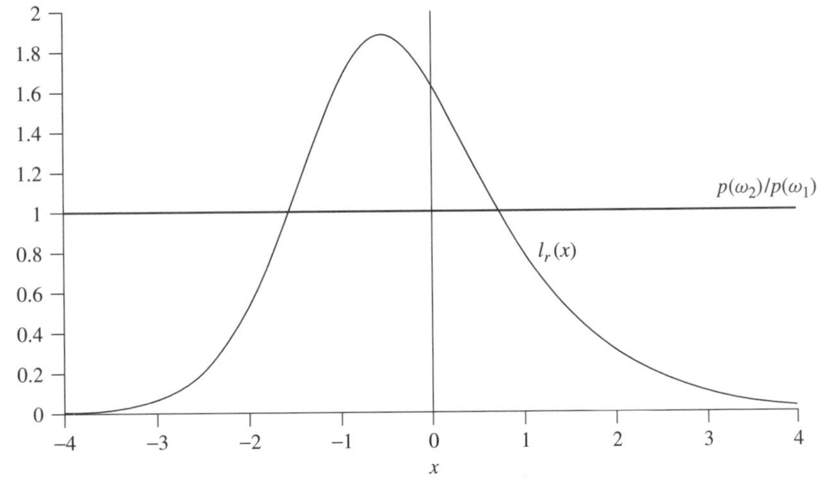

Figure 1.3 Likelihood function

where $p(\text{error}|\omega_i)$ is the probability of misclassifying patterns from class ω_i. This is given by

$$p(\text{error}|\omega_i) = \int_{\mathcal{C}[\Omega_i]} p(\boldsymbol{x}|\omega_i)\, d\boldsymbol{x} \tag{1.4}$$

the integral of the class-conditional density function over $\mathcal{C}[\Omega_i]$, the region of measurement space outside Ω_i ($\mathcal{C}$ is the complement operator), i.e. $\sum_{j=1, j\neq i}^{C} \Omega_j$. Therefore, we may write the probability of misclassifying a pattern as

$$\begin{aligned}
p(\text{error}) &= \sum_{i=1}^{C} \int_{\mathcal{C}[\Omega_i]} p(\boldsymbol{x}|\omega_i)p(\omega_i)\,d\boldsymbol{x} \\
&= \sum_{i=1}^{C} p(\omega_i)\left(1 - \int_{\Omega_i} p(\boldsymbol{x}|\omega_i)\,d\boldsymbol{x}\right) \\
&= 1 - \sum_{i=1}^{C} p(\omega_i) \int_{\Omega_i} p(\boldsymbol{x}|\omega_i)\,d\boldsymbol{x}
\end{aligned} \tag{1.5}$$

from which we see that minimising the probability of making an error is equivalent to maximising

$$\sum_{i=1}^{C} p(\omega_i) \int_{\Omega_i} p(\boldsymbol{x}|\omega_i)\,d\boldsymbol{x} \tag{1.6}$$

the probability of correct classification. Therefore, we wish to choose the regions Ω_i so that the integral given in (1.6) is a maximum. This is achieved by selecting Ω_i to be the region for which $p(\omega_i)p(\boldsymbol{x}|\omega_i)$ is the largest over all classes and the probability of correct classification, c, is

$$c = \int \max_i \, p(\omega_i)p(\boldsymbol{x}|\omega_i)\,d\boldsymbol{x} \tag{1.7}$$

where the integral is over the whole of the measurement space, and the Bayes error is

$$e_B = 1 - \int \max_i \, p(\omega_i)p(\boldsymbol{x}|\omega_i)\,d\boldsymbol{x} \tag{1.8}$$

This is illustrated in Figures 1.4 and 1.5. Figure 1.4 plots the two distributions $p(x|\omega_i), i = 1, 2$ (both normal with unit variance and means ± 0.5), and Figure 1.5 plots the functions $p(x|\omega_i)p(\omega_i)$ where $p(\omega_1) = 0.3$, $p(\omega_2) = 0.7$. The Bayes decision boundary is marked with a vertical line at x_B. The areas of the hatched regions in Figure 1.4 represent the probability of error: by equation (1.4), the area of the horizontal hatching is the probability of classifying a pattern from class 1 as a pattern from class 2 and the area of the vertical hatching the probability of classifying a pattern from class 2 as class 1. The sum of these two areas, weighted by the priors (equation (1.5)), is the probability of making an error.

Bayes decision rule for minimum error – reject option

As we have stated above, an error or misrecognition occurs when the classifier assigns a pattern to one class when it actually belongs to another. In this section we consider the reject option. Usually it is the uncertain classifications which mainly contribute to the error rate. Therefore, rejecting a pattern (withholding a decision) may lead to a reduction in the error rate. This rejected pattern may be discarded, or set aside until further information allows a decision to be made. Although the option to reject may alleviate or remove the problem of a high misrecognition rate, some otherwise correct

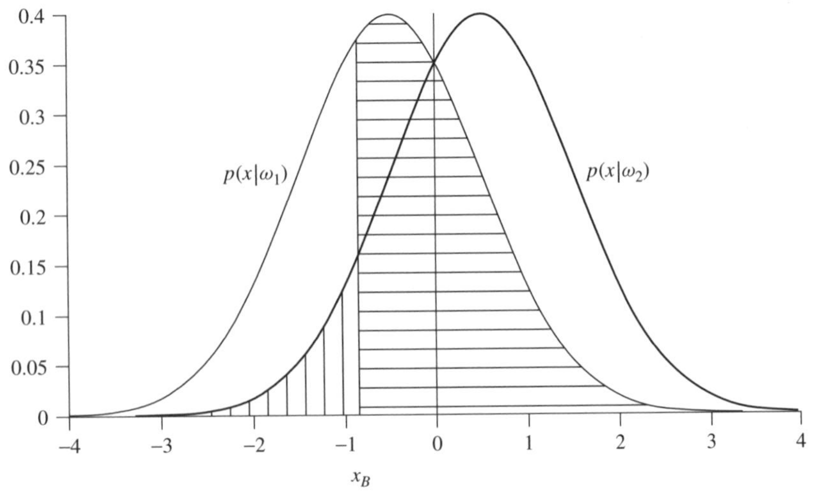

Figure 1.4 Class-conditional densities for two normal distributions

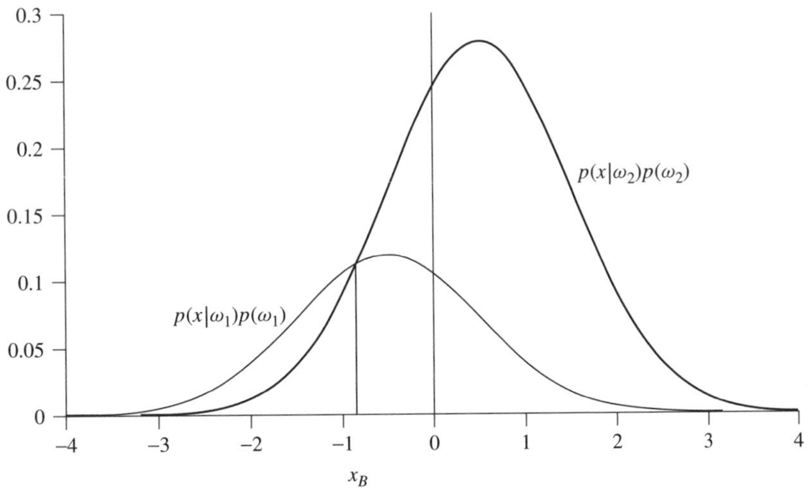

Figure 1.5 Bayes decision boundary for two normally distributed classes with unequal priors

classifications are also converted into rejects. Here we consider the trade-offs between error rate and reject rate.

Firstly, we partition the sample space into two complementary regions: R, a *reject region*, and A, an *acceptance* or *classification region*. These are defined by

$$R = \left\{ \boldsymbol{x} | 1 - \max_i p(\omega_i|\boldsymbol{x}) > t) \right\}$$

$$A = \left\{ \boldsymbol{x} | 1 - \max_i p(\omega_i|\boldsymbol{x}) \leq t) \right\}$$

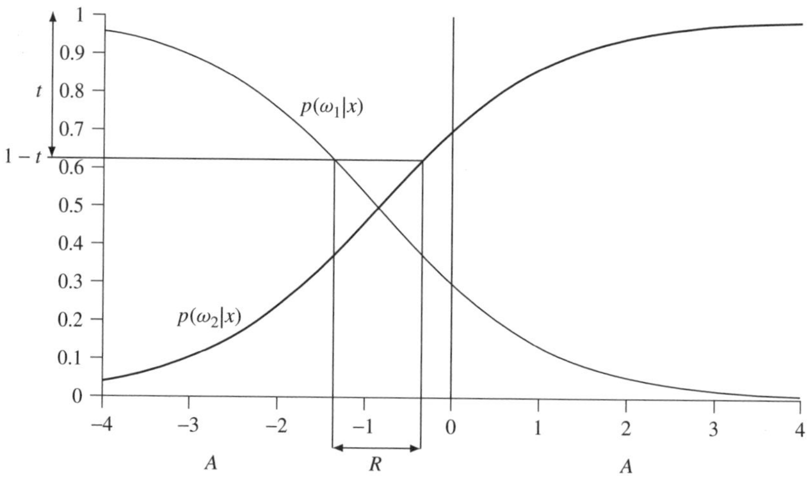

Figure 1.6 Illustration of acceptance and reject regions

where t is a threshold. This is illustrated in Figure 1.6 using the same distributions as those in Figures 1.4 and 1.5. The smaller the value of the threshold t, the larger is the reject region R. However, if t is chosen such that

$$1 - t < \frac{1}{C}$$

or equivalently,

$$t > \frac{C-1}{C}$$

where C is the number of classes, then the reject region is empty. This is because the minimum value which $\max_i p(\omega_i|\boldsymbol{x})$ can attain is $1/C$ (since $1 = \sum_{i=1}^{C} p(\omega_i|\boldsymbol{x}) \leq C \max_i p(\omega_i|\boldsymbol{x})$), when all classes are equally likely. Therefore, for the reject option to be activated, we must have $t \leq (C-1)/C$.

Thus, if a pattern $\boldsymbol{x}$ lies in the region A, we classify it according to the Bayes rule for minimum error (equation (1.2)). However, if $\boldsymbol{x}$ lies in the region R, we reject $\boldsymbol{x}$.

The probability of correct classification, $c(t)$, is a function of the threshold, t, and is given by equation (1.7), where now the integral is over the acceptance region, A, only

$$c(t) = \int_A \max_i \left[p(\omega_i) p(\boldsymbol{x}|\omega_i) \right] d\boldsymbol{x}$$

and the unconditional probability of rejecting a measurement $\boldsymbol{x}$, r, also a function of the threshold t, is

$$r(t) = \int_R p(\boldsymbol{x})\, d\boldsymbol{x} \tag{1.9}$$

Therefore, the error rate, e (the probability of accepting a point for classification and incorrectly classifying it), is

$$\begin{aligned} e(t) &= \int_A (1 - \max_i p(\omega_i|\boldsymbol{x}))p(\boldsymbol{x})\,d\boldsymbol{x} \\ &= 1 - c(t) - r(t) \end{aligned}$$

Thus, the error rate and reject rate are inversely related. Chow (1970) derives a simple functional relationship between $e(t)$ and $r(t)$ which we quote here without proof. Knowing $r(t)$ over the complete range of t allows $e(t)$ to be calculated using the relationship

$$e(t) = -\int_0^t s\,dr(s) \tag{1.10}$$

The above result allows the error rate to be evaluated from the reject function for the Bayes optimum classifier. The reject function can be calculated using unlabelled data and a practical application is to problems where labelling of gathered data is costly.

Bayes decision rule for minimum risk

In the previous section, the decision rule selected the class for which the *a posteriori* probability, $p(\omega_j|\boldsymbol{x})$, was the greatest. This minimised the probability of making an error. We now consider a somewhat different rule that minimises an expected *loss* or risk. This is a very important concept since in many applications the costs associated with misclassification depend upon the true class of the pattern and the class to which it is assigned. For example, in a medical diagnosis problem in which a patient has back pain, it is far worse to classify a patient with severe spinal abnormality as healthy (or having mild back ache) than the other way round.

We make this concept more formal by introducing a loss that is a measure of the cost of making the decision that a pattern belongs to class ω_i when the true class is ω_j. We define a loss matrix $\boldsymbol{\Lambda}$ with components

$$\lambda_{ji} = \text{cost of assigning a pattern } \boldsymbol{x} \text{ to } \omega_i \text{ when } \boldsymbol{x} \in \omega_j$$

In practice, it may be very difficult to assign costs. In some situations, λ may be measured in monetary units that are quantifiable. However, in many situations, costs are a combination of several different factors measured in different units – money, time, quality of life. As a consequence, they may be the subjective opinion of an expert. The *conditional risk* of assigning a pattern $\boldsymbol{x}$ to class ω_i is defined as

$$l^i(\boldsymbol{x}) = \sum_{j=1}^{C} \lambda_{ji}\, p(\omega_j|\boldsymbol{x})$$

The average risk over region Ω_i is

$$\begin{aligned} r^i &= \int_{\Omega_i} l^i(\boldsymbol{x})p(\boldsymbol{x})\,d\boldsymbol{x} \\ &= \int_{\Omega_i} \sum_{j=1}^{C} \lambda_{ji}\, p(\omega_j|\boldsymbol{x})p(\boldsymbol{x})\,d\boldsymbol{x} \end{aligned}$$

and the overall expected cost or *risk* is

$$r = \sum_{i=1}^{C} r^i = \sum_{i=1}^{C} \int_{\Omega_i} \sum_{j=1}^{C} \lambda_{ji} p(\omega_j|\boldsymbol{x}) p(\boldsymbol{x})\, d\boldsymbol{x} \qquad (1.11)$$

The above expression for the risk will be minimised if the regions Ω_i are chosen such that if

$$\sum_{j=1}^{C} \lambda_{ji} p(\omega_j|\boldsymbol{x}) p(\boldsymbol{x}) \leq \sum_{j=1}^{C} \lambda_{jk} p(\omega_j|\boldsymbol{x}) p(\boldsymbol{x}) \quad k = 1, \ldots, C \qquad (1.12)$$

then $\boldsymbol{x} \in \Omega_i$. This is the *Bayes decision rule for minimum risk*, with Bayes risk, r^*, given by

$$r^* = \int_{\boldsymbol{x}} \min_{i=1,\ldots,C} \sum_{j=1}^{C} \lambda_{ji} p(\omega_j|\boldsymbol{x}) p(\boldsymbol{x})\, d\boldsymbol{x}$$

One special case of the loss matrix $\boldsymbol{\Lambda}$ is the *equal cost* loss matrix for which

$$\lambda_{ij} = \begin{cases} 1 & i \neq j \\ 0 & i = j \end{cases}$$

Substituting into (1.12) gives the decision rule: assign $\boldsymbol{x}$ to class ω_i if

$$\sum_{j=1}^{C} p(\omega_j|\boldsymbol{x}) p(\boldsymbol{x}) - p(\omega_i|\boldsymbol{x}) p(\boldsymbol{x}) \leq \sum_{j=1}^{C} p(\omega_j|\boldsymbol{x}) p(\boldsymbol{x}) - p(\omega_k|\boldsymbol{x}) p(\boldsymbol{x}) \quad k = 1, \ldots, C$$

that is,

$$p(\boldsymbol{x}|\omega_i) p(\omega_i) \geq p(\boldsymbol{x}|\omega_k) p(\omega_k) \quad k = 1, \ldots, C$$

implies that $\boldsymbol{x} \in$ class ω_i; this is the Bayes rule for minimum error.

Bayes decision rule for minimum risk – reject option

As with the Bayes rule for minimum error, we may also introduce a reject option, by which the reject region, R, is defined by

$$R = \left\{ \boldsymbol{x} \;\middle|\; \min_i l^i(\boldsymbol{x}) > t \right\}$$

where t is a threshold. The decision is to accept a pattern $\boldsymbol{x}$ and assign it to class ω_i if

$$l^i(\boldsymbol{x}) = \min_j l^j(\boldsymbol{x}) \leq t$$

and to reject $\boldsymbol{x}$ if

$$l^i(\boldsymbol{x}) = \min_j l^j(\boldsymbol{x}) > t$$

This decision is equivalent to defining a reject region Ω_0 with a constant conditional risk

$$l^0(\boldsymbol{x}) = t$$

so that the Bayes decision rule is: assign $\boldsymbol{x}$ to class ω_i if

$$l^i(\boldsymbol{x}) \leq l^j(\boldsymbol{x}) \quad j = 0, 1, \ldots, C$$

with Bayes risk

$$r^* = \int_R t p(\boldsymbol{x})\, d\boldsymbol{x} + \int_A \min_{i=1,\ldots,C} \sum_{j=1}^{C} \lambda_{ji} p(\omega_j|\boldsymbol{x}) p(\boldsymbol{x})\, d\boldsymbol{x} \tag{1.13}$$

Neyman–Pearson decision rule

An alternative to the Bayes decision rules for a two-class problem is the Neyman–Pearson test. In a two-class problem there are two possible types of error that may be made in the decision process. We may classify a pattern of class ω_1 as belonging to class ω_2 or a pattern from class ω_2 as belonging to class ω_1. Let the probability of these two errors be ϵ_1 and ϵ_2 respectively, so that

$$\epsilon_1 = \int_{\Omega_2} p(\boldsymbol{x}|\omega_1)\, d\boldsymbol{x} = \text{error probability of Type I}$$

and

$$\epsilon_2 = \int_{\Omega_1} p(\boldsymbol{x}|\omega_2)\, d\boldsymbol{x} = \text{error probability of Type II}$$

The Neyman–Pearson decision rule is to minimise the error ϵ_1 subject to ϵ_2 being equal to a constant, ϵ_0, say.

If class ω_1 is termed the positive class and class ω_2 the negative class, then ϵ_1 is referred to as the *false negative rate*, the proportion of positive samples incorrectly assigned to the negative class; ϵ_2 is the *false positive rate*, the proportion of negative samples classed as positive.

An example of the use of the Neyman–Pearson decision rule is in radar detection where the problem is to detect a signal in the presence of noise. There are two types of error that may occur; one is to mistake noise for a signal present. This is called a *false alarm*. The second type of error occurs when a signal is actually present but the decision is made that only noise is present. This is a *missed detection*. If ω_1 denotes the signal class and ω_2 denotes the noise then ϵ_2 is the probability of false alarm and ϵ_1 is the probability of missed detection. In many radar applications, a threshold is set to give a fixed probability of false alarm and therefore the Neyman–Pearson decision rule is the one usually used.

We seek the minimum of

$$r = \int_{\Omega_2} p(\boldsymbol{x}|\omega_1)\, d\boldsymbol{x} + \mu \left\{ \int_{\Omega_1} p(\boldsymbol{x}|\omega_2)\, d\boldsymbol{x} - \epsilon_0 \right\}$$

where μ is a Lagrange multiplier[1] and ϵ_0 is the specified false alarm rate. The equation may be written

$$r = (1 - \mu\epsilon_0) + \int_{\Omega_1} \{\mu p(\boldsymbol{x}|\omega_2)\, d\boldsymbol{x} - p(\boldsymbol{x}|\omega_1)\, d\boldsymbol{x}\}$$

This will be minimised if we choose Ω_1 such that the integrand is negative, i.e.

$$\text{if } \mu p(\boldsymbol{x}|\omega_2) - p(\boldsymbol{x}|\omega_1) < 0, \quad \text{then } \boldsymbol{x} \in \Omega_1$$

or, in terms of the likelihood ratio,

$$\text{if } \frac{p(\boldsymbol{x}|\omega_1)}{p(\boldsymbol{x}|\omega_2)} > \mu, \quad \text{then } \boldsymbol{x} \in \Omega_1 \tag{1.14}$$

Thus the decision rule depends only on the within-class distributions and ignores the *a priori* probabilities.

The threshold μ is chosen so that

$$\int_{\Omega_1} p(\boldsymbol{x}|\omega_2)\, d\boldsymbol{x} = \epsilon_0,$$

the specified false alarm rate. However, in general μ cannot be determined analytically and requires numerical calculation.

Often, the performance of the decision rule is summarised in a receiver operating characteristic (ROC) curve, which plots the true positive against the false positive (that is, the probability of detection ($1 - \epsilon_1 = \int_{\Omega_1} p(\boldsymbol{x}|\omega_1)\, d\boldsymbol{x}$) against the probability of false alarm ($\epsilon_2 = \int_{\Omega_1} p(\boldsymbol{x}|\omega_2)\, d\boldsymbol{x}$)) as the threshold μ is varied. This is illustrated in Figure 1.7 for the univariate case of two normally distributed classes of unit variance and means separated by a distance, d. All the ROC curves pass through the $(0, 0)$ and $(1, 1)$ points and as the separation increases the curve moves into the top left corner. Ideally, we would like 100% detection for a 0% false alarm rate; the closer a curve is to this the better.

For the two-class case, the minimum risk decision (see equation (1.12)) defines the decision rules on the basis of the likelihood ratio ($\lambda_{ii} = 0$):

$$\text{if } \frac{p(\boldsymbol{x}|\omega_1)}{p(\boldsymbol{x}|\omega_2)} > \frac{\lambda_{21} p(\omega_2)}{\lambda_{12} p(\omega_1)}, \quad \text{then } \boldsymbol{x} \in \Omega_1 \tag{1.15}$$

The threshold defined by the right-hand side will correspond to a particular point on the ROC curve that depends on the misclassification costs and the prior probabilities.

In practice, precise values for the misclassification costs will be unavailable and we shall need to assess the performance over a range of expected costs. The use of the ROC curve as a tool for comparing and assessing classifier performance is discussed in Chapter 8.

---

[1]The method of Lagrange's undetermined multipliers can be found in most textbooks on mathematical methods, for example Wylie and Barrett (1995).

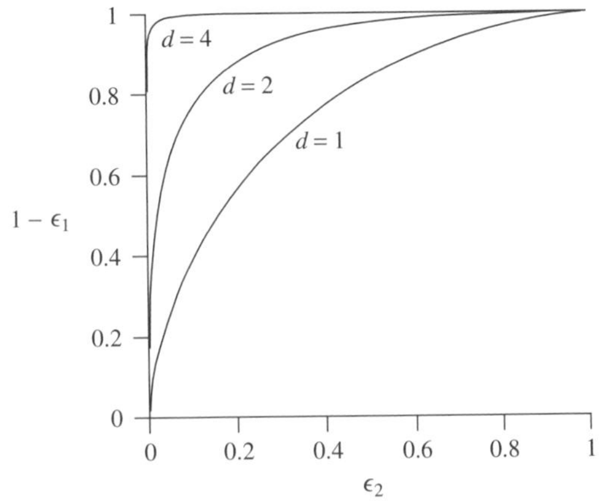

Figure 1.7 Receiver operating characteristic for two univariate normal distributions of unit variance and separation d; $1 - \epsilon_1 = \int_{\Omega_1} p(\boldsymbol{x}|\omega_1)\, d\boldsymbol{x}$ is the true positive (the probability of detection) and $\epsilon_2 = \int_{\Omega_1} p(\boldsymbol{x}|\omega_2)\, d\boldsymbol{x}$ is the false positive (the probability of false alarm)

Minimax criterion

The Bayes decision rules rely on a knowledge of both the within-class distributions and the prior class probabilities. However, situations may arise where the relative frequencies of new objects to be classified are unknown. In this situation a *minimax* procedure may be employed. The name *minimax* is used to refer to procedures for which either the maximum expected loss *or* the maximum of the error probability is a minimum. We shall limit our discussion below to the two-class problem and the minimum error probability procedure.

Consider the Bayes rule for minimum error. The decision regions Ω_1 and Ω_2 are defined by

$$p(\boldsymbol{x}|\omega_1)p(\omega_1) > p(\boldsymbol{x}|\omega_2)p(\omega_2) \text{ implies } \boldsymbol{x} \in \Omega_1 \tag{1.16}$$

and the Bayes minimum error, e_B, is

$$e_B = p(\omega_2)\int_{\Omega_1} p(\boldsymbol{x}|\omega_2)\, d\boldsymbol{x} + p(\omega_1)\int_{\Omega_2} p(\boldsymbol{x}|\omega_1)\, d\boldsymbol{x} \tag{1.17}$$

where $p(\omega_2) = 1 - p(\omega_1)$.

For *fixed* decision regions Ω_1 and Ω_2, e_B is a linear function of $p(\omega_1)$ (we denote this function $\tilde{e}_B$) attaining its maximum on the region $[0, 1]$ either at $p(\omega_1) = 0$ or $p(\omega_1) = 1$. However, since the regions Ω_1 and Ω_2 are also dependent on $p(\omega_1)$ through the Bayes decision criterion (1.16), the dependency of e_B on $p(\omega_1)$ is more complex, and not necessarily monotonic.

If Ω_1 and Ω_2 are fixed (determined according to (1.16) for some specified $p(\omega_i)$), the error given by (1.17) will only be the Bayes minimum error for a particular value of $p(\omega_1)$, say p_1^* (see Figure 1.8). For other values of $p(\omega_1)$, the error given by (1.17) must be greater than the minimum error. Therefore, the optimum curve touches the line at a tangent at p_1^* and is concave down at that point.

The minimax procedure aims to choose the partition Ω_1, Ω_2, or equivalently the value of $p(\omega_1)$ so that the maximum error (on a test set in which the values of $p(\omega_i)$ are unknown) is minimised. For example, in the figure, if the partition were chosen to correspond to the value p_1^* of $p(\omega_1)$, then the maximum error which could occur would be a value of b if $p(\omega_1)$ were actually equal to unity. The minimax procedure aims to minimise this maximum value, i.e. minimise

$$\max\{\tilde{e}_B(0), \tilde{e}_B(1)\}$$

or minimise

$$\max\left\{\int_{\Omega_2} p(\boldsymbol{x}|\omega_1)\,d\boldsymbol{x}, \int_{\Omega_1} p(\boldsymbol{x}|\omega_2)\,d\boldsymbol{x}\right\}$$

This is a minimum when

$$\int_{\Omega_2} p(\boldsymbol{x}|\omega_1)\,d\boldsymbol{x} = \int_{\Omega_1} p(\boldsymbol{x}|\omega_2)\,d\boldsymbol{x} \qquad (1.18)$$

which is when $a = b$ in Figure 1.8 and the line $\tilde{e}_B(p(\omega_1))$ is horizontal and touches the Bayes minimum error curve at its peak value.

Therefore, we choose the regions Ω_1 and Ω_2 so that the probabilities of the two types of error are the same. The minimax solution may be criticised as being over-pessimistic since it is a Bayes solution with respect to the least favourable prior distribution. The

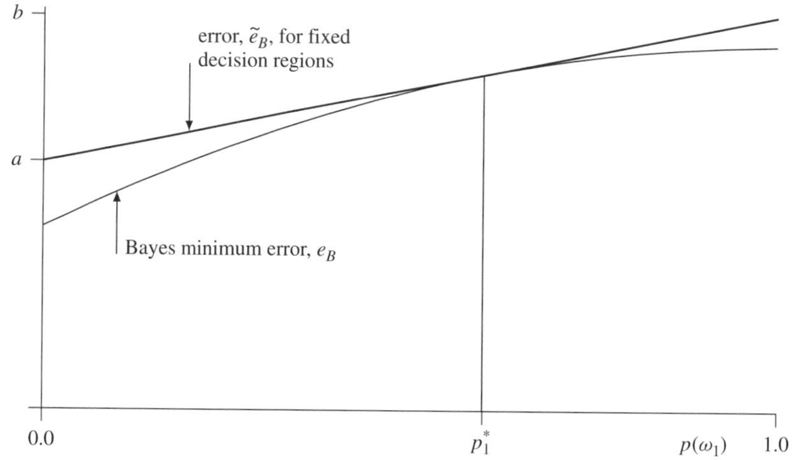

Figure 1.8 Minimax illustration

strategy may also be applied to minimising the maximum risk. In this case, the risk is

$$\begin{aligned}\int_{\Omega_1} [\lambda_{11} p(\omega_1|\boldsymbol{x}) + \lambda_{21} p(\omega_2|\boldsymbol{x})] p(\boldsymbol{x})\, d\boldsymbol{x} + \int_{\Omega_2} [\lambda_{12} p(\omega_1|\boldsymbol{x}) + \lambda_{22} p(\omega_2|\boldsymbol{x})] p(\boldsymbol{x})\, d\boldsymbol{x} \\ = p(\omega_1) \left[\lambda_{11} + (\lambda_{12} - \lambda_{11}) \int_{\Omega_2} p(\boldsymbol{x}|\omega_1)\, d\boldsymbol{x} \right] \\ + p(\omega_2) \left[\lambda_{22} + (\lambda_{21} - \lambda_{22}) \int_{\Omega_1} p(\boldsymbol{x}|\omega_2)\, d\boldsymbol{x} \right]\end{aligned}$$

and the boundary must therefore satisfy

$$\lambda_{11} - \lambda_{22} + (\lambda_{12} - \lambda_{11}) \int_{\Omega_2} p(\boldsymbol{x}|\omega_1)\, d\boldsymbol{x} - (\lambda_{21} - \lambda_{22}) \int_{\Omega_1} p(\boldsymbol{x}|\omega_2)\, d\boldsymbol{x} = 0$$

For $\lambda_{11} = \lambda_{22}$ and $\lambda_{21} = \lambda_{12}$, this reduces to condition (1.18).

Discussion

In this section we have introduced a decision-theoretic approach to classifying patterns. This divides up the measurement space into decision regions and we have looked at various strategies for obtaining the decision boundaries. The optimum rule in the sense of minimising the error is the Bayes decision rule for minimum error. Introducing the costs of making incorrect decisions leads to the Bayes rule for minimum risk. The theory developed assumes that the *a priori* distributions and the class-conditional distributions are known. In a real-world task, this is unlikely to be so. Therefore approximations must be made based on the data available. We consider techniques for estimating distributions in Chapters 2 and 3. Two alternatives to the Bayesian decision rule have also been described, namely the Neyman–Pearson decision rule (commonly used in signal processing applications) and the minimax rule. Both require knowledge of the class-conditional probability density functions. The receiver operating characteristic curve characterises the performance of a rule over a range of thresholds of the likelihood ratio.

We have seen that the error rate plays an important part in decision-making and classifier performance assessment. Consequently, estimation of error rates is a problem of great interest in statistical pattern recognition. For given fixed decision regions, we may calculate the probability of error using (1.5). If these decision regions are chosen according to the Bayes decision rule (1.2), then the error is the *Bayes error rate* or *optimal error rate*. However, regardless of how the decision regions are chosen, the error rate may be regarded as a measure of a given decision rule's performance.

The Bayes error rate (1.5) requires complete knowledge of the class-conditional density functions. In a particular situation, these may not be known and a classifier may be designed on the basis of a training set of samples. Given this training set, we may choose to form estimates of the distributions (using some of the techniques discussed in Chapters 2 and 3) and thus, with these estimates, use the Bayes decision rule and estimate the error according to (1.5).

However, even with accurate estimates of the distributions, evaluation of the error requires an integral over a multidimensional space and may prove a formidable task. An alternative approach is to obtain bounds on the optimal error rate or distribution-free estimates. Further discussion of methods of error rate estimation is given in Chapter 8.

1.5.2 Discriminant functions

In the previous subsection, classification was achieved by applying the Bayesian decision rule. This requires knowledge of the class-conditional density functions, $p(\boldsymbol{x}|\omega_i)$ (such as normal distributions whose parameters are estimated from the data – see Chapter 2), or nonparametric density estimation methods (such as kernel density estimation – see Chapter 3). Here, instead of making assumptions about $p(\boldsymbol{x}|\omega_i)$, we make assumptions about the forms of the *discriminant functions*.

A discriminant function is a function of the pattern $\boldsymbol{x}$ that leads to a classification rule. For example, in a two-class problem, a discriminant function $h(\boldsymbol{x})$ is a function for which

$$\begin{aligned} h(\boldsymbol{x}) &> k \Rightarrow \boldsymbol{x} \in \omega_1 \\ &< k \Rightarrow \boldsymbol{x} \in \omega_2 \end{aligned} \tag{1.19}$$

for constant k. In the case of equality ($h(\boldsymbol{x}) = k$), the pattern $\boldsymbol{x}$ may be assigned arbitrarily to one of the two classes. An optimal discriminant function for the two-class case is

$$h(\boldsymbol{x}) = \frac{p(\boldsymbol{x}|\omega_1)}{p(\boldsymbol{x}|\omega_2)}$$

with $k = p(\omega_2)/p(\omega_1)$. Discriminant functions are not unique. If f is a monotonic function then

$$\begin{aligned} g(\boldsymbol{x}) = f(h(\boldsymbol{x})) > k' \Rightarrow \boldsymbol{x} \in \omega_1 \\ g(\boldsymbol{x}) = f(h(\boldsymbol{x})) < k' \Rightarrow \boldsymbol{x} \in \omega_2 \end{aligned}$$

where $k' = f(k)$ leads to the same decision as (1.19).

In the C-group case we define C discriminant functions $g_i(\boldsymbol{x})$ such that

$$g_i(\boldsymbol{x}) > g_j(\boldsymbol{x}) \Rightarrow \boldsymbol{x} \in \omega_i \quad j = 1, \ldots, C; \quad j \neq i$$

That is, a pattern is assigned to the class with the largest discriminant. Of course, for two classes, a single discriminant function

$$h(\boldsymbol{x}) = g_1(\boldsymbol{x}) - g_2(\boldsymbol{x})$$

with $k = 0$ reduces to the two-class case given by (1.19).

Again, we may define an optimal discriminant function as

$$g_i(\boldsymbol{x}) = p(\boldsymbol{x}|\omega_i)p(\omega_i)$$

leading to the Bayes decision rule, but as we showed for the two-class case, there are other discriminant functions that lead to the same decision.

The essential difference between the approach of the previous subsection and the discriminant function approach described here is that the form of the discriminant function is specified and is not imposed by the underlying distribution. The choice of discriminant function may depend on prior knowledge about the patterns to be classified or may be a particular functional form whose parameters are adjusted by a training procedure. Many different forms of discriminant function have been considered in the literature, varying in complexity from the linear discriminant function (in which g is a linear combination of the x_i) to multiparameter nonlinear functions such as the multilayer perceptron.

Discrimination may also be viewed as a problem in *regression* (see Section 1.6) in which the dependent variable, y, is a class indicator and the regressors are the pattern vectors. Many discriminant function models lead to estimates of $E[y|\boldsymbol{x}]$, which is the aim of regression analysis (though in regression y is not necessarily a class indicator). Thus, many of the techniques we shall discuss for optimising discriminant functions apply equally well to regression problems. Indeed, as we find with feature extraction in Chapter 9 and also clustering in Chapter 10, similar techniques have been developed under different names in the pattern recognition and statistics literature.

Linear discriminant functions

First of all, let us consider the family of discriminant functions that are linear combinations of the components of $\boldsymbol{x} = (x_1, \ldots, x_p)^T$,

$$g(\boldsymbol{x}) = \boldsymbol{w}^T \boldsymbol{x} + w_0 = \sum_{i=1}^{p} w_i x_i + w_0 \tag{1.20}$$

This is a *linear discriminant function*, a complete specification of which is achieved by prescribing the *weight vector* $\boldsymbol{w}$ and *threshold weight* w_0. Equation (1.20) is the equation of a hyperplane with unit normal in the direction of $\boldsymbol{w}$ and a perpendicular distance $|w_0|/|\boldsymbol{w}|$ from the origin. The value of the discriminant function for a pattern $\boldsymbol{x}$ is a measure of the perpendicular distance from the hyperplane (see Figure 1.9).

A linear discriminant function can arise through assumptions of normal distributions for the class densities, with equal covariance matrices (see Chapter 2). Alternatively,

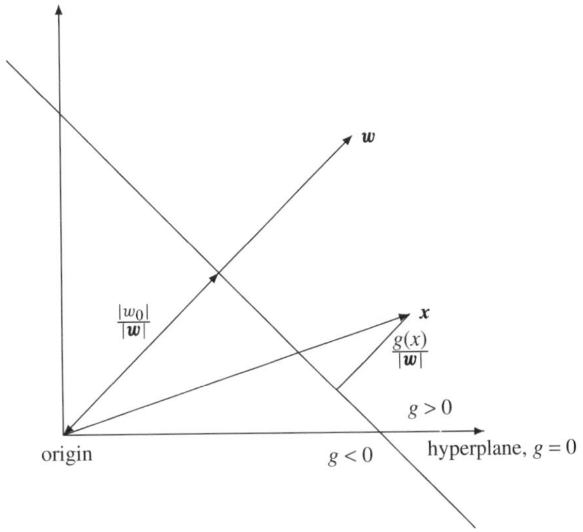

Figure 1.9 Geometry of linear discriminant function given by equation (1.20)

without making distributional assumptions, we may require the form of the discriminant function to be linear and determine its parameters (see Chapter 4).

A pattern classifier employing linear discriminant functions is termed a *linear machine* (Nilsson, 1965), an important special case of which is the *minimum-distance classifier* or nearest-neighbour rule. Suppose we are given a set of prototype points $\boldsymbol{p}_1, \ldots, \boldsymbol{p}_C$, one for each of the C classes $\omega_1, \ldots, \omega_C$. The minimum-distance classifier assigns a pattern $\boldsymbol{x}$ to the class ω_i associated with the nearest point $\boldsymbol{p}_i$. For each point, the squared Euclidean distance is

$$|\boldsymbol{x} - \boldsymbol{p}_i|^2 = \boldsymbol{x}^T\boldsymbol{x} - 2\boldsymbol{x}^T\boldsymbol{p}_i + \boldsymbol{p}_i^T\boldsymbol{p}_i$$

and minimum-distance classification is achieved by comparing the expressions $\boldsymbol{x}^T\boldsymbol{p}_i - \frac{1}{2}\boldsymbol{p}_i^T\boldsymbol{p}_i$ and selecting the largest value. Thus, the linear discriminant function is

$$g_i(\boldsymbol{x}) = \boldsymbol{w}_i^T\boldsymbol{x} + w_{i0}$$

where

$$\boldsymbol{w}_i = \boldsymbol{p}_i$$
$$w_{i0} = -\tfrac{1}{2}|\boldsymbol{p}_i|^2$$

Therefore, the minimum-distance classifier is a linear machine. If the prototype points, $\boldsymbol{p}_i$, are the class means, then we have the nearest class mean classifier. Decision regions for a minimum-distance classifier are illustrated in Figure 1.10. Each boundary is the perpendicular bisector of the lines joining the prototype points of regions that are contiguous. Also, note from the figure that the decision regions are convex (that is, two arbitrary points lying in the region can be joined by a straight line that lies entirely within the region). In fact, decision regions of a linear machine are always convex. Thus, the two class problems, illustrated in Figure 1.11, although separable, cannot be separated by a linear machine. Two generalisations that overcome this difficulty are piecewise linear discriminant functions and generalised linear discriminant functions.

Piecewise linear discriminant functions

This is a generalisation of the minimum-distance classifier to the situation in which there is more than one prototype per class. Suppose there are n_i prototypes in class ω_i, $\boldsymbol{p}_i^1, \ldots, \boldsymbol{p}_i^{n_i}, i = 1, \ldots, C$. We define the discriminant function for class ω_i to be

$$g_i(\boldsymbol{x}) = \max_{j=1,\ldots,n_i} g_i^j(\boldsymbol{x})$$

where g_i^j is a subsidiary discriminant function, which is linear and is given by

$$g_i^j(\boldsymbol{x}) = \boldsymbol{x}^T\boldsymbol{p}_i^j - \tfrac{1}{2}\boldsymbol{p}_i^{j\,T}\boldsymbol{p}_i^j \qquad j = 1, \ldots, n_i;\ i = 1, \ldots, C$$

A pattern $\boldsymbol{x}$ is assigned to the class for which $g_i(\boldsymbol{x})$ is largest; that is, to the class of the nearest prototype vector. This partitions the space into $\sum_{i=1}^{C} n_i$ regions known as

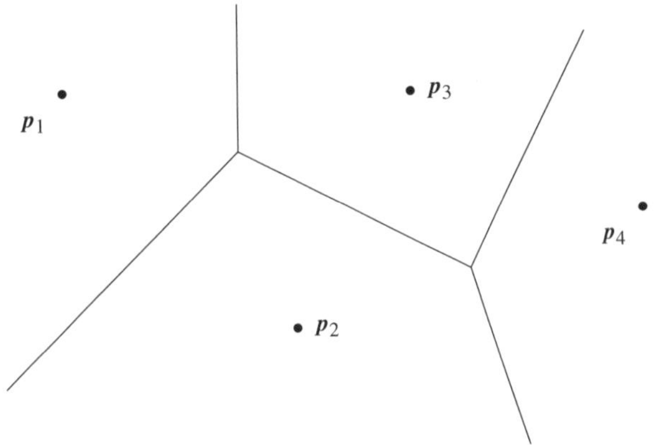

Figure 1.10 Decision regions for a minimum-distance classifier

Figure 1.11 Groups not separable by a linear discriminant

the Dirichlet tessellation of the space. When each pattern in the training set is taken as a prototype vector, then we have the nearest-neighbour decision rule of Chapter 3. This discriminant function generates a piecewise linear decision boundary (see Figure 1.12).

Rather than using the complete design set as prototypes, we may use a subset. Methods of reducing the number of prototype vectors (edit and condense) are described in Chapter 3, along with the nearest-neighbour algorithm. Clustering schemes may also be employed.

Generalised linear discriminant function

A *generalised linear discriminant function*, also termed a *phi machine* (Nilsson, 1965), is a discriminant function of the form

$$g(\boldsymbol{x}) = \boldsymbol{w}^T \boldsymbol{\phi} + w_0$$

where $\boldsymbol{\phi} = (\phi_1(\boldsymbol{x}), \dots, \boldsymbol{\phi}_D(\boldsymbol{x}))^T$ is a vector function of $\boldsymbol{x}$. If $D = p$, the number of variables, and $\phi_i(\boldsymbol{x}) = x_i$, then we have a linear discriminant function.

The discriminant function is linear in the functions ϕ_i, not in the original measurements x_i. As an example, consider the two-class problem of Figure 1.13. A linear discriminant function will not separate the classes, even though they are separable. However,

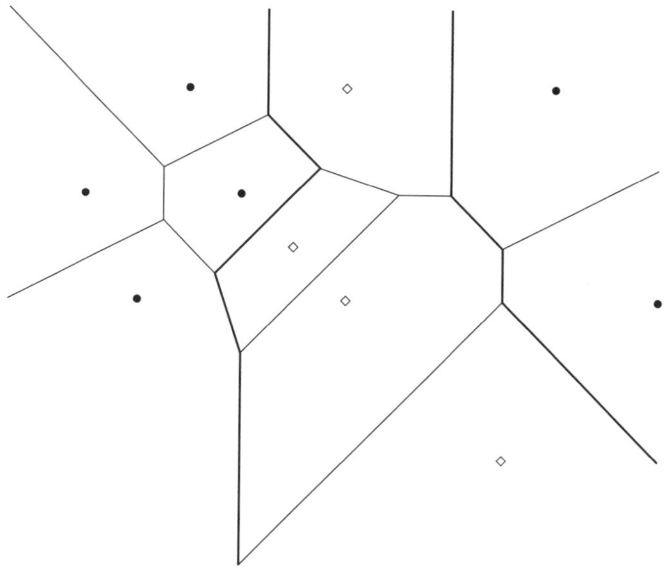

Figure 1.12 Dirichlet tessellation (comprising nearest-neighbour regions for a set of prototypes) and the decision boundary (thick lines) for two classes

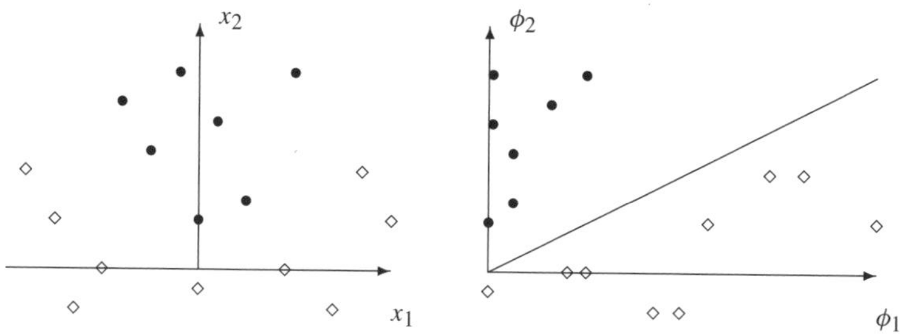

Figure 1.13 Nonlinear transformation of variables may permit linear discrimination

if we make the transformation

$$\phi_1(\boldsymbol{x}) = x_1^2$$
$$\phi_2(\boldsymbol{x}) = x_2$$

then the classes can be separated in the ϕ-space by a straight line. Similarly, disjoint classes can be transformed into a ϕ-space in which a linear discriminant function could separate the classes (provided that they are separable in the original space).

The problem, therefore, is simple. Make a good choice for the functions $\phi_i(\boldsymbol{x})$, then use a linear discriminant function to separate the classes. But, how do we choose ϕ_i? Specific examples are shown in Table 1.1.

Clearly there is a problem in that as the number of functions that are used as a basis set increases, so does the number of parameters that must be determined using the limited

Table 1.1 Discriminant functions, ϕ

Discriminant function	Mathematical form, $\phi_i(\boldsymbol{x})$
linear	$\phi_i(\boldsymbol{x}) = x_i,\ i = 1, \ldots, p$
quadratic	$\phi_i(\boldsymbol{x}) = x_{k_1}^{l_1} x_{k_2}^{l_2},\ i = 1, \ldots, (p+1)(p+2)/2 - 1$
	$l_1, l_2 = 0$ or 1; $k_1, k_2 = 1, \ldots, p$ l_1, l_2 not both zero
νth-order polynomial	$\phi_i(\boldsymbol{x}) = x_{k_1}^{l_1} \ldots x_{k_\nu}^{l_\nu},\ i = 1, \ldots, \binom{p+\nu}{\nu} - 1$
	$l_1, \ldots, l_\nu = 0$ or 1; $k_1, \ldots, k_\nu = 1, \ldots, p$
	l_i not all zero
radial basis function	$\phi_i(\boldsymbol{x}) = \phi(\lvert\boldsymbol{x} - \boldsymbol{v}_i\rvert)$ for centre $\boldsymbol{v}_i$ and function ϕ
multilayer perceptron	$\phi_i(\boldsymbol{x}) = f(\boldsymbol{x}^T \boldsymbol{v}_i + v_{i0})$ for direction $\boldsymbol{v}_i$ and offset v_{i0}. f is the logistic function, $f(z) = 1/(1 + \exp(-z))$

training set. A complete quadratic discriminant function requires $D = (p+1)(p+2)/2$ terms and so for C classes there are $C(p+1)(p+2)/2$ parameters to estimate. We may need to apply a constraint or 'regularise' the model to ensure that there is no over-fitting.

An alternative to having a set of different functions is to have a set of functions of the same parametric form, but which differ in the values of the parameters they take,

$$\phi_i(\boldsymbol{x}) = \phi(\boldsymbol{x};\ \boldsymbol{v}_i)$$

where $\boldsymbol{v}_i$ is a set of parameters. Different models arise depending on the way the variable $\boldsymbol{x}$ and the parameters $\boldsymbol{v}$ are combined. If

$$\phi(\boldsymbol{x};\ \boldsymbol{v}) = \phi(\lvert\boldsymbol{x} - \boldsymbol{v}\rvert)$$

that is, ϕ is a function only of the magnitude of the difference between the pattern $\boldsymbol{x}$ and the weight vector $\boldsymbol{v}$, then the resulting discriminant function is known as a *radial basis function*. On the other hand, if ϕ is a function of the scalar product of the two vectors

$$\phi(\boldsymbol{x};\ \boldsymbol{v}) = \phi(\boldsymbol{x}^T \boldsymbol{v} + v_0)$$

then the discriminant function is known as a *multilayer perceptron*. It is also a model known as projection pursuit. Both the radial basis function and the multilayer perceptron models can be used in regression.

In these latter examples, the discriminant function is no longer linear in the parameters. Specific forms for ϕ for radial basis functions and for the multilayer perceptron models will be given in Chapters 5 and 6.

Summary

In a multiclass problem, a pattern $\boldsymbol{x}$ is assigned to the class for which the discriminant function is the largest. A linear discriminant function divides the feature space by a hyperplane whose orientation is determined by the weight vector $\boldsymbol{w}$ and distance from the origin by the weight threshold w_0. The decision regions produced by linear discriminant functions are convex.

A piecewise linear discriminant function permits non-convex and disjoint decision regions. Special cases are the nearest-neighbour and nearest class mean classifier.

A generalised linear discriminant function, with fixed functions ϕ_i, is linear in its parameters. It permits non-convex and multiply connected decision regions (for suitable choices of ϕ_i). Radial basis functions and multilayer perceptrons can be regarded as generalised linear discriminant functions with flexible functions ϕ_i whose parameters must be determined or specified using the training set.

The Bayes decision rule is optimal (in the sense of minimising classification error) and with sufficient flexibility in our discriminant functions we ought to be able to achieve optimal performance in principle. However, we are limited by a finite number of training samples and also, once we start to consider parametric forms for the ϕ_i, we lose the simplicity and ease of computation of the linear functions.

1.6 Multiple regression

Many of the techniques and procedures described within this book are also relevant to problems in *regression*, the process of investigating the relationship between a dependent (or response) variable Y and independent (or predictor) variables $X_1, \ldots, X_p$; a regression function expresses the expected value of Y in terms of $X_1, \ldots, X_p$ and model parameters. Regression is an important part of statistical pattern recognition and, although the emphasis of the book is on discrimination, practical illustrations are sometimes given on problems of a regression nature.

The discrimination problem itself is one in which we are attempting to predict the values of one variable (the class variable) given measurements made on a set of independent variables (the pattern vector, $\boldsymbol{x}$). In this case, the response variable is categorical. Posing the discrimination problem as one in regression is discussed in Chapter 4.

Regression analysis is concerned with predicting the mean value of the response variable given measurements on the predictor variables and assumes a model of the form

$$E[y|\boldsymbol{x}] \stackrel{\triangle}{=} \int y p(y|\boldsymbol{x})\, dy = f(\boldsymbol{x}; \boldsymbol{\theta})$$

where f is a (possibly nonlinear) function of the measurements $\boldsymbol{x}$ and $\boldsymbol{\theta}$, a set of parameters of f. For example,

$$f(\boldsymbol{x}; \boldsymbol{\theta}) = \theta_0 + \boldsymbol{\theta}^T \boldsymbol{x}$$

where $\boldsymbol{\theta} = (\theta_1, \ldots, \theta_p)^T$, is a model that is linear in the parameters and the variables. The model

$$f(\boldsymbol{x}; \boldsymbol{\theta}) = \theta_0 + \boldsymbol{\theta}^T \boldsymbol{\phi}(\boldsymbol{x})$$

where $\boldsymbol{\theta} = (\theta_1, \ldots, \theta_D)^T$ and $\boldsymbol{\phi} = (\phi_1(\boldsymbol{x}), \ldots, \phi_D(\boldsymbol{x}))^T$ is a vector of nonlinear functions of $\boldsymbol{x}$, is linear in the parameters but nonlinear in the variables. *Linear regression*

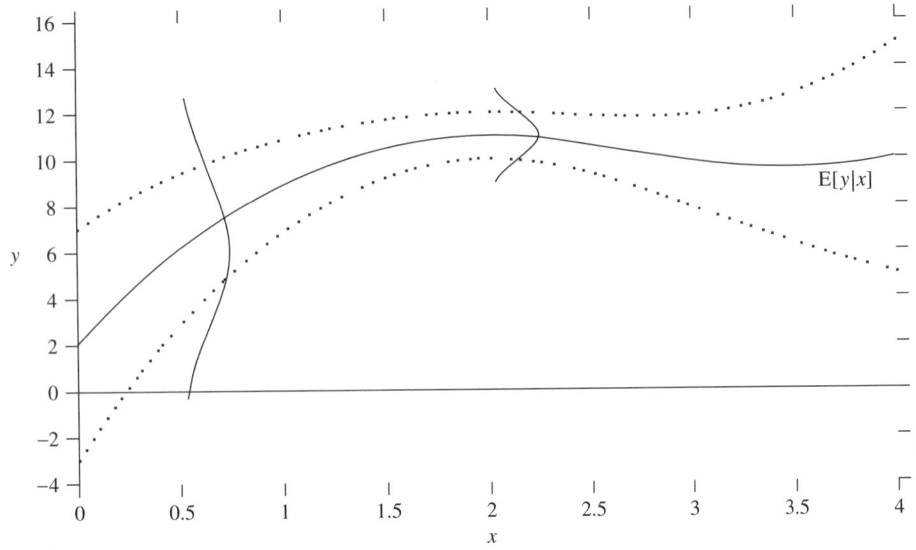

Figure 1.14 Population regression line (solid line) with representation of spread of conditional distribution (dotted lines) for normally distributed error terms, with variance depending on x

refers to a regression model that is linear in the parameters, but not necessarily in the variables.

Figure 1.14 shows a regression summary for some hypothetical data. For each value of x, there is a population of y values that varies with x. The solid line connecting the conditional means, $E[y|x]$, is the *regression line*. The dotted lines either side represent the spread of the conditional distribution (± 1 standard deviation from the mean).

It is assumed that the difference (commonly referred to as an error or residual), ϵ_i, between the measurement on the response variable and its predicted value conditional on the measurements on the predictors,

$$\epsilon_i = y_i - E[y|\boldsymbol{x}_i]$$

is an unobservable random variable. A normal model for the errors (see Appendix E) is often assumed,

$$p(\epsilon) = \frac{1}{\sqrt{2\pi}\sigma} \exp\left(-\frac{1}{2}\frac{\epsilon^2}{\sigma^2}\right)$$

That is,

$$p(y_i|\boldsymbol{x}_i, \boldsymbol{\theta}) = \frac{1}{\sqrt{2\pi}\sigma} \exp\left(-\frac{1}{2\sigma^2}(y_i - f(\boldsymbol{x}_i; \boldsymbol{\theta}))^2\right)$$

Given a set of data $\{(y_i, \boldsymbol{x}_i), i = 1, \ldots, n\}$, the maximum likelihood estimate of the model parameters (the value of the parameters for which the data are 'most likely', discussed further in Appendix B), $\boldsymbol{\theta}$, is that for which

$$p(\{(y_i, \boldsymbol{x}_i)\}|\boldsymbol{\theta})$$

is a maximum. Assuming independent samples, this amounts to determining the value of $\boldsymbol{\theta}$ for which the commonly used least squares error,

$$\sum_{i=1}^{n}(y_i - f(\boldsymbol{x}_i; \boldsymbol{\theta}))^2 \tag{1.21}$$

is a minimum (see the exercises at the end of the chapter).

For the linear model, procedures for estimating the parameters are described in Chapter 4.

1.7 Outline of book

The aim in writing this volume is to provide a comprehensive account of statistical pattern recognition techniques with emphasis on methods and algorithms for discrimination and classification. In recent years there have been many developments in multivariate analysis techniques, particularly in nonparametric methods for discrimination and classification. These are described in this book as extensions to the basic methodology developed over the years.

This chapter has presented some basic approaches to statistical pattern recognition. Supplementary material on probability theory and data analysis can be found in the appendices.

A road map to the book is given in Figure 1.15, which describes the basic pattern recognition cycle. The numbers in the figure refer to chapters and appendices of this book.

Chapters 2 and 3 describe basic approaches to supervised classification via Bayes' rule and estimation of the class-conditional densities. Chapter 2 considers normal-based models. Chapter 3 addresses nonparametric approaches to density estimation.

Chapters 4–7 take a discriminant function approach to supervised classification. Chapter 4 describes algorithms for linear discriminant functions. Chapter 5 considers kernel-based approaches for constructing nonlinear discriminant functions, namely radial basis functions and support vector machine methods. Chapter 6 describes alternative, projection-based methods, including the multilayer perceptron neural network. Chapter 7 describes tree-based approaches.

Chapter 8 addresses the important topic of performance assessment: how good is your designed classifier and how well does it compare with competing techniques? Can improvement be achieved with an ensemble of classifiers?

Chapters 9 and 10 consider techniques that may form part of an exploratory data analysis. Chapter 9 describes methods of feature selection and extraction, both linear and nonlinear. Chapter 10 addresses *unsupervised* classification or clustering.

Finally, Chapter 11 covers additional topics on pattern recognition including model selection.

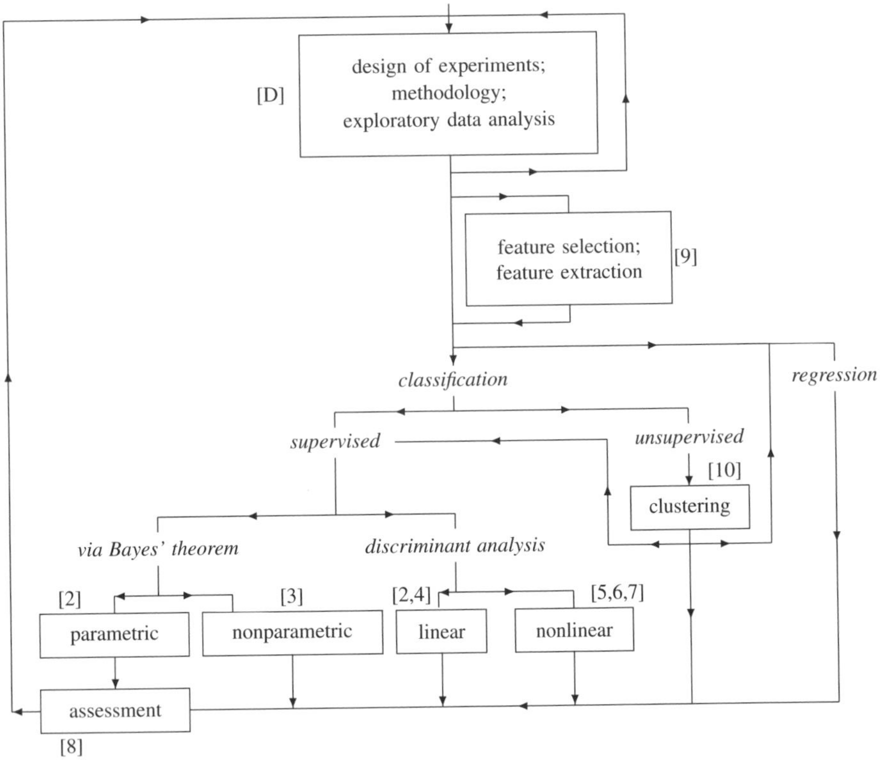

Figure 1.15 The pattern recognition cycle; numbers in parentheses refer to chapters and appendices of this book

1.8 Notes and references

There was a growth of interest in techniques for automatic pattern recognition in the 1960s. Many books appeared in the early 1970s, some of which are still very relevant today and have been revised and reissued. More recently, there has been another flurry of books on pattern recognition, particularly incorporating developments in neural network methods.

A very good introduction is provided by the book of Hand (1981a). Perhaps a little out of date now, it provides nevertheless a very readable account of techniques for discrimination and classification written from a statistical point of view and is to be recommended. Two of the main textbooks on statistical pattern recognition are those by Fukunaga (1990) and Devijver and Kittler (1982). Written perhaps with an engineering emphasis, Fukunaga's book provides a comprehensive account of the most important aspects of pattern recognition, with many examples, computer projects and problems. Devijver and Kittler's book covers the nearest-neighbour decision rule and feature selection and extraction in some detail, though not at the neglect of other important areas of statistical pattern recognition. It contains detailed mathematical accounts of techniques and algorithms, treating some areas in depth.

Another important textbook is that by Duda *et al.* (2001). Recently revised, this presents a thorough account of the main topics in pattern recognition, covering many recent developments. Other books that are an important source of reference material are those by Young and Calvert (1974), Tou and Gonzales (1974) and Chen (1973). Also, good accounts are given by Andrews (1972), a more mathematical treatment, and Therrien (1989), an undergraduate text.

Recently, there have been several books that describe the developments in pattern recognition that have taken place over the last decade, particularly the 'neural network' aspects, relating these to the more traditional methods. A comprehensive treatment of neural networks is provided by Haykin (1994). Bishop (1995) provides an excellent introduction to neural network methods from a statistical pattern recognition perspective. Ripley's (1996) account provides a thorough description of pattern recognition from within a statistical framework. It includes neural network methods, approaches developed in the field of machine learning, recent advances in statistical techniques as well as development of more traditional pattern recognition methods and gives valuable insights into many techniques gained from practical experience. Hastie *et al.* (2001) provide a thorough description of modern techniques in pattern recognition. Other books that deserve a mention are those by Schalkoff (1992) and Pao (1989).

Hand (1997) gives a short introduction to pattern recognition techniques and the central ideas in discrimination and places emphasis on the comparison and assessment of classifiers.

A more specialised treatment of discriminant analysis and pattern recognition is the book by McLachlan (1992a). This is a very good book. It is not an introductory textbook, but provides a thorough account of recent advances and sophisticated developments in discriminant analysis. Written from a statistical perspective, the book is a valuable guide to theoretical and practical work on statistical pattern recognition and is to be recommended for researchers in the field.

Comparative treatments of pattern recognition techniques (statistical, neural and machine learning methods) are provided in the volume edited by Michie *et al.* (1994) who report on the outcome of the *Statlog* project. Technical descriptions of the methods are given, together with the results of applying those techniques to a wide range of problems. This volume provides the most extensive comparative study available. More than 20 different classification procedures were considered for about 20 data sets.

The book by Watanabe (1985), unlike the books above, is not an account of statistical methods of discrimination, though some are included. Rather, it considers a wider perspective of human cognition and learning. There are many other books in this latter area. Indeed, in the early days of pattern recognition, many of the meetings and conferences covered the humanistic and biological side of pattern recognition in addition to the mechanical aspects. Although these non-mechanical aspects are beyond the scope of this book, the monograph by Watanabe provides one unifying treatment that we recommend for background reading.

There are many other books on pattern recognition. Some of those treating more specific parts (such as clustering) are cited in the appropriate chapter of this book. In addition, most textbooks on multivariate analysis devote some attention to discrimination and classification. These provide a valuable source of reference and are cited elsewhere in the book.

The website `www.statistical-pattern-recognition.net` contains references and links to further information on techniques and applications.

Exercises

In some of the exercises, it will be necessary to generate samples from a multivariate density with mean $\boldsymbol{\mu}$ and covariance matrix $\boldsymbol{\Sigma}$. Many computer packages offer routines for this. However, it is a simple matter to generate samples from a normal distribution with unit variance and zero mean (for example, Press *et al.*, 1992). Given a vector $\boldsymbol{Y}_i$ of such samples, then the vector $\boldsymbol{U}\boldsymbol{\Lambda}^{1/2}\boldsymbol{Y}_i + \boldsymbol{\mu}$ has the required distribution, where $\boldsymbol{U}$ is the matrix of eigenvectors of the covariance matrix and $\boldsymbol{\Lambda}^{1/2}$ is the diagonal matrix whose diagonal elements are the square roots of the corresponding eigenvalues (see Appendix C).

1. Consider two multivariate normally distributed classes,

$$p(\boldsymbol{x}|\omega_i) = \frac{1}{(2\pi)^{p/2}|\boldsymbol{\Sigma}_i|^{1/2}} \exp\left\{-\frac{1}{2}(\boldsymbol{x} - \boldsymbol{\mu}_i)^T \boldsymbol{\Sigma}_i^{-1}(\boldsymbol{x} - \boldsymbol{\mu}_i)\right\}$$

with means $\boldsymbol{\mu}_1$ and $\boldsymbol{\mu}_2$ and equal covariance matrices, $\boldsymbol{\Sigma}_1 = \boldsymbol{\Sigma}_2 = \boldsymbol{\Sigma}$. Show that the logarithm of the likelihood ratio is linear in the feature vector $\boldsymbol{x}$. What is the equation of the decision boundary?

2. Determine the equation of the decision boundary for the more general case of $\boldsymbol{\Sigma}_1 = \alpha\boldsymbol{\Sigma}_2$, for scalar α (normally distributed classes as in Exercise 1). In particular, for two univariate distributions, $N(0, 1)$ and $N(1, 1/4)$, show that one of the decision regions is bounded and determine its extent.

3. For the distributions in Exercise 1, determine the equation of the minimum risk decision boundary for the loss matrix

$$\boldsymbol{\Lambda} = \begin{pmatrix} 0 & 2 \\ 1 & 0 \end{pmatrix}$$

4. Consider two multivariate normally distributed classes (ω_2 with mean $(-1, 0)^T$ and ω_1 with mean $(1, 0)^T$, and identity covariance matrix). For a given threshold μ (see equation (1.14)) on the likelihood ratio, determine the regions Ω_1 and Ω_2 in a Neyman–Pearson rule.

5. Consider three bivariate normal distributions, ω_1, ω_2, ω_3 with identity covariance matrices and means $(-2, 0)^T$, $(0, 0)^T$ and $(0, 2)^T$. Show that the decision boundaries are piecewise linear. Now define a class A as the *mixture* of ω_1 and ω_3,

$$p_A(\boldsymbol{x}) = 0.5p(\boldsymbol{x}|\omega_1) + 0.5p(\boldsymbol{x}|\omega_3)$$

and class B as bivariate normal with identity covariance matrix and mean $(a, b)^T$, for some a, b. What is the equation of the Bayes decision boundary? Under what conditions is it piecewise linear?

6. Consider two uniform distributions with equal priors

$$p(x|\omega_1) = \begin{cases} 1 & \text{when } 0 \le x \le 1 \\ 0 & \text{otherwise} \end{cases}$$

$$p(x|\omega_2) = \begin{cases} \frac{1}{2} & \text{when } \frac{1}{2} \le x \le \frac{5}{2} \\ 0 & \text{otherwise} \end{cases}$$

   Show that the reject function is given by

$$r(t) = \begin{cases} \frac{3}{8} & \text{when } 0 \le t \le \frac{1}{3} \\ 0 & \text{when } \frac{1}{3} < t \le 1 \end{cases}$$

   Hence calculate the error rate by integrating (1.10).

7. Reject option. Consider two classes, each normally distributed with means $x = 1$ and $x = -1$ and unit variances; $p(\omega_1) = p(\omega_2) = 0.5$. Generate a test set and use it (without using class labels) to estimate the reject rate as a function of the threshold t. Hence, estimate the error rate for no rejection. Compare with the estimate based on a labelled version of the test set. Comment on the use of this procedure when the true distributions are unknown and the densities have to be estimated.

8. The area of a sphere of radius r in p dimensions, S_p, is

$$S_p = \frac{2\pi^{\frac{p}{2}} r^{p-1}}{\Gamma(p/2)}$$

   where Γ is the gamma function ($\Gamma(1/2) = \pi^{1/2}$, $\Gamma(1) = 1$, $\Gamma(x+1) = x\Gamma(x)$). Show that the probability of a sample, $\boldsymbol{x}$, drawn from a zero-mean normal distribution with covariance matrix $\sigma^2 \boldsymbol{I}$ ($\boldsymbol{I}$ is the identity matrix) and having $|\boldsymbol{x}| \le R$ is

$$\int_0^R S_p(r) \frac{1}{(2\pi\sigma^2)^{p/2}} \exp\left(-\frac{r^2}{2\sigma^2}\right) dr$$

   Evaluate this numerically for $R = 2\sigma$ and for $p = 1, \ldots, 10$. What do the results tell you about the distribution of normal samples in high-dimensional spaces?

9. In a two-class problem, let the cost of misclassifying a class ω_1 pattern be C_1 and the cost of misclassifying a class ω_2 pattern be C_2. Show that the point on the ROC curve that minimises the risk has gradient

$$\frac{C_2 p(\omega_2)}{C_1 p(\omega_1)}$$

10. Show that under the assumption of normally distributed residuals, the maximum likelihood solution for the parameters of a linear model is equivalent to minimising the sum-square error (1.21).

2

Density estimation – parametric

Overview

A discrimination rule may be constructed through explicit estimation of the class-conditional density functions and the use of Bayes' rule. One approach is to assume a simple parametric model for the density functions and to estimate the parameters of the model using an available training set.

2.1 Introduction

In Chapter 1 we considered the basic theory of pattern classification. All the information regarding the density functions $p(\boldsymbol{x}|\omega_i)$ was assumed known. In practice, this knowledge is often not or only partially available. Therefore, the next question that we must address is the estimation of the density functions themselves. If we can assume some parametric form for the distribution, perhaps from theoretical considerations, then the problem reduces to one of estimating a finite number of parameters. In this chapter, special consideration is given to the normal distribution which leads to algorithms for the Gaussian classifier.

We described in Chapter 1 how the minimum error decision is based on the probability of class membership $p(\omega_i|\boldsymbol{x})$, which may be written

$$p(\omega_i|\boldsymbol{x}) = p(\omega_i)\frac{p(\boldsymbol{x}|\omega_i)}{p(\boldsymbol{x})}$$

Assuming that the prior probability $p(\omega_i)$ is known, then in order to make a decision we need to estimate the class-conditional density $p(\boldsymbol{x}|\omega_i)$ (the probability density function $p(\boldsymbol{x})$ is independent of ω_i and therefore is not required in the decision-making process). The estimation of the density is based on a sample of observations $\mathcal{D}_i = \{\boldsymbol{x}^i_1, \ldots, \boldsymbol{x}^i_{n_i}\}$ ($\boldsymbol{x}^i_j \in \mathbb{R}^p$) from class ω_i. In this chapter and the following one we consider two basic approaches to density estimation: the parametric and nonparametric approaches. In the parametric approach, we assume that the class-conditional density for class ω_i is of a known form but has an unknown parameter, or set of parameters, θ_i, and we write this as $p(\boldsymbol{x}|\theta_i)$. In the *estimative approach* we use an estimate of the parameter θ_i, based on the samples $\mathcal{D}_i$ in the density. Thus we take

$$p(\boldsymbol{x}|\omega_i) = p(\boldsymbol{x}|\hat{\theta}_i)$$

where $\hat{\theta}_i = \hat{\theta}_i(\mathcal{D}_i)$ is an estimate of the parameter θ_i based on the sample. A different data sample, $\mathcal{D}_i$, would give rise to a different estimate $\hat{\theta}_i$, but the estimative approach does not take into account this sampling variability.

In the *predictive* or *Bayesian approach*, we write

$$p(\boldsymbol{x}|\omega_i) = \int p(\boldsymbol{x}|\theta_i)p(\theta_i|\mathcal{D}_i)\,d\theta_i$$

where $p(\theta_i|\mathcal{D}_i)$ can be regarded as a weighting function based on the data set $\mathcal{D}_i$, or as a full Bayesian posterior density function for θ_i based on a prior $p(\theta_i)$ and the data (Aitchison *et al.*, 1977). Thus, we admit that we do not know the true value of θ_i and instead of taking a single estimate, we take a weighted sum of the densities $p(\boldsymbol{x}|\theta_i)$, weighted by the distribution $p(\theta_i|\mathcal{D}_i)$. This approach may be regarded as making allowance for the sampling variability of the estimate of θ_i.

The alternative nonparametric approach to density estimation that we consider in this book does not assume a functional form for the density and is discussed in Chapter 3.

2.2 Normal-based models

2.2.1 Linear and quadratic discriminant functions

Perhaps the most widely used classifier is that based on the normal distribution (Appendix E),

$$p(\boldsymbol{x}|\omega_i) = \frac{1}{(2\pi)^{\frac{p}{2}}|\boldsymbol{\Sigma}_i|^{\frac{1}{2}}} \exp\left\{-\frac{1}{2}(\boldsymbol{x}-\boldsymbol{\mu}_i)^T\boldsymbol{\Sigma}_i^{-1}(\boldsymbol{x}-\boldsymbol{\mu}_i)\right\}$$

Classification is achieved by assigning a pattern to a class for which the posterior probability, $p(\omega_i|\boldsymbol{x})$, is the greatest, or equivalently $\log(p(\omega_i|\boldsymbol{x}))$. Using Bayes' rule and the normal assumption for the conditional densities above, we have

$$\begin{aligned}\log(p(\omega_i|\boldsymbol{x})) &= \log(p(\boldsymbol{x}|\omega_i)) + \log(p(\omega_i)) - \log(p(\boldsymbol{x}))\\ &= -\frac{1}{2}(\boldsymbol{x}-\boldsymbol{\mu}_i)^T\boldsymbol{\Sigma}_i^{-1}(\boldsymbol{x}-\boldsymbol{\mu}_i) - \frac{1}{2}\log(|\boldsymbol{\Sigma}_i|)\\ &\quad - \frac{p}{2}\log(2\pi) + \log(p(\omega_i)) - \log(p(\boldsymbol{x}))\end{aligned}$$

Since $p(\boldsymbol{x})$ is independent of class, the discriminant rule is: assign $\boldsymbol{x}$ to ω_i if $g_i > g_j$, for all $j \neq i$, where

$$g_i(\boldsymbol{x}) = \log(p(\omega_i)) - \tfrac{1}{2}\log(|\boldsymbol{\Sigma}_i|) - \tfrac{1}{2}(\boldsymbol{x}-\boldsymbol{\mu}_i)^T\boldsymbol{\Sigma}_i^{-1}(\boldsymbol{x}-\boldsymbol{\mu}_i) \qquad (2.1)$$

Classifying a pattern $\boldsymbol{x}$ on the basis of the values of $g_i(\boldsymbol{x}), i = 1, \ldots, C$, gives the *normal-based quadratic discriminant function* (McLachlan, 1992a).

In the *estimative approach*, the quantities $\boldsymbol{\mu}_i$ and $\boldsymbol{\Sigma}_i$ in the above are replaced by estimates based on a training set. Consider a set of samples, $\{\boldsymbol{x}_1, \ldots, \boldsymbol{x}_n\}$, $\boldsymbol{x}_j \in \mathbb{R}^p$, and a normal distribution characterised by $\boldsymbol{\theta} = (\boldsymbol{\mu}, \boldsymbol{\Sigma})$. Then the likelihood function, $L(\boldsymbol{x}_1, \ldots, \boldsymbol{x}_n|\boldsymbol{\theta})$, is

$$L(\boldsymbol{x}_1, \ldots, \boldsymbol{x}_n|\boldsymbol{\theta}) = \prod_{i=1}^{n} \frac{1}{(2\pi)^{\frac{p}{2}}|\boldsymbol{\Sigma}|^{\frac{1}{2}}} \exp\left\{-\frac{1}{2}(\boldsymbol{x}_i - \boldsymbol{\mu})^T \boldsymbol{\Sigma}^{-1}(\boldsymbol{x}_i - \boldsymbol{\mu})\right\}$$

Differentiating $\log(L)$ with respect to $\boldsymbol{\theta}$ gives the equations[1]

$$\frac{\partial \log(L)}{\partial \boldsymbol{\mu}} = \frac{1}{2}\sum_{i=1}^{n} \boldsymbol{\Sigma}^{-1}(\boldsymbol{x}_i - \boldsymbol{\mu}) + \frac{1}{2}\sum_{i=1}^{n}(\boldsymbol{\Sigma}^{-1})^T(\boldsymbol{x}_i - \boldsymbol{\mu})$$

and

$$\frac{\partial \log(L)}{\partial \boldsymbol{\Sigma}} = -\frac{n}{2}\boldsymbol{\Sigma}^{-1} + \frac{1}{2}\sum_{i=1}^{n}(\boldsymbol{x}_i - \boldsymbol{\mu})(\boldsymbol{x}_i - \boldsymbol{\mu})^T \boldsymbol{\Sigma}^{-1}\boldsymbol{\Sigma}^{-1}$$

where we have used the result that

$$\frac{\partial |\boldsymbol{A}|}{\partial \boldsymbol{A}} = [\mathrm{adj}(\boldsymbol{A})]^T = |\boldsymbol{A}|(\boldsymbol{A}^{-1})^T$$

Equating the above two equations to zero gives the maximum likelihood estimate of the mean as

$$\boldsymbol{m} = \frac{1}{n}\sum_{i=1}^{n} \boldsymbol{x}_i$$

the sample mean vector, and the covariance matrix estimate as

$$\hat{\boldsymbol{\Sigma}} = \frac{1}{n}\sum_{i=1}^{n}(\boldsymbol{x}_i - \boldsymbol{m})(\boldsymbol{x}_i - \boldsymbol{m})^T$$

the sample covariance matrix.

Substituting the estimates of the means and the covariance matrices (termed the 'plug-in estimates') of each class into (2.1) gives the *Gaussian classifier* or quadratic discrimination rule: assign $\boldsymbol{x}$ to ω_i if $g_i > g_j$, for all $j \neq i$, where

$$g_i(\boldsymbol{x}) = \log(p(\omega_i)) - \tfrac{1}{2}\log(|\hat{\boldsymbol{\Sigma}}_i|) - \tfrac{1}{2}(\boldsymbol{x} - \boldsymbol{m}_i)^T \hat{\boldsymbol{\Sigma}}_i^{-1}(\boldsymbol{x} - \boldsymbol{m}_i) \qquad (2.2)$$

If the training data have been gathered by sampling from the classes, then a plug-in estimate for the prior probability, $p(\omega_i)$, is $n_i / \sum_j n_j$, where n_i is the number of patterns in class ω_i.

[1] Differentiation with respect to a p-dimensional vector means differentiating with respect to each component of the vector. This gives a set of p equations which may be expressed as a vector equation (see Appendix C).

In the above, we may apply the discrimination rule to all members of the design set and the separate test set, if available. Problems will occur in the Gaussian classifier if any of the matrices $\hat{\boldsymbol{\Sigma}}_i$ is singular. There are several alternatives commonly employed. One is simply to use diagonal covariance matrices; that is, set all off-diagonal terms of $\hat{\boldsymbol{\Sigma}}_i$ to zero. Another approach is to project the data onto a space in which $\hat{\boldsymbol{\Sigma}}_i$ is nonsingular, perhaps using a principal components analysis (see Chapter 9), and then to use the Gaussian classifier in the reduced dimension space. Such an approach is assessed by Schott (1993) and linear transformations for reducing dimensionality are discussed in Chapter 9. A further alternative is to assume that the class covariance matrices $\boldsymbol{\Sigma}_1, \ldots, \boldsymbol{\Sigma}_C$ are all the same, in which case the discriminant function (2.1) simplifies and the discriminant rule becomes: assign $\boldsymbol{x}$ to ω_i if $g_i > g_j$, for all $j \neq i$, where g_i is the linear discriminant

$$g_i(\boldsymbol{x}) = \log(p(\omega_i)) - \tfrac{1}{2}\boldsymbol{m}_i^T \boldsymbol{S}_W^{-1} \boldsymbol{m}_i + \boldsymbol{x}^T \boldsymbol{S}_W^{-1} \boldsymbol{m}_i \tag{2.3}$$

in which $\boldsymbol{S}_W$ is the common group covariance matrix. This is the *normal-based linear discriminant function*. The maximum likelihood estimate is the pooled within-group sample covariance matrix

$$\boldsymbol{S}_W = \sum_{i=1}^{C} \frac{n_i}{n} \hat{\boldsymbol{\Sigma}}_i$$

The unbiased estimate is given by

$$\frac{n}{n-C}\boldsymbol{S}_W$$

A special case of (2.3) occurs when the matrix $\boldsymbol{S}_W$ is taken to be the identity and the class priors $p(\omega_i)$ are equal. This is the *nearest class mean classifier*: assign $\boldsymbol{x}$ to class ω_i if

$$-2\boldsymbol{x}^T \boldsymbol{m}_j + \boldsymbol{m}_j^T \boldsymbol{m}_j > -2\boldsymbol{x}^T \boldsymbol{m}_i + \boldsymbol{m}_i^T \boldsymbol{m}_i \text{ for all } i \neq j$$

For the special case of two classes, the rule (2.3) may be written: assign $\boldsymbol{x}$ to class ω_1 if

$$\boldsymbol{w}^T \boldsymbol{x} + w_0 > 0 \tag{2.4}$$

else assign $\boldsymbol{x}$ to class ω_2, where in the above

$$\begin{aligned} \boldsymbol{w} &= \boldsymbol{S}_W^{-1}(\boldsymbol{m}_1 - \boldsymbol{m}_2) \\ w_0 &= -\log\left(\frac{p(\omega_2)}{p(\omega_1)}\right) - \frac{1}{2}(\boldsymbol{m}_1 + \boldsymbol{m}_2)^T \boldsymbol{w} \end{aligned} \tag{2.5}$$

In problems where the data are from multivariate normal distributions with different covariance matrices, there may be insufficient data to obtain good estimates of class covariance matrices. Sampling variability may mean that it is better to assume equal covariance matrices (leading to the linear discriminant rule) rather than different covariance matrices. However, there are several intermediate covariance matrix structures (Flury, 1987) that may be considered without making the restrictive assumption of equality of covariance matrices. These include diagonal, but different, covariance matrices; common principal components; and proportional covariance matrices models. These models have been considered in the context of multivariate mixtures (see Section 2.3) and are discussed further in Chapter 9.

The linear discriminant rule (2.3) is quite robust to departures from the equal covariance matrix assumptions (Wahl and Kronmal, 1977; O'Neill, 1992), and may give better performance than the optimum quadratic discriminant rule for normally distributed classes when the true covariance matrices are unknown and the sample sizes are small. However, it is better to use the quadratic rule if the sample size is sufficient. The linear discriminant function can be greatly affected by non-normality and, if possible, variables should be transformed to approximate normality before applying the rule. Nonlinear transformations of data are discussed in Chapter 9.

2.2.2 Regularised discriminant analysis

Regularised discriminant analysis (RDA) was proposed by Friedman (1989) for small-sample, high-dimensional data sets as a means of overcoming the degradation in performance of the quadratic discriminant rule. Two parameters are involved: λ, a complexity parameter providing an intermediate between a linear and a quadratic discriminant rule; and γ, a shrinkage parameter for covariance matrix updates.

Specifically, $\hat{\boldsymbol{\Sigma}}_i$ is replaced by a linear combination, $\boldsymbol{\Sigma}_i^{\lambda}$, of the sample covariance matrix $\hat{\boldsymbol{\Sigma}}_i$ and the pooled covariance matrix $\boldsymbol{S}_W$,

$$\boldsymbol{\Sigma}_i^{\lambda} = \frac{(1-\lambda)\boldsymbol{S}_i + \lambda \boldsymbol{S}}{(1-\lambda)n_i + \lambda n} \tag{2.6}$$

where $0 \leq \lambda \leq 1$ and

$$\boldsymbol{S}_i = n_i \hat{\boldsymbol{\Sigma}}_i, \quad \boldsymbol{S} = n\boldsymbol{S}_W$$

At the extremes of $\lambda = 0$ and $\lambda = 1$ we have covariance matrix estimates that lead to the quadratic discriminant rule and the linear discriminant rule respectively:

$$\boldsymbol{\Sigma}_i^{\lambda} = \begin{cases} \hat{\boldsymbol{\Sigma}}_i & \lambda = 0 \\ \boldsymbol{S}_W & \lambda = 1 \end{cases}$$

The second parameter γ is used to *regularise* the sample class covariance matrix further beyond that provided by (2.6),

$$\boldsymbol{\Sigma}_i^{\lambda,\gamma} = (1-\gamma)\boldsymbol{\Sigma}_i^{\lambda} + \gamma c_i(\lambda)\boldsymbol{I}_p \tag{2.7}$$

where $\boldsymbol{I}_p$ is the $p \times p$ identity matrix and

$$c_i(\lambda) = \mathrm{Tr}\{\boldsymbol{\Sigma}_i^{\lambda}\}/p,$$

the average eigenvalue of $\boldsymbol{\Sigma}_i^{\lambda}$.

The matrix $\boldsymbol{\Sigma}_i^{\lambda,\gamma}$ is then used as the plug-in estimate of the covariance matrix in the normal-based discriminant rule: assign $\boldsymbol{x}$ to ω_i if $g_i > g_j$, for all $j \neq i$, where

$$g_i(\boldsymbol{x}) = -\tfrac{1}{2}(\boldsymbol{x} - \boldsymbol{m}_i)^T[\boldsymbol{\Sigma}_i^{\lambda,\gamma}]^{-1}(\boldsymbol{x} - \boldsymbol{m}_i) - \tfrac{1}{2}\log(|\boldsymbol{\Sigma}_i^{\lambda,\gamma}|) + \log(p(\omega_i))$$

Friedman's RDA approach involves choosing the values of λ and γ to minimise a *cross-validated* estimate of future misclassification cost (that is, a robust estimate of the error rate using a procedure called cross-validation – see Chapter 8). The strategy is to evaluate the misclassification risk on a grid of points ($0 \leq \lambda, \gamma \leq 1$) and to choose the optimal values of λ and γ to be those grid point values with smallest estimated risk.

Robust estimates of the covariance matrices (see Chapter 11) may also be incorporated in the analysis. Instead of using $\boldsymbol{\Sigma}_i^\lambda$ in (2.7), Friedman proposes $\tilde{\boldsymbol{\Sigma}}_i^\lambda$, given by

$$\tilde{\boldsymbol{\Sigma}}_i^\lambda = \frac{(1-\lambda)\tilde{\boldsymbol{S}}_i + \lambda\tilde{\boldsymbol{S}}}{W_i^\lambda}$$

where

$$\tilde{\boldsymbol{S}}_i = \sum_{j=1}^{n} z_{ij} w_j (\boldsymbol{x}_j - \tilde{\boldsymbol{m}}_i)(\boldsymbol{x}_j - \tilde{\boldsymbol{m}}_i)^T$$

$$\tilde{\boldsymbol{S}} = \sum_{k=1}^{C} \tilde{\boldsymbol{S}}_k$$

$$\tilde{\boldsymbol{m}}_i = \sum_{j=1}^{n} z_{ij} w_j \boldsymbol{x}_j / W_i$$

$$W_i = \sum_{j=1}^{n} z_{ij} w_j$$

$$W = \sum_{i=1}^{C} W_i$$

$$W_i^\lambda = (1-\lambda)W_i + \lambda W$$

in which the w_j ($0 \leq w_j \leq 1$) are weights associated with each observation and $z_{ij} = 1$ if $\boldsymbol{x}_j \in$ class ω_i and 0 otherwise. For $w_j = 1$ for all j, $\tilde{\boldsymbol{\Sigma}}_i^\lambda = \boldsymbol{\Sigma}_i^\lambda$.

In order to reduce the computation cost, a strategy that updates a covariance matrix when one observation is removed from a data set is employed.

Friedman (1989) assesses the effectiveness of RDA on simulated and real data sets. He finds that model selection based on the cross-validation procedure performs well, and that fairly accurate classification can be achieved with RDA for a small ratio of the number of observations (n) to the number of variables (p). He finds that RDA has the potential for dramatically increasing the power of discriminant analysis when sample sizes are small and the number of variables is large.

In conclusion, RDA can improve classification performance when the covariance matrices are not close to being equal and/or the sample size is too small for quadratic discriminant analysis to be viable.

2.2.3 Example application study

The problem The purpose of this study is to investigate the feasibility of predicting the degree of recovery of patients entering hospital with severe head injury using data collected shortly after injury (Titterington *et al.*, 1981).

Summary Titterington *et al.* (1981) report the results of several classifiers, each designed using different training sets. In this example, results of the application of a quadratic discriminant rule (2.2) are presented.

The data The data set comprises measurements on patients entering hospital with a head injury involving a minimum degree of brain damage. Measurements are made on six categorical variables: `Age`, grouped into decades 0–9, 10–19, ... , 60–69, 70+; `EMV` score, relating to eye, motor and verbal responses to stimulation, grouped into seven categories; `MRP`, a summary of the motor responses in all four limbs, graded 1 (nil) to 7 (normal); `Change`, the change in neurological function over the first 24 hours, graded 1 (deteriorating) to 3 (improving); `Eye Indicant`, a summary of eye movement scores, graded 1 (bad) to 3 (good); `Pupils`, the reaction of pupils to light, graded 1 (non-reacting) or 2 (reacting). There are 500 patients in the training and test sets, distributed over three classes related to the predicted outcome (dead or vegetative; severe disability; and moderate disability or good recovery). The number of patterns in each of the three classes for the training and the test sets are: training – 259, 52, 189; test – 250, 48, 202. Thus there is an uneven class distribution. Also, there are many missing values. These have been substituted by class means on training and population means on test. Further details of the data are given by Titterington *et al.* (1981).

The model The data in each class are modelled using a normal distribution leading to the discriminant rule (2.2).

Training procedure Training consists of estimating the quantities $\{\boldsymbol{m}_i, \hat{\boldsymbol{\Sigma}}_i, p(\omega_i), i = 1, \dots, C\}$, the sample mean, sample covariance matrix and prior class probability for each class from the data. The prior class probability is taken to be $p(\omega_i) = n_i/n$. Once $\hat{\boldsymbol{\Sigma}}_i$ has been estimated, a numerical procedure must be used to calculate the inverse, $\hat{\boldsymbol{\Sigma}}_i^{-1}$, and the determinant, $|\boldsymbol{\Sigma}_i|$. Once calculated, these quantities are substituted into equation (2.2) to give C functions, $g_i(\boldsymbol{x})$.

For each pattern, $\boldsymbol{x}$, in the training and test set, $g_i(\boldsymbol{x})$ is calculated and $\boldsymbol{x}$ assigned to the class for which the corresponding discriminant function, $g_i(\boldsymbol{x})$, is the largest.

Results Results on training and test sets for a Gaussian classifier (quadratic rule) are given in Table 2.1 as misclassification matrices or *confusion matrices* (see also Exercise 2). Note that class 2 is nearly always classified incorrectly as class 1 or class 3.

Table 2.1 Left: confusion matrix for training data; right: results for the test data

		True class		
		1	2	3
Predicted	1	209	22	15
class	2	0	1	1
	3	50	29	173

		True class		
		1	2	3
Predicted	1	188	19	29
class	2	3	1	2
	3	59	28	171

2.2.4 Further developments

There have also been several investigations of the robustness of the linear and quadratic discriminant rules to certain types of non-normality (for example, Lachenbruch *et al.*, 1973; Chingánda and Subrahmaniam, 1979; Ashikaga and Chang, 1981; Balakrishnan and Subrahmaniam, 1985).

Robustness of the discrimination rule to outliers is discussed by Todorov *et al.* (1994); see Krusińska (1988) for a review and also Chapter 11.

Aeberhard *et al.* (1994) report an extensive simulation study on eight statistical classification methods applied to problems when the number of observations is less than the number of variables. They found that out of the techniques considered, RDA was the most powerful, being outperformed by linear discriminant analysis only when the class covariance matrices were identical and for a large training set size. Reducing the dimensionality by feature extraction methods generally led to poorer results. However, Schott (1993) finds that dimension reduction prior to quadratic discriminant analysis can substantially reduce misclassification rates for small sample sizes. It also decreases the sample sizes necessary for quadratic discriminant analysis to be preferred over linear discriminant analysis. Alternative approaches to the problem of discriminant analysis with singular covariance matrices are described by Krzanowski *et al.* (1995).

Further simulations have been carried out by Rayens and Greene (1991) who compare RDA with an approach based on an empirical Bayes framework for addressing the problem of unstable covariance matrices (see also Greene and Rayens, 1989). Aeberhard *et al.* (1993) propose a modified model selection procedure for RDA and Celeux and Mkhadri (1992) present a method of regularised discriminant analysis for discrete data. Expressions for the shrinkage parameter are proposed by Loh (1995) and Mkhadri (1995).

An alternative regularised Gaussian discriminant analysis approach is proposed by Bensmail and Celeux (1996). Termed eigenvalue decomposition discriminant analysis, it is based on the reparametrisation of the covariance matrix of a class in terms of its eigenvalue decomposition. Fourteen different models are assessed, and results compare favourably with RDA. Raudys (2000) considers a similar development.

Hastie *et al.* (1995) cast the discrimination problem as one of regression using *optimal scaling* and use a penalised regression procedure (regularising the within-class covariance matrix). In situations where there are many highly correlated variables, their procedure offers promising results.

Extensions of linear and quadratic discriminant analysis to data sets where the patterns are curves or functions are developed by James and Hastie (2001).

2.2.5 Summary

Linear and quadratic discriminants (or equivalently, Gaussian classifiers) are widely used methods of supervised classification and are supported by many statistical packages. Problems occur when the covariance matrices are close to singular and when class boundaries are nonlinear. The former can be overcome by regularisation. This can be achieved by imposing structure on the covariance matrices, pooling/combining covariance matrices or adding a penalty term to the within-class scatter. Friedman (1989) proposes a scheme that includes a combination of matrices and the addition of a penalty term.

2.3 Normal mixture models

Finite mixture models have received wide application, being used to model distributions where the measurements arise from separate groups, but individual membership is unknown. As methods of density estimation, mixture models are more flexible than the simple normal-based models of Section 2.2, providing improved discrimination in some circumstances. Applications of mixture models include (see Section 2.5) textile flaw detection (where the data comprise measurements from a background 'noise' and a flaw), waveform classification (where the signal may comprise a sample from a waveform or noise) and target classification (in which the radar target density is approximated by a sum of simple component densities).

There are several issues associated with mixture models that are of interest. The most important for density estimation concerns the estimation of the model parameters. We may also be interested in how many components are present and whether there are any 'natural' groupings in the data that may be identified. This is the problem of clustering (or unsupervised classification) that we return to in Chapter 10.

2.3.1 Maximum likelihood estimation via EM

A finite mixture model is a distribution of the form

$$p(\boldsymbol{x}) = \sum_{j=1}^{g} \pi_j p(\boldsymbol{x}; \boldsymbol{\theta}_j)$$

where g is the number of mixture components, $\pi_j \geq 0$ are the mixing proportions ($\sum_{j=1}^{g} \pi_j = 1$) and $p(\boldsymbol{x}; \boldsymbol{\theta}_j)$, $j = 1, \ldots, g$, are the *component density* functions which depend on a parameter vector $\boldsymbol{\theta}_j$. There are three sets of parameters to estimate: the values of π_j, the components of $\boldsymbol{\theta}_j$ and the value of g. The component densities may be of different parametric forms and are specified using knowledge of the data generation process, if available. In the normal mixture model, $p(\boldsymbol{x}; \boldsymbol{\theta}_j)$ is the multivariate normal distribution, with $\boldsymbol{\theta}_j = \{\boldsymbol{\mu}_j, \boldsymbol{\Sigma}_j\}$.

Given a set of n observations $(\boldsymbol{x}_1, \ldots, \boldsymbol{x}_n)$, the likelihood function is

$$L(\boldsymbol{\Psi}) = \prod_{i=1}^{n} \sum_{j=1}^{g} \pi_j p(\boldsymbol{x}_i | \boldsymbol{\theta}_j) \tag{2.8}$$

where $\boldsymbol{\Psi}$ denotes the set of parameters $\{\pi_1, \ldots, \pi_g; \boldsymbol{\theta}_1, \ldots, \boldsymbol{\theta}_g\}$ and we now denote the dependence of the component densities on their parameters as $p(\boldsymbol{x}|\boldsymbol{\theta}_j)$. In general, it is not possible to solve $\partial L / \partial \boldsymbol{\Psi} = 0$ explicitly for the parameters of the model and iterative schemes must be employed. One approach for maximising the likelihood $L(\boldsymbol{\Psi})$ is to use a general class of iterative procedures known as EM (expectation – maximisation) algorithms, introduced in the context of missing data estimation by Dempster *et al.* (1977), though it had appeared in many forms previously.

The basic procedure is as follows. We suppose that we have a set of 'incomplete' data vectors $\{\boldsymbol{x}\}$ and we wish to maximise the likelihood $L(\boldsymbol{\Psi}) = p(\{\boldsymbol{x}\}|\boldsymbol{\Psi})$. Let $\{\boldsymbol{y}\}$ denote a typical 'complete' version of $\{\boldsymbol{x}\}$, that is, each vector $\boldsymbol{x}_i$ is augmented by the 'missing' values so that $\boldsymbol{y}_i^T = (\boldsymbol{x}_i^T, \boldsymbol{z}_i^T)$. There may be many possible vectors $\boldsymbol{y}_i$ in which we can embed $\boldsymbol{x}_i$, though there may be a natural choice for some problems. In the finite mixture case, $\boldsymbol{z}_i$ is a class indicator vector $\boldsymbol{z}_i = (z_{1i}, \ldots, z_{gi})^T$, where $z_{ji} = 1$ if $\boldsymbol{x}_i$ belongs to the jth component and zero otherwise.

General EM procedure

Let the likelihood of $\{\boldsymbol{y}\}$ be $g(\{\boldsymbol{y}\}|\boldsymbol{\Psi})$ whose form we know explicitly so that the likelihood $p(\{\boldsymbol{x}\}|\boldsymbol{\Psi})$ is obtained from $g(\{\boldsymbol{y}\}|\boldsymbol{\Psi})$ by integrating over all possible $\{\boldsymbol{y}\}$ in which the set $\{\boldsymbol{x}\}$ is embedded:

$$L(\boldsymbol{\Psi}) = p(\{\boldsymbol{x}\}|\boldsymbol{\Psi}) = \int \prod_{i=1}^{n} g(\boldsymbol{x}_i, \boldsymbol{z}|\boldsymbol{\Psi})\, d\boldsymbol{z}$$

The EM procedure generates a sequence of estimates of $\boldsymbol{\Psi}$, $\{\boldsymbol{\Psi}^{(m)}\}$, from an initial estimate $\boldsymbol{\Psi}^{(0)}$ and consists of two steps:

1. E-step: Evaluate $Q(\boldsymbol{\Psi}, \boldsymbol{\Psi}^{(m)}) \triangleq \mathrm{E}[\log(g(\{\boldsymbol{y}\}|\boldsymbol{\Psi}))|\{\boldsymbol{x}\}, \boldsymbol{\Psi}^{(m)}]$, that is,

$$Q(\boldsymbol{\Psi}, \boldsymbol{\Psi}^{(m)}) = \int \sum_i \log(g(\boldsymbol{x}_i, \boldsymbol{z}_i|\boldsymbol{\Psi}))\, p(\{\boldsymbol{z}\}|\{\boldsymbol{x}\}, \boldsymbol{\Psi}^{(m)})\, d\boldsymbol{z}_1 \ldots d\boldsymbol{z}_n$$

   the expectation of the complete data log-likelihood, conditional on the observed data, $\{\boldsymbol{x}\}$, and the current value of the parameters, $\boldsymbol{\Psi}^{(m)}$.

2. M-step: Find $\boldsymbol{\Psi} = \boldsymbol{\Psi}^{(m+1)}$ that maximises $Q(\boldsymbol{\Psi}, \boldsymbol{\Psi}^{(m)})$. Often the solution for the M-step may be obtained in closed form.

The likelihoods of interest satisfy

$$L\{\boldsymbol{\Psi}^{(m+1)}\} \geq L\{\boldsymbol{\Psi}^{(m)}\}$$

so they are monotonically increasing (see the exercises).

An illustration of the EM iterative scheme is shown in Figure 2.1. It is one of a class of *iterative majorisation* schemes (de Leeuw 1977; de Leeuw and Heiser 1977, 1980) that find a local maximum of a function $f(\boldsymbol{\Psi})$ by defining an auxiliary function, $Q(\boldsymbol{\Psi}, \boldsymbol{\Psi}^{(m)})$, that touches the function f at the point $(\boldsymbol{\Psi}^{(m)}, f(\boldsymbol{\Psi}^{(m)}))$ and lies everywhere else below it (strictly, iterative majorisation refers to a procedure for finding the minimum of a function by iteratively defining a majorising function). The auxiliary function is maximised and the position of the maximum, $\boldsymbol{\Psi}^{(m+1)}$, gives a value of the original function f that is greater than at the previous iteration. This process is repeated, with a new auxiliary function being defined that touches the curve at $(\boldsymbol{\Psi}^{(m+1)}, f(\boldsymbol{\Psi}^{(m+1)}))$, and continues until convergence. The shape as well as the position of the auxiliary function will also change as the iteration proceeds. In the case of the EM algorithm, $Q(\boldsymbol{\Psi}, \boldsymbol{\Psi}^{(m)})$ differs from the log-likelihood at $\boldsymbol{\Psi}$ by $H(\boldsymbol{\Psi}, \boldsymbol{\Psi}^{(m)})$ (see Dempster *et al.*, 1977, for details):

$$\log(L(\boldsymbol{\Psi})) = Q(\boldsymbol{\Psi}, \boldsymbol{\Psi}^{(m)}) - H(\boldsymbol{\Psi}, \boldsymbol{\Psi}^{(m)})$$

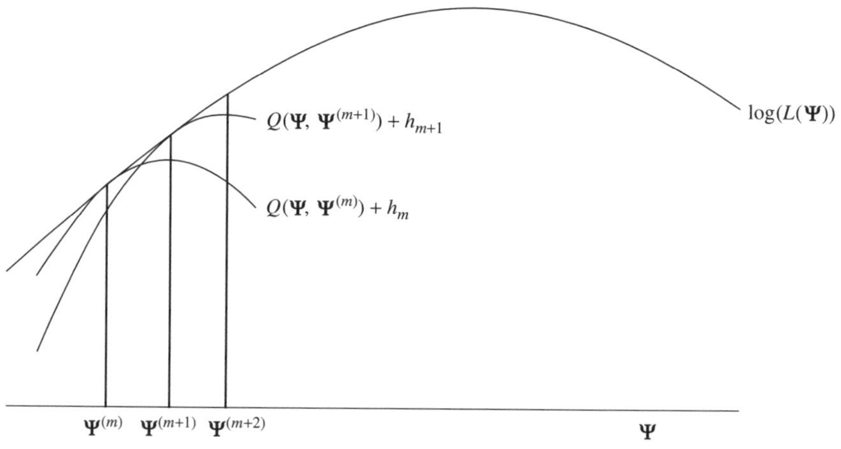

Figure 2.1 EM illustration: successive maximisation of the function $Q(\Psi, \Psi^{(m)})$ leads to increases in the log-likelihood

where

$$\begin{aligned} H(\Psi, \Psi^{(m)}) &= \mathrm{E}[\log(g(\{y\}|\Psi)/p(\{x\}|\Psi))|\{x\}, \Psi^{(m)}] \\ &= \int \sum_i \log(p(z_i|\{x\}, \Psi))p(z_i|\{x\}, \Psi^{(m)})\, dz_1 \ldots dz_n \end{aligned} \tag{2.9}$$

and in Figure 2.1, $h_m = -H(\Psi^{(m)}, \Psi^{(m)})$.

EM algorithm for mixtures

Let us now consider the application of the EM algorithm to mixture distributions. For fully labelled data, we define the complete data vector y to be the observation augmented by a class label; that is, $y^T = (x^T, z^T)$, where z is an indicator vector of length g with a 1 in the kth position if x is in category k and zeros elsewhere. The likelihood of y is

$$\begin{aligned} g(y|\Psi) &= p(x|z, \Psi)p(z|\Psi) \\ &= p(x|\theta_k)\pi_k \end{aligned}$$

which may be written as

$$g(y|\Psi) = \prod_{j=1}^{g} \left[p(x|\theta_j)\pi_j\right]^{z_j}$$

since z_j is zero except for $j = k$. The likelihood of x is

$$\begin{aligned} p(x|\Psi) &= \sum_{\text{all possible } z \text{ values}} g(y|\Psi) \\ &= \sum_{j=1}^{g} \pi_j p(x|\theta_j) \end{aligned}$$

which is a mixture distribution. Thus, we may interpret mixture data as incomplete data where the missing values are the class labels.

For n observations we have

$$g(\boldsymbol{y}_1, \ldots, \boldsymbol{y}_n|\boldsymbol{\Psi}) = \prod_{i=1}^{n}\prod_{j=1}^{g}[p(\boldsymbol{x}_i|\boldsymbol{\theta}_j)\pi_j]^{z_{ji}}$$

with

$$\log(g(\boldsymbol{y}_1, \ldots, \boldsymbol{y}_n|\boldsymbol{\Psi})) = \sum_{i=1}^{n} \boldsymbol{z}_i^T \boldsymbol{l} + \sum_{i=1}^{n} \boldsymbol{z}_i^T \boldsymbol{u}_i(\boldsymbol{\theta})$$

where the vector $\boldsymbol{l}$ has jth component $\log(\pi_j)$, $\boldsymbol{u}_i$ has jth component $\log(p(\boldsymbol{x}_i|\boldsymbol{\theta}_j))$ and $\boldsymbol{z}_i$ has components z_{ji}, $j = 1, \ldots, g$, where z_{ji} are the indicator variables taking value one if pattern $\boldsymbol{x}_i$ is in group j, and zero otherwise. The likelihood of $(\boldsymbol{x}_1, \ldots, \boldsymbol{x}_n)$ is $L_0(\boldsymbol{\Psi})$, as follows. given by (2.8). The steps in the basic iteration are

1. E-step: Form

$$Q(\boldsymbol{\Psi}, \boldsymbol{\Psi}^{(m)}) = \sum_{i=1}^{n} \boldsymbol{w}_i^T \boldsymbol{l} + \sum_{i=1}^{n} \boldsymbol{w}_i^T \boldsymbol{u}_i(\boldsymbol{\theta})$$

where

$$\boldsymbol{w}_i = \mathrm{E}(\boldsymbol{z}_i|\boldsymbol{x}_i, \boldsymbol{\Psi}^{(m)})$$

with jth component, the probability that $\boldsymbol{x}_i$ belongs to group j given the current estimates $\boldsymbol{\Psi}^{(m)}$, given by

$$w_{ij} = \frac{\pi_j^{(m)} p(\boldsymbol{x}_i|\boldsymbol{\theta}_j^{(m)})}{\sum_k \pi_k^{(m)} p(\boldsymbol{x}_i|\boldsymbol{\theta}_k^{(m)})} \tag{2.10}$$

2. M-step: This consists of maximising Q with respect to $\boldsymbol{\Psi}$. Consider the parameters π_i, $\boldsymbol{\theta}_i$ in turn. Maximising Q with respect to π_i (subject to the constraint that $\sum_{j=1}^{g} \pi_j = 1$) leads to the equation

$$\sum_{i=1}^{n} w_{ij} \frac{1}{\pi_j} - \lambda = 0$$

obtained by differentiating $Q - \lambda(\sum_{j=1}^{g} \pi_j - 1)$ with respect to π_j, where λ is a Lagrange multiplier. The constraint $\sum \pi_j = 1$ gives $\lambda = \sum_{j=1}^{g} \sum_{i=1}^{n} w_{ij} = n$ and we have the estimate of π_j as

$$\hat{\pi}_j = \frac{1}{n} \sum_{i=1}^{n} w_{ij} \tag{2.11}$$

For *normal* mixtures, $\boldsymbol{\theta}_i = (\boldsymbol{\mu}_i, \boldsymbol{\Sigma}_i)$ and we consider the mean and covariance matrix re-estimation separately. Differentiating Q with respect to $\boldsymbol{\mu}_j$ and equating to zero gives

$$\sum_{i=1}^{n} w_{ij}(\boldsymbol{x}_i - \boldsymbol{\mu}_j) = 0$$

which gives the re-estimation for $\boldsymbol{\mu}_j$ as

$$\hat{\boldsymbol{\mu}}_j = \frac{\sum_{i=1}^n w_{ij}\boldsymbol{x}_i}{\sum_{i=1}^n w_{ij}} = \frac{1}{n\hat{\pi}_j}\sum_{i=1}^n w_{ij}\boldsymbol{x}_i \tag{2.12}$$

Differentiating Q with respect to $\boldsymbol{\Sigma}_j$ and equating to zero gives

$$\begin{aligned}\hat{\boldsymbol{\Sigma}}_j &= \frac{\sum_{i=1}^n w_{ij}(\boldsymbol{x}_i - \hat{\boldsymbol{\mu}}_j)(\boldsymbol{x}_i - \hat{\boldsymbol{\mu}}_j)^T}{\sum_{i=1}^n w_{ij}} \\ &= \frac{1}{n\hat{\pi}_j}\sum_{i=1}^n w_{ij}(\boldsymbol{x}_i - \hat{\boldsymbol{\mu}}_j)(\boldsymbol{x}_i - \hat{\boldsymbol{\mu}}_j)^T\end{aligned} \tag{2.13}$$

Thus, the EM algorithm for normal mixtures alternates between the E-step of estimating the $\boldsymbol{w}_i$ (equation (2.10)) and the M-step of calculating $\hat{\pi}_j$, $\hat{\boldsymbol{\mu}}_j$ and $\hat{\boldsymbol{\Sigma}}_j$ $(j = 1, \ldots, g)$ given the values of $\boldsymbol{w}_i$ (equations (2.11), (2.12) and (2.13)). These estimates become the estimates at stage $m+1$ and are substituted into the right-hand side of (2.10) for the next stage of the iteration. The process iterates until convergence of the likelihood.

Discussion

The EM procedure is very easy to implement, but the convergence rate can be poor depending on the data distribution and the initial estimates for the parameters. Optimisation procedures, in addition to the EM algorithm, include Newton–Raphson iterative schemes (Hasselblad, 1966) and simulated annealing (Ingrassia, 1992).

One of the main problems with likelihood optimisation is that there is a multitude of 'useless' global maxima (Titterington *et al.*, 1985). For example, if the mean of one of the groups is taken as one of the sample points, then the likelihood tends to infinity as the variance of the component centred on that sample point tends to zero. Similarly, if sample points are close together, then there will be high local maxima of the likelihood function and it appears that the maximum likelihood procedure fails for this class of mixture models. However, provided that we do not allow the variances to tend to zero, perhaps by imposing an equality constraint on the covariance matrices, then the maximum likelihood method is still viable (Everitt and Hand, 1981). Equality of covariance matrices may be a rather restrictive assumption in many applications. Convergence to parameter values associated with singularities is more likely to occur with small sample sizes and when components are not well separated.

2.3.2 Mixture models for discrimination

If we wish to use the normal mixture model in a discrimination problem, one approach is to obtain the parameters $\boldsymbol{\Psi}_1, \ldots \boldsymbol{\Psi}_C$ for each of the classes in turn. Then, given a set of patterns $\boldsymbol{x}_i, i = 1, \ldots, n$, that we wish to classify, we calculate $L(\boldsymbol{x}_i, |\boldsymbol{\Psi}_j)$, $i = 1, \ldots, n$; $j = 1, \ldots, C$, the likelihood of the observation given each of the models, and combine this with the class priors to obtain a probability of class membership.

Hastie and Tibshirani (1996) consider the use of mixture models for discrimination, allowing each group within a class to have its own mean vector, but the covariance matrix is common across all mixture components and across all classes. This is one way of restricting the number of parameters to be estimated.

If we let π_{jr} be the mixing probabilities for the rth subgroup within class ω_j, $j = 1, \ldots, C$, $\sum_{r=1}^{R_j} \pi_{jr} = 1$, where class ω_j has R_j subgroups; $\boldsymbol{\mu}_{jr}$ be the mean of the rth subgroup within class ω_j; and $\boldsymbol{\Sigma}$ be the common covariance matrix; then the re-estimation equations are (Hastie and Tibshirani, 1996)

$$w_{ijr} = \frac{\pi_{jr} p(\boldsymbol{x}_i | \boldsymbol{\theta}_{jr})}{\sum_{k=1}^{R_j} \pi_{jk} p(\boldsymbol{x}_i | \boldsymbol{\theta}_{jk})} \tag{2.14}$$

where $p(\boldsymbol{x}_i | \boldsymbol{\theta}_{jr})$ is the density of the rth subgroup of class ω_j evaluated at $\boldsymbol{x}_i$ ($\boldsymbol{\theta}_{jr}$ denotes the parameters $\boldsymbol{\mu}_{jr}$ and $\boldsymbol{\Sigma}$) and

$$\begin{aligned} \hat{\pi}_{jr} &\propto \sum_{g_i=j} w_{ijr}, \quad \sum_{r=1}^{R_j} \hat{\pi}_{jr} = 1 \\ \hat{\boldsymbol{\mu}}_{jr} &= \frac{\sum_{g_i=j} w_{ijr} \boldsymbol{x}_i}{\sum_{g_i=j} w_{ijr}} \\ \hat{\boldsymbol{\Sigma}} &= \frac{1}{n} \sum_{j=1}^{C} \sum_{g_i=j} \sum_{r=1}^{R_j} w_{ijr} (\boldsymbol{x}_i - \hat{\boldsymbol{\mu}}_{jr})(\boldsymbol{x}_i - \hat{\boldsymbol{\mu}}_{jr})^T \end{aligned} \tag{2.15}$$

$\sum_{g_i=j}$ denoting the sum over observations $\boldsymbol{x}_i$ where $\boldsymbol{x}_i$ belongs to class ω_j.

Other constraints on the covariance matrix structure may be considered. Within the context of clustering using normal mixtures, Celeux and Govaert (1995) propose a parametrisation of the covariance matrix that covers several different conditions, ranging from equal spherical clusters (covariance matrices equal and proportional to the identity matrix) to different covariance matrices for each cluster. Such an approach can be developed for discrimination.

2.3.3 How many components?

Several authors have considered the problem of testing for the number of components, g, of a normal mixture. This is not a trivial problem and depends on many factors including shape of clusters, separation, relative sizes, sample size and dimension of data. Wolfe (1971) proposes a modified likelihood ratio test in which the null hypothesis $g = g_0$ is tested against the alternative hypothesis that $g = g_1$. The quantity

$$-\frac{2}{n}\left(n - 1 - p - \frac{g_1}{2}\right) \log(\lambda)$$

where λ is the likelihood ratio, is tested as a chi-square with the degrees of freedom, d, being twice the difference in the number of parameters in the two hypotheses (Everitt, *et al.*, 2001), excluding mixing proportions. For components of a normal mixture with arbitrary covariance matrices,

$$d = 2(g_1 - g_0)\frac{p(p+3)}{2}$$

and with common covariance matrices (the case that was studied by Wolfe, 1971), $d = 2(g_1 - g_0)p$.

This test has been investigated by Everitt (1981) and Anderson (1985). For the common covariance structure, Everitt finds that, for testing $g = 1$ against $g = 2$ in a two-component mixture, the test is appropriate if the number of observations is at least ten times the number of variables. McLachlan and Basford (1988) recommend that Wolfe's modified likelihood ratio test be used as a guide to structure rather than rigidly interpreted.

In a discrimination context, we may monitor performance on a separate test set (see Chapter 11) and choose the model that gives best performance on this test set. Note that the test set is part of the training procedure (more properly termed a *validation set*) and error rates quoted using these data will be optimistically biased.

2.3.4 Example application study

The problem The practical application concerns the automatic recognition of ships using high-resolution radar measurements of the radar cross-section of targets (Webb, 2000).

Summary This is a straightforward mixture model approach to discrimination, with maximum likelihood estimates of the parameters obtained via the EM algorithm. The mixture component distributions are taken to be gamma distributions.

The data The data consist of radar range profiles (RRPs) of ships of seven class types. An RRP describes the magnitude of the radar reflections of the ship as a function of distance from the radar. The profiles are sampled at 3 m spacing and each RRP comprises 130 measurements. RRPs are recorded from all aspects of a ship as the ship turns through 360 degrees. There are 19 data files and each data file comprises between 1700 and 8800 training patterns. The data files are divided into train and test sets. Several classes have more than one rotation available for training and testing.

The model The density of each class is modelled using a mixture model. Thus, we have

$$p(\boldsymbol{x}) = \sum_{i=1}^{g} \pi_i p(\boldsymbol{x}|\boldsymbol{\theta}_i) \tag{2.16}$$

where $\boldsymbol{\theta}_i$ represents the set of parameters of mixture component i. An independence model is assumed for each mixture component, $p(\boldsymbol{x}|\boldsymbol{\theta}_i)$, therefore

$$p(\boldsymbol{x}|\boldsymbol{\theta}_i) = \prod_{j=1}^{130} p(x_j|\boldsymbol{\theta}_{ij})$$

and the univariate factor, $p(x_j|\boldsymbol{\theta}_{ij})$, is modelled as a gamma distribution[2] with parameters $\boldsymbol{\theta}_{ij} = (m_{ij}, \mu_{ij})$,

$$p(x_j|\boldsymbol{\theta}_{ij}) = \frac{m_{ij}}{(m_{ij}-1)!\mu_{ij}} \left(\frac{m_{ij}x_j}{\mu_{ij}}\right)^{m_{ij}-1} \exp\left(-\frac{m_{ij}x_j}{\mu_{ij}}\right)$$

where m_{ij} is the order parameter for variable x_j of mixture component i and μ_{ij} is the mean. Thus for each mixture component, there are two parameters associated with each dimension. We denote by $\boldsymbol{\theta}_i$ the set $\{\boldsymbol{\theta}_{ij}, j = 1, \ldots, 130\}$, the parameters of component i. The gamma distribution is chosen from physical considerations – it has special cases of Rayleigh scattering and a non-fluctuating target. Also, empirical measurements have been found to be gamma-distributed.

Training procedure Given a set of n observations (in a given class), the likelihood function is

$$L(\boldsymbol{\Psi}) = \prod_{i=1}^{n} \sum_{j=1}^{g} \pi_j p(\boldsymbol{x}_i|\boldsymbol{\theta}_j)$$

where $\boldsymbol{\Psi}$ represents all the parameters of the model, $\boldsymbol{\Psi} = \{\boldsymbol{\theta}_j, \pi_j, j = 1, \ldots, g\}$.

An EM approach to maximum likelihood is taken. If $\{\boldsymbol{\theta}_k^{(m)}, \pi_k^{(m)}\}$ denotes the estimate of the parameters of the kth component at the mth stage of the iteration, then the E-step estimates w_{ij}, the probability that $\boldsymbol{x}_i$ belongs to group j given the current estimates of the parameters (2.10),

$$w_{ij} = \frac{\pi_j^{(m)} p(\boldsymbol{x}_i|\boldsymbol{\theta}_j^{(m)})}{\sum_k \pi_k^{(m)} p(\boldsymbol{x}_i|\boldsymbol{\theta}_k^{(m)})} \tag{2.17}$$

The M-step leads to the estimate of the mixture weights, π_j, as in (2.11),

$$\hat{\pi}_j = \frac{1}{n}\sum_{i=1}^{n} w_{ij} \tag{2.18}$$

The equation for the mean is given by (2.12),

$$\hat{\boldsymbol{\mu}}_j = \frac{\sum_{i=1}^{n} w_{ij}\boldsymbol{x}_i}{\sum_{i=1}^{n} w_{ij}} = \frac{1}{n\hat{\pi}_j}\sum_{i=1}^{n} w_{ij}\boldsymbol{x}_i \tag{2.19}$$

but the equation for the gamma order parameters, m_{ij}, cannot be solved in closed form,

$$\nu(m_{jk}) = -\frac{\sum_{i=1}^{n} w_{ik}\log(x_{ij}/\mu_{jk})}{\sum_{i=1}^{n} w_{ik}}$$

where $\nu(m) = \log(m) - \psi(m)$ and $\psi(m)$ is the digamma function.

Thus, for the gamma mixture problem, an EM approach may be taken, but a numerical root-finding routine must be used within the EM loop for the gamma distribution order parameters.

[2]There are several parametrisations of a gamma distribution. We present the one used in the study.

The number of mixture components per class was varied between 5 and 110 and the model for each ship determined by minimising a penalised likelihood criterion (the likelihood penalised by a complexity term – see Chapter 11). This resulted in between 50 and 100 mixture components per ship. The density function for each class was constructed using (2.16). The class priors were taken to be equal. The model was then applied to a separate test set and the error rate estimated.

2.3.5 Further developments

Jamshidian and Jennrich (1993) propose an approach for accelerating the EM algorithm based on a generalised conjugate gradients numerical optimisation scheme (see also Jamshidian and Jennrich, 1997). Other competing numerical schemes for normal mixtures are described by Everitt and Hand (1981) and Titterington *et al.* (1985). Lindsay and Basak (1993) describe a method of moments approach to multivariate normal mixtures that may be used to initialise an EM algorithm.

Further extensions to the EM algorithm are given by Meng and Rubin (1992, 1993). The SEM (supplemented EM) algorithm is a procedure for computing the asymptotic variance-covariance matrix. The ECM (expectation/conditional maximisation) algorithm is a procedure for implementing the M-step when a closed-form solution is not available, replacing each M-step by a sequence of conditional maximisation steps. An alternative gradient algorithm for approximating the M-step is presented by Lange (1995) and the algorithm is further generalised to the ECME (ECM either) algorithm by Liu and Rubin (1994).

Developments for data containing groups of observations with longer than normal tails are described by Peel and McLachlan (2000), who develop a mixture of t distributions model, with parameters determined using the ECM algorithm.

Approaches for choosing the number of components of a normal mixture include that of Bozdogan (1993), who has compared several information-theoretic criteria on simulated data consisting of overlapping and non-overlapping clusters of different shape and compactness. Celeux and Soromenho (1996) propose an entropy criterion, evaluated as a by-product of the EM algorithm, and compare its performance with several other criteria.

2.3.6 Summary

Modelling using normal mixtures is a simple way of developing the normal model to nonlinear discriminant functions. Even if we assume a common covariance matrix for mixture components, the decision boundary is not linear. The EM algorithm provides an appealing scheme for parameter estimation, and there have been various extensions accelerating the technique. A mixture model may also be used to partition a given data set by modelling the data set using a mixture and assigning data samples to the group for which the probability of membership is the greatest. The use of mixture models in this context is discussed further in Chapter 10. Bayesian approaches to mixture modelling have also received attention and are considered in Section 2.4.3 in a discrimination context and in Chapter 10 for clustering.

2.4 Bayesian estimates

2.4.1 Bayesian learning methods

Here we seek to estimate some quantity such as the density at $\boldsymbol{x}$

$$p(\boldsymbol{x}|\mathcal{D})$$

where $\mathcal{D} = \{\boldsymbol{x}_1, \dots, \boldsymbol{x}_n\}$ is the set of training patterns characterising the distribution. The dependence of the density at $\boldsymbol{x}$ on $\mathcal{D}$ is through the parameters of the model assumed for the density. If we assume a particular model, $p(\boldsymbol{x}|\boldsymbol{\theta})$, then the Bayesian approach does not base the density estimate on a single estimate of the parameters, $\boldsymbol{\theta}$, of the probability density function $p(\boldsymbol{x}|\boldsymbol{\theta})$, but admits that we do not know the true value of $\boldsymbol{\theta}$ and we write

$$p(\boldsymbol{x}|\mathcal{D}) = \int p(\boldsymbol{x}|\boldsymbol{\theta})p(\boldsymbol{\theta}|\mathcal{D})\,d\boldsymbol{\theta} \tag{2.20}$$

where by Bayes' theorem the posterior density of $\boldsymbol{\theta}$ may be expressed as

$$p(\boldsymbol{\theta}|\mathcal{D}) = \frac{p(\mathcal{D}|\boldsymbol{\theta})p(\boldsymbol{\theta})}{\int p(\mathcal{D}|\boldsymbol{\theta})p(\boldsymbol{\theta})\,d\boldsymbol{\theta}} \tag{2.21}$$

Bayes' theorem allows us to combine any prior, $p(\boldsymbol{\theta})$, with any likelihood, $p(\mathcal{D}|\boldsymbol{\theta})$, to give the posterior. However, it is convenient for particular likelihood functions to take special forms of the prior that lead to simple, or at least tractable, solutions for the posterior. For a given model, $p(\boldsymbol{x}|\boldsymbol{\theta})$, the family of prior distributions for which the posterior density, $p(\boldsymbol{\theta}|\mathcal{D})$, is of the same functional form is called *conjugate* with respect to $p(\boldsymbol{x}|\boldsymbol{\theta})$. Some of the more common forms of conjugate priors are given by Bernardo and Smith (1994).

The posterior density may also be calculated in a recursive manner. If the measurements, $\boldsymbol{x}_i$, are given successively, we may write (2.21) as

$$p(\boldsymbol{\theta}|\boldsymbol{x}_1, \dots, \boldsymbol{x}_n) = \frac{p(\boldsymbol{x}_n|\boldsymbol{\theta})p(\boldsymbol{\theta}|\boldsymbol{x}_1, \dots, \boldsymbol{x}_{n-1})}{\int p(\boldsymbol{x}_n|\boldsymbol{\theta})p(\boldsymbol{\theta}|\boldsymbol{x}_1, \dots, \boldsymbol{x}_{n-1})\,d\boldsymbol{\theta}} \tag{2.22}$$

for $\boldsymbol{x}_1, \dots, \boldsymbol{x}_n$ conditionally independent. This expresses the posterior distribution of $\boldsymbol{\theta}$ given n measurements in terms of the posterior distribution given $n-1$ measurements. Starting with $p(\boldsymbol{\theta})$, we may perform the operation (2.22) n times to obtain the posterior.

For situations when there is no conjugate prior distribution, or the denominator in (2.21) cannot be evaluated analytically, we must resort to numerical methods, discussed in the following section.

To illustrate the Bayesian learning approach we shall consider the problem of estimating the mean of a univariate normal distribution with known variance, and quote the result for the multivariate normal distribution with unknown mean and covariance matrix. Further details on estimating the parameters of normal models are given in the books by Fukunaga (1990) and Fu (1968).

Example 1 Estimating the mean of a normal distribution with known variance, σ^2.

Let the model for the density be normal with mean μ and variance σ^2, denoted $p(x|\mu)$ (σ^2 known),

$$p(x|\mu) = \frac{1}{\sqrt{2\pi}\sigma} \exp\left(-\frac{1}{2\sigma^2}(x-\mu)^2\right)$$

Assume a prior density for the mean μ that is also normal with mean μ_0 and variance σ_0^2,

$$p(\mu) = \frac{1}{\sqrt{2\pi}\sigma_0} \exp\left\{-\frac{1}{2}\left(\frac{\mu-\mu_0}{\sigma_0}\right)^2\right\}$$

Now,

$$\begin{aligned} p(x_1, \ldots, x_n|\mu)p(\mu) & \\ &= p(\mu)\prod_{i=1}^{n} p(x_i|\mu) \\ &= \frac{1}{\sqrt{2\pi}\sigma_0} \exp\left\{-\frac{1}{2}\left(\frac{\mu-\mu_0}{\sigma_0}\right)^2\right\} \prod_{i=1}^{n}\left[\frac{1}{\sqrt{2\pi}\sigma}\exp\left\{-\frac{1}{2}\left(\frac{x_i-\mu}{\sigma}\right)^2\right\}\right] \end{aligned}$$

This may be written in the form

$$p(x_1, \ldots, x_n|\mu)p(\mu) = \frac{1}{\sigma^n\sigma_0}\frac{1}{(2\pi)^{(n+1)/2}} \exp\left\{-\frac{1}{2}\left(\frac{\mu-\mu_n}{\sigma_n}\right)^2\right\}\exp\left(-\frac{k_n}{2}\right)$$

where

$$\frac{1}{\sigma_n^2} = \frac{1}{\sigma_0^2} + \frac{n}{\sigma^2}$$

$$\mu_n = \sigma_n^2\left(\frac{\mu_0}{\sigma_0^2} + \frac{\sum x_i}{\sigma^2}\right)$$

$$k_n = \frac{\mu_0^2}{\sigma_0^2} - \frac{\mu_n^2}{\sigma_n^2} + \frac{\sum x_i^2}{\sigma^2}$$

Substituting into equation (2.21) gives the posterior distribution as

$$p(\mu|x_1, \ldots, x_n) = \frac{1}{\sqrt{2\pi}\sigma_n} \exp\left\{-\frac{1}{2}\left(\frac{\mu-\mu_n}{\sigma_n}\right)^2\right\} \tag{2.23}$$

which is normal with mean μ_n and variance σ_n^2.

As $n \to \infty$, $\mu_n \to$ the sample mean, $m = \sum_i x_i/n$, and the variance of μ, namely σ_n^2, tends to zero as $1/n$. Thus, as more samples are used to obtain the distribution (2.23), the contribution of the initial guesses σ_0 and μ_0 becomes smaller. This is illustrated in Figure 2.2. Samples, x_i, from a normal distribution with unit mean and unit variance are generated. A normal model is assumed with mean μ and unit variance.

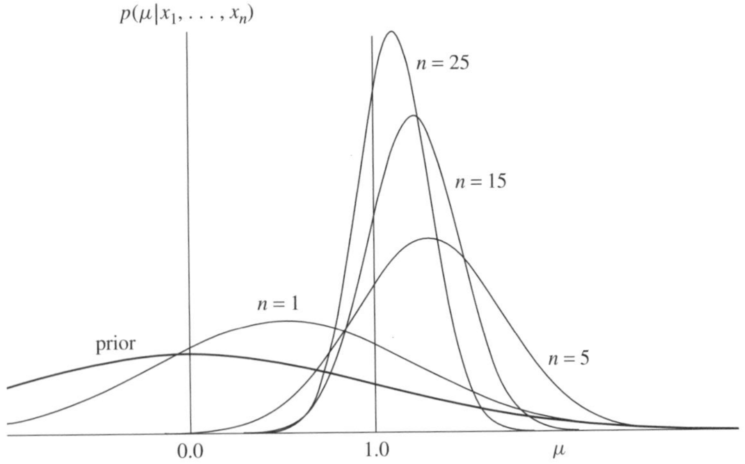

Figure 2.2 Bayesian learning of the mean of a normal distribution of known variance

Figure 2.2 plots the posterior distribution of the mean, μ, given different numbers of samples (using (2.23)) for a prior distribution of the mean that is normal with $\mu_0 = 0$ and $\sigma_0^2 = 1$. As the number of samples increases, the posterior distribution narrows about the true mean.

Finally, substituting the density into (2.20) gives the conditional distribution

$$p(x|x_1, \ldots, x_n) = \frac{1}{(\sigma^2 + \sigma_n^2)^{1/2}\sqrt{2\pi}} \exp\left\{-\frac{1}{2}\frac{(x-\mu_n)^2}{\sigma^2 + \sigma_n^2}\right\}$$

which is normal with mean μ_n and variance $\sigma^2 + \sigma_n^2$. □

Example 2 Estimating the mean and the covariance matrix of a multivariate normal distribution.

In this example, we consider the multivariate problem in which the mean and the covariance matrix of a normal distribution are unknown. Let the model for the data be normal with mean $\boldsymbol{\mu}$ and covariance matrix $\boldsymbol{\Sigma}$:

$$p(\boldsymbol{x}|\boldsymbol{\mu}, \boldsymbol{\Sigma}) = \frac{1}{(2\pi)^{p/2}|\boldsymbol{\Sigma}|^{\frac{1}{2}}} \exp\left\{-\frac{1}{2}(\boldsymbol{x}-\boldsymbol{\mu})^T \boldsymbol{\Sigma}^{-1}(\boldsymbol{x}-\boldsymbol{\mu})\right\}$$

We wish to estimate the distribution of $\boldsymbol{\mu}$ and $\boldsymbol{\Sigma}$ given measurements $\boldsymbol{x}_1, \ldots, \boldsymbol{x}_n$.

When the mean $\boldsymbol{\mu}$ and the covariance matrix $\boldsymbol{\Sigma}$ of a normal distribution are to be estimated, an appropriate choice of prior density is the Gauss–Wishart, or normal – Wishart, probability density function in which the mean is normally distributed with mean $\boldsymbol{\mu}_0$ and covariance matrix $\boldsymbol{K}^{-1}/\lambda$, and $\boldsymbol{K}$ (the inverse of the covariance matrix $\boldsymbol{\Sigma}$) is distributed according to a Wishart distribution (see Appendix E for a definition of some of the commonly used distributions) with parameters α and $\boldsymbol{\beta}$:

$$\begin{aligned} p(\boldsymbol{\mu}, \boldsymbol{K}) &= \mathrm{N}_p(\boldsymbol{\mu}|\boldsymbol{\mu}_0, \lambda\boldsymbol{K})\mathrm{Wi}_p(\boldsymbol{K}|\alpha, \boldsymbol{\beta}) \\ &= \frac{|\lambda\boldsymbol{K}|^{1/2}}{(2\pi)^{p/2}} \exp\left\{-\frac{1}{2}\lambda(\boldsymbol{\mu}-\boldsymbol{\mu}_0)^T\boldsymbol{K}(\boldsymbol{\mu}-\boldsymbol{\mu}_0)\right\} \\ &\quad \times c(p,\alpha)|\boldsymbol{\beta}|^{\alpha}|\boldsymbol{K}|^{(\alpha-(p+1)/2)} \exp\{-\mathrm{Tr}(\boldsymbol{\beta}\boldsymbol{K})\} \end{aligned} \tag{2.24}$$

where

$$c(p,\alpha) = \left[\pi^{p(p-1)/4} \prod_{i=1}^{p} \Gamma\left(\frac{2\alpha+1-i}{2}\right)\right]^{-1}$$

The term λ expresses the confidence in $\boldsymbol{\mu}_0$ as the initial value of the mean and α ($2\alpha > p-1$) the initial confidence in the covariance matrix. It can be shown that the posterior distribution

$$p(\boldsymbol{\mu}, \boldsymbol{K}|\boldsymbol{x}_1, \dots, \boldsymbol{x}_n) = \frac{p(\boldsymbol{x}_1, \dots, \boldsymbol{x}_n|\boldsymbol{\mu}, \boldsymbol{K})p(\boldsymbol{\mu}, \boldsymbol{K})}{\int p(\boldsymbol{x}_1, \dots, \boldsymbol{x}_n|\boldsymbol{\mu}, \boldsymbol{K})p(\boldsymbol{\mu}, \boldsymbol{\Sigma})\, d\boldsymbol{\mu}\, d\boldsymbol{K}}$$

is also Gauss–Wishart with the parameters $\boldsymbol{\mu}_0$, $\boldsymbol{\beta}$, λ and α replaced by (Fu, 1968)

$$\begin{aligned} \lambda_n &= \lambda + n \\ \alpha_n &= \alpha + n/2 \\ \boldsymbol{\mu}_n &= (\lambda\boldsymbol{\mu}_0 + n\boldsymbol{m})/(\lambda + n) \\ 2\boldsymbol{\beta}_n &= 2\boldsymbol{\beta} + (n-1)\boldsymbol{S} + \frac{n\lambda}{n+\lambda}(\boldsymbol{\mu}_0 - \boldsymbol{\mu})(\boldsymbol{\mu}_0 - \boldsymbol{\mu})^T \end{aligned} \tag{2.25}$$

where

$$\boldsymbol{S} = \frac{1}{n-1}\sum_{i=1}^{n}(\boldsymbol{x}_i - \boldsymbol{m})(\boldsymbol{x}_i - \boldsymbol{m})^T$$

and $\boldsymbol{m}$ is the sample mean. That is,

$$p(\boldsymbol{\mu}, \boldsymbol{K}|\boldsymbol{x}_1, \dots, \boldsymbol{x}_n) = \mathrm{N}_p(\boldsymbol{\mu}|\boldsymbol{\mu}_n, \lambda_n\boldsymbol{K})\mathrm{Wi}_p(\boldsymbol{K}|\alpha_n, \boldsymbol{\beta}_n) \tag{2.26}$$

The conditional distribution of $\boldsymbol{\mu}$ given $\boldsymbol{K}$ is normal. Marginalising (integrating with respect to $\boldsymbol{\mu}$) gives the posterior for $\boldsymbol{K}$ as $\mathrm{Wi}_p(\boldsymbol{K}|\alpha_n, \boldsymbol{\beta}_n)$. The posterior for $\boldsymbol{\mu}$ (integrating with respect to $\boldsymbol{K}$) is

$$p(\boldsymbol{\mu}|\boldsymbol{x}_1, \dots, \boldsymbol{x}_n) = \mathrm{St}_p(\boldsymbol{\mu}|\boldsymbol{\mu}_n, \lambda_n(\alpha_n - (p-1)/2)\boldsymbol{\beta}_n^{-1}, 2(\alpha_n - (p-1)/2))$$

which is the p-dimensional generalisation of the univariate Student distribution (see Appendix E for a definition). Finally, we may substitute into (2.20) to obtain the density for the case of the normal distribution with unknown mean and covariance matrix,

$$p(\boldsymbol{x}|\boldsymbol{x}_1,\ldots,\boldsymbol{x}_n) = \mathrm{St}_p(\boldsymbol{x}|\boldsymbol{\mu}_n, (\lambda_n+1)^{-1}\lambda_n(\alpha_n-(p-1)/2)\boldsymbol{\beta}_n^{-1}, 2(\alpha_n-(p-1)/2)) \tag{2.27}$$

which is also a Student distribution. □

The parameters α_n, λ_n, $\boldsymbol{\mu}_n$ and $\boldsymbol{\beta}_n$ may be calculated for each class and the conditional density (2.27) may be used as a basis for discrimination: assign $\boldsymbol{x}$ to class ω_i for which $g_i > g_j$, $j = 1, \ldots, C$, $j \neq i$, where

$$g_i = p(\boldsymbol{x}|\boldsymbol{x}_1,\ldots,\boldsymbol{x}_{n_i} \in \omega_i)p(\omega_i) \tag{2.28}$$

Unknown priors

In equation (2.28), $p(\omega_i)$ represents the prior probabilities for class ω_i. It may happen that the prior class probabilities, $p(\omega_i)$, are unknown. In this case, we may treat them as parameters of the model that may be updated using the data. We write

$$\begin{aligned} p(\omega_i|\boldsymbol{x},\mathcal{D}) &= \int p(\omega_i,\boldsymbol{\pi},|\boldsymbol{x},\mathcal{D})\,d\boldsymbol{\pi} \\ &\propto \int p(\boldsymbol{x}|\omega_i,\boldsymbol{\pi},\mathcal{D})p(\omega_i,\boldsymbol{\pi}|\mathcal{D})\,d\boldsymbol{\pi} \\ &\propto \int p(\boldsymbol{x}|\omega_i,\mathcal{D})p(\omega_i,\boldsymbol{\pi}|\mathcal{D})\,d\boldsymbol{\pi} \end{aligned} \tag{2.29}$$

where we use $\boldsymbol{\pi} = (\pi_1,\ldots,\pi_C)$ to represent the prior class probabilities.

The term $p(\omega_i,\boldsymbol{\pi}|\mathcal{D})$ may be written

$$\begin{aligned} p(\omega_i,\boldsymbol{\pi}|\mathcal{D}) &= p(\boldsymbol{\pi}|\mathcal{D})p(\omega_i|\boldsymbol{\pi},\mathcal{D}) \\ &= p(\boldsymbol{\pi}|\mathcal{D})\pi_i \end{aligned} \tag{2.30}$$

The aspects of the measurements that influence the distribution of the class probabilities, $\boldsymbol{\pi}$, are the numbers of patterns in each class, $n_i, i = 1,\ldots,C$. A suitable prior for $\pi_i = p(\omega_i)$ is a *Dirichlet prior* (see Appendix E), with parameters $\boldsymbol{a}_0 = (a_{01},\ldots,a_{0C})$,

$$p(\pi_1,\ldots,\pi_C) = k\prod_{j=1}^{C}\pi_j^{a_{0j}-1}$$

where $k = \Gamma(\sum_i a_{0i})/\prod_i \Gamma(a_{0i})$ (Γ is the gamma function) and $\pi_C = 1 - \sum_{i=1}^{C-1}\pi_i$. This is written $\boldsymbol{\pi} \sim \mathrm{Di}_C(\boldsymbol{\pi}|\boldsymbol{a}_0)$ for $\boldsymbol{\pi} = (\pi_1,\ldots,\pi_C)^T$. Assuming that the distribution of the data (the n_i) given the priors is multinomial,

$$p(\mathcal{D}|\boldsymbol{\pi}) = \frac{n!}{\prod_{l=1}^{C} n_l!}\prod_{l=1}^{C}\pi_l^{n_l}$$

then the posterior

$$p(\boldsymbol{\pi}|\mathcal{D}) \propto p(\mathcal{D}|\boldsymbol{\pi})p(\boldsymbol{\pi}) \tag{2.31}$$

is also distributed as $\text{Di}_C(\boldsymbol{\pi}|\boldsymbol{a})$, where $\boldsymbol{a} = \boldsymbol{a}_0 + \boldsymbol{n}$ and $\boldsymbol{n} = (n_1, \ldots, n_C)^T$, the vector of numbers of patterns in each class.

Substituting the Dirichlet distribution for $p(\boldsymbol{\pi}|\mathcal{D})$ into equation (2.30), and then $p(\omega_i, \boldsymbol{\pi}|\mathcal{D})$ from (2.30) into equation (2.29), gives

$$p(\omega_i|\boldsymbol{x}, \mathcal{D}) \propto p(\boldsymbol{x}|\omega_i, \mathcal{D}) \int \pi_i \, \text{Di}_C(\boldsymbol{\pi}|\boldsymbol{a}) \, d\boldsymbol{\pi} \tag{2.32}$$

which replaces π_i in (2.28) by its expected value, $a_i / \sum_j a_j$. Thus, the posterior probability of class membership now becomes

$$p(\omega_i|\boldsymbol{x}, \mathcal{D}) = \frac{(n_i + a_{0i}) p(\boldsymbol{x}|\omega_i)}{\sum_i (n_i + a_{0i}) p(\boldsymbol{x}|\omega_i)} \tag{2.33}$$

This treatment has assumed that the π_is are unknown but can be estimated from the training data as well as prior information. Whether the training data can be used depends on the sampling scheme used to gather the data. There are various modifications to (2.33) depending on the assumed knowledge concerning the π_i (see Geisser, 1964).

Summary

The Bayesian approach described above involves two stages. The first stage is concerned with learning about the parameters of the distribution, $\boldsymbol{\theta}$, through the recursive calculation of the posterior density $p(\boldsymbol{\theta}|\boldsymbol{x}_1, \ldots, \boldsymbol{x}_n)$ for a specified prior

$$p(\boldsymbol{\theta}|\boldsymbol{x}_1, \ldots, \boldsymbol{x}_n) \propto p(\boldsymbol{x}_n|\boldsymbol{\theta}) p(\boldsymbol{\theta}|\boldsymbol{x}_1, \ldots, \boldsymbol{x}_{n-1})$$

For a suitable prior and choice of class-conditional densities, the posterior distribution for $\boldsymbol{\theta}$ is of the same form as the prior. The second stage is the integration over $\boldsymbol{\theta}$ to obtain the conditional density $p(\boldsymbol{x}|\boldsymbol{x}_1, \ldots, \boldsymbol{x}_n)$ which may be viewed as making allowance for the variability in the estimate due to sampling. Although it is relatively straightforward to perform the integrations for the normal case that has been considered here, it may be necessary to perform two multivariate numerical integrations for more complicated probability density functions. This is the case for the normal mixture model for which there exist no reproducing (conjugate) densities.

2.4.2 Markov chain Monte Carlo

Introduction

In the previous section we developed the Bayesian approach to density estimation and illustrated it using two normal distribution examples, for which the integral in the denominator of (2.21) may be evaluated analytically for suitable choices of prior distribution. We now consider some of the computational machinery for practical implementation of Bayesian methods to problems for which the normalising integral cannot be evaluated analytically and numerical integration over possibly high-dimensional spaces is infeasible. The following section will illustrate these ideas on an application to a discrimination problem.

Let $\mathcal{D}$ denote the observed data. In a classification problem, $\mathcal{D}$ comprises the training set of patterns and their class labels $\{(\boldsymbol{x}_i, z_i), i = 1, \ldots, n\}$, where $\boldsymbol{x}_i$ are the patterns and z_i are the class labels, together with $\{\boldsymbol{x}_i^t, i = 1, \ldots, n_t\}$, a set of patterns for which the class labels are assumed unknown. Let $\boldsymbol{\theta}$ denote the model parameters and the 'missing data'. In a classification problem, the missing data are the class labels of the patterns of unknown class and the model parameters would be, for example, the means and covariance matrices for normally distributed classes.

The posterior distribution of $\boldsymbol{\theta}$ conditional on the observed data, $\mathcal{D}$, may be written, using Bayes' theorem, as the normalised product of the likelihood, $p(\mathcal{D}|\boldsymbol{\theta})$, and the prior distribution, $p(\boldsymbol{\theta})$:

$$p(\boldsymbol{\theta}|\mathcal{D}) = \frac{p(\mathcal{D}|\boldsymbol{\theta})p(\boldsymbol{\theta})}{\int p(\mathcal{D}|\boldsymbol{\theta})p(\boldsymbol{\theta})\,d\boldsymbol{\theta}} \tag{2.34}$$

This posterior distribution tells us all that we need to know about $\boldsymbol{\theta}$ and can be used to calculate summary statistics. The posterior expectation of a function $h(\boldsymbol{\theta})$ is

$$\mathrm{E}[h(\boldsymbol{\theta})|\mathcal{D}] = \frac{\int h(\boldsymbol{\theta})p(\mathcal{D}|\boldsymbol{\theta})p(\boldsymbol{\theta})\,d\boldsymbol{\theta}}{\int p(\mathcal{D}|\boldsymbol{\theta})p(\boldsymbol{\theta})\,d\boldsymbol{\theta}} \tag{2.35}$$

However, the integrals in equations (2.34) and (2.35) have led to practical difficulties for the implementation of Bayesian methods. The normalising constant in (2.34) is often unknown because analytic evaluation of the integral can only be performed for simple models.

A group of methods known as Markov chain Monte Carlo (MCMC) has proven to be effective at generating samples asymptotically from the posterior distribution (without knowing the normalising constant) from which inference about model parameters may be made by forming sample averages. MCMC methodology may be used to analyse complex problems, no longer requiring users to force the problem into an oversimplified framework for which analytic treatment is possible.

The Gibbs sampler

We begin with a description of the Gibbs sampler, one of the most popular MCMC methods, and discuss some of the issues that must be addressed for practical implementation. Both the Gibbs sampler and a more general algorithm, the Metropolis–Hastings algorithm, have formed the basis for many variants.

Let $f(\boldsymbol{\theta})$ denote the posterior distribution from which we wish to draw samples; $\boldsymbol{\theta}$ is a p-dimensional parameter vector, $(\theta_1, \ldots, \theta_p)^T$. We may not know f exactly, but we know a function $g(\boldsymbol{\theta})$, where $f(\boldsymbol{\theta}) = g(\boldsymbol{\theta}) / \int g(\boldsymbol{\theta})\,d\boldsymbol{\theta}$.

Let $\boldsymbol{\theta}_{(i)}$ be the set of parameters with the ith parameter removed; that is, $\boldsymbol{\theta}_{(i)} = \{\theta_1, \ldots, \theta_{i-1}, \theta_{i+1}, \ldots, \theta_p\}$. We assume that we are able to draw samples from the one-dimensional conditional distributions, $f(\theta_i|\boldsymbol{\theta}_{(i)})$, derived from the normalisation of $g(\theta_i|\boldsymbol{\theta}_{(i)})$, the function g regarded as a function of θ_i alone, all other parameters being fixed.

Gibbs sampling is a simple algorithm that consists of drawing samples from these distributions in a cyclical way as follows.

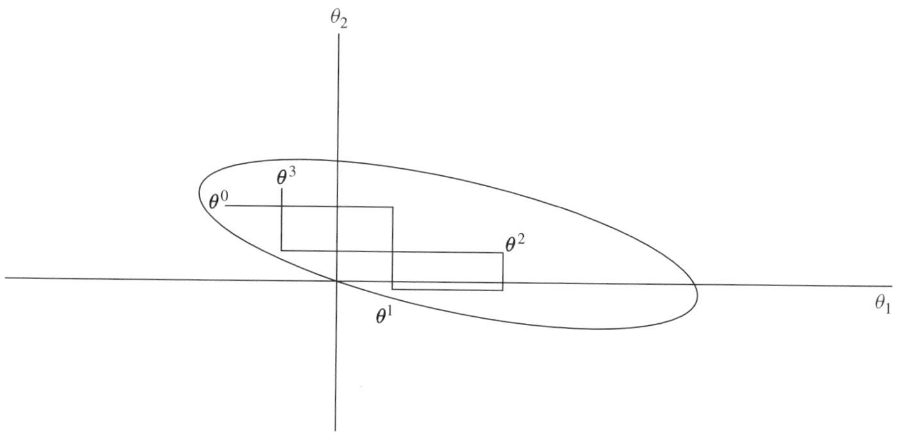

Figure 2.3 Gibbs sampler illustration

Gibbs sampling

To generate a sequence from $f(\theta_1, \ldots, \theta_p)$, choose an arbitrary starting value for $\boldsymbol{\theta}$, $\boldsymbol{\theta}^0 = (\theta_1^0, \ldots, \theta_p^0)^T$ from the support of the prior/posterior distribution. At stage t of the iteration,

- draw a sample, θ_1^{t+1} from $f(\theta_1|\theta_2^t, \ldots, \theta_p^t)$;
- draw a sample, θ_2^{t+1} from $f(\theta_2|\theta_1^{t+1}, \theta_3^t, \ldots, \theta_p^t)$;
- and continue through the variables, finally drawing a sample, θ_p^{t+1} from $f(\theta_p|\theta_1^{t+1}, \ldots, \theta_{p-1}^{t+1})$;

After a large number of iterations, the vectors $\boldsymbol{\theta}^t$ behave like a random draw from the joint density $f(\boldsymbol{\theta})$ (Bernardo and Smith, 1994).

Figure 2.3 illustrates the Gibbs sampler for a bivariate distribution. The θ_1 and θ_2 components are updated alternately, producing moves in the horizontal and vertical directions.

In the Gibbs sampling algorithm, the distribution of $\boldsymbol{\theta}^t$ given all previous values $\boldsymbol{\theta}^0, \boldsymbol{\theta}^1, \ldots, \boldsymbol{\theta}^{t-1}$ depends only on $\boldsymbol{\theta}^{t-1}$. This is the Markov property and the sequence generated is termed a *Markov chain*.

For the distribution of $\boldsymbol{\theta}^t$ to converge to a stationary distribution (a distribution that does not depend on $\boldsymbol{\theta}^0$ or t), the chain must be *aperiodic*, *irreducible* and *positive recurrent*. A Markov chain is aperiodic if it does not oscillate between different subsets in a regular periodic way. Recurrence is the property that all sets of $\boldsymbol{\theta}$ values will be reached infinitely often at least from almost all starting points.

It is irreducible if it can reach all possible $\boldsymbol{\theta}$ values from any starting point. Figure 2.4 illustrates a distribution that is uniform on $([0, 1] \times [0, 1]) \bigcup ([1, 2] \times [1, 2])$, the union of

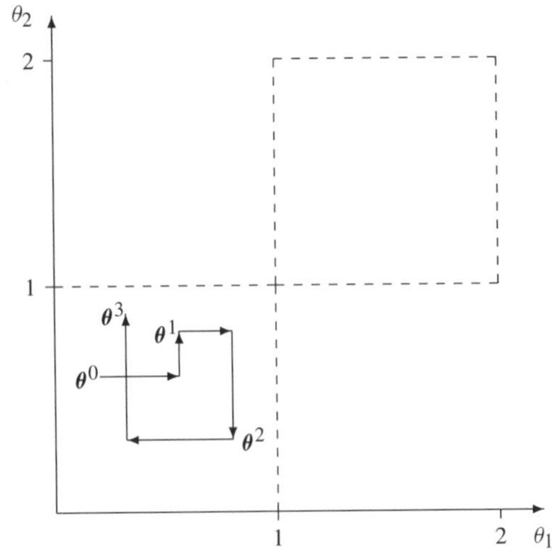

Figure 2.4 Illustration of a reducible Gibbs sampler

two non-overlapping unit squares. Consider the Gibbs sampler that uses the coordinate directions as sampling directions. For a point $\theta_1 \in [0, 1]$, the conditional distribution of θ_2 given θ_1 is uniform over [0, 1]. Similarly for a point $\theta_2 \in [0, 1]$, the conditional distribution of θ_1 given θ_2 is uniform over [0, 1]. We can see that if the starting point for θ_1 is in [0, 1], then the Gibbs sampler will generate a point θ_2, also in [0, 1]. The next step of the algorithm will generate a value for θ_1, also in [0, 1], and so on. Therefore, successive values of $\boldsymbol{\theta}^t$ will be uniformly distributed on the square $([0, 1] \times [0, 1])$. The square $([1, 2] \times [1, 2])$ will not be visited. Conversely, a starting point in $([1, 2] \times [1, 2])$ will yield a limiting distribution uniform on $([1, 2] \times [1, 2])$. Thus the limiting distribution depends on the starting value and therefore is not irreducible.

By designing a *transition kernel* $\mathrm{K}(\boldsymbol{\theta}, \boldsymbol{\theta}')$ (the probability of moving from $\boldsymbol{\theta}$ to $\boldsymbol{\theta}'$) that satisfies *detailed balance* (time-reversibility)

$$f(\boldsymbol{\theta})\mathrm{K}(\boldsymbol{\theta}, \boldsymbol{\theta}') = f(\boldsymbol{\theta}')\mathrm{K}(\boldsymbol{\theta}', \boldsymbol{\theta})$$

for all pairs of states $(\boldsymbol{\theta}, \boldsymbol{\theta}')$ in the support of f, then the stationary distribution is the target distribution of interest, $f(\boldsymbol{\theta})$.

Summarisation

After a sufficiently large number of iterations (referred to as the *burn-in* period), the samples $\{\boldsymbol{\theta}^t\}$ will be dependent samples from the posterior distribution $f(\boldsymbol{\theta})$. These samples may be used to obtain estimators of expectations using *ergodic averages*. Evaluation of the expectation of a function, h, of interest is achieved by the approximation

$$\mathrm{E}[h(\boldsymbol{\theta})] \approx \frac{1}{N - M} \sum_{t=M+1}^{N} h(\boldsymbol{\theta}^t) \tag{2.36}$$

where N is the number of iterations of the Gibbs sampler and M is the number of iterations in the burn-in period.

Other summaries include plots of the marginal densities using some of the general nonparametric methods of density estimation, such as *kernel methods* (discussed in detail in Chapter 3). The kernel density estimate of θ_i given samples $\{\theta_i^t, t = M+1, \ldots, N\}$ is

$$p(\theta_i) = \frac{1}{N-M} \sum_{t=M+1}^{N} K(\theta_i, \theta_i^t) \tag{2.37}$$

where the *kernel* $K(\theta, \theta^*)$ is a density centred at θ^*. Choices for kernels and their widths are discussed in Chapter 3.

An alternative estimator, due to Gelfand and Smith (1990) and termed the Rao-Blackwellised estimator, makes use of the conditional densities $f(\theta_i|\boldsymbol{\theta}^t_{(i)})$,

$$p(\theta_i) = \frac{1}{N-M} \sum_{t=M+1}^{N} f(\theta_i|\boldsymbol{\theta}^t_{(i)}) \tag{2.38}$$

This estimates the tails of the distribution better than more general methods of density estimation (O'Hagan, 1994).

The Rao-Blackwellised estimator of $\mathrm{E}[h(\theta_i)]$ is then

$$\mathrm{E}[h(\theta_i)] \approx \frac{1}{N-M} \sum_{t=M+1}^{N} \mathrm{E}[h(\theta_i)|\boldsymbol{\theta}^t_{(i)}] \tag{2.39}$$

The difference between (2.39) and (2.36) is that (2.39) requires an analytic expression for the conditional expectation so that it may be evaluated at each step of the iteration. For reasonably long runs, the improvement in using (2.39) over (2.36) is small.

If $\boldsymbol{\theta}$ are the parameters of a density and we require $p(\boldsymbol{x}|\mathcal{D})$, we may estimate this by approximating the integral in (2.20) using a Monte Carlo integration:

$$p(\boldsymbol{x}|\mathcal{D}) = \frac{1}{N-M} \sum_{t=M+1}^{N} p(\boldsymbol{x}|\boldsymbol{\theta}^t)$$

Convergence

In an implementation of Gibbs sampling, there are a number of practical considerations to be addressed. These include the length of the burn-in period, M; the length of the sequence, N; and the spacing between samples taken from the final sequence of iterations (the final sequence may be subsampled in an attempt to produce approximately independent samples and to reduce the amount of storage required).

The length of the chain should be long enough for it to 'forget' its starting value and such that all regions of the parameter space have been traversed by the chain. The limiting distribution should not depend on its starting value, $\boldsymbol{\theta}^0$, but the length of the sequence will depend on the correlation between the variables. Correlation between the θ_is will tend to slow convergence. It can be difficult to know when a sequence has converged as the Gibbs sampler can spend long periods in a relatively small region, thus giving the impression of convergence.

The most commonly used method for determining the burn-in period is by visually inspecting plots of the output values, $\boldsymbol{\theta}^t$, and making a subjective judgement. More formal tools, *convergence diagnostics*, exist and we refer to Raftery and Lewis (1996) Gelman (1996) and Mengersen *et al.* (1999) for further details of the most popular methods. However, convergence diagnostics do not tell when a chain has converged, but tell when it has not converged – sometimes.

There are various approaches for reducing correlation (and hence speeding up convergence) including *reparametrisation* and *grouping variables*.

Reparametrisation transforms the set $\boldsymbol{\theta}$ using a linear transformation to a new set $\boldsymbol{\phi}$ with zero correlation between the variables. The linear transformation is calculated using an estimate of the covariance matrix based on a short initial sequence (in pattern recognition, the process of deriving a set of uncorrelated variables that is a linear combination of the original variables is *principal components analysis*, which we shall describe in Chapter 9). The Gibbs sampler then proceeds using the variables $\boldsymbol{\phi}$, provided that it is straightforward to sample from the new conditionals $f(\phi_i|\boldsymbol{\phi}_{(i)})$. The process may be repeated until the correlation in the final sequence is small, hopefully leading to more rapid convergence.

Grouping variables means that at each step of the iteration a sample from a multivariate distribution $f(\boldsymbol{\theta}_i|\boldsymbol{\theta}_{(i)})$ is generated, where $\boldsymbol{\theta}_i$ is a subvector of $\boldsymbol{\theta}$ and $\boldsymbol{\theta}_{(i)}$ is the set of remaining variables. Provided correlations between variables are caused primarily by correlations between elements of the subvectors, with low correlations between subvectors, we can hope for more rapid convergence. A method for sampling from $f(\boldsymbol{\theta}_i|\boldsymbol{\theta}_{(i)})$ (which may be complex) is required.

Starting point

The starting point is any point you do not mind having in the sequence. Preliminary runs, started where the last one ended, will give you some feel for suitable starting values. There is some argument to say that since the starting point is a legitimate point from the sequence (although perhaps in the tail of the distribution), it would be visited anyway by the Markov chain, at some stage; hence there is no need for burn-in. However, using a burn-in period and removing initial samples will make estimators approximately unbiased.

Parallel runs

Instead of running one chain until convergence, it is possible to run multiple chains (with different starting values) as an approach to monitoring convergence, although more formal methods exist (Roberts, 1996; Raftery and Lewis, 1996), as well as to obtaining independent observations from $f(\boldsymbol{\theta})$. This is a somewhat controversial issue since independent samples are not required in many cases, and certainly not for ergodic averaging (equation (2.37)). Comparing several chains may help in identifying convergence. For example, are estimates of quantities of interest consistent between runs? In such cases, it is desirable to choose different starting values, $\boldsymbol{\theta}^0$, for each run, widely dispersed.

In practice, you will probably do several runs if computational resources permit, either to compare related probability models or to gain information about a chosen model such as burn-in length. Then, you would perform a long run in order to obtain samples for computing statistics.

Sampling from conditional distributions

The Gibbs sampler requires ways of sampling from the conditional distributions $f(\boldsymbol{\theta}_i|\boldsymbol{\theta}_{(i)})$ and it is essential that sampling from these distributions is computationally efficient. If $f(\boldsymbol{\theta}_i|\boldsymbol{\theta}_{(i)})$ is a standard distribution, then it is likely that algorithms exist for drawing samples from it. For algorithms for sampling from some of the more common distributions, see, for example, Devroye (1986) and Ripley (1987).

As an example, consider the univariate equivalent of the distribution given by equation (2.26):

$$
\begin{aligned}
p(\mu, 1/\sigma^2) &= \mathrm{N}_1(\mu|\mu_n, \lambda_n/\sigma^2)\mathrm{Ga}(1/\sigma^2|\alpha_n, \beta_n) \\
&\propto \frac{1}{\sigma}\exp\left\{-\frac{\lambda+n}{2\sigma^2}(\mu-\mu_n)^2\right\}\left(\frac{1}{\sigma^2}\right)^{\alpha_n-1}\exp\left\{-\frac{\beta_n}{\sigma^2}\right\}
\end{aligned}
$$

Given σ^2, μ is normally distributed with mean μ_n (the prior updated by the data samples – see (2.25)) and variance $\sigma^2/(\lambda+n)$; given the mean, $1/\sigma^2$ has a gamma distribution (see Appendix E) $\mathrm{Ga}(1/\sigma^2|\alpha_n+\frac{1}{2}, \beta_n+\lambda_n(\mu-\mu_n)^2/2))$, with marginal (integrating over μ) $\mathrm{Ga}(1/\sigma^2|\alpha_n, \beta_n)$. The mean of σ^2 is $\beta_n/(\alpha_n-1)$, $\alpha_n > 1$, and variance $\beta_n^2/[(\alpha_n-1)^2(\alpha_n-2)]$, $\alpha_n > 2$.

The data, x_i, comprise 20 points from a normal distribution with zero mean and unit variance. The priors are $\mu \sim \mathrm{N}_1(\mu_0, \lambda/\sigma^2)$, $1/\sigma^2 \sim \mathrm{Ga}(\alpha, \beta)$. The parameters of the prior distribution of μ and σ^2 are taken to be $\lambda = 1, \mu_0 = 1, \alpha = 1/2, \beta = 2$. For this example, the true posteriors of μ and σ^2 may be calculated; μ has a t distribution and the inverse of σ^2 has a gamma distribution. The mean and the variance of the true posteriors of μ and σ^2 are given in Table 2.2.

A Gibbs sampling approach is taken for generating samples from the joint posterior density of μ and σ^2. The steps are to initialise σ^2 (in this example, a sample is taken from the prior distribution) and then sample μ; then for a given μ, sample σ^2, and so on, although this is not necessarily the best approach for a diffuse prior.

Figure 2.5 shows the first 1000 samples in a sequence of μ and σ^2 samples. Taking the first 500 samples as burn-in, and using the remainder to calculate summary statistics,

Table 2.2 Summary statistics for μ and σ^2. The true values are calculated from the known marginal posterior densities. The short-run values are calculated from a Gibbs sampler run of 1000 samples less a burn-in of 500 samples. The long-run values are calculated after a run of 100 000 samples, less a burn-in of 500 samples

	True	Short run	Long run
mean μ	−0.10029	−0.10274	−0.10042
var μ	0.05796	0.05738	0.05781
mean σ^2	1.217	1.236	1.2166
var σ^2	0.17428	0.20090	0.17260

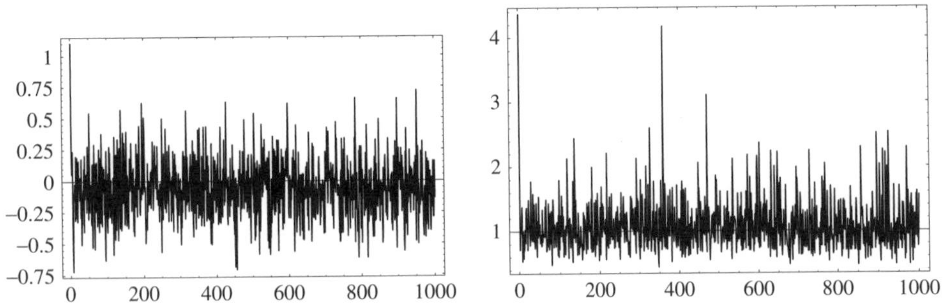

Figure 2.5 One thousand iterations from Gibbs sampler for a normal-inverse gamma posterior density; left: samples of μ; right: samples of σ^2

gives values for the mean and variance of μ and σ^2 that are close to the true values (calculated analytically); see Table 2.2. A longer sequence gives values closer to the truth.

Rejection sampling

If the conditional distribution is not recognised to be of a standard form for which efficient sampling exists, then other sampling schemes must be employed. Let $f(\boldsymbol{\theta}) = g(\boldsymbol{\theta})/\int g(\boldsymbol{\theta})\,d\boldsymbol{\theta}$ be the density from which we wish to sample. Rejection sampling uses a density $s(\boldsymbol{\theta})$ from which we can conveniently sample (cheaply) and requires that $g(\boldsymbol{\theta})/s(\boldsymbol{\theta})$ is bounded. Let an upper bound of $g(\boldsymbol{\theta})/s(\boldsymbol{\theta})$ be A.

The rejection sampling algorithm is as follows.

Rejection sampling algorithm

Repeat

- sample a point $\boldsymbol{\theta}$ from the known distribution $s(\boldsymbol{\theta})$;
- sample y from the uniform distribution on $[0, 1]$;
- if $Ay \leq g(\boldsymbol{\theta})/s(\boldsymbol{\theta})$ then accept $\boldsymbol{\theta}$;

until one $\boldsymbol{\theta}$ is accepted.

Depending on the choice of s, many samples could be rejected before one is accepted. If s is close to the shape of $g(\boldsymbol{\theta})$, so that $g(\boldsymbol{\theta})/s(\boldsymbol{\theta}) \sim A$ for all $\boldsymbol{\theta}$, then the acceptance condition is almost always accepted.

The distribution of the samples, $\boldsymbol{\theta}$, generated is $f(\boldsymbol{\theta})$.

Ratio of uniforms

Let D denote the region in $\mathbb{R}^2$ such that

$$D = \{(u, v);\ 0 \leq u \leq \sqrt{g(v/u)}\},$$

then sampling a point uniformly from D and taking $\theta = v/u$ gives a sample from the density proportional to $g(\theta)$, namely $g(\theta)/\int g(\theta)\,d\theta$.

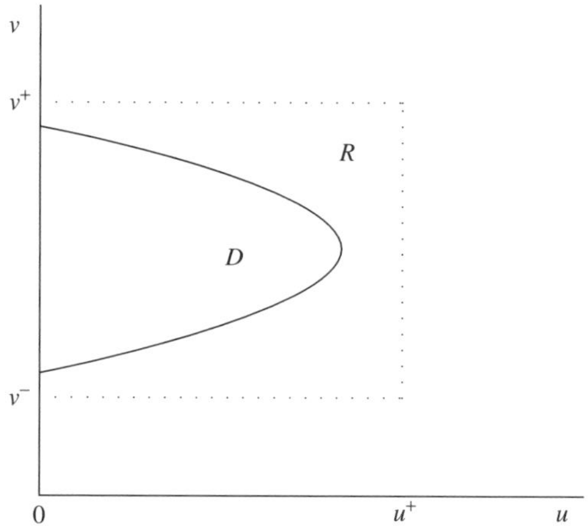

Figure 2.6 Envelope rectangle, R, for the region D defined by $D = \{(u, v);\ 0 \le u \le \sqrt{g(v/u)}\}$

A sample from D could be drawn by a simple application of rejection sampling: sample uniformly from the rectangle, R, bounding D (see Figure 2.6) and if (u, v) is in D then accept.

Metropolis–Hastings algorithm

The Metropolis–Hastings algorithm is a widely used technique for sampling from distributions for which the conditional densities cannot be computed or are of a form from which it is difficult to sample. It uses a *proposal distribution*, from which sampling is easy, and accepts a sample with a probability that depends on the proposal distribution and the (unnormalised) density from which we wish to sample.

Let $\boldsymbol{\theta}^t$ be the current sample. In the Metropolis–Hastings algorithm, a proposal distribution, which may depend on $\boldsymbol{\theta}^t$, is specified. We denote this $q(\boldsymbol{\theta}|\boldsymbol{\theta}^t)$. The Metropolis–Hastings algorithm is as follows.

Metropolis–Hastings algorithm

- Sample a point $\boldsymbol{\theta}$ from the proposal distribution $q(\boldsymbol{\theta}|\boldsymbol{\theta}^t)$.
- Sample y from the uniform distribution on $[0, 1]$.
- If

$$y \le \min\left(1, \frac{g(\boldsymbol{\theta})q(\boldsymbol{\theta}^t|\boldsymbol{\theta})}{g(\boldsymbol{\theta}^t)q(\boldsymbol{\theta}|\boldsymbol{\theta}^t)}\right)$$

then accept $\boldsymbol{\theta}$ and set $\boldsymbol{\theta}^{t+1} = \boldsymbol{\theta}$, else reject $\boldsymbol{\theta}$ and set $\boldsymbol{\theta}^{t+1} = \boldsymbol{\theta}^t$.

It produces a different Markov chain than Gibbs sampling, but with the same limiting distribution, $g(\boldsymbol{\theta})/\int g(\boldsymbol{\theta})\,d\boldsymbol{\theta}$.

The proposal distribution can take any sensible form and the stationary distribution will still be $g(\boldsymbol{\theta})/\int g(\boldsymbol{\theta})\,d\boldsymbol{\theta}$. For example, $q(\boldsymbol{X}|\boldsymbol{Y})$ may be a multivariate normal distribution with mean $\boldsymbol{Y}$ and fixed covariance matrix, $\boldsymbol{\Sigma}$. However, the scale of $\boldsymbol{\Sigma}$ will need to be chosen carefully. If it is too small, then there will be a high acceptance rate, but poor *mixing*; that is, the chain may not move rapidly throughout the support of the target distribution and will have to be run for longer than necessary to obtain good estimates from equation (2.36). If the scale of $\boldsymbol{\Sigma}$ is too large, then there will be a poor acceptance rate, and so the chain may stay at the same value for some time, again leading to poor mixing.

For symmetric proposal distributions, $q(\boldsymbol{X}|\boldsymbol{Y}) = q(\boldsymbol{Y}|\boldsymbol{X})$, the acceptance probability reduces to

$$\min\left(1, \frac{g(\boldsymbol{\theta})}{g(\boldsymbol{\theta}^t)}\right)$$

and, in particular, for $q(\boldsymbol{X}|\boldsymbol{Y}) = q(|\boldsymbol{X} - \boldsymbol{Y}|)$, so that q is a function of the difference between $\boldsymbol{X}$ and $\boldsymbol{Y}$ only, the algorithm is the *random-walk* Metropolis algorithm.

Single-component Metropolis–Hastings

The Metropolis–Hastings algorithm given above updates all components of the parameter vector, $\boldsymbol{\theta}$, in one step. An alternative approach is to update a single component at a time.

Single-component Metropolis–Hastings algorithm

At stage t of the iteration, do the following.

- Draw a sample, Y from the proposal distribution $q(\theta|\theta_1^t, \ldots, \theta_p^t)$.
  - Accept the sample with probability

$$\alpha = \min\left(1, \frac{g(Y|\theta_2^t, \ldots, \theta_p^t)q(\theta_1^t|Y, \theta_2^t, \ldots, \theta_p^t)}{g(\theta_1^t|\theta_2^t, \ldots, \theta_p^t)q(Y|\theta_1^t, \theta_2^t, \ldots, \theta_p^t)}\right)$$

  - If Y is accepted, then $\theta_1^{t+1} = Y$, else $\theta_1^{t+1} = \theta_1^t$.
- Continue through the variables as in the Gibbs sampler, finally drawing a sample, Y, from the proposal distribution $q(\theta|\theta_1^{t+1}, \ldots, \theta_{p-1}^{t+1}, \theta_p^t)$.
  - Accept the sample with probability

$$\alpha = \min\left(1, \frac{g(Y|\theta_1^{t+1}, \ldots, \theta_{p-1}^{t+1})q(\theta_p^t|\theta_1^{t+1}, \ldots, \theta_{p-1}^{t+1}, Y)}{g(\theta_p^t|\theta_1^{t+1}, \ldots, \theta_{p-1}^{t+1})q(Y|\theta_1^{t+1}, \ldots, \theta_{p-1}^{t+1}, \theta_p^t)}\right)$$

  - If Y is accepted, then $\theta_p^{t+1} = Y$, else $\theta_p^{t+1} = \theta_p^t$.

In this single-component update case, for proposal distributions that are the conditionals of the multivariate distribution that we wish to sample,

$$q(\theta_i|\boldsymbol{\theta}_{(i)}) = f(\theta_i|\boldsymbol{\theta}_{(i)})$$

then the sample is always accepted and the algorithm is identical to Gibbs sampling.

Choice of proposal distribution in Metropolis–Hastings

If the distribution that we wish to approximate, f, is unimodal, and is not heavy-tailed (loosely, heavy-tailed means that it tends to zero more slowly than the exponential, but also the term is used to describe distributions with infinite variance), then an appropriate choice for the proposal distribution might be normal, with parameters chosen to be a best fit of $\log(q)$ to $\log(g)$ $(f = g/\int g)$. For more complex distributions, the proposal could be a multivariate normal, or mixtures of multivariate normal, but for distributions with heavy tails, Student t distributions (see Appendix E) might be used. For computational efficiency, q should be chosen so that it can be easily sampled and evaluated. Often a random-walk algorithm is used (symmetric proposal distribution), and can give good results.

Data augmentation

Introducing auxiliary variables can often lead to more simple and efficient MCMC sampling methods, with improved mixing. If we require samples from a posterior $p(\boldsymbol{\theta}|\mathcal{D})$, then the basic idea is to notice that it may be easier to sample from $p(\boldsymbol{\theta}, \boldsymbol{\phi}|\mathcal{D})$, where $\boldsymbol{\phi}$ is a set of auxiliary variables. In some applications, the choice of $\boldsymbol{\phi}$ may be obvious, in others some experience is necessary to recognise suitable choices. The distribution $p(\boldsymbol{\theta}|\mathcal{D})$ is then simply a marginal of the augmented distribution, $p(\boldsymbol{\theta}, \boldsymbol{\phi}|\mathcal{D})$, and the method of sampling is termed the *data augmentation method*. Statistics concerning the distribution of $p(\boldsymbol{\theta}|\mathcal{D})$ can be obtained by using the $\boldsymbol{\theta}$ components of the samples of the augmented parameter vector $(\boldsymbol{\theta}, \boldsymbol{\phi})$ and ignoring the $\boldsymbol{\phi}$ components.

One type of problem where data augmentation is used is that involving missing data. Suppose that we have a data set $\mathcal{D}$ and some 'missing values', $\boldsymbol{\phi}$. In a classification problem, where there are some unlabelled data available for training the classifier, $\boldsymbol{\phi}$ represents the class labels of these data. Alternatively, there may be incomplete pattern vectors; that is, for some patterns, measurements on some of the variables may be absent.

The posterior distribution of parameters $\boldsymbol{\theta}$ is given by

$$p(\boldsymbol{\theta}|\mathcal{D}) \propto \int p(\mathcal{D}, \boldsymbol{\phi}|\boldsymbol{\theta})\, d\boldsymbol{\phi}\, p(\boldsymbol{\theta})$$

However, it may be difficult to marginalise the joint density $p(\mathcal{D}, \boldsymbol{\phi}|\boldsymbol{\theta})$ and it is simpler to obtain samples of the augmented vector $(\boldsymbol{\theta}, \boldsymbol{\phi})$. In this case,

$p(\boldsymbol{\theta}|\mathcal{D}, \boldsymbol{\phi})$ is the posterior based on the complete data, which is easy to sample, either directly or by use of MCMC methods (for example, Metropolis–Hastings);

$p(\boldsymbol{\phi}|\boldsymbol{\theta}, \mathcal{D})$ is the sampling distribution for the missing data; again, typically easy to sample.

Examples of missing data problems are given in Section 2.4.3.

Example

In this example, we seek to model a time series as a sum of k sinusoids of unknown amplitude, frequency and phase (ψ, ω, ϕ). The approach and example are based on work by Andrieu and Doucet (1999). We assume a model of the form

$$y = h(x; \boldsymbol{\xi}) + \epsilon = \sum_{j=1}^{k} \psi_j \cos(\omega_j x + \phi_j) + \epsilon, \tag{2.40}$$

where $\epsilon \sim N(0, \sigma^2)$ and $\boldsymbol{\xi} = \{(\psi_j, \omega_j, \phi_j), j = 1, \ldots, k\}$; thus

$$p(y|x; \boldsymbol{\theta}) = \frac{1}{\sqrt{2\pi\sigma^2}} \exp\{-(y - h(x; \boldsymbol{\xi}))^2/(2\sigma^2)\}$$

where the parameters of the density are $\boldsymbol{\theta} = (\boldsymbol{\xi}, \sigma^2)$. The training data, $\mathcal{D} = \{y_i, i = 1, \ldots, n\}$, comprise n measurements of y at regular intervals, $x_i = i$, for $i = 0, 1, \ldots, n-1$. Assuming independent noise samples, we have

$$p(\mathcal{D}|\boldsymbol{\theta}) \propto \prod_{i=1}^{n} \frac{1}{\sigma} \exp\{-(y_i - h(x_i; \boldsymbol{\xi}))^2/(2\sigma^2)\}$$

What we would like now is information about the parameters given the data set and the model for predicting y given a new sample x_n. Information about the parameters $\boldsymbol{\theta}$ requires specification of a prior distribution, $p(\boldsymbol{\theta})$. Then using Bayes' theorem, we have

$$p(\boldsymbol{\theta}|\mathcal{D}) \propto p(\mathcal{D}|\boldsymbol{\theta}) P(\boldsymbol{\theta})$$

For predicting a new sample, we take

$$p(y|x_n) \approx \sum_{t=M+1}^{N} p(y|x_n; \boldsymbol{\theta}^t)$$

where M is the burn-in period; N is the length of the sequence; $\boldsymbol{\theta}^t$ are the parameters at stage t.

It is convenient to reparametrise the model as

$$y_i = \sum_{j=1}^{k} \{g_j \cos(\omega_j x_i) + h_j \sin(\omega_j x_i)\} + \epsilon_i,$$

where $g_j = \psi_j \cos(\phi_j)$ and $h_j = -\psi_j \sin(\phi_j)$ represent the new amplitudes of the problem, which lie in the range $(-\infty, \infty)$. This may be written as

$$\boldsymbol{y} = \boldsymbol{D}\boldsymbol{a} + \boldsymbol{\epsilon},$$

where $\boldsymbol{y}^T = (y_1, \ldots, y_n)$; $\boldsymbol{a}^T = (g_1, h_1, \ldots, g_k, h_k)$ is the $2k$-dimensional vector of amplitudes, and $\boldsymbol{D}$ is an $n \times 2k$ matrix, defined by:

$$D_{i,j} = \begin{cases} \cos(\omega_j x_i) & j \text{ odd} \\ \sin(\omega_j x_i) & j \text{ even} \end{cases}$$

Data Data are generated according to the model (2.40), with $k = 3$, $n = 64$, $\{\omega_j\} = 2\pi(0.2, 0.2 + 1/n, 0.2 + 2/n)$, $\{\psi_j\} = (\sqrt{20}, \sqrt{2\pi}, \sqrt{20})$, $\sigma = 2.239$ and $\{\phi_j\} = (0, \pi/4, \pi/3)$; the time series is shown in Figure 2.7.

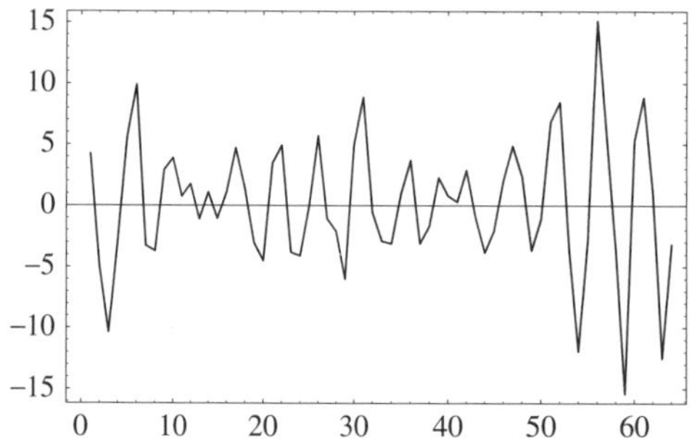

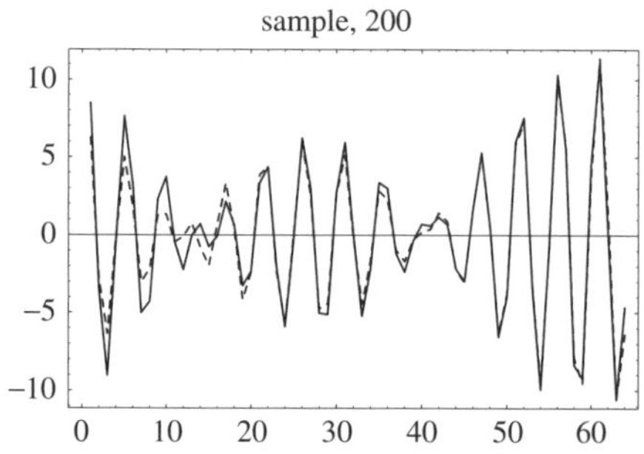

Figure 2.7 Data for sinusoids estimation problem (left); underlying model (solid line) with reconstruction, based on the 200th set of MCMC samples (dashed line, right)

Prior The prior distribution for the random variables, (ω, σ^2, a), is written as

$$p(\omega, \sigma^2, a) = p(\omega)p(\sigma^2)p(a|\omega, \sigma^2),$$

where $\omega = \{\omega_j\}$. Specifically,

$$p(\omega) = \frac{1}{\pi^k}\mathrm{I}[\omega \in [0, \pi]^k]$$
$$\boldsymbol{a}|\omega, \sigma^2 \sim \mathrm{N}_{2k}(0, (\sigma^2\boldsymbol{\Sigma})^{-1}), \quad \text{where} \quad \boldsymbol{\Sigma}^{-1} = \delta^{-2}\boldsymbol{D}^T\boldsymbol{D}$$
$$\sigma^2 \sim \mathrm{Ig}(\nu_0/2, \gamma_0/2)$$

for parameters, δ^2, ν_0 and γ_0. Values of $\delta^2 = 50$, $\nu_0 = 0.01$ and $\gamma_0 = 0.01$ have been used in the illustrated example.

Posterior The posterior distribution can be rearranged to:

$$p(\boldsymbol{a}, \omega, \sigma^2|\mathcal{D}) \propto \frac{1}{\sigma^{2\left(\frac{n+\nu_0}{2}+k+1\right)}} \exp\left[\frac{-(\gamma_0 + \boldsymbol{y}^T\boldsymbol{P}\boldsymbol{y})}{2\sigma^2}\right]\mathrm{I}[\omega \in [0, \pi]^k]$$
$$\times |\boldsymbol{\Sigma}|^{-1/2} \exp\left[\frac{-(\boldsymbol{a}-\boldsymbol{m})^T\boldsymbol{M}^{-1}(\boldsymbol{a}-\boldsymbol{m})}{2\sigma^2}\right] \tag{2.41}$$

where

$$\boldsymbol{M}^{-1} = \boldsymbol{D}^T\boldsymbol{D} + \boldsymbol{\Sigma}^{-1}, \quad \boldsymbol{m} = \boldsymbol{M}\boldsymbol{D}^T\boldsymbol{y}, \quad \text{and} \quad \boldsymbol{P} = \boldsymbol{I}_n - \boldsymbol{D}\boldsymbol{M}\boldsymbol{D}^T$$

The amplitude, $\boldsymbol{a}$, and variance, σ^2, can be integrated out analytically, giving:

$$p(\omega|\mathcal{D}) \propto (\gamma_0 + \boldsymbol{y}^T\boldsymbol{P}\boldsymbol{y})^{-\frac{n+\nu_0}{2}}$$

This cannot be dealt with analytically so samples are drawn, by sampling from the conditional distributions of the individual components, ω_j, using Metropolis–Hastings sampling, which uses

$$p(\omega_j|\omega_{(j)}, \mathcal{D}) \propto p(\omega|\mathcal{D})$$

where $\omega_{(j)}$ is the set of variables with the jth one omitted.

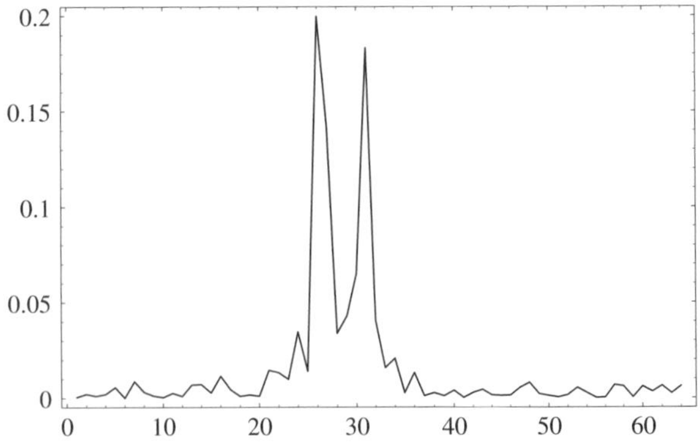

Figure 2.8 Proposal distribution

Proposal At each update choose randomly between two possible proposal distributions. The first, chosen with probability 0.2, is given by:

$$q_j(\omega'_j|\omega) \propto \sum_{l=0}^{n-1} p_l \mathrm{I}\left[\frac{l\pi}{n} < \omega'_j < \frac{(l+1)\pi}{n}\right],$$

where p_l is the squared modulus of the Fourier transform of the data (see Figure 2.8) at frequency $l\pi/n$ (this proposal aims to prevent the Markov chain getting stuck with one solution for ω_j). The second is the normally distributed random walk, $N(0, \pi/(2n))$ (ensuring irreducibility of the Markov chain).

Results Having drawn samples for ω, the amplitudes can be sampled using

$$\boldsymbol{a}|\omega, \sigma^2, \mathcal{D} \sim N(\boldsymbol{m}, \sigma^2\boldsymbol{M})$$

which comes directly from (2.41). The noise variance can also be sampled using

$$\sigma^2|\omega, \mathcal{D} \sim \mathrm{Ig}\left(\frac{n+\nu_0}{2}, \frac{\gamma_0 + \boldsymbol{y}^T\boldsymbol{P}\boldsymbol{y}}{2}\right)$$

which comes from (2.41) after analytical integration of $\boldsymbol{a}$.

The algorithm is initialised with a sample from the prior for ω. Convergence for the illustrated example was very quick, with a burn-in of less than 100 iterations required. Figure 2.9 gives plots of samples of the noise, σ (true value, 2.239), and frequencies, ω_j. Figure 2.7 shows the reconstruction of the data using 200th set of MCMC samples.

Summary

MCMC methods can provide effective approaches to inference problems in situations where analytic evaluation of posterior probabilities is not feasible. Their main strength is in their flexibility. They enable Bayesian approaches to be adapted to real-world problems without having to make unnecessarily restrictive assumptions regarding prior

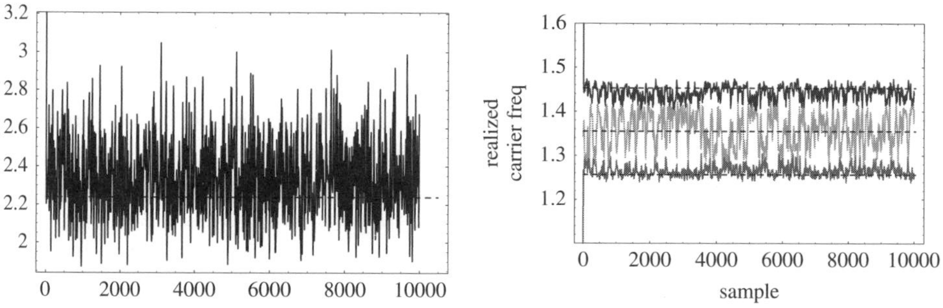

Figure 2.9 Ten thousand iterations from MCMC sampler; left: samples of the noise; right: samples of the frequencies

distributions which may make the mathematics tractable. Originating in the statistical physics literature, the development in Bayesian statistics has been driven by the difficulty in performing numerical integration. The main disadvantage concerns uncertainty over convergence, and hence over the accuracy of estimates computed from the samples.

In many respects, the implementation of these methods is still something of an art, with several trial runs being performed in order to explore models and parameter values. Techniques are required for reducing the amount of computation per iteration. Run times can be long, caused by poor mixing.

The main features of the MCMC method are as follows.

1. It performs iterative sampling from a proposal distribution. The samples may be univariate, multivariate or a subvector of the parameter vector. A special case is Gibbs sampling when samples from conditional probability density functions are made.

2. The samples provide a summary of the posterior probability distribution. They may be used to calculate summary statistics either by averaging functions of the samples (equation (2.36)) or by averaging conditional expectations (Rao-Blackwellisation, equation (2.39)).

3. Correlated variables lead to longer convergence.

4. The parameters of the method are N, the sequence length; M, the burn-in period; $q(.|.)$, the proposal distribution; and s, the subsampling factor.

5. In practice, you would run several chains to estimate parameter values and then one long chain to calculate statistics.

6. Subsampling of the final chain may be performed to reduce the amount of storage required to represent the distribution.

7. Sampling from standard distributions is readily performed using algorithms in the books by Devroye (1986) and Ripley (1987), for example. For non-standard distributions, the rejection methods and ratio-of-uniforms methods may be used as well as Metropolis–Hastings, but there are other possibilities (Gilks *et al.*, 1996).

2.4.3 Bayesian approaches to discrimination

In this section we apply the Bayesian learning methods of Section 2.4.1 to the discrimination problem, making use of analytic solutions where we can, but using the numerical techniques of the previous section where that is not possible.

Let $\mathcal{D}$ denote the data set used to train the classifier. In the first instance, let it comprise a set of labelled patterns $\{(\boldsymbol{x}_i, z_i), i = 1, \ldots, n\}$, where $z_i = j$ implies that the pattern $\boldsymbol{x}_i$ is in class ω_j. Given pattern $\boldsymbol{x}$ from an unknown class, we would like to predict its class membership; that is, we require

$$p(z = j|\mathcal{D}, \boldsymbol{x}) \quad j = 1, \ldots, C$$

where z is the class indicator variable corresponding to $\boldsymbol{x}$. The Bayes decision rule for minimum error is to assign $\boldsymbol{x}$ to the class for which $p(z = j|\mathcal{D}, \boldsymbol{x})$ is the greatest. The above may be written (compare with equation (2.32))

$$p(z = j|\mathcal{D}, \boldsymbol{x}) \propto p(\boldsymbol{x}|\mathcal{D}, z = j)p(z = j|\mathcal{D}) \tag{2.42}$$

where the constant of proportionality does not depend on class. The first term, $p(\boldsymbol{x}|\mathcal{D}, z = j)$, is the probability density of class ω_j evaluated at $\boldsymbol{x}$. If we assume a model for the density, with parameters $\boldsymbol{\Phi}_j$, then by (2.20) this may be written

$$p(\boldsymbol{x}|\mathcal{D}, z = j) = \int p(\boldsymbol{x}|\boldsymbol{\Phi}_j)p(\boldsymbol{\Phi}_j|\mathcal{D}, z = j)\,d\boldsymbol{\Phi}_j$$

For certain special cases of the density model, $p(\boldsymbol{x}|\boldsymbol{\Phi}_j)$, we may evaluate this analytically. For example, as we have seen in Section 2.4.1, a normal model with parameters $\boldsymbol{\mu}$ and $\boldsymbol{\Sigma}$ with Gauss–Wishart priors leads to a posterior distribution of the parameters that is also Gauss–Wishart and a multivariate Student t distribution for the density $p(\boldsymbol{x}|\mathcal{D}, z = j)$.

If we are unable to obtain an analytic solution, then a numerical approach will be required. For example, if we use one of the MCMC methods of the previous section to draw samples from the posterior density of the parameters, $p(\boldsymbol{\Phi}_j|\mathcal{D}, z = j)$, we may approximate $p(\boldsymbol{x}|\mathcal{D}, z = j)$ by

$$p(\boldsymbol{x}|\mathcal{D}, z = j) \approx \frac{1}{N - M} \sum_{t=M+1}^{N} p(\boldsymbol{x}|\boldsymbol{\Phi}_j^t) \tag{2.43}$$

where $\boldsymbol{\Phi}_j^t$ are samples generated from the MCMC process and M and N are the burn-in period and run length, respectively.

The second term in (2.42) is the probability of class ω_j given the data set, $\mathcal{D}$. Thus, it is the prior probability updated by the data. We saw in Section 2.4.1 that if we assume Dirichlet priors for π_i, the prior probability of class ω_i, that is,

$$p(\boldsymbol{\pi}) = \text{Di}_C(\boldsymbol{\pi}|\boldsymbol{a}_0)$$

then

$$p(z = j|\mathcal{D}) = \mathrm{E}[\pi_j|\mathcal{D}] = \frac{a_{0j} + n_j}{\sum_{j=1}^{C}(a_{0j} + n_j)} \tag{2.44}$$

Unlabelled training data

The case of classifier design using a normal model when the training data comprise both labelled and unlabelled patterns is considered by Lavine and West (1992). It provides an example of the Gibbs sampling methods that uses some of the analytic results of Section 2.4.1. We summarise the approach here.

Let the data set $\mathcal{D} = \{(\boldsymbol{x}_i, z_i), i = 1, \ldots, n;\ \boldsymbol{x}_i^u, i = 1, \ldots, n_u\}$, where $\boldsymbol{x}_i^u$ are the unlabelled patterns. Let $\boldsymbol{\mu} = \{\boldsymbol{\mu}_i, i = 1, \ldots, C\}$ and $\boldsymbol{\Sigma} = \{\boldsymbol{\Sigma}_i, i = 1, \ldots, C\}$ be the set of class means and covariance matrices and $\boldsymbol{\pi}$ the class priors. Denote by $z^u = \{z_i^u, i = 1, \ldots, n_u\}$, the set of unknown class labels.

The parameters of the model are $\theta = \{\boldsymbol{\mu}, \boldsymbol{\Sigma}, \boldsymbol{\pi}, z^u\}$. Taking a Gibbs sampling approach, we successively draw samples from three conditional distributions.

1. Sample from $p(\boldsymbol{\mu}, \boldsymbol{\Sigma}|\boldsymbol{\pi}, z^u, \mathcal{D})$. This density may be written

$$p(\boldsymbol{\mu}, \boldsymbol{\Sigma}|\boldsymbol{\pi}, z^u, \mathcal{D}) = \prod_{i=1}^{C} p(\boldsymbol{\mu}_i, \boldsymbol{\Sigma}_i|z^u, \mathcal{D}),$$

   the product of C independent Gauss–Wishart distributions given by (2.26).

2. Sample from $p(\boldsymbol{\pi}|\boldsymbol{\mu}, \boldsymbol{\Sigma}, z^u, \mathcal{D})$. This is Dirichlet $\mathrm{Di}_C(\boldsymbol{\pi}|\boldsymbol{a})$, $\boldsymbol{a} = \boldsymbol{a}_0 + \boldsymbol{n}$, independent of $\boldsymbol{\mu}$ and $\boldsymbol{\Sigma}$, with $\boldsymbol{n} = (n_1, \ldots, n_C)$, where n_j is the number of patterns in class ω_j as determined by $\mathcal{D}$ and z^u.

3. Sample from $p(z^u|\boldsymbol{\mu}, \boldsymbol{\Sigma}, \boldsymbol{\pi}, \mathcal{D})$. Since the samples z_i^u are conditionally independent, we require samples from

$$p(z_i^u = j|\boldsymbol{\mu}, \boldsymbol{\Sigma}, \boldsymbol{\pi}, \mathcal{D}) \propto \pi_j p(\boldsymbol{x}_i|\boldsymbol{\mu}_j, \boldsymbol{\Sigma}_j, z_i^u = j) \tag{2.45}$$

   the product of the prior and the normal density of class ω_j at $\boldsymbol{x}_i$. The constant of proportionality is chosen so that the sum over classes is unity. Sampling a value for z_i^u is then trivial.

The Gibbs sampling procedure produces a set of samples $\{\boldsymbol{\mu}^t, \boldsymbol{\Sigma}^t, \boldsymbol{\pi}^t, (z^u)^t, t = 1, \ldots, N\}$, which may be used to classify the unlabelled patterns and future observations. To classify the unlabelled patterns in the training set, we use

$$p(z_i^u = j|\mathcal{D}) = \frac{1}{N - M} \sum_{t=M+1}^{N} p(z_i^u = j|\boldsymbol{\mu}_j^t, \boldsymbol{\Sigma}_j^t, \pi_j^t, \mathcal{D})$$

where the terms in the summation are, by (2.45), products of the prior and class-conditional density (normalised), evaluated for each set of parameters in the Markov chain. To classify a new pattern $\boldsymbol{x}$, we require $p(z = j|\boldsymbol{x}, \mathcal{D})$, given by

$$p(z = j|\boldsymbol{x}, \mathcal{D}) \propto p(z = j|\mathcal{D}) p(\boldsymbol{x}|\mathcal{D}, z = j) \tag{2.46}$$

where the first term in the product, $p(z = j|\mathcal{D})$, is

$$p(z = j|\mathcal{D}) = \mathrm{E}[\pi_j|\mathcal{D}] = \frac{1}{N-M}\sum_{t=M+1}^{N} \mathrm{E}[\pi_j|\mathcal{D}, (z^u)^t] \tag{2.47}$$

The expectation in the summation can be evaluated using (2.44). The second term can be written, by (2.43), as

$$p(\boldsymbol{x}|\mathcal{D}, z = j) \approx \frac{1}{N-M}\sum_{t=M+1}^{N} p(\boldsymbol{x}|\mathcal{D}, (z^u)^t, z = j)$$

the sum of Student t distributions.

Illustration

The illustration given here is based on that of Lavine and West (1992). Two-dimensional data from three classes are generated from equally weighted non-normal distributions. Defining matrices

$$\boldsymbol{C}_1 = \begin{pmatrix} 5 & 1 \\ 3 & 5 \end{pmatrix} \quad \boldsymbol{C}_2 = \begin{pmatrix} 0 & 1 \\ 1 & 5 \end{pmatrix} \quad \boldsymbol{C}_3 = \begin{pmatrix} 5 & 0 \\ 3 & 1 \end{pmatrix}$$

then an observation from class ω_i is generated according to

$$\boldsymbol{x}_j = \boldsymbol{C}_i \begin{pmatrix} w_j \\ 1 - w_j \end{pmatrix} + \boldsymbol{\epsilon}_j$$

where w_j is uniform over [0, 1] and $\boldsymbol{\epsilon}_j$ is normally distributed with zero mean and diagonal covariance matrix, $\boldsymbol{I}/2$.

The labelled training data are shown in Figure 2.10.

Using a normal model, with diagonal covariance matrix, for the density of each class, an MCMC approach using Gibbs sampling is taken. The training set consists of 1200 labelled and 300 unlabelled patterns. The priors for the mean and variances are normal and inverse gamma, respectively. The parameter values are initialised as samples from the prior. Figure 2.11 shows components of the mean and covariance matrix that are produced from the chain. The package WinBugs (Lunn *et al.*, 2000) has been used to complete the MCMC sampling.

2.4.4 Example application study

The problem This application concerns the classification of mobile ground targets from inverse synthetic aperture radar (ISAR) images (Copsey and Webb, 2001).

Summary This study follows the approach above, with the addition that the class-conditional densities are themselves mixtures.

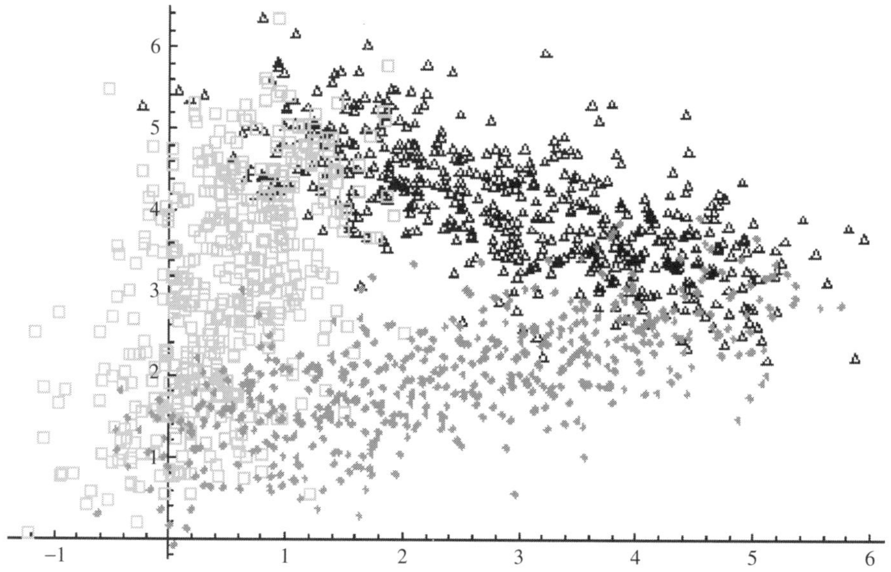

Figure 2.10 Three-class, two-dimensional training data

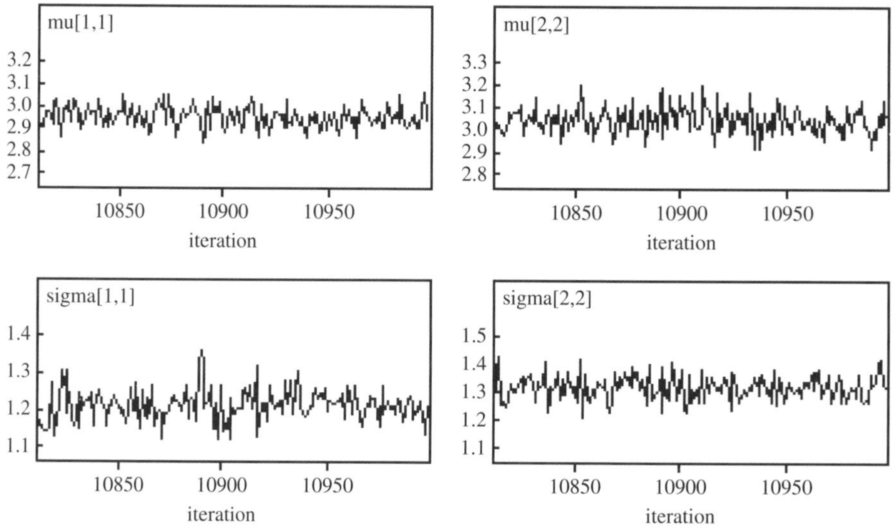

Figure 2.11 MCMC samples of $\boldsymbol{\mu}$ and $\boldsymbol{\Sigma}$. Top left: μ_1 component of class 1; top right: μ_2 component of class 2; bottom left: $\Sigma_{1,1}$ component of class 1; bottom right: $\Sigma_{2,2}$ component of class 2

The data The data comprise ISAR images of three types of vehicle, gathered over a complete rotation of the vehicle on a turntable. An ISAR image is an image generated by processing the signals received by the radar. One axis corresponds to range (distance from the radar); the second axis corresponds to cross-range. The training data consist of approximately equal numbers of images (about 2000) per class, collected over single complete rotations at a constant depression angle. The test data comprise six sets of approximately 400 ISAR images collected from single rotations of six vehicles. The images are 38 pixels in range by 26 pixels in cross-range.

The model The probability density function of the data is written as

$$p(x) = \sum_{j=1}^{C} \pi_j p(x|j)$$

where $\pi = (\pi_1, \dots, \pi_C)$ is the set of prior class probabilities and $p(x|j)$ is the class-conditional probability density of class j, which is also modelled as a mixture, with the jth class having R_j mixture components (termed subclasses)

$$p(x|j) = \sum_{r=1}^{R_j} \lambda_{j,r} p(x|j, r)$$

where $\lambda_j = (\lambda_{j,1}, \dots, \lambda_{j,R_j})$ represents the prior subclass probabilities within class j; that is, $\lambda_{j,r}$ is the mixing probability for the rth subclass of the jth class, satisfying $\sum_{r=1}^{R_j} \lambda_{j,r} = 1$. The probability density of the data for the subclass r of class j, $p(x|j, r)$, is taken to be normal with mean $\mu_{j,r}$ and covariance matrix $\Sigma_{j,r}$. Let $\mu = \{\mu_{j,r}\}$; $\Sigma = \{\Sigma_{j,r}\}$.

Training procedure A Gibbs sampling approach is taken. The random variable set, $\{\pi, \lambda, \mu, \Sigma\}$ is augmented by allocation variables $\{z, Z\}$ such that $(z_i = j, Z_i = r)$ implies that observation x_i is modelled as being drawn from subclass r of class j; z_i is known for labelled training data; Z_i is always unknown.

Let $\mathcal{D}$ denote the measurements and known allocations; z^u, the set of unknown class labels; $Z = (Z_1, \dots, Z_n)$ the subclass allocation labels. The stages in the Gibbs sampling iterations are as follows.

1. Sample from $p(\mu, \Sigma|\pi, \lambda, z^u, Z, \mathcal{D})$.
2. Sample from $p(\pi, \lambda|\mu, \Sigma, z^u, Z, \mathcal{D})$.
3. Sample from $p(z^u, Z|\mu, \Sigma, \pi, \lambda, \mathcal{D})$.

Future observations, x are classified by evaluating $p(z = j|x\mathcal{D})$,

$$p(z = j|\boldsymbol{x}, \mathcal{D}) = p(z = j|\mathcal{D}) p(\boldsymbol{x}|\mathcal{D}, z = j)$$

where the first term in the product, $p(z = j|\mathcal{D})$, is evaluated using equation (2.47) and the second term is approximated as

$$p(\boldsymbol{x}|\mathcal{D}, z = j) \approx \frac{1}{N - M} \sum_{t=M+1}^{N} p(\boldsymbol{x}|\mathcal{D}, Z^t, (z^u)^t, z = j)$$

where $p(\boldsymbol{x}|\mathcal{D}, Z^t, (z^u)^t, z=j)$ is written as a mixture

$$p(\boldsymbol{x}|\mathcal{D}, Z^t, (z^u)^t, z=j) = \sum_{r=1}^{R_j} p(\boldsymbol{x}|\mathcal{D}, Z^t, (z^u)^t, z=j, Z=r) \times p(Z=r|\mathcal{D}, Z^t, (z^u)^t, z=j)$$

in which $p(Z=r|\mathcal{D}, Z^t, (z^u)^t, z=j) = \mathrm{E}[\lambda_{j,r}|\mathcal{D}, (z^u)^t, Z^t]$ and $p(\boldsymbol{x}|\mathcal{D}, Z^t, (z^u)^t, z=j, Z=r)$ is the predictive density for a data point drawn from subclass r of class j, the parameters being determined using the MCMC algorithm outputs. This predictive distribution is also shown to be a product of Student t distributions.

A fixed model order is adopted ($R_j = 12$ for all classes); the data are preprocessed to produce 35 values (on the principal components); the burn-in period is 10 000 iterations; 1000 samples are drawn to calculate statistics; the decorrelation gap is 10 iterations.

2.4.5 Further developments

There are many developments of the basic methodology presented in this section, particularly with respect to computational implementation of the Bayesian approach. These include strategies for improving MCMC; monitoring convergence; and adaptive MCMC methods. A good starting point is the book by Gilks *et al* (1996).

Developments of the MCMC methodology to problems when observations arrive sequentially and one is interested in performing inference on-line are described by Doucet *et al.* (2001).

A Bayesian methodology for *univariate* normal mixtures that jointly models the number of components and the mixture component parameters is presented by Richardson and Green (1997).

2.4.6 Summary

A Bayesian approach to density estimation can only be treated analytically for simple distributions. For problems in which the normalising integral in the denominator of the expression for a posterior density cannot be evaluated analytically, Bayesian computational methods must be employed.

Monte Carlo methods, including the Gibbs sampler, can be applied routinely, allowing efficient practical application of Bayesian methods, at least for some problems. Approaches to discrimination can make use of unlabelled test samples to refine models. The procedure described in Section 2.4.3 implements an iterative procedure to classify test data. Although the procedure is attractive in that it uses the test data to refine knowledge about the parameters, its iterative nature may prevent its application in problems with real-time requirements.

2.5 Application studies

The application of the normal-based linear and quadratic discriminant rules covers a wide range of problems. These include the areas of:

- Medical research. Aitchison *et al.* (1977) compare predictive and estimative approaches to discrimination. Harkins *et al.* (1994) use a quadratic rule for the classification of red cell disorders. Hand (1992) reviews statistical methodology in medical research, including discriminant analysis (see also Jain and Jain, 1994). Stevenson (1993) discusses the role of discriminant analysis in psychiatric research.
- Machine vision. Magee *et al.* (1993) use a Gaussian classifier to discriminate bottles based on five features derived from images of the bottle tops.
- Target recognition. Kreithen *et al.* (1993) develop a target and clutter discrimination algorithm based on multivariate normal assumptions for the class distributions.
- Spectroscopic data. Krzanowski *et al.* (1995) consider ways of estimating linear discriminant functions when covariance matrices are singular and analyse data consisting of infrared reflectance measurements.
- Radar. Haykin *et al.* (1991) evaluate a Gaussian classifier on a clutter classification problem. Lee *et al.* (1994) develop a classifier for polarimetric synthetic aperture radar imagery based on the Wishart distribution.
- As part of a study by Aeberhard *et al.* (1994), regularised discriminant analysis was one of eight discrimination techniques (including linear and quadratic discriminant analysis) applied to nine real data sets. An example is the wine data set – the results of a chemical analysis of wines from the same region of Italy, but derived from different varieties of grape. The feature vectors were 13-dimensional, and the training set was small, comprising only 59, 71 and 48 samples in each of three classes. Since there is no separate test set, a leave-one-out procedure (see Chapter 8) was used to estimate error rate. On all the real data sets, RDA performed best overall.

Comparative studies of normal-based models with other discriminant methods can be found in the papers by Curram and Mingers (1994); Bedworth *et al.* (1989) on a speech recognition problem; and Aeberhard *et al.* (1994).

Applications of mixture models include:

- Plant breeding. Jansen and Den Nijs (1993) use a mixture of normals to model the distribution of pollen grain size.
- Image processing. Luttrell (1994) uses a *partitioned mixture distribution* for low-level image processing operations.
- Speech recognition. Rabiner *et al.* (1985), and Juang and Rabiner (1985) describe a hidden Markov model approach to isolated digit recognition in which the probability density function associated with each state of the Markov process is a normal mixture model.
- Handwritten character recognition. Revow *et al.* (1996) use a development of conventional mixture models (in which the means are constrained to lie on a spline) for handwritten digit recognition. Hastie and Tibshirani (1996) apply their mixture discriminant analysis approach to the classification of handwritten 3s, 5s and 8s.
- Motif discovery in biopolymers. Bailey and Elkan (1995) use a two-component mixture model to identify motifs (a pattern common to a set of nucleic or amino acid subsequences which share some biological property of interest) in a set of unaligned genetic or protein sequences.

- Face detection and tracking. In a study of face recognition (McKenna *et al.*, 1998), data characterising each subject's face (20- and 40-dimensional feature vectors) are modelled as a Gaussian mixture, with component parameters estimated using the EM procedure. Classification is performed by using these density estimates in Bayes' rule.

A compilation of examples of applications of Bayesian methodology is given in the book by French and Smith (1997). This includes applications in clinical medicine, flood damage analysis, nuclear plant reliability and asset management.

2.6 Summary and discussion

The approaches developed in this chapter towards discrimination have been based on estimation of the class-conditional density functions using parametric and *semiparametric* techniques. It is certainly true that we cannot design a classifier that performs better than the Bayes discriminant rule. No matter how sophisticated a classifier is, or how appealing it may be in terms of reflecting a model of human decision processes, it cannot achieve a lower error rate than the Bayes classifier. Therefore a natural step is to estimate the components of the Bayes rule from the data, namely the class-conditional probability density functions and the class priors.

In Section 2.2, we gave a short introduction to discrimination based on normal models. The models are easy to use and have been widely applied in discrimination problems. In Section 2.3 we introduced the normal mixture model and the EM algorithm. We will return to this in Chapter 10 when we shall consider such models for clustering. Section 2.4 considered Bayesian approaches to discrimination (which take into account parameter variability due to sampling), and Bayesian computational procedures that produce samples from the posterior distributions of interest were described. Such techniques remove the mathematical nicety of conjugate prior distributions in a Bayesian analysis, allowing models to be tailored to the beliefs and needs of the user.

2.7 Recommendations

An approach based on density estimation is not without its dangers of course. If incorrect assumptions are made about the form of the distribution in the parametric approach (and in many cases we will not have a physical model of the data generation process to use) or data points are sparse leading to poor estimates, then we cannot hope to achieve optimal performance. However, the linear and quadratic rules are widely used, simple to implement and have been used with success in many applications. Therefore, it is worth applying such techniques to provide at least a baseline performance on which to build. It may prove to be sufficient.

2.8 Notes and references

A comparison of the predictive and estimative approaches is found in the articles by Aitchison *et al.* (1977) and Moran and Murphy (1979). McLachlan (1992a) gives a very thorough account of normal-based discriminant rules and is an excellent source of reference material. Simple procedures for correcting the bias of the discriminant rule are also given. Mkhadri *et al.* (1997) provide a review of regularisation in discriminant analysis.

Mixture distributions, and in particular the normal mixture model, are discussed in a number of texts. The book by Everitt and Hand (1981) provides a good introduction, and a more detailed treatment is given by Titterington *et al.* (1985) (see also McLachlan and Basford, 1988). A thorough treatment, with recent methodological and computational developments, applications and software description is presented by McLachlan and Peel (2000). Lavine and West (1992) discuss Bayesian approaches to normal mixture models for discrimination and classification, with posterior probabilities obtained using an iterative resampling technique (see also West, 1992). Several approaches for determining the number of components of a normal mixture have been proposed and are discussed further in the context of clustering in Chapter 10. A review of mixture densities and the EM algorithm is given by Redner and Walker (1984). A thorough description of the EM algorithm and its extensions is provided in the book by McLachlan and Krishnan (1996). See also the review by Meng and van Dyk (1997), where the emphasis is on strategies for faster convergence. Software for the fitting of mixture models is publicly available.

Bayesian learning is discussed in many of the standard pattern recognition texts including Fu (1968), Fukunaga (1990), Young and Calvert (1974) and Hand (1981a). Geisser (1964) presents methods for Bayesian learning of means and covariance matrices under various assumptions on the parameters. Bayesian methods for discrimination are described by Lavine and West (1992) and West (1992).

A more detailed treatment of Bayesian inference, with descriptions of computational procedures, is given by O'Hagan (1994) and Bernardo and Smith (1994).

Gelfand (2000) gives a review of the Gibbs sampler and its origins (see also Casella and George, 1992). Monte Carlo techniques for obtaining characteristics of posterior distributions are also reviewed by Tierney (1994).

The website `www.statistical-pattern-recognition.net` contains references and links to further information on techniques and applications.

Exercises

1. In the example application study of Section 2.2.3, is it appropriate to use a Gaussian classifier for the head injury data? Justify your answer.

2. Suppose that $\boldsymbol{B} = \boldsymbol{A} + \boldsymbol{u}\boldsymbol{u}^T$, where $\boldsymbol{A}$ is a nonsingular $(p \times p)$ matrix and $\boldsymbol{u}$ is a vector. Show that $\boldsymbol{B}^{-1} = \boldsymbol{A}^{-1} - k\boldsymbol{A}^{-1}\boldsymbol{u}\boldsymbol{u}^T\boldsymbol{A}^{-1}$, where $k = 1/(1 + \boldsymbol{u}^T\boldsymbol{A}^{-1}\boldsymbol{u})$. (Krzanowski and Marriott, 1996)

3. Show that the estimate of the covariance matrix given by

$$\hat{\boldsymbol{\Sigma}} = \frac{1}{n-1}\sum_{i=1}^{n}(\boldsymbol{x}_i - \boldsymbol{m})(\boldsymbol{x}_i - \boldsymbol{m})^T$$

where $\boldsymbol{m}$ is the sample mean, is unbiased.

4. Suppose that the p-element $\boldsymbol{x}$ is normally distributed $N(\boldsymbol{\mu}, \boldsymbol{\Sigma}_i)$ in population i ($i = 1, 2$), where $\boldsymbol{\Sigma}_i = \sigma_i^2[(1-\rho_i)\boldsymbol{I} + \rho_i \boldsymbol{1}\boldsymbol{1}^T]$ and $\boldsymbol{1}$ denotes the p-vector all of whose elements are 1. Show that the optimal (i.e. Bayes) discriminant function is given, apart from an additive constant, by

$$-\tfrac{1}{2}(c_{11} - c_{12})Q_1 + \tfrac{1}{2}(c_{21} - c_{22})Q_2$$

where $Q_1 = \boldsymbol{x}\boldsymbol{x}^T$, $Q_2 = (\boldsymbol{1}^T\boldsymbol{x})^2$, $c_{1i} = [\sigma_i^2 + (1-\rho_i)]^{-1}$ and $c_{2i} = \rho_i[\sigma_i^2(1-\rho_i)\{1 + (p-1)\rho_i\}]^{-1}$. (Krzanowski and Marriott, 1996)

5. Verify that the simple one-pass algorithm

   (a) Initialise $\boldsymbol{S} = \boldsymbol{0}$, $\boldsymbol{m} = \boldsymbol{0}$.

   (b) For $r = 1$ to n do

      i. $\boldsymbol{d}_r = \boldsymbol{x}_r - \boldsymbol{m}$

      ii. $\boldsymbol{S} = \boldsymbol{S} + \left(1 - \dfrac{1}{r}\right)\boldsymbol{d}_r\boldsymbol{d}_r^T$

      iii. $\boldsymbol{m} = \boldsymbol{m} + \dfrac{\boldsymbol{d}_r}{r}$

   results in $\boldsymbol{m}$ as the sample mean and $\boldsymbol{S}$ as n times the sample covariance matrix.

6. Derive the EM update equations (2.14) and (2.15).

7. Consider a gamma distribution of the form

$$p(x|\mu, m) = \frac{m}{\Gamma(m)\mu}\left(\frac{mx}{\mu}\right)^{m-1}\exp\left(-\frac{mx}{\mu}\right)$$

for mean μ and order parameter m. Derive the EM update equations for the π_i, μ_i and m_i for the gamma mixture

$$p(x) = \sum_{i=1}^{g} \pi_i p(x|\mu_i, m_i)$$

8. Given that the log-likelihood in the EM procedure is given by

$$\log(L(\boldsymbol{\Psi})) = Q(\boldsymbol{\Psi}, \boldsymbol{\Psi}^{(m)}) - H(\boldsymbol{\Psi}, \boldsymbol{\Psi}^{(m)})$$

where H is given by (2.9), and using the result that $\log(x) \leq x - 1$, show that

$$H(\boldsymbol{\Psi}^{(m+1)}, \boldsymbol{\Psi}^{(m)}) - H(\boldsymbol{\Psi}^{(m)}, \boldsymbol{\Psi}^{(m)}) \leq 0$$

where $\boldsymbol{\Psi}^{(m+1)}$ is chosen to maximise $Q(\boldsymbol{\Psi}, \boldsymbol{\Psi}^{(m)})$. Hence, show that the log-likelihood is increased: $\log(L(\boldsymbol{\Psi}^{(m+1)})) \geq \log(L(\boldsymbol{\Psi}^{(m)}))$.

9. Generate three data sets (train, validation and test sets) for the three-class, 21-variable, waveform data (Breiman *et al.*, 1984):

$$\begin{aligned} x_i &= uh_1(i) + (1-u)h_2(i) + \epsilon_i \quad \text{(class 1)} \\ x_i &= uh_1(i) + (1-u)h_3(i) + \epsilon_i \quad \text{(class 2)} \\ x_i &= uh_2(i) + (1-u)h_3(i) + \epsilon_i \quad \text{(class 3)} \end{aligned}$$

   where $i = 1, \ldots, 21$; u is uniformly distributed on $[0, 1]$; ϵ_i are normally distributed with zero mean and unit variance; and the h_i are shifted triangular waveforms: $h_1(i) = \max(6 - |i - 11|, 0)$, $h_2(i) = h_1(i-4)$, $h_3(i) = h_1(i+4)$. Assume equal class priors. Construct a three-component mixture model for each class using a common covariance matrix across components and classes (Section 2.3.2). Investigate starting values for the means and covariance matrix and choose a model based on the validation set error rate. For this model, evaluate the classification error on the test set.

   Compare the results with a linear discriminant classifier and a quadratic discriminant classifier constructed using the training set and evaluated on the test set.

10. For the distribution illustrated by Figure 2.4, show that a suitable linear transformation of the coordinate system, to new variables ϕ_1 and ϕ_2, will lead to an irreducible chain.

11. For a transformation of variables from $(X_1, \ldots, X_p)$ to $(Y_1, \ldots, Y_p)$, given by

$$\boldsymbol{Y} = \boldsymbol{g}(\boldsymbol{X})$$

   where $\boldsymbol{g} = (g_1, g_2, \ldots, g_p)^T$, the density functions of $\boldsymbol{X}$ and $\boldsymbol{Y}$ are related by

$$p_Y(\boldsymbol{y}) = \frac{p_X(\boldsymbol{x})}{|\boldsymbol{J}|}$$

   where $|\boldsymbol{J}|$ is the absolute value of the Jacobian determinant

$$\boldsymbol{J}(x_1, \ldots, x_p) = \begin{vmatrix} \dfrac{\partial g_1}{\partial x_1} & \cdots & \dfrac{\partial g_1}{\partial x_p} \\ \vdots & \ddots & \vdots \\ \dfrac{\partial g_p}{\partial x_1} & \cdots & \dfrac{\partial g_p}{\partial x_p} \end{vmatrix}$$

   Let D denote the region in $\mathbb{R}^2$

$$D = \{(u, v);\ 0 \le u \le \sqrt{g(v/u)}\}$$

   Show that if (u, v) is uniformly distributed over the region D, then the change of variables $(U, V) \to (U, X = V/U)$ gives

$$p(u, x) = ku \quad 0 < u < \sqrt{g(x)}$$

   where k is a constant. Determine the value of k, and then by marginalising with respect to u show that

$$p(x) = \frac{g(x)}{\int g(x)\,dx}$$

3

Density estimation – nonparametric

Overview

Nonparametric methods of density estimation can provide class-conditional density estimates for use in Bayes' rule. Three main methods are introduced: a histogram approach with generalisation to include Bayesian networks; k-nearest-neighbour methods and variants; and kernel methods of density estimation.

3.1 Introduction

Many of the classification methods discussed in this book require knowledge of the class-conditional probability density functions. Given these functions, we can apply the likelihood ratio test (see Chapter 1) and decide the class to which a pattern $\boldsymbol{x}$ can be assigned. In some cases we may be able to make simplifying assumptions regarding the form of the density function; for example, that it is normal or a normal mixture (see Chapter 2). In these cases we are left with the problem of estimating the parameters that describe the densities from available data samples.

In many cases, however, we cannot assume that the density is characterised by a set of parameters and we must resort to *nonparametric* methods of density estimation; that is, there is no formal structure for the density prescribed. There are many methods that have been used for statistical density estimation and in the following paragraphs we shall consider four of them, namely the histogram approach, k-nearest-neighbour, expansion by basis functions and kernel-based methods. First, we shall consider some basic properties of density estimators.

Unbiasedness

If $\boldsymbol{X}_1, \ldots, \boldsymbol{X}_n$ are independent and identically distributed p-dimensional random variables with continuous density $p(\boldsymbol{x})$,

$$p(\boldsymbol{x}) \geq 0 \quad \int_{\mathbb{R}^p} p(\boldsymbol{x})\, d\boldsymbol{x} = 1 \tag{3.1}$$

the problem is to estimate $p(\boldsymbol{x})$ given measurements on these variables. If the estimator $\hat{p}(\boldsymbol{x})$ also satisfies (3.1), then it is not unbiased (Rosenblatt, 1956). That is, if we impose the condition that our estimator is itself a density (in that it satisfies (3.1)), it is biased:

$$\mathrm{E}[\hat{p}(\boldsymbol{x})] \neq p(\boldsymbol{x})$$

where $\mathrm{E}[\hat{p}(\boldsymbol{x})] = \int \hat{p}(\boldsymbol{x}|\boldsymbol{x}_1 \ldots \boldsymbol{x}_n) p(\boldsymbol{x}_1) \ldots p(\boldsymbol{x}_n)\, d\boldsymbol{x}_1 \ldots d\boldsymbol{x}_n$, the expectation over the random variables $\boldsymbol{X}_1, \ldots, \boldsymbol{X}_n$. Although estimators can be derived that are asymptotically unbiased, $\mathrm{E}[\hat{p}(\boldsymbol{x})] \rightarrow p(\boldsymbol{x})$ as $n \rightarrow \infty$, in practice we are limited by the number of samples that we have.

Consistency

There are other measures of discrepancy between the density and its estimate. The mean squared error (MSE) is defined by

$$\mathrm{MSE}_x(\hat{p}) = \mathrm{E}[(\hat{p}(\boldsymbol{x}) - p(\boldsymbol{x}))^2]$$

where the subscript x is used to denote that MSE is a function of $\boldsymbol{x}$. The above equation may be written

$$\mathrm{MSE}_x(\hat{p}) = \mathrm{var}(\hat{p}(\boldsymbol{x})) + \{\mathrm{bias}(\hat{p}(\boldsymbol{x}))\}^2$$

If $\mathrm{MSE}_x \rightarrow 0$ for all $\boldsymbol{x} \in \mathbb{R}^p$, then $\hat{p}$ is a *pointwise consistent estimator of* p *in the quadratic mean.* A global measure of accuracy is given by the integrated squared error (ISE)

$$\mathrm{ISE} = \int [\hat{p}(\boldsymbol{x}) - p(\boldsymbol{x})]^2\, d\boldsymbol{x}$$

and by the mean integrated squared error (MISE)

$$\mathrm{MISE} = \mathrm{E}\left[\int [\hat{p}(\boldsymbol{x}) - p(\boldsymbol{x})]^2\, d\boldsymbol{x}\right]$$

which represents an average over all possible data sets. Since the order of the expectation and the integral may be reversed, the MISE is equivalent to the integral of the MSE, that is the sum of the integrated squared bias and the integrated variance.

Density estimates

Although one might naïvely expect that density estimates have to satisfy the property (3.1), this need not be the case. We shall want them to be pointwise consistent, so that we can get arbitrarily close to the true density given enough samples. Consideration has been given to density estimates that may be negative in parts in order to improve the convergence properties. Also, as we shall see in a later section, the integral constraint may be relaxed. The k-nearest-neighbour density estimate has an infinite integral.

3.2 Histogram method

The histogram method is perhaps the oldest method of density estimation. It is the classical method by which a probability density is constructed from a set of samples.

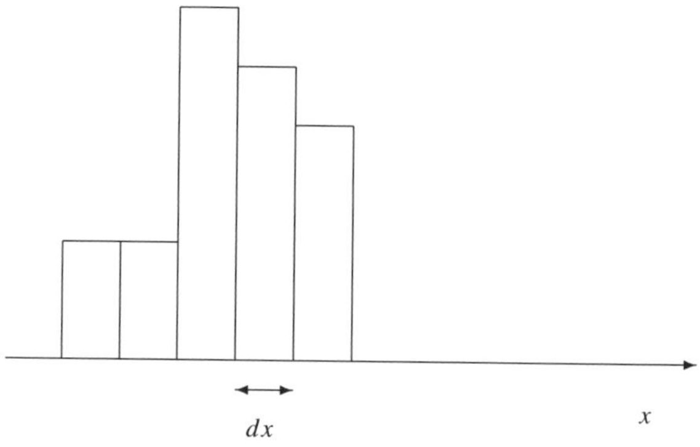

Figure 3.1 Histogram

In one dimension, the real line is partitioned into a number of equal-sized cells (see Figure 3.1) and the estimate of the density at a point x is taken to be

$$\hat{p}(x) = \frac{n_j}{\sum_j^N n_j \, dx}$$

where n_j is the number of samples in the cell of width dx that straddles the point x, N is the number of cells and dx is the size of the cell. This generalises to

$$\hat{p}(\boldsymbol{x}) = \frac{n_j}{\sum_j n_j dV}$$

for a multidimensional observation space, where dV is the volume of bin j.

Although this is a very simple concept and easy to implement, and it has the advantage of not needing to retain the sample points, there are several problems with the basic histogram approach. First of all, it is seldom practical in high-dimensional spaces. In one dimension, there are N cells; in two dimensions, there are N^2 cells (assuming that each variable is partitioned into N cells). For data samples $\boldsymbol{x} \in \mathbb{R}^p$ (p-dimensional vector $\boldsymbol{x}$) there are N^p cells. This exponential growth in the number of cells means that in high dimensions a very large amount of data is required to estimate the density. For example, where the data samples are six-dimensional, then dividing each variable range into 10 cells (a not unreasonable figure) gives a million cells. In order to prevent the estimate being zero over a large region, many observations will be required. A second problem with the histogram approach is that the density estimate is discontinuous and falls abruptly to zero at the boundaries of the region. We shall now consider some of the proposed approaches for overcoming these difficulties.

3.2.1 Data-adaptive histograms

One approach to the problem of constructing approximations to probability density functions from a limited number of samples using p-dimensional histograms is to allow the

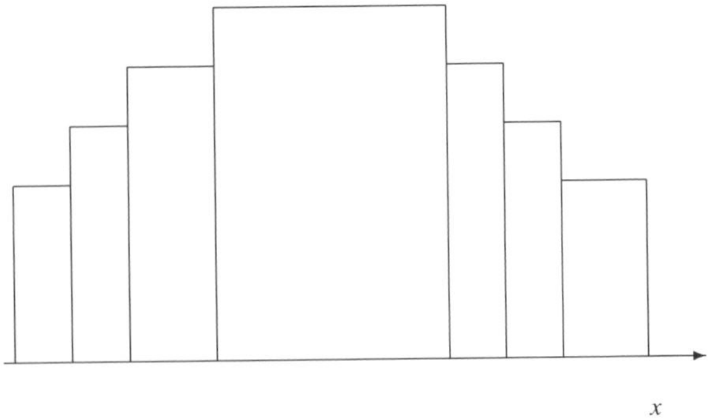

Figure 3.2 Variable cell size histogram

histogram descriptors – location, shape and size – to adapt to the data. This is illustrated in Figure 3.2.

An early approach was by Sebestyen and Edie (1966) who described a sequential method to multivariate density estimation using cells that are hyperellipsoidal in shape.

3.2.2 Independence assumption

Another approach for reducing the number of cells in high-dimensional problems is to make some simplifying assumptions regarding the form of the probability density function. We may assume that the variables are independent so that $p(\boldsymbol{x})$ may be written in the form

$$p(\boldsymbol{x}) = \prod_{i=1}^{p} p(x_i)$$

where $p(x_i)$ are the individual (one-dimensional) densities of the components of $\boldsymbol{x}$. Various names have been used to describe such a model including *naïve Bayes*, *idiot's Bayes* and *independence Bayes*. A histogram approach may be used for each density individually, giving pN cells (assuming an equal number of cells, N, per variable), rather than N^p. A particular implementation of the independence model is (Titterington *et al.*, 1981)

$$p(\boldsymbol{x}) \sim \left\{ \prod_{r=1}^{p} \frac{n(x_r) + \frac{1}{C_r}}{N(r) + 1} \right\}^B \tag{3.2}$$

where

x_r	is the rth component of $\boldsymbol{x}$;
$n(x_r)$	is the number of samples with value x_r on variable r;
$N(r)$	is the number of observations on variable r (this may vary due to missing data);
C_r	is the number of cells in variable r;
B	is an 'association factor' representing the 'proportion of non-redundant information' in the variables.

Note that the above expression takes account of missing data (which may be a problem in some categorical data problems). It has a non-constant number of cells per variable.

3.2.3 Lancaster models

Lancaster models are a means of representing the joint distribution in terms of the marginal distributions, assuming all interactions higher than a certain order vanish. For example, if we assume that all interactions higher than order $s = 1$ vanish, then a Lancaster model is equivalent to the independence assumption. If we take $s = 2$, then the probability density function is expressed in terms of the marginals $p(x_i)$ and the joint distributions $p(x_i, x_j), i \neq j$, as (Zentgraf, 1975)

$$p(\boldsymbol{x}) = \left\{ \sum_{i,j,i<j} \frac{p(x_i, x_j)}{p(x_i)p(x_j)} - \left[\binom{p}{2} - 1 \right] \right\} p_{\text{indep}}(\boldsymbol{x})$$

where $p_{\text{indep}}(\boldsymbol{x})$ is the density function obtained by the independence assumption,

$$p_{\text{indep}}(\boldsymbol{x}) = \prod_{k=1}^{p} p(x_k)$$

Lancaster models permit a range of models from the independence assumption to the full multinomial, but do have the disadvantage that some of the probability density estimates may be negative. Titterington *et al.* (1981) take the two-dimensional marginal estimates as

$$p(x_i, x_j) = \frac{n(x_i, x_j) + 1/(C_i C_j)}{N(i, j) + 1}$$

where the definitions of $n(x_i, x_j)$ and $N(i, j)$ are analogous to the definitions of $n(x_i)$ and $N(i)$ given above for the independence model, and

$$p(x_i) = \left(\frac{n(x_i) + 1/C_i}{N(i) + 1} \right)^B$$

Titterington *et al.* adopt the independence model whenever the estimate of the joint distribution is negative.

3.2.4 Maximum weight dependence trees

Lancaster models are one way to capture dependencies between variables without making the sometimes unrealistic assumption of total independence, yet having a model that does not require an unrealistic amount of storage or number of observations. Chow and Liu (1968) propose a tree-dependent model in which the probability distribution $p(\boldsymbol{x})$ is modelled as a tree-dependent distribution $p^t(\boldsymbol{x})$ that can be written as the product of $p - 1$ pairwise conditional probability distributions

$$p^t(\boldsymbol{x}) = \prod_{i=1}^{p} p(x_i | x_{j(i)}) \tag{3.3}$$

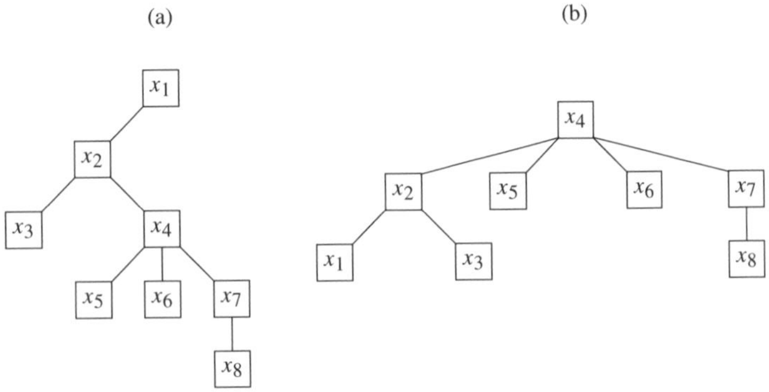

Figure 3.3 Tree representations

where $x_{j(i)}$ is the variable designated as the parent of x_i, with the root x_1 chosen arbitrarily and characterised by the prior probability $p(x_1|x_0) = p(x_1)$. For example, the density

$$p^t(\boldsymbol{x}) = p(x_1)p(x_2|x_1)p(x_3|x_2)p(x_4|x_2)p(x_5|x_4)p(x_6|x_4)p(x_7|x_4)p(x_8|x_7) \qquad (3.4)$$

has the tree depicted in Figure 3.3a, with root x_1. An alternative tree representation is Figure 3.3b, with root x_4, since (3.4) may be written using Bayes' theorem as

$$p^t(\boldsymbol{x}) = p(x_4)p(x_1|x_2)p(x_3|x_2)p(x_2|x_4)p(x_5|x_4)p(x_6|x_4)p(x_7|x_4)p(x_8|x_7)$$

Indeed, any node may be taken as the root node. If each variable can take N values, then the density (3.3) has $N(N-1)$ parameters for each of the conditional densities and $N-1$ parameters for the prior probability, giving a total of $N(N-1)(p-1)+N-1$ parameters to estimate.

The approach of Chow and Liu (1968) is to seek the tree-dependent distribution, $p^t(\boldsymbol{x})$, that best approximates the distribution $p(\boldsymbol{x})$. They use the Kullback–Leibler cross-entropy measure as the measure of closeness in approximating $p(\boldsymbol{x})$ by $p^t(\boldsymbol{x})$,

$$D(p, p^t) = \int p(\boldsymbol{x}) \log\left(\frac{p(\boldsymbol{x})}{p^t(\boldsymbol{x})}\right) d\boldsymbol{x}$$

or, for discrete variables,

$$D = \sum_{\boldsymbol{x}} p(\boldsymbol{x}) \log\left(\frac{p(\boldsymbol{x})}{p^t(\boldsymbol{x})}\right)$$

where the sum is over all values that the variable $\boldsymbol{x}$ can take, and they seek the tree-dependent distribution $p^\tau(\boldsymbol{x})$ such that $D(p(\boldsymbol{x}), p^\tau(\boldsymbol{x})) \leq D(p(\boldsymbol{x}), p^t(\boldsymbol{x}))$ over all t in the set of possible first-order dependence trees. Using the mutual information between variables

$$I(X_i, X_j) = \sum_{x_i, x_j} p(x_i, x_j) \log\left(\frac{p(x_i, x_j)}{p(x_i)p(x_j)}\right)$$

to assign weights to every branch of the dependence tree, Chow and Liu (1968) show (see the exercises) that the tree-dependent distribution $p^t(\boldsymbol{x})$ that best approximates $p(\boldsymbol{x})$ is the one with maximum weight defined by

$$W = \sum_{i=1}^{p} I(X_i, X_{j(i)})$$

This is termed a *maximum weight dependence tree* (MWDT) or a *maximum weight spanning tree*. The steps in the algorithm to find a MWDT are as follows.

1. Compute the branch weights for all $p(p-1)/2$ variable pairs and order them in decreasing magnitude.
2. Assign the branches corresponding to the two largest branches to the tree.
3. Consider the next largest value and add the corresponding branch to the tree if it does not form a cycle, otherwise discard it.
4. Repeat this procedure until $p-1$ branches have been selected.
5. The probability distribution may be computed by selecting an arbitrary root node and computing (3.3).

Applying the above procedure to the six-dimensional head injury data of Titterington *et al.* (1981) produces (for class 1) the tree illustrated in Figure 3.4. The labels for the variables are described in Section 2.2.3 (x_1 – `Age`; x_2 – `EMV`; x_3 – `MRP`; x_4 – `Change`; x_5 – `Eye Indicant`; x_6 – `Pupils`). To determine the tree-dependent distribution, select a root node (say node 1) and write the density using the figure as

$$p^t(\boldsymbol{x}) = p(x_1)p(x_2|x_1)p(x_3|x_2)p(x_5|x_2)p(x_4|x_3)p(x_6|x_5)$$

The first stage in applying MWDTs to a classification problem is to apply the algorithm to the data set for each class individually to give C trees.

There are several features that make MWDTs attractive. The algorithm requires only second-order distributions but, unlike the second-order Lancaster model, it need store

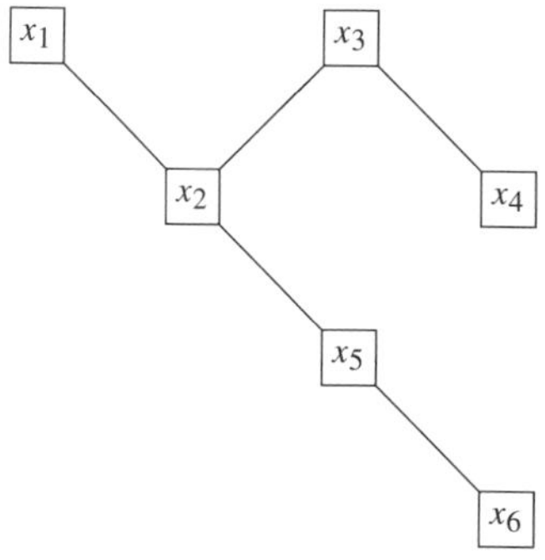

Figure 3.4 MWDT applied to head injury patient data (class 1)

only $p-1$ of these. The tree is computed in $\mathcal{O}(p^2)$ steps (though additional computation is required in order to obtain the mutual information) and if $p(\boldsymbol{x})$ is indeed tree-dependent then the approximation $p^t(\boldsymbol{x})$ is a consistent estimate in the sense that

$$\max_{\boldsymbol{x}} |p_n^{t(n)}(\boldsymbol{x}) - p(\boldsymbol{x})| \rightarrow 0 \text{ with probability 1 as } n \rightarrow \infty$$

where $p_n^{t(n)}$ is the tree-dependent distribution estimated from n independent samples of the distribution $p(\boldsymbol{x})$.

3.2.5 Bayesian networks

In Section 3.2.4 a development of the naïve Bayes model (which assumes independence between variables) was described. This was the MWDT model, which allows pairwise dependence between variables and is a compromise between approaches that specify all relationships between variables and the rather restrictive independence assumption. Bayesian networks also provide an intermediate model between these two extremes and have the tree-based models as a special case.

We introduce Bayesian networks by considering a graphical representation of a multivariate density. The chain rule allows a joint density, $p(x_1, \ldots, x_p)$, to be expressed in the form

$$p(x_1, \ldots, x_p) = p(x_p|x_1, \ldots, x_{p-1})p(x_{p-1}|x_1, \ldots, x_{p-2}) \ldots p(x_2|x_1)p(x_1)$$

We may depict such a representation of the density graphically. This is illustrated in Figure 3.5 for $p = 6$ variables. Each node in the graph represents a variable and the directed links denote the dependencies of a given variable. The *parents* of a given variable are those variables with directed links towards it. For example, the parents of x_5 are x_1, x_2, x_3 and x_4. The *root node* is the node without parents (the node corresponding to variable x_1). The probability density that such a figure depicts is the product of conditional densities

$$p(x_1, \ldots, x_p) = \prod_{i=1}^{p} p(x_i|\boldsymbol{\pi}_i) \tag{3.5}$$

where $\boldsymbol{\pi}_i$ is the set of parents of x_i (cf. equation (3.3)). If $\boldsymbol{\pi}_i$ is empty, $p(x_i|\boldsymbol{\pi}_i)$ is set to $p(x_i)$.

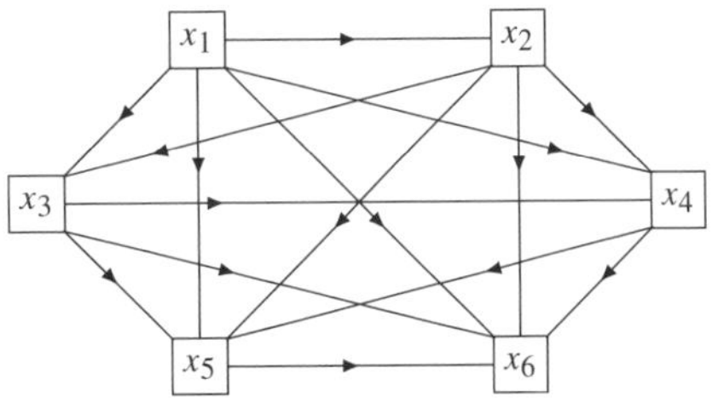

Figure 3.5 Graphical representation of the multivariate density $p(x_1, \ldots, x_6)$

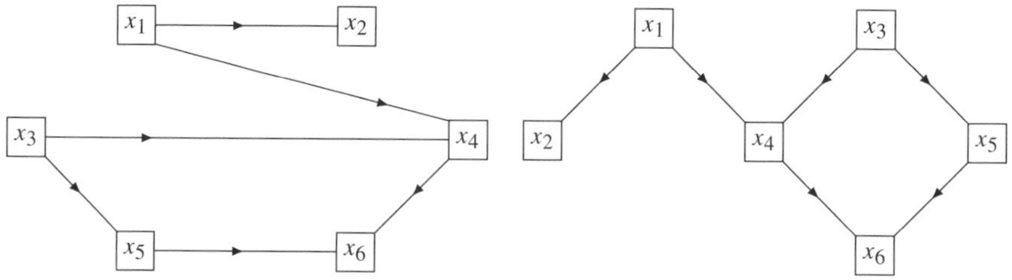

Figure 3.6 Graphical representations of the multivariate density $p(x_1, \ldots, x_6) = p(x_6|x_4, x_5)$ $p(x_5|x_3)p(x_4|x_1, x_3)p(x_3)p(x_2|x_1)p(x_1)$

Figure 3.5 is the graphical part of the Bayesian network and equation (3.5) is the density associated with the graphical representation. However, there is little to be gained in representing a full multivariate density $p(x_1, \ldots, x_6)$ as a product using the chain rule with the corresponding graph (Figure 3.5), unless we can make some simplifying assumptions concerning the dependence of variables. For example, suppose

$$p(x_6|x_1, \ldots, x_5) = p(x_6|x_4, x_5)$$

that is, x_6 is independent of x_1, x_2, x_3 given x_4, x_5; also

$$p(x_5|x_1, \ldots, x_4) = p(x_5|x_3)$$
$$p(x_4|x_1, x_2, x_3) = p(x_4|x_1, x_3)$$
$$p(x_3|x_1, x_2) = p(x_3)$$

Then the multivariate density may be expressed as the product

$$p(x_1, \ldots, x_6) = p(x_6|x_4, x_5)p(x_5|x_3)p(x_4|x_1, x_3)p(x_3)p(x_2|x_1)p(x_1) \tag{3.6}$$

This is depicted graphically in Figure 3.6; the left-hand graph is obtained by removing links in Figure 3.5 and the right-hand graph is an equivalent graph to make the 'parentage' more apparent. This figure, with the general probability interpretation (3.5), gives (3.6). Note that there are two root nodes.

Definition

A *graph* is a pair (V, E), where V is a set of vertices and E a set of edges (connections between vertices).

A directed graph is a graph in which all edges are directed: the edges are ordered pairs; if $(\alpha, \beta) \in E$, for vertices α and β, then $(\beta, \alpha) \notin E$.

A *directed acyclic graph* (DAG) is one in which there are no cycles: there is no path $\alpha_1 \rightarrow \alpha_2 \rightarrow \cdots \rightarrow \alpha_1$, for any vertex α_1.

A Bayesian network is a DAG where the vertices correspond to variables and associated with a variable X with parents $Y_1, \ldots, Y_p$ is a conditional probability density function, $p(X|Y_1, \ldots, Y_p)$.

The Bayesian network, together with the conditional densities, specifies a joint probability density function by (3.5).

The graphical representation of Bayesian networks is convenient for visualising dependencies. The factorisation of a multivariate density into a product of densities defined on perhaps only a few variables allows better nonparametric density estimates. For example, the product (3.6) requires densities defined on at most three variables.

Classification

As with the MWDT, we may construct a different Bayesian network to model the probability density function of each class $p(\boldsymbol{x}|\omega_i)$ separately. These densities may be substituted into Bayes' rule to obtain estimates of the posterior probabilities of class membership. Alternatively, we may construct a Bayesian network to model the joint density $p(\boldsymbol{x}, \omega)$, where ω is a class label. Again, we may evaluate this for each of the classes and use Bayes' rule to obtain $p(\omega|\boldsymbol{x})$. Separate networks for each class allow a more flexible model. Such a set of networks has been termed a *Bayesian multinet* (Friedman *et al.*, 1997).

Specifying the network

Specifying the structure of a Bayesian network consists of two parts: specifying the network topology and estimating the parameters of the conditional probability density functions. The topology may be specified by someone with an expert knowledge of the problem domain and who is able to make some statements about dependencies of variables. Hence, an alternative name for Bayesian networks is probabilistic expert systems: an *expert system* because the network encodes expert knowledge of a problem in its structure and *probabilistic* because the dependencies between variables are probabilistic. Acquiring expert knowledge can be a lengthy process. An alternative approach is to learn the graph from data if they are available, perhaps in a similar manner to the MWDT algorithm. In some applications, sufficient data may not be available.

In learning structure from data, the aim is to find the Bayesian network that best characterises the dependencies in the data. There are many approaches. Buntine (1996) reviews the literature on learning networks. Ideally, we would want to combine expert knowledge where available and statistical data. Heckerman *et al.* (1995; see also Cooper and Herskovits, 1992) discuss Bayesian approaches to network learning.

The probability density functions are usually specified as a conditional probability table, with continuous variables discretised, perhaps as part of the structure learning process. For density estimation, it is not essential to discretise the variables and some nonparametric density estimate, perhaps based on *product kernels* (see Section 3.5), could be used.

Discussion

Bayesian networks provide a graphical representation of the variables in a problem and the relationship between them. This representation needs to be specified or learned from data. This structure, together with the conditional density functions, allows the multivariate density function to be specified through the product rule (3.5). In a classification problem, a density may be estimated for each class and Bayes' rule used to obtain the posterior probabilities of class membership.

Bayesian networks have been used to model many complex problems, other than ones in classification, with the structure being used to calculate the conditional density of a variable given measurements made on some (or all) of the remaining variables. Such a situation may arise when measurements are made sequentially and we wish to update our *belief* that a variable takes a particular value as measurements arrive. For example, suppose we model the joint density $p(y, x_1, \ldots, x_p)$ as a Bayesian network, and we are interested in the quantity $p(y|\boldsymbol{e})$, where $\boldsymbol{e}$ comprises measurements made on a subset of the variables, $x_1, \ldots, x_p$. Then, using Bayes' theorem, we may write

$$p(y|\boldsymbol{e}) = \frac{\int p(y, \boldsymbol{e}, \tilde{\boldsymbol{e}})\, d\tilde{\boldsymbol{e}}}{\iint p(y, \boldsymbol{e}, \tilde{\boldsymbol{e}})\, d\tilde{\boldsymbol{e}}\, dy}$$

where $\tilde{\boldsymbol{e}}$ is the set of variables $x_1, \ldots, x_p$ not instantiated. Efficient algorithms that make use of the graphical structure have been developed for computing $p(y|\boldsymbol{e})$ in the case of discrete variables (Lauritzen and Spiegelhalter, 1988).

3.2.6 Example application study

The problem Prediction of the occurrence of rainfall using daily observations of meteorological data (Liu *et al.*, 2001).

Summary The study used the naïve Bayes (that is, independence assumption) classifier with marginal densities estimated through a histogram approach.

The data The data comprise daily observations from May to October for the years 1984–1992 from Hong Kong Observatory. There are mixed continuous and categorical data including wind direction and speed, daily mean atmospheric pressure, five-day mean pressure, temperature, rainfall and so on. There are 38 basic input variables and three classes of rainfall. There is some data preprocessing to handle missing values and standardisation of variables.

The model The model is a naïve Bayes classifier, with priors calculated from the data and class-conditional densities estimated using histograms.

Training procedure Although training for a histogram-based naïve Bayes classifier is minimal in general, the degree of discretisation must be specified (here, a class-dependent method was used). Also, some variable selection (see Chapter 9) was carried out. Another difference from the standard histogram approach is that the density estimates were updated using past tested data. Thus, the size of the training set is variable; it increases as we apply the method to test patterns.

Results For this application, this simple approach worked well and performed better that other methods also assessed.

3.2.7 Further developments

The MWDT tree-dependence approximation can also be derived by minimising an upper bound on the Bayes error rate under certain circumstances (Wong and Poon, 1989).

Computational improvements in the procedure have been proposed by Valiveti and Oommen (1992, 1993), who suggest a chi-square metric in place of the expected mutual information measure.

A further development is to impose a common tree structure across all classes. The mutual information between variables is then written as

$$I(X_i, X_j) = \sum_{x_i, x_j, \omega} p(x_i, x_j, \omega) \log \left(\frac{p(x_i, x_j|\omega)}{p(x_i|\omega)p(x_j|\omega)} \right)$$

It is termed a tree-augmented naïve Bayesian network by Friedman *et al.* (1997), who provide a thorough evaluation of the model on 23 data sets from the UCI repository (Murphy and Aha, 1995) and two artificial data sets. Continuous attributes are discretised and patterns with missing values omitted from the analysis. Performance is measured in terms of classification accuracy with the holdout method used to estimate error rates. See Chapter 8 for performance assessment and Chapter 11 for a discussion of the missing data problem. The model was compared to one in which separate trees were constructed for each class and the naïve Bayes model. Both tree models performed well in practice.

One of the disadvantages of histogram procedures for continuous data that we noted in the introduction to this section was that the density is discontinuous at cell boundaries. Procedures based on splines have been proposed for overcoming this difficulty. More recent developments in the use of splines for density estimation are described by Gu and Qiu (1993).

3.2.8 Summary

In the development of the basic histogram approach described in this section, we have concentrated on methods for reducing the number of cells for high-dimensional data. The approaches described assume that the data are categorical with integer-labelled categories, though the categories themselves may or may not be ordered. The simplest models are the independence models, and on the head injury data of Titterington *et al.* (1981) they gave consistently good performance over a range of values for B, the association factor (0.8 to 1.0). The independence assumption results in a very severe factorisation of the probability density function – clearly one that is unrealistic for many practical problems. Yet, as discussed by Hand and Yu (2001), it is a model that has had a long and successful history (see also Domingos and Pazzani, 1997). Practical studies, particularly in medical areas, have shown it to perform surprisingly well. Hand (1992) provides some reasons why this may be so: its intrinsic simplicity means low variance in its estimates; although its probability density estimates are biased, this may not matter in supervised classification so long as $\hat{p}(\omega_1|\boldsymbol{x}) > \hat{p}(\omega_2|\boldsymbol{x})$ when $p(\omega_1|\boldsymbol{x}) > p(\omega_2|\boldsymbol{x})$; in many cases, variables have undergone a selection process to reduce dependencies.

More complex interactions between variables may be represented using Lancaster models and MWDTs. Introduced by Chow and Liu (1968), MWDTs provide an efficient means of representing probability density functions using only second-order statistics.

Dependence trees are a special case of Bayesian networks which model a multivariate density as a product of conditional densities defined on a smaller number of variables. These networks may be specified by an expert, or learned from data.

3.3 *k*-nearest-neighbour method

The k-nearest-neighbour method is a simple method of density estimation. The probability that a point x' falls within a volume V centred at a point $\boldsymbol{x}$ is given by

$$\theta = \int_{V(\boldsymbol{x})} p(\boldsymbol{x})\, d\boldsymbol{x}$$

where the integral is over the volume V. For a small volume

$$\theta \sim p(\boldsymbol{x}) V \tag{3.7}$$

The probability, θ, may be approximated by the proportion of samples falling within V. If k is the number of samples, out of a total of n, falling within V (k is a function of $\boldsymbol{x}$) then

$$\theta \sim \frac{k}{n} \tag{3.8}$$

Equations (3.7) and (3.8) combine to give an approximation for the density,

$$\hat{p}(\boldsymbol{x}) = \frac{k}{nV} \tag{3.9}$$

The k-nearest-neighbour approach is to fix the probability k/n (or, equivalently, for a given number of samples n, to fix k) and to determine the volume V which contains k samples centred on the point $\boldsymbol{x}$. For example, if x_k is the kth nearest-neighbour point to $\boldsymbol{x}$, then V may be taken to be a sphere, centred at $\boldsymbol{x}$, of radius $\|\boldsymbol{x} - \boldsymbol{x}_k\|$ (the volume of a sphere of radius r in n dimensions is $2r^n\pi^{\frac{n}{2}}/n\Gamma(n/2)$, where $\Gamma(x)$ is the gamma function). The ratio of the probability to this volume gives the density estimate. This is in contrast to the basic histogram approach which is to fix the cell size and to determine the number of points lying within it.

One of the parameters to choose is the value of k. If it is too large, then the estimate will be smoothed and fine detail averaged out. If it is too small, then the probability density estimate is likely to be spiky. This is illustrated in Figures 3.7 (peaks truncated) and 3.8, where 13 samples are plotted on the x-axis, and the k-nearest-neighbour density estimate shown for $k = 1$ and 2. One thing to note about the density estimate is that it is not in fact a density. The integral under the curve is infinite. This is because for large enough $|\boldsymbol{x}|$, the estimate varies as $1/|\boldsymbol{x}|$. However, it can be shown that the density estimator is asymptotically unbiased and consistent if

$$\lim_{n\to\infty} k(n) = \infty$$

$$\lim_{n\to\infty} \frac{k(n)}{n} = 0$$

3.3.1 *k*-nearest-neighbour decision rule

Having obtained an expression for a density estimate, we can now use this in a decision rule. Suppose that in the first k samples there are k_m in class ω_m (so that $\sum_{m=1}^{C} k_m = k$).

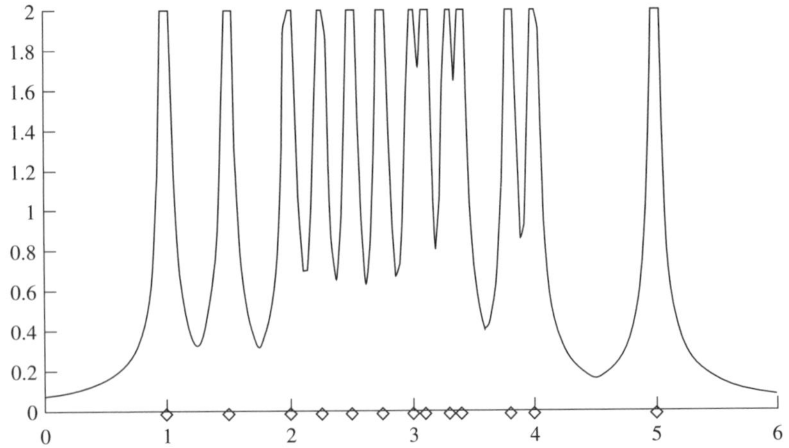

Figure 3.7 Nearest-neighbour density estimates for $k = 1$

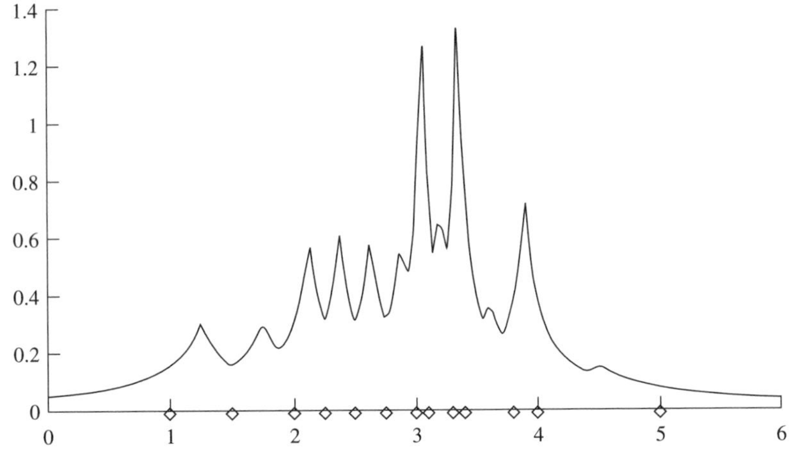

Figure 3.8 Nearest-neighbour density estimates for $k = 2$

Let the total number of samples in class ω_m be n_m (so that $\sum_{m=1}^{C} n_m = n$). Then we may estimate the class-conditional density, $p(\boldsymbol{x}|\omega_m)$, as

$$\hat{p}(\boldsymbol{x}|\omega_m) = \frac{k_m}{n_m V} \tag{3.10}$$

and the prior probability, $p(\omega_m)$, as

$$\hat{p}(\omega_m) = \frac{n_m}{n}$$

Then the decision rule is to assign $\boldsymbol{x}$ to ω_m if

$$\hat{p}(\omega_m|\boldsymbol{x}) \geq \hat{p}(\omega_i|\boldsymbol{x}) \quad \text{for all } i$$

or, using Bayes' theorem,

$$\frac{k_m}{n_m V}\frac{n_m}{n} \geq \frac{k_i}{n_i V}\frac{n_i}{n} \quad \text{for all } i$$

that is, assign $\boldsymbol{x}$ to ω_m if

$$k_m \geq k_i \quad \text{for all } i$$

Thus, the decision rule is to assign $\boldsymbol{x}$ to the class that receives the largest vote amongst the k nearest neighbours. There are several ways of breaking ties. Ties may be broken arbitrarily. Alternatively, $\boldsymbol{x}$ may be assigned to the class, out of the classes with tying values of k_i, that has nearest mean vector to $\boldsymbol{x}$ (with the mean vector calculated over the k_i samples). Another method is to assign $\boldsymbol{x}$ to the most compact class – that is, to the one for which the distance to the k_ith member is the smallest. This does not require any extra computation. Dudani (1976) proposes a distance-weighted rule in which weights are assigned to the k nearest neighbours, with closest neighbours being weighted more heavily. A pattern is assigned to that class for which the weights of the representatives among the k neighbours sum to the greatest value.

3.3.2 Properties of the nearest-neighbour rule

The asymptotic misclassification rate of the nearest-neighbour rule, e, satisfies the condition (Cover and Hart, 1967)

$$e^* \leq e \leq e^*\left(2 - \frac{Ce^*}{C-1}\right)$$

where e^* is the Bayes probability of error and C is the number of classes. Thus in the large-sample limit, the nearest-neighbour error rate is bounded above by twice the Bayes error rate. The inequality may be inverted to give

$$\frac{C-1}{C} - \sqrt{\frac{C-1}{C}}\sqrt{\frac{C-1}{C} - e} \leq e^* \leq e$$

The leftmost quantity is a lower bound on the Bayes error rate. Therefore, any classifier must have an error rate greater than this value.

3.3.3 Algorithms

Identifying the nearest neighbour of a given observation vector from among a set of training vectors is conceptually straightforward with n distance calculations to be performed. However, as the number n in the training set becomes large, this computational overhead may become excessive.

Many algorithms for reducing the nearest-neighbour search time involve the significant computational overhead of preprocessing the prototype data set in order to form a distance matrix (see Dasarathy, 1991, for a summary). There is also the overhead of storing $n(n-1)/2$ distances. There are many approaches to this problem. The linear approximating and eliminating search algorithm (LAESA), a development of the AESA algorithm of Vidal (1986, 1994), has a preprocessing stage that computes a number of *base prototypes* that are in some sense maximally separated from among the set of training vectors (Micó *et al*., 1994). Although it does not guarantee an optimal solution (in the sense that the sum of all pairwise distances between members of the set of base prototypes is a maximum), it can be achieved in linear preprocessing time. However, the LAESA algorithm requires the storage of an array of size n by n_b, the number of base prototypes. This will place an upper bound on the permitted number of base prototypes. Figure 3.9 illustrates a set of 21 training samples in two dimensions and four base prototypes chosen by the base prototype algorithm of Micó *et al*. (1994). The algorithm begins by first selecting a base prototype, $\boldsymbol{b}_1$, arbitrarily from the set of prototypes and the distance to every member of the remaining prototypes is calculated and stored in an array A. The second base prototype, $\boldsymbol{b}_2$, is the prototype that is furthest from $\boldsymbol{b}_1$. The distance of this prototype to every remaining prototype is calculated and the array A incremented. Thus, A represents the accumulated distances of non-base prototypes to base prototypes. The third base prototype is the one for which the accumulated distance is the greatest. This process of selecting a base prototype, calculating distances and accumulating the distances continues until the required number of base prototypes has been selected.

The LAESA searching algorithm uses the set of base prototypes and the interpoint distances between these vectors and those in the training set as follows. The algorithm uses the metric properties of the data space. Let $\boldsymbol{x}$ be the test sample (whose nearest neighbour from the set of prototypes we seek), $\boldsymbol{n}$ be its current nearest neighbour at a distance d_{xn}, and $\boldsymbol{q}$ be a base prototype whose distance to $\boldsymbol{x}$ has been computed at an earlier stage of the algorithm (see Figure 3.10). The condition for a prototype $\boldsymbol{p}$ to be rejected as a nearest-neighbour candidate is

$$d_{xp} \geq d_{xn}$$

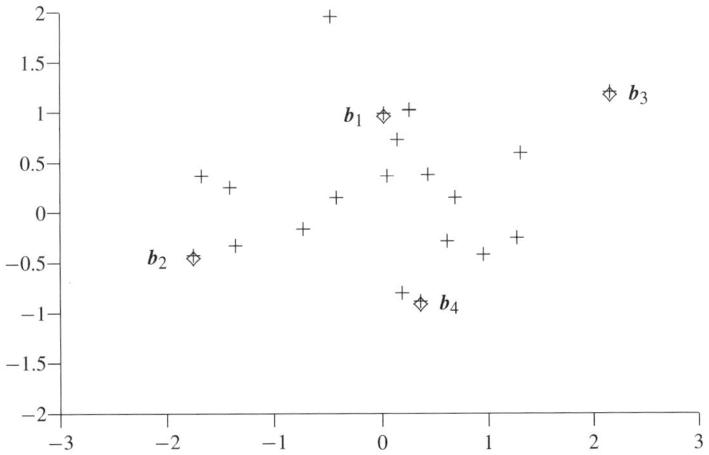

Figure 3.9 Selection of base prototypes (◇) from the data set (+)

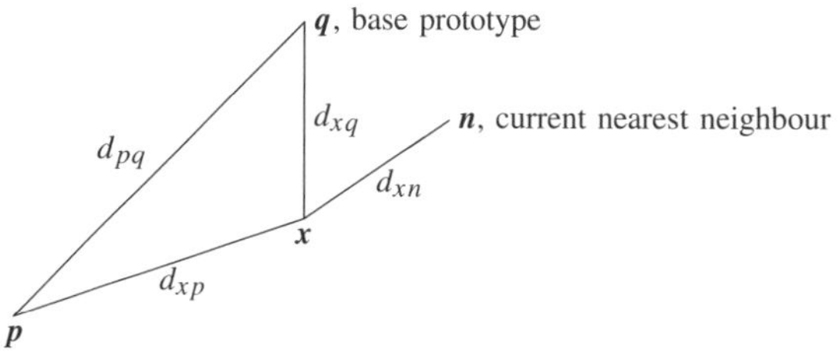

Figure 3.10 Nearest-neighbour selection using LAESA

This requires the calculation of the distance d_{xp}. However, a lower bound on the distance d_{xp} is given by

$$d_{xp} \geq |d_{pq} - d_{xq}|$$

and if this lower bound (which does not require any additional distance calculation) exceeds the current nearest-neighbour distance then clearly we may reject $\boldsymbol{p}$. We may go further by stating

$$d_{xp} \geq G(p) \triangleq \max_{\boldsymbol{q}} |d_{pq} - d_{xq}| \tag{3.11}$$

where the maximum is over all base prototypes considered so far in the iteration process, and therefore if $G(p) \geq d_{xn}$, we may reject $\boldsymbol{p}$ without computing d_{xp}.

Given $\boldsymbol{x}$, the algorithm selects a base prototype as an initial candidate $\boldsymbol{s}$ for a nearest neighbour and removes this from the set of prototypes. It then searches through the remaining set and rejects all samples from the set of prototypes whose lower bound on the distance to $\boldsymbol{x}$ exceeds the distance $|\boldsymbol{x} - \boldsymbol{s}|$. At the initial stage, the lower bound (3.11) is based on the selected base prototype, $\boldsymbol{s}$, only. A record of this lower bound is stored in an array. Out of the remaining prototypes, the $EC\infty$ version of the algorithm selects as the next candidate $\boldsymbol{s}$ for the nearest neighbour the base prototype for which this lower bound is a minimum (assuming the base prototypes have not been eliminated, otherwise a non-base prototype sample is chosen).

The candidate vector $\boldsymbol{s}$ need not necessarily be nearer than the previous choice, though if it is the nearest neighbour is updated. The data set is searched through again, rejecting vectors that are greater than the lower bound (which is updated if $\boldsymbol{s}$ is a base prototype). This process is repeated and is summarised in the five following steps.

1. Distance computing – calculate the distance of $\boldsymbol{x}$ to the candidate $\boldsymbol{s}$ for a nearest neighbour.

2. Update the prototype nearest to $\boldsymbol{x}$ if necessary.

3. Update the lower bounds, $G(p)$, if $\boldsymbol{s}$ is a base prototype.

4. Eliminate the prototypes with lower bounds greater than d_{xs}.

5. Approximating – select the candidates for the next nearest neighbour.

Further details are given by Micó *et al.* (1994). Figure 3.11 gives an illustration of the LAESA procedure (using the $EC\infty$ condition). The test sample $\boldsymbol{x}$ is at the origin $(0, 0)$

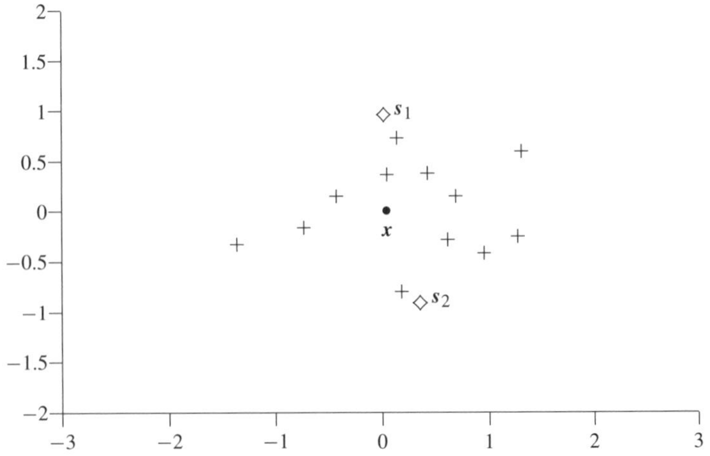

Figure 3.11 Nearest-neighbour selection using LAESA; s_1 and s_2 are the first two candidates for nearest neighbour

and the first two choices for $\boldsymbol{s}$ are shown. The remaining samples are those left after two passes through the data set (two distance calculations).

There are two factors governing the choice of the number of base prototypes, n_b. One is the amount of storage available. An array of size $n \times n_b$ must be stored. This could become prohibitive if n and n_b are large. At the other extreme, too small a value of n_b will result in a large number of distance calculations if n is large. We suggest that you choose a value as large as possible without placing constraints on memory since the number of distance calculations decreases monotonically (approximately) with n_b for the $EC\infty$ model of Micó *et al.* (1994) given above.

3.3.4 Editing techniques

One of the disadvantages of the k-nearest-neighbour rule is that it requires the storage of all n data samples. If n is large, then this may mean an excessive amount of storage. However, a major disadvantage may be the computation time for obtaining the k nearest neighbours. There have been several studies concerned with reducing the number of class prototypes with the joint aims of increasing computational efficiency and increasing the generalisation error rate.

We shall consider algorithms for two procedures for reducing the number of prototypes. The first of these belongs to the family of *editing* techniques. These techniques process the design set with the aim of removing prototypes that contribute to the misclassification rate. This is illustrated in Figure 3.12 for a two-class problem. Figure 3.12 plots samples from two overlapping distributions, together with the Bayes decision boundary. Each region, to the left and right of the boundary, contains prototypes that are misclassified by a Bayes classifier. Removing these to form Figure 3.13 gives two homogeneous sets of prototypes with a nearest-neighbour decision boundary that approximates the Bayes decision boundary. The second technique that we

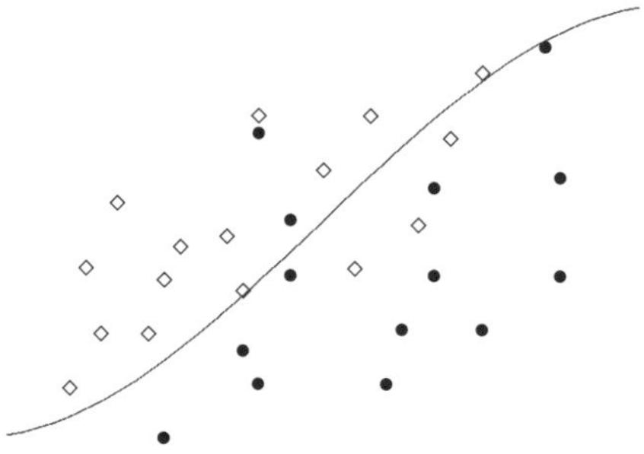

Figure 3.12 Editing illustration – samples and decision boundary

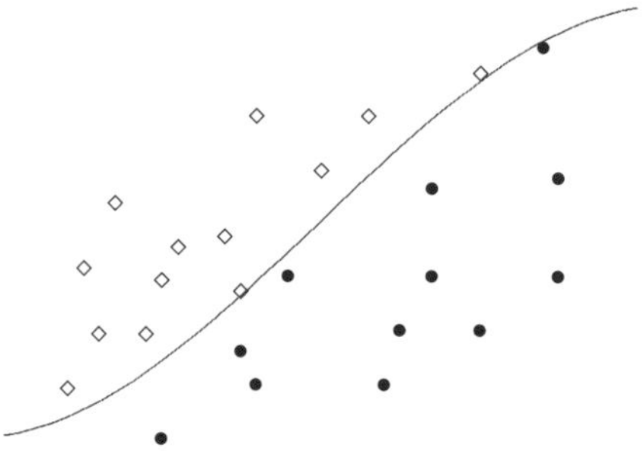

Figure 3.13 Editing illustration – edited data set

consider is that of *condensing*. This aims to reduce the number of prototypes per class without changing the nearest-neighbour approximation to the Bayes decision boundary substantially.

The basic editing procedure is as follows. Given a design set R (of known classification), together with a classification rule η, let S be the set of samples misclassified by the classification rule η. Remove these from the design set to form $R = R - S$ and repeat the procedure until a stopping criterion is met. Thus, we end up with a set of samples correctly classified by the rule.

One implementation of applying this scheme to the k-nearest-neighbour rule is as follows.

1. Make a random partition of the data set, R, into N groups $R_1, \dots, R_N$.

2. Classify the samples in the set R_i using the k-nearest-neighbour rule with the union of the 'next' M sets $R_{(i+1)\mathrm{modN}} \cup \cdots \cup R_{(i+M-1)\mathrm{modN}}$ as the design set, for $i = 1, \dots, N$ where $1 \leq M \leq N - 1$. Let S be the set of misclassified samples.

3. Remove all misclassified samples from the data set to form a new data set, $R = R - S$.

4. If the last I iterations have resulted in no samples being removed from the design set, then terminate the algorithm, otherwise go back to step 1.

If $M = 1$, then we have a *modified holdout* method of error estimation, and taking $k = 1$ gives the multiedit algorithm of Devijver and Kittler (1982). Taking $M = N - 1$ (so that all remaining sets are used) gives an N-fold cross-validation error estimate. If N is equal to the number of samples in the design set (and $M = N - 1$), then we have the leave-one-out error estimate which is Wilson's method of editing (Wilson, 1972). Note that after the first iteration, the number of design samples has reduced and the number of partitions cannot exceed the number of samples. For small data sets, an editing procedure using a cross-validation method of error estimation is preferred to multiedit (Ferri and Vidal, 1992a; see also Ferri *et al.*, 1999).

The editing algorithm above creates homogeneous sets of clusters of samples. The basic idea behind condensing is to remove those samples deeply embedded within each cluster that do not contribute significantly to the nearest-neighbour approximation to the Bayes decision region. The procedure that we describe is due to Hart (1968). We begin with two areas of store, labelled A and B. One sample is placed in A and the remaining samples in B. Each sample point in B is classified using the nearest-neighbour rule with the contents of A (initially a single vector) as prototypes. If a sample is classified correctly, it is returned to B, otherwise it is added to A. The procedure terminates when a complete pass through the set B fails to transfer any points.

The final contents of A constitute the condensed subset to be used with the nearest-neighbour rule.

The result of a condensing procedure is shown in Figure 3.14. There can be considerable reduction in the number of training samples.

Ferri and Vidal (1992b) apply both editing and condensing techniques to image data gathered for a robotic harvesting application. The problem consists of detecting the location of pieces of fruit within a scene. The data comprise six images, captured into arrays of 128×128 pixels; two are used for training, four for test. From the training images, a training set of 1513 10-dimensional feature vectors (obtained from RGB values at a pixel location and its neighbours) spanning three classes (fruit, leaves, sky) is constructed.

The editing algorithm is run with a value of N (the number of partitions of the data set) chosen randomly from $\{3, 4, 5\}$ at each iteration of the editing process. Results are reported for values of I (the number of iterations in the editing procedure with no

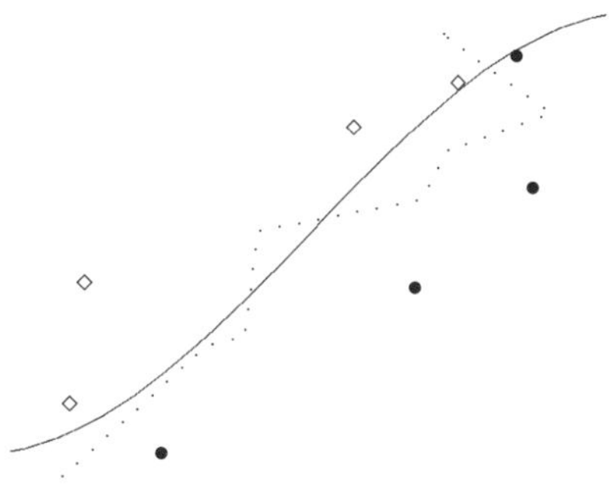

Figure 3.14 Condensing illustration: the solid line represents the Bayes decision boundary; the dotted line is the approximation based on condensing

Table 3.1 Results for multiedit and condensing for image segmentation (after Ferri and Vidal, 1992b)

	Original	$I = 6$		$I = 10$	
		Multiedit	Condense	Multiedit	Condense
Size	1513	1145	22	1139	28
Average errors over four test images	18.26	10.55	10.10	10.58	8.95

samples removed from the design set) of 6 and 10. Results for $I = 6$ are summarised in Table 3.1.

Condensing reduces the size of the data set considerably. Editing reduces the error rate, with a further smaller reduction after condensing.

3.3.5 Choice of distance metric

The most commonly used metric in measuring the distance of a new sample from a prototype is Euclidean distance. Therefore, since all variables are treated equally, the input variables must be scaled to ensure that the *k*-nearest-neighbour rule is independent of measurement units. The generalisation of this is

$$d(\boldsymbol{x}, \boldsymbol{n}) = \left\{(\boldsymbol{x} - \boldsymbol{n})^T \boldsymbol{A}(\boldsymbol{x} - \boldsymbol{n})\right\}^{\frac{1}{2}} \tag{3.12}$$

for a matrix $\boldsymbol{A}$. Choices for $\boldsymbol{A}$ have been discussed by Fukunaga and Flick (1984). Todeschini (1989) assesses six global metrics on ten data sets after four ways of standardising the data. Another development of the Euclidean rule is proposed by van der Heiden and Groen (1997),

$$d_p(\boldsymbol{x}, \boldsymbol{n}) = \left\{(\boldsymbol{x}^{(p)} - \boldsymbol{n}^{(p)})^T (\boldsymbol{x}^{(p)} - \boldsymbol{n}^{(p)})\right\}^{\frac{1}{2}}$$

where $\boldsymbol{x}^{(p)}$ is a transformation of the vector $\boldsymbol{x}$ defined for each element, x_i, of the vector $\boldsymbol{x}$ by

$$x_i^{(p)} = \begin{cases} (x_i^p - 1)/p & \text{if } 0 < p \leq 1 \\ \log(x_i) & \text{if } p = 0 \end{cases}$$

In experiments on radar range profiles of aircraft, van der Heiden and Groen (1997) evaluate the classification error as a function of p.

Friedman (1994) considers basic extensions to the *k*-nearest-neighbour method and presents a hybrid between a *k*-nearest-neighbour rule and a recursive partitioning method (see Chapter 7) in which the metric depends on position in the data space. In some classification problems (those in which there is unequal influence of the input variables on the classification performance), this can offer improved performance. Myles and Hand (1990) assess *local* metrics where the distance between $\boldsymbol{x}$ and $\boldsymbol{n}$ depends on local estimates of the posterior probability.

In the discriminant adaptive nearest-neighbour approach (Hastie and Tibshirani, 1996), a local metric is defined in which, loosely, the nearest-neighbour region is parallel to the

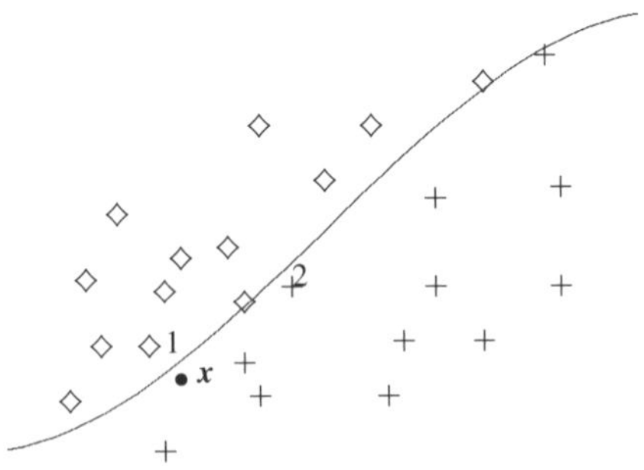

Figure 3.15 Discriminant adaptive nearest-neighbour illustration

decision boundary. It is in the region of the decision boundary where most misclassifications occur. As an illustration, consider the edited data set of Figure 3.13 shown in Figure 3.15. The nearest neighbour to the point $\boldsymbol{x}$ is that labelled 1, and $\boldsymbol{x}$ is classified as $\diamond$. However, if we measure in a coordinate system orthogonal to the decision boundary, then the distance between two points is the difference between their distances from the decision boundary, and the point labelled 2 is the nearest neighbour. In this case, $\boldsymbol{x}$ is classified as $+$. The procedure of Hastie and Tibshirani uses a local definition of the matrix $\boldsymbol{A}$ (based on local estimates of the *within-* and *between-class scatter matrices* – see Section 9.1 for definitions, for example). This procedure can offer substantial improvements in some problems.

3.3.6 Example application study

The problem To develop credit scoring techniques for assessing the creditworthiness of consumer loan applicants (Henley and Hand, 1996).

Summary The approach developed is based on a k-nearest-neighbour method, with an adjusted version of the Euclidean distance metric that attempts to incorporate knowledge of class separation contained in the data.

The data The data comprised measurements on 16 variables (resulting from a variable selection procedure), usually nominal or ordinal, for a set of credit applicants, split into training and test sets of sizes 15 054 and 4132 respectively. The data were preprocessed into a ratio form, so that the jth value on the ith feature or variable is replaced by $\log(p_{ij}/q_{ij})$, where p_{ij} is the proportion of those classified good in attribute j of variable i and q_{ij} is the proportion characterised as bad in attribute j of variable i.

The model A k-nearest-neighbour classifier is used. The quadratic metric (3.12) is used, where the positive definite symmetric matrix, $\boldsymbol{A}$, is given by

$$\boldsymbol{A} = (\boldsymbol{I} + D\boldsymbol{w}\boldsymbol{w}^T)$$

Here D is a distance parameter and $\boldsymbol{w}$ is in the direction of the equiprobability contours of $p(g|\boldsymbol{x})$, the posterior probability of class g; that is, $\boldsymbol{w}$ is a vector parallel to the decision boundary. This is similar to the discriminant adaptive nearest-neighbour classifier of Section 3.3.5.

Training procedure The value of D was chosen to minimise the *bad risk rate*, the proportion of bad risk applicants accepted for a prespecified proportion accepted, based on the design set. The value of k was also based on the design set. The performance assessment criterion used in this study is not the usual one of error rate used in most studies involving k-nearest-neighbour classifiers. In this investigation, the proportion to be accepted is prespecified and the aim is to minimise the number of bad-risk applicants accepted.

Results The main conclusion of the study was that the k-nearest-neighbour approach was fairly insensitive to the choice of parameters and it is a practical classification rule for credit scoring.

3.3.7 Further developments

There are many varieties of the k-nearest-neighbour method. The probability density estimator on which the k-nearest-neighbour decision rule is based has been studied by Buturović (1993), who proposes modifications to (3.10) for reducing the bias and the variance of the estimator.

There are other preprocessing schemes to reduce nearest-neighbour search time. Approximation–elimination algorithms for fast nearest-neighbour search are given by Vidal (1994) and reviewed and compared by Ramasubramanian and Paliwal (2000). Fukunaga and Narendra (1975) structure the design set as a tree. This avoids the computation of some of the distances. Jiang and Zhang (1993) use an efficient branch and bound technique. Increased speed of finding neighbours is usually bought at increased preprocessing or storage (for example, Djouadi and Bouktache, 1997). Dasarathy (1994a) proposes an algorithm for reducing the number of prototypes in nearest-neighbour classification using a scheme based on the concept of an optimal subset selection. This 'minimal consistent set' is derived using an iterative procedure, and on the data sets tested gives improved performance over condensed nearest-neighbour. The approach by Friedman *et al.* (1977) for finding nearest neighbours using a k–d tree algorithm does not rely on the triangle inequality (as the LAESA algorithm does). It can be applied to data with a wide variety of dissimilarity measures. The expected number of samples examined is independent of the number of prototypes. A review of the literature on computational procedures is given by Dasarathy (1991).

Hamamoto *et al.* (1997) propose generating bootstrap samples by linearly combining local training samples. This increases the design set, rather than reducing it, but gives improved performance over conventional k-nearest-neighbour, particularly in high dimensions.

There are theoretical bounds on the Bayes error rate for the k-nearest-neighbour method. We have given those for the nearest-neighbour rule. For small samples, the true error rate may be very different from the Bayes error rate. The effects on the error rates of k-nearest-neighbour rules of finite sample size have been investigated by Fukunaga and Hummels (1987a, 1987b) who demonstrate that the bias in the nearest-neighbour error decreases slowly with sample size, particularly when the dimensionality of the data is high. This indicates that increasing the sample size is not an effective means of reducing bias when dimensionality is high. However, a means of compensating for the bias is to obtain an expression for the rate of convergence of the error rate and to predict the asymptotic limit by evaluating the error rate on training sets of different sample size. This has been explored further by Psaltis *et al.* (1994) who characterise the error rate as an asymptotic series expansion. Leave-one-out procedures for error rate estimation are presented by Fukunaga and Hummels (1987b, 1989), who find that sensitivity of error rate to the choice of k (in a two-class problem) can be reduced by appropriate selection of a threshold for the likelihood function.

3.3.8 Summary

Nearest-neighbour methods have received considerable attention over the years. The simplicity of the approach has made it very popular with researchers. A comprehensive review is given by Dasarathy (1991), and many of the important contributions to the literature are included in the same volume. It is perhaps conceptually the simplest of the classification rules that we present, and the decision rule has been summed up as ‘judge a person by the company he keeps’ (Dasarathy, 1991). The approach requires a set of labelled templates that are used to classify a set of test patterns. The simple-minded implementation of the k-nearest-neighbour rule (calculating the distance of a test pattern from every member of the training set and retaining the class of the k nearest patterns for a decision) is likely to be computationally expensive for a large data set, but for many applications it may well prove acceptable. If you get the answer in minutes, rather than seconds, you may probably not be too worried. The LAESA algorithm trades off storage requirements against computation. LAESA relies on the selection of base prototypes and computes distances of the stored prototypes from these base prototypes.

Additional means of reducing the search time for classifying a pattern using a nearest-neighbour method (and increasing generalisation) are given by the editing and condensing procedures. Both reduce the number of prototypes; editing with the purpose of increasing generalisation performance and condensing with the aim of reducing the number of prototypes without significant degradation of the performance. Experimental studies have been performed by Hand and Batchelor (1978).

Improvements may be obtained through the use of alternative metrics, either local metrics that use local measures of within-class and between-class distance or non-Euclidean distance.

One question that we have not addressed is the choice of k. The larger the value of k, the more robust is the procedure. Yet k must be much smaller than the minimum of n_i, the number of samples in class i, otherwise the neighbourhood is no longer the local neighbourhood of the sample (Dasarathy, 1991). In a limited study, Enas and Choi (1986) give a rule $k \approx N^{2/8}$ or $k \approx N^{3/8}$, where N is the population size. The approach described by Dasarathy is to use a cross-validation procedure to classify each sample in the design set using the remaining samples for various values of k and to determine overall performance. Take the optimum value of k as the one giving the smallest error rate, though the lower the value of k the better from a computational point of view. Keep this value fixed in subsequent editing and condensing procedures if used.

3.4 Expansion by basis functions

The method of density estimation based on an orthogonal expansion by basis functions was first introduced by Čencov (1962). The basic approach is to approximate a density function, $p(x)$, by a weighted sum of orthogonal basis functions. We suppose that the density admits the expansion

$$p(x) = \sum_{i=1}^{\infty} a_i \phi_i(x) \tag{3.13}$$

where the $\{\phi_i\}$ form a complete orthonormal set of functions satisfying

$$\int k(x)\phi_i(x)\phi_j(x)\,dx = \lambda_i \delta_{ij} \tag{3.14}$$

for a *kernel* or *weighting function* $k(x)$, and $\delta_{ij} = 1$ if $i = j$ and zero otherwise. Thus, multiplying (3.13) by $k(x)\phi_i(x)$ and integrating gives

$$\lambda_i a_i = \int k(x) p(x) \phi_i(x)\,dx$$

Given $\{x_1, x_2, \dots, x_n\}$, a set of independently and identically distributed samples from $p(x)$, then the a_i can be estimated in an unbiased manner by

$$\lambda_i \hat{a}_i = \frac{1}{n} \sum_{j=1}^{n} k(x_j)\phi_i(x_j)$$

The orthogonal series estimator based on the sample $\{x_1, x_2, \dots, x_n\}$ of $p(x)$ is then given by

$$\hat{p}_n(x) = \sum_{i=1}^{s} \frac{1}{n\lambda_i} \sum_{j=1}^{n} k(x_j)\phi_i(x_j)\phi_i(x) \tag{3.15}$$

where s is the number of terms retained in the expansion. The coefficients $\hat{a}_i$ may be computed sequentially from

$$\lambda_i \hat{a}_i(r+1) = \frac{r}{r+1}\lambda_i \hat{a}_i(r) + \frac{1}{r+1} k(x_{r+1})\phi_i(x_{r+1})$$

where $\hat{a}_i(r+1)$ is the value obtained using $r+1$ data samples. This means that, given an extra sample point, it is a simple matter to update the coefficients. Also, a large number of data vectors could be used to calculate the coefficients without storing the data in memory.

A further advantage of the series estimator method is that the final estimate is easy to store. It is not in the form of a complicated analytic function, but a set of coefficients.

But what are the disadvantages? First of all, the method is limited to low-dimensional data spaces. Although, in principle, the method may be extended to estimate multivariate probability density functions in a straightforward manner, the number of coefficients in the series increases exponentially with dimensionality. It is not an easy matter to calculate the coefficients. Another disadvantage is that the density estimate is not necessarily a density (as in the nearest-neighbour method described earlier in this chapter). This may or may not be a problem, depending on the application. Also, the density estimate is not necessarily non-negative.

Many different functions have been used as basis functions. These include Fourier and trigonometric functions on [0, 1], and Legendre polynomials on [−1, 1]; and those with unbounded support such as Laguerre polynomials on [0, ∞] and Hermite functions on the real line. If we have no prior knowledge as to the form of $p(x)$, then the basis functions are chosen for their simplicity of implementation. The most popular orthogonal series estimator for densities with unbounded support is the Hermite series estimator. The normalised Hermite functions are given by

$$\phi_k(x) = \frac{\exp(-x^2/2)}{(2^k k! \pi^{\frac{1}{2}})^{\frac{1}{2}}} H_k(x)$$

where $H_k(x)$ is the kth Hermite polynomial

$$H_k(x) = (-1)^k \exp(x^2) \frac{d^k}{dx^k} \exp(-x^2)$$

The performance and the smoothness of the density estimator depends on the number of terms used in the expansion. Too few terms leads to over-smoothed densities. Different stopping rules (rules for choosing the number of terms, s, in expansion (3.15)) have been proposed and are briefly reviewed by Izenman (1991). Kronmal and Tarter (1962) propose a stopping rule based on minimising a mean integrated squared error. Termination occurs when the test

$$\hat{a}_j^2 > \frac{2}{n+1} \hat{b}_j^2$$

fails, where

$$\hat{b}_j^2 = \frac{1}{n} \sum_{k=1}^{n} \phi_j^2(x_k)$$

or, alternatively, when t or more successive terms fail the test. Practical and theoretical difficulties are encountered with this test, particularly with sharply peaked or multimodal densities, and it could happen that an infinite number of terms pass the test. Alternative procedures for overcoming these difficulties have been proposed (Diggle and Hall, 1986; Hart, 1985).

3.5 Kernel methods

One of the problems with the histogram approach, as discussed earlier in the chapter, is that for a fixed cell dimension, the number of cells increases exponentially with dimension of the data vectors. This problem can be overcome somewhat by having a variable cell size. The k-nearest-neighbour method (in its simplest form) overcomes the problem by estimating the density using a cell in which the number of design samples is fixed and finds the cell volume that contains the nearest k. The kernel method (also known as the Parzen method of density estimation, after Parzen, 1962) fixes the cell volume and finds the number of samples within the cell and uses this to estimate the density.

Let us consider a one-dimensional example and let $\{x_1, \dots, x_n\}$ be the set of observations or data samples that we shall use to estimate the density. We may easily write down an estimate of the cumulative distribution function as

$$\hat{P}(x) = \frac{\text{number of observations} \leq x}{n}$$

The density function, $p(x)$, is the derivative of the distribution, but the distribution is discontinuous (at observation values – see Figure 3.16) and its derivative results in a set of spikes at the sample points, x_i, and a value zero elsewhere. However, we may define an estimate of the density as

$$\hat{p}(x) = \frac{\hat{P}(x+h) - \hat{P}(x-h)}{2h}$$

where h is a positive number. This is the proportion of observations falling within the interval $(x-h, x+h)$ divided by $2h$. This may be written as

$$\hat{p}(x) = \frac{1}{hn}\sum_{i=1}^{n} K\left(\frac{x - x_i}{h}\right) \tag{3.16}$$

where

$$K(z) = \begin{cases} 0 & |z| > 1 \\ \frac{1}{2} & |z| \leq 1 \end{cases} \tag{3.17}$$

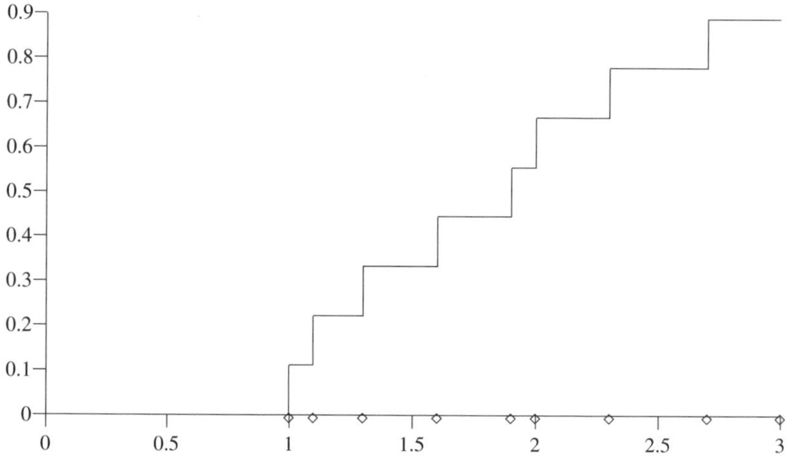

Figure 3.16 Cumulative distribution

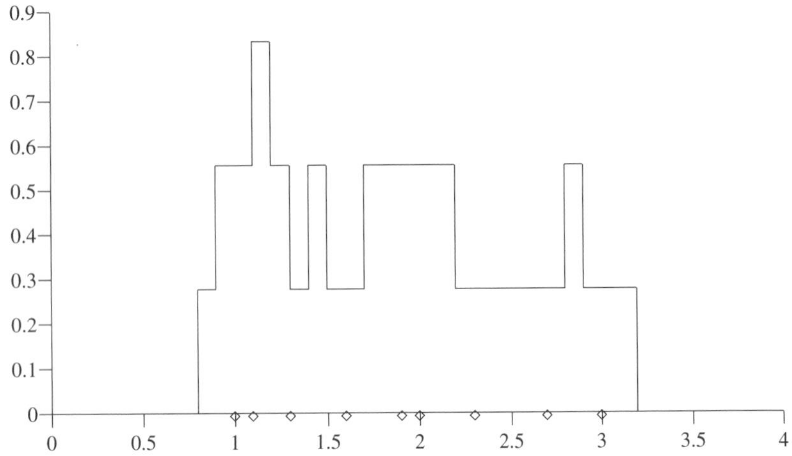

Figure 3.17 Probability density estimate with top hat kernel, $h = 0.2$

since for sample points, x_i, within h of x, the summation gives a value of half the number of observations within the interval. Thus each point within the interval contributes equally to the summation. Figure 3.17 gives an estimate of the density using equations (3.16) and (3.17) for the data used to form the cumulative distribution in Figure 3.16.

Figure 3.17 shows us that the density estimate is itself discontinuous. This arises from the fact that points within a distance h of x contribute a value $\frac{1}{2hn}$ to the density and points further away a value of zero. It is this jump from $\frac{1}{2hn}$ to zero that gives rise to the discontinuities. We can remove this, and generalise the estimator, by using a smoother weighting function than that given by (3.17). For example, we could have a weighting function $K_1(z)$ (also with the property that the integral over the real line is unity) that decreases as $|z|$ increases. Figure 3.18 plots the density estimate for a weighting given by

$$K_1(z) = \frac{1}{\sqrt{2\pi}} \exp\left\{-\frac{z^2}{2}\right\}$$

and a value of h of 0.2. This gives a smoother density estimate. Of course, it does not mean that this estimate is necessarily more 'correct' than that of Figure 3.17, but we might suppose that the underlying density is a smooth function and want a smooth estimate.

The above derivation, together with the three figures, provides a motivation for the kernel method of density estimation, which we formulate as follows. Given a set of observations $\{x_1, \ldots, x_n\}$, an estimate of a density function, in one dimension, is taken to be

$$\hat{p}(x) = \frac{1}{nh} \sum_{i=1}^{n} K\left(\frac{x - x_i}{h}\right) \tag{3.18}$$

where $K(z)$ is termed the *kernel function* and h is the *spread* or *smoothing parameter* (sometimes termed the *bandwidth*). Examples of popular univariate kernel functions are given in Table 3.2.

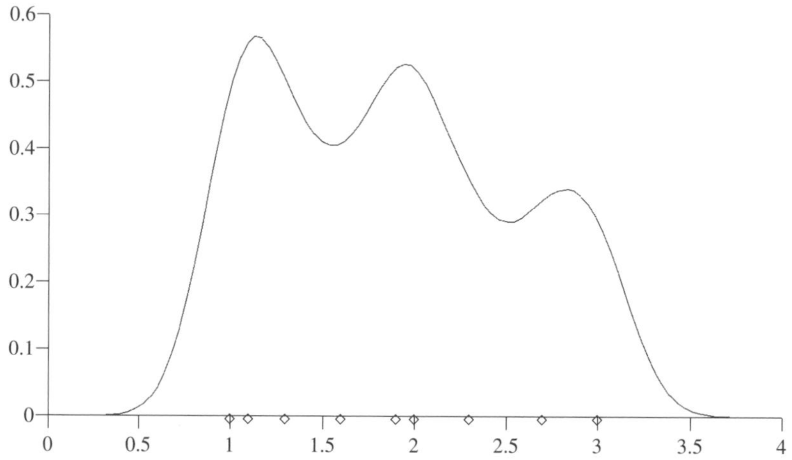

Figure 3.18 Probability density figure with Gaussian kernel and $h = 0.2$

Table 3.2 Commonly used kernel functions for univariate data

Kernel function	Analytic form, $K(x)$
Rectangular	$\frac{1}{2}$ for $\lvert x\rvert < 1$, 0 otherwise
Triangular	$1 - \lvert x\rvert$ for $\lvert x\rvert < 1$, 0 otherwise
Biweight	$\frac{15}{16}(1 - x^2)^2$ for $\lvert x\rvert < 1$, 0 otherwise
Normal	$\frac{1}{\sqrt{2\pi}} \exp(-x^2/2)$
Bartlett–Epanechnikov	$\frac{3}{4}(1 - x^2/5)/\sqrt{5}$ for $\lvert x\rvert < \sqrt{5}$, 0 otherwise

Multivariate kernels (p variables) are usually radially symmetric univariate densities such as the Gaussian kernel $K(\boldsymbol{x}) = (2\pi)^{-p/2} \exp(-\boldsymbol{x}^T\boldsymbol{x}/2)$ and the Bartlett–Epanechnikov kernel $K(\boldsymbol{x}) = (1 - \boldsymbol{x}^T\boldsymbol{x})(p+2)/(2c_p)$ for $|\boldsymbol{x}| < 1$ (0 otherwise) where $c_p = \pi^{p/2}/\ \Gamma((p/2)+1)$ is the volume of the p-dimensional unit sphere.

If we impose the conditions that the kernel $K(z) \geq 0$ and $\int K(z)dz = 1$, then the density estimate $\hat{p}(x)$ given by (3.18) also satisfies the necessary conditions for a probability density function, $p(x) \geq 0$ and $\int p(x)\,dx = 1$.

The theorem due to Rosenblatt (1956) implies that for positive kernels the density estimate will be biased for any finite sample size (see Section 3.1). That is, the estimate of the density averaged over an ensemble of data sets is a biased estimate of the true probability density function. In order to obtain an unbiased estimate, we would be required to relax the condition of positivity of the kernels. Thus the estimate of the density function would not necessarily be a density function itself since it may have negative values. To some people, that would not matter. After all, why should the properties of the *estimate* be the same as the *true* density? On the other hand, there are some who could not live with an estimate of a probability that had negative values, and so would readily accept a bias. They would also point out that asymptotically the estimate is unbiased (it is unbiased as the number of samples used in the estimate tends to infinity) and asymptotically consistent if certain conditions on the smoothing parameter and the kernel hold. These conditions on the kernel are

$$\int |K(z)|\,dz < \infty$$

$$\int K(z)dz = 1$$

$$\sup_z |K(z)| < \infty \qquad K(z) \text{ is finite everywhere}$$

$$\lim_{z\to\infty} |zK(z)| = 0$$

and the conditions on the smoothing parameter are

$$\lim_{n\to\infty} h(n) = 0 \quad \text{for an asymptotic unbiased estimate}$$

$$\lim_{n\to\infty} nh(n) = \infty \quad \text{for an asymptotic consistent estimate}$$

The effect of changing the smoothing parameter is shown in Figure 3.19. For a large value of h, the density is smoothed and detail is obscured. As h becomes smaller, the density estimate shows more structure and becomes spiky as h approaches zero.

The extension to multivariate data is straightforward, with the multivariate kernel density estimate defined as

$$\hat{p}(\boldsymbol{x}) = \frac{1}{nh^p} \sum_{i=1}^{n} K\left(\frac{1}{h}(\boldsymbol{x} - \boldsymbol{x}_i)\right)$$

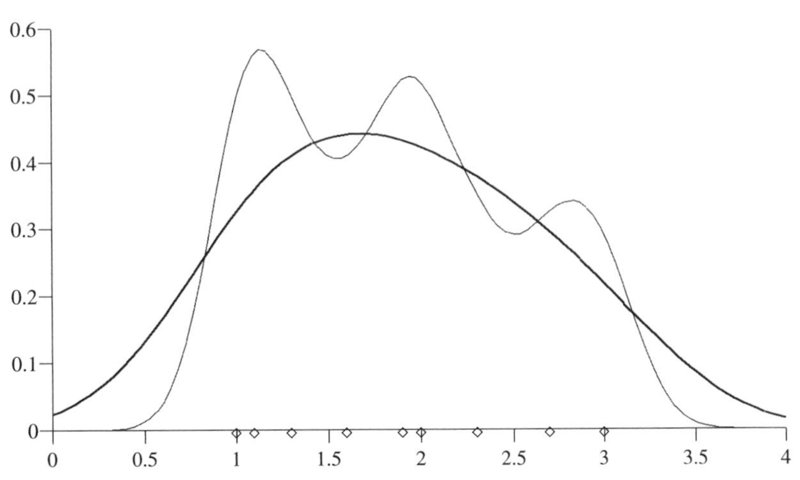

Figure 3.19 Probability density with different levels of smoothing ($h = 0.2$ and $h = 0.5$)

where $K(\boldsymbol{x})$ is a multivariate kernel defined for p-dimensional $\boldsymbol{x}$

$$\int_{\mathbb{R}^p} K(\boldsymbol{x})\,d\boldsymbol{x} = 1$$

and h is the window width. One form of the probability density function estimate commonly used is a sum of *product kernels* (note that this does not imply independence of the variables)

$$\hat{p}(\boldsymbol{x}) = \frac{1}{n}\frac{1}{h_1 \ldots h_p}\sum_{i=1}^{n}\prod_{j=1}^{p} K_j\left(\frac{[\boldsymbol{x} - \boldsymbol{x}_i]_j}{h_j}\right)$$

where there is a different smoothing parameter associated with each variable. The K_j can take any of the univariate forms in Table 3.2. Usually, the K_j are taken to be the same form.

More generally, we may take

$$\hat{p}(\boldsymbol{x}) = \hat{p}(\boldsymbol{x}, \boldsymbol{H}) = \frac{1}{n}\sum_{i=1}^{n}|\boldsymbol{H}|^{-1/2}K(\boldsymbol{H}^{-1/2}(\boldsymbol{x} - \boldsymbol{x}_i))$$

where K is a p-variate spherically symmetric density function and $\boldsymbol{H}$ is a symmetric positive definite matrix. In a classification context, $\boldsymbol{H}$ is commonly taken to be $h_k^2\hat{\boldsymbol{\Sigma}}_k$ for class ω_k, where h_k is a scaling for class ω_k and $\hat{\boldsymbol{\Sigma}}_k$ is the sample covariance matrix. Various approximations to the covariance are evaluated by Hamamoto *et al.* (1996) in situations where the sample size is small and the dimensionality is high.

3.5.1 Choice of smoothing parameter

One of the problems with this 'nonparametric' method is the choice of the smoothing parameter, h. If h is too small, the density estimator is a collection of n sharp peaks, positioned at the sample points. If h is too large, the density estimate is smoothed and structure in the probability density estimate is lost. The optimal choice of h depends on several factors. Firstly, it depends on the data: the number of data points and their distribution. It also depends on the choice of the kernel function and on the optimality criterion used for its estimation. The maximum likelihood estimation of h that maximises the likelihood

$$p(\boldsymbol{x}_1, \ldots, \boldsymbol{x}_n|h)$$

is given by $h = 0$ – an estimate that consists of a spike at each data point and zero elsewhere. Therefore, some other technique must be used for estimating h. There are many possible methods. Surveys are given in the articles by Jones *et al.* (1996) and Marron (1988).

1. Find the average distance between samples and their kth nearest neighbour and use this for h. A value of $k = 10$ has been suggested (Hand, 1981a).

2. Find the value of h that minimises the mean integrated squared error between the density and its approximation. For a radially symmetric normal kernel, Silverman

(1986) suggests

$$h = \sigma \left(\frac{4}{p+2} \right)^{\frac{1}{p+4}} n^{-\frac{1}{p+4}} \tag{3.19}$$

where a choice for σ is

$$\sigma^2 = \frac{1}{p} \sum_{i=1}^{p} s_{ii}$$

and s_{ii} are the diagonal elements of a sample covariance matrix, possibly a *robust* estimate (see Chapter 11). The above estimate will work well if the data come from a population that is normally distributed, but may over-smooth the density if the population is multimodal. A slightly smaller value may be appropriate. You could try several values and assess the misclassification rate.

3. There are more sophisticated ways of choosing kernel widths based on least squares cross-validation and likelihood cross-validation. In likelihood cross-validation the value of h is chosen to maximise (Duin, 1976)

$$\prod_{i=1}^{n} \hat{p}_i(\boldsymbol{x}_i)$$

where $\hat{p}_i(\boldsymbol{x}_i)$ is the density estimate based on $n-1$ samples (all samples but the ith). However, a major problem with this method is its reported poor performance in the heavy-tailed case.

4. Many bandwidth estimators have been considered for the univariate case. The basic idea behind the 'plug-in' estimate is to plug an estimate for the unknown curvature, $S \triangleq \int (p^{ii})^2\, dx$ (the integral of the square of the second derivative of the density), into the expression for h that minimises the asymptotic mean integrated squared error,

$$h = \left[\frac{c}{d^2 S n} \right]^{1/5}$$

where $c = \int K^2(t)dt$ and $d = \int t^2 K(t)dt$. Jones and Sheather (1991) propose a kernel estimate for the curvature, but this, in turn, requires an estimate of the bandwidth, which will be different from that used to estimate the density. Cao *et al.* (1994), in simulations, use

$$S = n^{-2} g^{-5} \sum_{i,j} K^{iv} \left(\frac{x_i - x_j}{g} \right)$$

where K^{iv} is the fourth derivative of K and the smoothing parameter g is given by

$$g = \left(\frac{2K^{iv}(0)}{d} \right)^{1/7} \hat{T}^{-1/7} n^{-1/7}$$

in which $\hat{T}$ is a parametric estimator of $\int p^{iii}(t)^2\, dt$. Development of the 'plug-in' ideas to bandwidth selectors for multivariate data has been considered by Wand and Jones (1994).

Cao *et al.* (1994) perform comparative studies on a range of smoothing methods for univariate densities. Although there is no uniformly best estimator, they find that the plug-in estimator of Sheather and Jones (1991) shows satisfactory performance over a range of problems.

The previous discussion has assumed that h is a fixed value over the whole of the space of data samples. The 'optimal' value of h may in fact be location-dependent, giving a large value in regions where the data samples are sparse and a small value where the data samples are densely packed. There are two main approaches: (i) $h = h(x)$; that is, h depends on the location of the sample in the data space. Such approaches are often based on nearest-neighbour ideas. (ii) $h = h(x_i)$; that is, h is fixed for each kernel and depends on the local density. These are termed *variable kernel* methods.

One particular choice for h is (Breiman *et al.*, 1977)

$$h_j = \alpha_k d_{jk}$$

where α_k is a constant multiplier and d_{jk} is the distance from $\boldsymbol{x}_j$ to its kth nearest neighbour in the training/design set. However, we still have the problem of parameter estimation – namely that of estimating α_k and k.

Breiman *et al.* (1977) find that good fits can be obtained over a wide range of k provided α_k satisfies

$$\beta_k \stackrel{\triangle}{=} \frac{\alpha_k \overline{d_k}^2}{\sigma(d_k)} = \text{constant}$$

where $\overline{d_k}$ is the mean of the kth nearest-neighbour distances ($\frac{1}{n}\sum_{j=1}^{n} d_{jk}$) and $\sigma(d_k)$ is their standard deviation. In their simulations, this constant was 3–4 times larger than the best value of h obtained for the fixed kernel estimator.

Other approaches that have a different bandwidth for the kernel associated with each data point employ a 'pilot' density estimate to set the bandwidth. Abramson (1982) has proposed a bandwidth inversely proportional to the square root of the density, $hp^{-1/2}(x)$, which may lead to $\mathcal{O}(h^4)$ bias under certain conditions on the density (see Terrell and Scott, 1992; Hall *et al.* 1995). Although a pilot estimate of $p(z)$ is required, the method is insensitive to the fine detail of the pilot (Silverman, 1986).

3.5.2 Choice of kernel

Another choice which we have to make in the form of our density estimate is the kernel function. In practice, the most widely used kernel is the normal form

$$K\left(\frac{x}{h}\right) = \frac{1}{h\sqrt{2\pi}}\exp\left\{-\frac{x^2}{2h^2}\right\}$$

with product kernels being used for multivariate density estimation. Alternatively, radially symmetric unimodal probability density functions such as the multivariate normal density function are used. There is evidence that the form is relatively unimportant, though the product form may not be ideal for the multivariate case. There are some arguments in favour of kernels that are not themselves densities and admit negative values (Silverman, 1986).

3.5.3 Example application study

The problem The practical problem addressed relates to one in the oil industry: to predict the type of subsurface (for example, sand, shale, coal – the lithofacies class) from physical properties such as electron density, velocity of sound and electrical resistivity obtained from a logging instrument lowered into a well.

Summary In some practical problems, the probability density function of data may vary with time – termed *population drift* (Hand, 1997). Thus, the test conditions differ from those used to define the training data. Kraaijveld (1996) addresses the problem of classification when the training data are only approximately representative of the test conditions using a kernel approach to discriminant analysis.

The data Twelve standard data sets were generated from data gathered from two fields and 18 different well sites. The data sets comprised different-sized feature sets (4 to 24 features) and 2, 3 and 4 classes.

The model A Gaussian kernel is used, with the width estimated by maximising a modified likelihood function (solved using a numerical root-finding procedure and giving a width, s_1). An approximation, based on the assumption that the density at a point is determined by the nearest kernel only, is given by

$$s_2 = \sqrt{\frac{1}{pn} \sum_{i=1}^{n} |\boldsymbol{x}_i^* - \boldsymbol{x}_i|^2}$$

for a p-dimensional data set $\{\boldsymbol{x}_i, i = 1, \dots, n\}$, where $\boldsymbol{x}_i^*$ is the nearest sample to $\boldsymbol{x}_i$.

A robust estimate is also derived using a test data set, again using a modified likelihood criterion, to give s_3. The nearest-neighbour approximation is

$$s_4 = \sqrt{\frac{1}{pn_t} \sum_{i=1}^{n_t} |\boldsymbol{x}_i^* - \boldsymbol{x}_i^t|^2}$$

where the test set is $\{\boldsymbol{x}_i^t, i = 1, \dots, n_t\}$. This provides a modification to the kernel width using the test set distribution (but not the test set labels). Thus, the test set is used as part of the training procedure.

Training procedure Twelve experiments were defined by using different combinations of data sets as train and test and five classifiers assessed (kernels with bandwidth estimators s_1 to s_4 and nearest neighbour).

Results The robust methods led to an increase in performance (measured in terms of error rate) in all but one of the 12 experiments. The nearest-neighbour approximation to the bandwidth tended to underestimate the bandwidth by about 20%.

3.5.4 Further developments

There have been several papers comparing the variable kernel approach with the fixed kernel method. Breiman *et al.* (1977) find superior performance compared to the fixed kernel approach. It seems that a variable kernel method is potentially advantageous when the underlying density is heavily skewed or long-tailed (Remme *et al.*, 1980; Bowman, 1985). Terrell and Scott (1992) report good performance of the Breiman *et al.* model for small to moderate sample sizes, but performance deteriorates as sample size grows compared to the fixed kernel approach.

Further investigations into variable kernel approaches include those of Krzyzak (1983) (who examines classification rules) and Terrell and Scott (1992). Terrell and Scott conclude that it is 'surprisingly difficult to do significantly better than the original fixed kernel scheme'. An alternative to the variable bandwidth kernel is the variable location kernel (Jones, *et al.*, 1994), in which the location of the kernel is perturbed to reduce bias.

For multivariate data, procedures for approximating the kernel density using a reduced number of kernels are described by Fukunaga and Hayes (1989a) and Babich and Camps (1996). Fukunaga and Hayes select a set from the given data set by minimising an entropy expression. Babich and Camps use an agglomerative clustering procedure to find a set of prototypes and weight the kernels appropriately (cf. mixture modelling). Jeon and Landgrebe (1994) use a clustering procedure, together with a branch and bound method, to eliminate data samples from the density calculation.

There are several approaches to density estimation that combine parametric and nonparametric approaches (*semiparametric* density estimators). Hjort and Glad (1995) describe an approach that multiplies an initial parametric start with a nonparametric kernel-type estimate for the necessary correction factor. Hjort and Jones (1996) find the best local parametric approximation to a density $p(x, \boldsymbol{\theta})$, where the parameter values, $\boldsymbol{\theta}$, depend on x.

The kernel methods described in this chapter apply to real-valued continuous quantities. We have not considered issues such as dealing with missing data (see Titterington and Mill, 1983; Pawlak, 1993) or kernel methods for discrete data (see McLachlan, 1992a, for a summary of kernel methods for other data types).

3.5.5 Summary

Kernel methods, both for multivariate density estimation and regression, have been extensively studied. The idea behind kernel density estimation is very simple – put a 'bump' function over each data point and then add them up to form a density. One of the disadvantages of the kernel approach is the high computational requirements for large data sets – there is a kernel at every data point that contributes to the density at a given point, x. Computation can be excessive and for large data sets it may be appropriate to use kernels other than the normal density in an effort to reduce computation. Also, since kernels are localised, only a small proportion will contribute to the density at a given point. Some preprocessing of the data will enable non-contributing kernels to be identified and omitted from a density calculation. For univariate density estimates, based on the normal kernel, Silverman (1982) proposes an efficient algorithm for computation based on the fact that the density estimate is a convolution of the data with the kernel and the Fourier transform is used to perform the convolution (see also Silverman, 1986; Jones and Lotwick, 1984). Speed improvements can be obtained in a similar manner to those used for k-nearest-neighbour methods – by reducing the number of prototypes (as in the edit and condense procedures).

The k-nearest-neighbour methods may also be viewed as kernel approaches to density estimation in which the kernel has uniform density in the sphere centred at a point $\boldsymbol{x}$ and of radius equal to the distance to the kth nearest neighbour. The attractiveness of k-nearest-neighbour methods is that the kernel width varies according to the local density, but is discontinuous. The work of Breiman *et al.* described earlier in this chapter is an attempt to combine the best features of k-nearest-neighbour methods with the fixed kernel approaches.

There may be difficulties in applying the kernel method in high dimensions. Regions of high density may contain few samples, even for moderate sample sizes. For example, in the 10-dimensional unit multivariate normal distribution (Silverman, 1986), 99% of the mass of the distribution is at points at a distance greater than 1.6, whereas in one dimension, 90% of the distribution lies between ± 1.6. Thus, reliable estimates of the density can only be made for extremely large samples in high dimensions. As an indication of the sample sizes required to obtain density estimates, Silverman considers again the special case of a unit multivariate normal distribution, and a kernel density estimate with normal kernels where the window width is chosen to minimise the mean squared error at the origin. In order that the relative mean squared error, $E[(\hat{p}(0) - p(0))^2 / p^2(0)]$, is less than 0.1, a small number of samples is required in one and two dimensions (see Table 3.3).

However, for 10 dimensions, over 800 000 samples are necessary. Thus, in order to obtain accurate density estimates in high dimensions, an enormous sample size is needed. Further, these results are likely to be optimistic and more samples would be required to estimate the density at other points in the distribution to the same accuracy.

Kernel methods are motivated by the asymptotic results and as such are only really relevant to low-dimensional spaces due to sample size considerations. However, as far as discrimination is concerned, we may not necessarily be interested in accurate estimates of the densities themselves, but rather the Bayes decision region for which approximate estimates of the densities may suffice. In practice, kernel methods do work well on multivariate data, in the sense that error rates similar to other classifiers can be achieved.

3.6 Application studies

The nonparametric methods of density estimation described in this chapter have been applied to a wide range of problems. Applications of Bayesian networks include the following.

- Drug safety. Cowell *et al.* (1991) develop a Bayesian network for analysing a specific adverse drug reaction problem (drug-induced pseudomembranous colitis). The algorithm of Lauritzen and Spiegelhalter (1988) was used for manipulating the probability

Table 3.3 Required sample size as a function of dimension for a relative mean squared error at the origin of less than 0.1 when estimating a standard multivariate normal density using normal kernels with width chosen so that the mean squared error at the origin is minimised (Silverman, 1986)

Dimensionality	Required sample size
1	4
2	19
3	67
4	223
5	768
6	2 790
7	10 700
8	43 700
9	187 000
10	842 000

density functions in Bayesian networks; see also Spiegelhalter *et al.* (1991) for a clear account of the application of probabilistic expert systems.

- Endoscope navigation. In a study on the use of computer vision techniques for automatic guidance and advice in colon endoscopy, Kwoh and Gillies (1996) construct a Bayesian network (using subjective knowledge from an expert) and compare performance with a maximum weight dependence tree learned from data. The latter gave better performance.
- Geographic information processing. Stassopoulou *et al.* (1996) compare Bayesian networks and neural networks (see Chapters 5 and 6) in a study to combine remote sensing and other data for assessing the risk of desertification of burned forest areas in the Mediterranean region. A Bayesian network is constructed using information provided by experts. An equivalent neural network is trained and its parameters used to set the conditional probability tables in the Bayesian network.
- Image segmentation. Williams and Feng (1998) use a tree-structured network, in conjunction with a neural network, as part of an image labelling scheme. The conditional probability tables are estimated from training data using a maximum likelihood procedure based on the EM algorithm (see Chapter 2).

A comparative study on 25 data sets of the naïve Bayes classifier and a tree-structured classifier is performed by Friedman *et al.* (1997). The tree-structured classifiers outperform the naïve Bayes while maintaining simplicity.

Examples of applications of k-nearest-neighbour methods include the following.

- Target classification. Chen and Walton (1986) apply nearest-neighbour methods to ship and aircraft classification using radar cross-section measurements. Drake *et al.* (1994) compare several methods, including k-nearest-neighbour, to multispectral imagery.
- Handwritten character recognition. There have been many studies in which nearest-neighbour methods have been applied to handwritten characters. Smith *et al.* (1994) use three distance metrics for a k-nearest-neighbour application to handwritten digits. Yan (1994) uses nearest-neighbour with a multilayer perceptron to refine prototypes.
- Credit scoring. Von Stein and Ziegler (1984) compare several discriminant analysis methods, including nearest-neighbour, to the problem of credit assessment for commercial borrowers.
- Nuclear reactors. In a small data set study, Dubuisson and Lavison (1980) use both k-nearest-neighbour and kernel density estimators to discriminate between two classes of power signals in the monitoring of a high-flux isotope reactor.
- Astronomy. Murtagh (1994) uses linear methods and k-nearest-neighbour to classify stellar objects using a variety of feature sets (up to 48 variables).

Example applications of kernel methods are as follows.

- Philatelic mixtures. Izenman and Sommer (1988) use nonparametric density estimates (and finite mixture models) to model postage stamp paper thicknesses.
- Chest pain. Scott *et al.* (1978) use a quartic kernel to estimate the density of plasma lipids in two groups (diseased and normal). The aim of the investigation was to ascertain the dependence of the risk of coronary artery disease on the joint variation of plasma lipid concentrations.
- Fruit fly classification. Sutton and Steck (1994) use Epanechnikov kernels in a two-class fruit fly discrimination problem.
- Forecasting abundance. Rice (1993) uses kernel methods based on Cauchy distributions in a study of abundance of fish.
- Astronomy. Studies involving kernel methods in astronomy include those of (i) De Jager *et al.* (1986) who compare histogram and kernel approaches to estimate a light curve, which is characteristic of periodic sources of gamma rays; (ii) Merritt and Tremblay (1994) who estimate surface and space density profiles of a spherical stellar system; and (iii) Vio *et al.* (1994) who consider the application of kernel methods to several problems in astronomy – bimodality, rectification and intrinsic shape determination of galaxies.
- Lithofacies recognition from wireline logs. Kraaijveld (1996) considers the problem of subsurface classification from well measurements. In particular, he develops the kernel classification method to the case where the test data differ from the training data by noise.

A comparative study of different kernel methods applied to multivariate data is reported by Breiman *et al.* (1977) and Bowman (1985). Hwang *et al.* (1994a) compare kernel density estimators with projection pursuit density estimation which interprets multidimensional density through several one-dimensional projections (see Chapter 6). Jones and Signorini (1997) perform a comparison of 'improved' univariate kernel density estimators; see also Cao *et al.* (1994) and Titterington (1980) for a comparison of kernels for categorical data.

The *Statlog* project (Michie *et al.*, 1994) provides a thorough comparison of a wide range of classification methods, including k-nearest-neighbour and kernel discriminant analysis (algorithm ALLOC80). ALLOC80 has a slightly lower error rate than k-nearest-neighbour (for the special case of $k = 1$ at any rate), but had longer training and test times. More recent comparisons include Liu and White (1995).

3.7 Summary and discussion

The approaches to discrimination developed in this chapter have been based on estimation of the class-conditional density functions using nonparametric techniques. It is certainly true that we cannot design a classifier that performs better than the Bayes discriminant rule. No matter how sophisticated a classifier is, or how appealing it may be in terms of reflecting a model of human decision processes, it cannot outperform the Bayes classifier for any proper performance criterion; in particular, it cannot achieve a lower error rate. Therefore a natural step is to estimate the components of the Bayes rule from the data, namely the class-conditional probability density functions and the class priors. We shall see in later chapters that we do not need to model the density explicitly to get good estimates of the posterior probabilities of class membership.

We have described four nonparametric methods of density estimation: the histogram approach and developments to reduce the number of parameters (naïve Bayes, tree-structured density estimators and Bayesian networks); the k-nearest-neighbour method leading to the k-nearest-neighbour classifier; series methods; and finally, kernel methods of density estimation. With advances in computing in recent years, these methods have now become viable and nonparametric methods of density estimation have had an impact on nonparametric approaches to discrimination and classification. Of the methods described, for discrete data the developments of the histogram – the independence model, the Lancaster models and maximum weight dependence trees – are easy to implement. Learning algorithms for Bayesian networks can be computationally demanding. For continuous data, the kernel method is probably the most popular, with normal kernels with the same window width for each dimension. However, it is reported by Terrell and Scott (1992) that nearest-neighbour methods are superior to fixed kernel approaches to density estimation beyond four dimensions. The kernel method has also been applied to discrete data.

In conclusion, an approach based on density estimation is not without its dangers of course. If incorrect assumptions are made about the form of the distribution in the parametric approach (and in many cases we will not have a physical model of the data generation process to use) or data points are sparse leading to poor kernel density estimates in the nonparametric approach, then we cannot hope to achieve good density estimates. However, the performance of a classifier, in terms of error rate, may not deteriorate too dramatically. Thus, it is a strategy worth trying.

3.8 Recommendations

1. Nearest-neighbour methods are easy to implement and are recommended as a starting point for a nonparametric approach. In the *Statlog* project (Michie *et al.*, 1994), the k-nearest-neighbour method came out best on the image data sets (top in four and runner-up in two of the six image data sets) and did very well on the whole.

2. For large data sets, some form of data reduction in the form of condensing/editing is advised.

3. As density estimators, kernel methods are not appropriate for high-dimensional data, but if smooth estimates of the density are required they are to be preferred over k-nearest-neighbour. Even poor estimates of the density may still give good classification performance.

4. For multivariate data sets, it is worth trying a simple independence model as a baseline. It is simple to implement, handles missing values easily and can give good performance.

5. Domain-specific and expert knowledge should be used where available. Bayesian networks are an attractive scheme for encoding such knowledge.

3.9 Notes and references

There is a large literature on nonparametric methods of density estimation. A good starting point is the book by Silverman (1986), placing emphasis on the practical aspects of density estimation. The book by Scott (1992) provides a blend of theory and applications, placing some emphasis on the visualisation of multivariate density estimates. The article by Izenman (1991) is to be recommended, covering some of the more recent developments in addition to providing an introduction to nonparametric density estimation. Other texts are Devroye and Györfi (1985) and Nadaraya (1989). The book by Wand and Jones (1995) presents a thorough treatment of kernel smoothing.

The literature on kernel methods for regression and density estimation is vast. A treatment of kernel density estimation can be found in most textbooks on density estimation. Silverman (1986) gives a particularly lucid account and the book by Hand (1982) provides a very good introduction and considers the use of kernel methods for discriminant analysis. Other treatments, more detailed than that presented here, may be found in the books by Scott (1992), McLachlan (1992a) and Nadaraya (1989). A thorough treatment of kernel smoothing is given in the book by Wand and Jones (1995).

Good introductions to Bayesian networks are provided by Jensen (1996), Pearl (1988) and Neapolitan (1990) and the article by Heckerman (1999).

The website `www.statistical-pattern-recognition.net` contains references and links to further information on techniques and applications.

Exercises

Data set 1: Generate p-dimensional multivariate data (500 samples in train and test sets, equal priors) for two classes: for ω_1, $\boldsymbol{x} \sim N(\boldsymbol{\mu}_1, \boldsymbol{\Sigma}_1)$ and for ω_2, $\boldsymbol{x} \sim 0.5N(\boldsymbol{\mu}_2, \boldsymbol{\Sigma}_1) + 0.5N(\boldsymbol{\mu}_3, \boldsymbol{\Sigma}_3)$ where $\boldsymbol{\mu}_1 = (0, \ldots, 0)^T$, $\boldsymbol{\mu}_2 = (2, \ldots, 2)^T$, $\boldsymbol{\mu}_3 = (-2, \ldots, -2)^T$ and $\boldsymbol{\Sigma}_1 = \boldsymbol{\Sigma}_2 = \boldsymbol{\Sigma}_3 = \boldsymbol{I}$, the identity matrix.

Data set 2: Generate data from p-dimensional multivariate data (500 samples in train and test sets, equal priors) for three normally distributed classes with $\boldsymbol{\mu}_1 = (0, \ldots, 0)^T$, $\boldsymbol{\mu}_2 = (2, \ldots, 2)^T$, $\boldsymbol{\mu}_3 = (-2, \ldots, -2)^T$ and $\boldsymbol{\Sigma}_1 = \boldsymbol{\Sigma}_2 = \boldsymbol{\Sigma}_3 = \boldsymbol{I}$, the identity matrix.

1. For three variables, X_1, X_2 and X_3 taking one of two values, 1 or 2, denote by P^{ij}_{ab} the probability that $X_i = a$ and $X_j = b$. Specifying the density as

$$P^{12}_{12} = P^{13}_{11} = P^{23}_{11} = P^{23}_{12} = P^{12}_{22} = P^{13}_{21} = P^{23}_{21} = P^{23}_{22} = \tfrac{1}{4}$$
$$P^{12}_{11} = P^{13}_{12} = \tfrac{7}{16};\ P^{12}_{21} = P^{13}_{22} = \tfrac{1}{16}$$

   show that the Lancaster density estimate of the probability $p(X_1 = 2, X_2 = 1, X_3 = 2)$ is negative $(= -1/64)$.

2. For a tree-dependent distribution (equation (3.3)):

$$p^t(\boldsymbol{x}) = \prod_{i=1}^{p} p(x_i | x_{j(i)})$$

   and noting that $p(\boldsymbol{x}) \log(p(\boldsymbol{x}))$ does not depend on the tree structure, show that minimising the Kullback–Leibler distance

$$D = \sum_{\boldsymbol{x}} p(\boldsymbol{x}) \log\left(\frac{p(\boldsymbol{x})}{p^t(\boldsymbol{x})}\right)$$

   is equivalent to finding the tree that minimises

$$\sum_{i=1}^{p} \sum_{x_i, x_{j(i)}} p(x_i, x_{j(i)}) \log\left(\frac{p(x_i, x_{j(i)})}{p(x_i)p(x_{j(i)})}\right)$$

3. Verify that the Bartlett–Epanechnikov kernel satisfies the properties

$$\int K(t)\,dt = 1$$
$$\int tK(t)\,dt = 0$$
$$\int t^2 K(t)\,dt = k_2 \neq 0$$

4. Compare and contrast the k-nearest-neighbour classifier with the Gaussian classifier. What assumptions do the models make? Also, consider such issues as training requirements, computation time, and storage and performance in high dimensions.

5. Consider a sample of n observations $(x_1, \ldots, x_n)$ from a density p. An estimate $\hat{p}$ is calculated using a kernel density estimate with Gaussian kernels for various bandwidths h. How would you expect the number of relative maxima of $\hat{p}$ to vary as h increases? Suppose that the x_i's are drawn from the Cauchy density $p(x) = \pi/(1+x^2)$. Show that the variance of X is infinite. Does this mean that the variance of the Gaussian kernel density estimate $\hat{p}$ is infinite?

6. Consider the multivariate kernel density estimate ($\boldsymbol{x} \in \mathbb{R}^p$),

$$\hat{p}(\boldsymbol{x}) = \frac{1}{nh^p} \sum_{i=1}^{n} K\left(\frac{1}{h}(\boldsymbol{x} - \boldsymbol{x}_i)\right)$$

Show that the k-nearest-neighbour density estimate given by (3.9) is a special case of the above for suitable choice of K and h (which varies with position, $\boldsymbol{x}$).

7. Implement a k-nearest-neighbour classifier using data set 1 and investigate its performance as a function of dimensionality $p = 1, 3, 5, 10$ and k. Comment on the results.

8. For the data of data set 1, implement a Gaussian kernel classifier. Construct a separate validation set to obtain a value of the kernel bandwidth (initialise at the value given by (3.19) and vary from this). Describe the results.

9. Implement a base prototype selection algorithm to select n_b base prototypes using data set 2. Implement the LAESA procedure and plot the number of distance calculations in classifying the test data as a function of the number of base prototypes, n_b, for $p = 2, 4, 6, 8$ and 10.

10. Using data set 2, implement a nearest-neighbour classifier with edit and condensing. Calculate the nearest-neighbour error rate, the error rate after editing, and the error rate after editing and condensing.

11. Again, for the three-class data above, investigate procedures for choosing k in the k-nearest-neighbour method.

12. Consider nearest-neighbour with edit and condense. Suggest ways of reducing the final number of prototypes by careful initialisation of the condensing algorithm. Plan a procedure to test your hypotheses. Implement it and describe your results.

4

Linear discriminant analysis

Overview

Discriminant functions that are linear in the features are constructed, resulting in (piecewise) linear decision boundaries. Different optimisation schemes give rise to different methods including the perceptron, Fisher's linear discriminant function and support vector machines. The relationship between these methods is discussed.

4.1 Introduction

This chapter deals with the problem of finding the weights of a linear discriminant function. Techniques for performing this task have sometimes been referred to as *learning algorithms*, and we retain some of the terminology here even though the methods are ones of optimisation or training rather than learning. A linear discriminant function has already appeared in Chapter 2. In that chapter, it arose as a result of a normal distribution assumption for the class densities in which the class covariance matrices were equal. In this chapter, we make no distributional assumptions, but start from the assumption that the decision boundaries are linear. The algorithms have been extensively treated in the literature, but they are included here as an introduction to the nonlinear models discussed in the following chapter, since a stepping stone to the nonlinear models is the generalised linear model in which the discriminant functions are linear combinations of nonlinear functions.

The treatment is divided into two parts: the binary classification problem and the multiclass problem. Although the two-class problem is clearly a special case of the multiclass situation (and in fact all the algorithms in the multiclass section can be applied to two classes), the two-class case is of sufficient interest in its own right to warrant a special treatment. It has received a considerable amount of attention and many different algorithms have been proposed.

4.2 Two-class algorithms

4.2.1 General ideas

In Chapter 1 we introduced the discriminant function approach to supervised classification; here we briefly restate that approach for linear discriminant functions.

Suppose we have a set of training patterns $\boldsymbol{x}_1, \ldots, \boldsymbol{x}_n$, each of which is assigned to one of two classes, ω_1 or ω_2. Using this design set, we seek a weight vector $\boldsymbol{w}$ and a threshold w_0 such that

$$\boldsymbol{w}^T\boldsymbol{x} + w_0 \begin{cases} > & 0 \\ < & 0 \end{cases} \Rightarrow \boldsymbol{x} \in \begin{cases} \omega_1 \\ \omega_2 \end{cases}$$

or

$$\boldsymbol{v}^T\boldsymbol{z} \begin{cases} > & 0 \\ < & 0 \end{cases} \Rightarrow \boldsymbol{x} \in \begin{cases} \omega_1 \\ \omega_2 \end{cases}$$

where $\boldsymbol{z} = (1, x_1, \ldots, x_p)^T$ is the *augmented pattern vector* and $\boldsymbol{v}$ is a $(p+1)$-dimensional vector $(w_0, w_1, \ldots, w_p)^T$. In what follows, $\boldsymbol{z}$ could also be $(1, \phi_1(\boldsymbol{x}), \ldots, \phi_D(\boldsymbol{x}))^T$, with $\boldsymbol{v}$ a $(D+1)$-dimensional vector of weights, where $\{\phi_i, i = 1, \ldots, D\}$ is a set of D functions of the original variables. Thus, we may apply these algorithms in a transformed feature space.

A sample in class ω_2 is classified correctly if $\boldsymbol{v}^T\boldsymbol{z} < 0$. If we were to redefine all samples in class ω_2 in the design set by their negative values and denote these redefined samples by $\boldsymbol{y}$, then we seek a value for $\boldsymbol{v}$ which satisfies

$$\begin{aligned} \boldsymbol{v}^T\boldsymbol{y} > 0 \qquad & \text{for all } \boldsymbol{y}_i \text{ corresponding to } \boldsymbol{x}_i \text{ in the design set} \\ & [\boldsymbol{y}_i^T = (1, \boldsymbol{x}_i^T), \boldsymbol{x}_i \in \omega_1;\ \boldsymbol{y}_i^T = (-1, -\boldsymbol{x}_i^T), \boldsymbol{x}_i \in \omega_2] \end{aligned} \tag{4.1}$$

Ideally, we would like a solution for $\boldsymbol{v}$ that makes $\boldsymbol{v}^T\boldsymbol{y}$ positive for as many samples in the design set as possible. This minimises the misclassification error on the design set. If $\boldsymbol{v}^T\boldsymbol{y}_i > 0$ for all members of the design set then the data are said to be *linearly separable*.

However, it is difficult to minimise the number of misclassifications. Usually some other criterion is employed. The sections that follow introduce a range of criteria adopted for discrimination between two classes. Some are suitable if the classes are separable, others for overlapping classes. Some lead to algorithms that are deterministic, others can be implemented using stochastic algorithms.

4.2.2 Perceptron criterion

Perhaps the simplest criterion to minimise is the perceptron criterion function

$$J_P(\boldsymbol{v}) = \sum_{\boldsymbol{y}_i \in \mathcal{Y}} (-\boldsymbol{v}^T\boldsymbol{y}_i)$$

where $\mathcal{Y} = \{\boldsymbol{y}_i | \boldsymbol{v}^T\boldsymbol{y}_i < 0\}$ (the set of misclassified samples). J_P is proportional to the sum of the distances of the misclassified samples to the decision boundary.

Error-correction procedure

Since the criterion function J_P is continuous, we can use a gradient-based procedure, such as the method of steepest descent (Press *et al.*, 1992), to determine its minimum:

$$\frac{\partial J_P}{\partial \boldsymbol{v}} = \sum_{\boldsymbol{y}_i \in \mathcal{Y}} (-\boldsymbol{y}_i)$$

which is the sum of the misclassified patterns, and the method of steepest descent gives a movement along the negative of the gradient with update rule

$$\boldsymbol{v}_{k+1} = \boldsymbol{v}_k + \rho_k \sum_{\boldsymbol{y}_i \in \mathcal{Y}} \boldsymbol{y}_i \qquad (4.2)$$

where ρ_k is a scale parameter that determines the step size. If the sample sets are separable, then this procedure is guaranteed to converge to a solution that separates the sets. Algorithms of the type (4.2) are sometimes referred to as *many-pattern adaptation* or *batch update* since all given pattern samples are used in the update of $\boldsymbol{v}$. The corresponding single-pattern adaptation scheme is

$$\boldsymbol{v}_{k+1} = \boldsymbol{v}_k + \rho_k \boldsymbol{y}_i \qquad (4.3)$$

where $\boldsymbol{y}_i$ is a training sample that has been misclassified by $\boldsymbol{v}_k$. This procedure cycles through the training set, modifying the weight vector whenever a sample is misclassified. There are several types of *error-correction* procedure of the form of (4.3). The *fixed increment* rule takes $\rho_k = \rho$, a constant, and is the simplest algorithm for solving systems of linear inequalities.

The error-correction procedure is illustrated geometrically in weight space in Figures 4.1 and 4.2. In Figure 4.1, the plane is partitioned by the line $\boldsymbol{v}^T \boldsymbol{y}_k = 0$. Since the current estimate of the weight vector $\boldsymbol{v}_k$ has $\boldsymbol{v}_k \boldsymbol{y}_k < 0$, the weight vector is

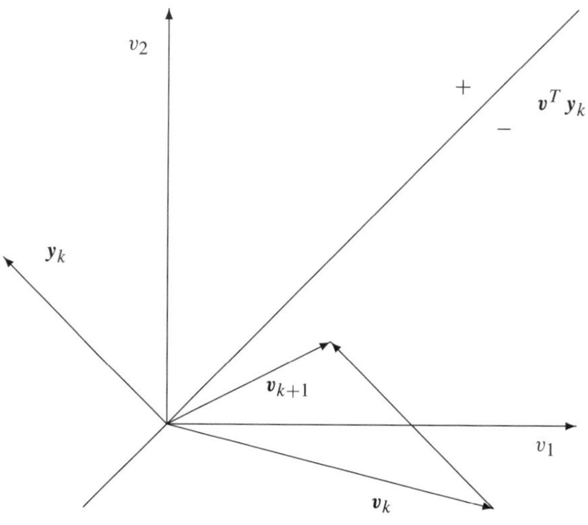

Figure 4.1 Perceptron training

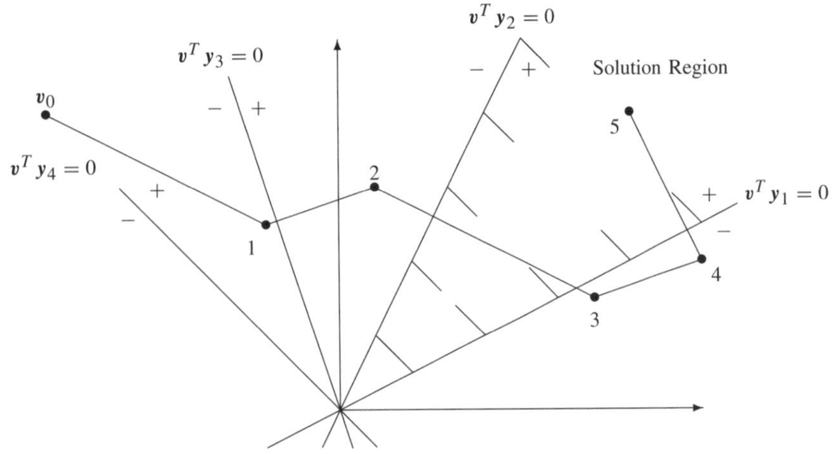

Figure 4.2 Perceptron training

updated by (for $\rho = 1$) adding on the pattern vector $\boldsymbol{y}_k$. This moves the weight vector towards and possibly into the region $\boldsymbol{v}^T \boldsymbol{y}_k > 0$. In Figure 4.2, the lines $\boldsymbol{v}^T \boldsymbol{y}_k = 0$ are plotted for four separate training patterns. If the classes are separable, the solution for $\boldsymbol{v}$ must lie in the shaded region (the *solution region* for which $\boldsymbol{v}^T \boldsymbol{y}_k > 0$ for all patterns $\boldsymbol{y}_k$). A solution path is also shown starting from an initial estimate $\boldsymbol{v}_0$. By presenting the patterns $\boldsymbol{y}_1, \boldsymbol{y}_2, \boldsymbol{y}_3, \boldsymbol{y}_4$ cyclically, a solution for $\boldsymbol{v}$ is obtained in five steps: the first change to $\boldsymbol{v}_0$ occurs when $\boldsymbol{y}_2$ is presented ($\boldsymbol{v}_0$ is already on the positive side of $\boldsymbol{v}^T \boldsymbol{y}_1 = 0$, so presenting $\boldsymbol{y}_1$ does not update $\boldsymbol{v}$). The second step adds on $\boldsymbol{y}_3$; the third $\boldsymbol{y}_2$ ($\boldsymbol{y}_4$ and $\boldsymbol{y}_1$ are not used because $\boldsymbol{v}$ is on the positive side of both hyperplanes $\boldsymbol{v}^T \boldsymbol{y}_4 = 0$ and $\boldsymbol{v}^T \boldsymbol{y}_1 = 0$); the final two stages are the addition of $\boldsymbol{y}_3$, then $\boldsymbol{y}_1$. Thus, from the sequence

$$\boldsymbol{y}_1, \hat{\boldsymbol{y}}_2, \hat{\boldsymbol{y}}_3, \boldsymbol{y}_4, \boldsymbol{y}_1, \hat{\boldsymbol{y}}_2, \hat{\boldsymbol{y}}_3, \boldsymbol{y}_4, \hat{\boldsymbol{y}}_1$$

only those vectors with a caret are used. Note that it is possible for an adjustment to undo a correction previously made. In this example, although the iteration started on the right (positive) side of $\boldsymbol{v}^T \boldsymbol{y}_1 = 0$, successive iterations of $\boldsymbol{v}$ gave an estimate with $\boldsymbol{v}^T \boldsymbol{y}_1 < 0$ (at stages 3 and 4). Eventually a solution with $J_P(\boldsymbol{v}) = 0$ will be obtained for separable patterns.

Variants

There are many variants on the fixed increment rule given in the previous section. We consider just a few of them here.

(1) Absolute correction rule Choose the value of ρ so that the value of $\boldsymbol{v}_{k+1}^T \boldsymbol{y}_i$ is positive. Thus

$$\rho > |\boldsymbol{v}_k^T \boldsymbol{y}_i| / |\boldsymbol{y}_i|^2$$

where $\boldsymbol{y}_i$ is the misclassified pattern presented at the kth step. This means that the iteration corrects for each misclassified pattern as it is presented. For example, ρ may be taken to be the smallest integer greater than $|\boldsymbol{v}_k^T \boldsymbol{y}_i| / |\boldsymbol{y}_i|^2$.

(2) Fractional correction rule This sets ρ to be a function of the distance to the hyperplane $\boldsymbol{v}^T \boldsymbol{y}_i = 0$, i.e.

$$\rho = \lambda |\boldsymbol{v}_k^T \boldsymbol{y}_i| / |\boldsymbol{y}_i|^2$$

where λ is the fraction of the distance to the hyperplane $\boldsymbol{v}^T \boldsymbol{y}_i = 0$, traversed in going from $\boldsymbol{v}_k$ to $\boldsymbol{v}_{k+1}$. If $\lambda > 1$, then pattern $\boldsymbol{y}_i$ will be classified correctly after the adjustment to $\boldsymbol{v}$.

(3) Introduction of a margin, b A margin, $b > 0$, is introduced (see Figure 4.3) and the weight vector is updated whenever $\boldsymbol{v}^T \boldsymbol{y}_i \leq b$. Thus, the solution vector $\boldsymbol{v}$ must lie at a distance greater than $b/|\boldsymbol{y}_i|$ from each hyperplane $\boldsymbol{v}^T \boldsymbol{y}_i = 0$. The training procedures given above are still guaranteed to produce a solution when the classes are separable. One of the reasons often given for the introduction of a threshold is to aid generalisation. Without the threshold, some of the points in the data space may lie close to the separating boundary. Viewed in data space, all points $\boldsymbol{x}_i$ lie at a distance greater than $b/|\boldsymbol{w}|$ from the separating hyperplane. Clearly, the solution is not unique and in Section 4.2.5 we address the problem of seeking a 'maximal margin' classifier.

(4) Variable increment ρ One of the problems with the above procedures is that, although they will converge if the classes are separable, the solution for $\boldsymbol{v}$ will oscillate if the classes overlap. The error-correction procedure also converges (for linearly separable classes) if ρ_k satisfies the following conditions

$$\rho_k \geq 0$$

$$\sum_{k=1}^{\infty} \rho_k = \infty$$

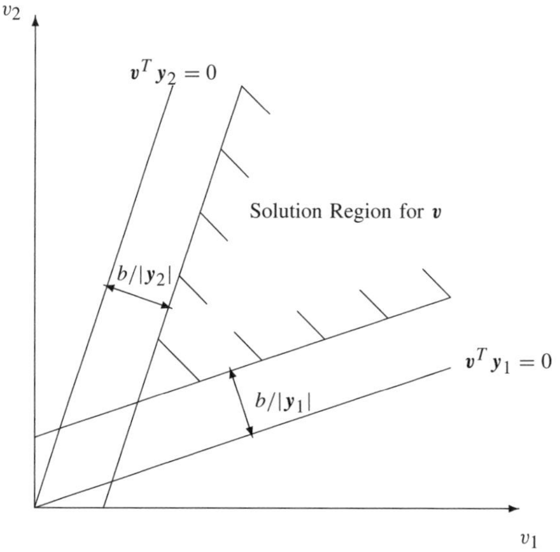

Figure 4.3 Solution region for a margin

and

$$\lim_{m\to\infty} \frac{\sum_{k=1}^{m} \rho_k^2}{\left(\sum_{k=1}^{m} \rho_k\right)^2} = 0$$

In many problems, we do not know *a priori* whether the samples are separable or not. If they are separable, then we would like a procedure that yields a solution that separates the classes. On the other hand, if they are not separable then a method with $\rho_k \to 0$ will decrease the effects of misclassified samples as the iteration proceeds. One possible choice is $\rho_k = 1/k$.

(5) Relaxation algorithm The *relaxation* or *Agmon–Mays algorithm* minimises the criterion

$$J_r = \frac{1}{2} \sum_{\boldsymbol{y}_i \in \mathcal{Y}} (\boldsymbol{v}^T \boldsymbol{y}_i - b)^2 / |\boldsymbol{y}_i|^2$$

where $\mathcal{Y}$ is $\{\boldsymbol{y}_i | \boldsymbol{y}_i^T \boldsymbol{v} \leq b\}$. Thus not only do the misclassified samples contribute to J_r, but so also do those correctly classified samples lying closer than $b/|\boldsymbol{v}|$ to the boundary $\boldsymbol{v}^T \boldsymbol{y} = 0$. The basic algorithm is

$$\boldsymbol{v}_{k+1} = \boldsymbol{v}_k + \rho_k \sum_{\boldsymbol{y}_i \in \mathcal{Y}_k} \frac{b - \boldsymbol{v}_k^T \boldsymbol{y}_i}{|\boldsymbol{y}_i|^2} \boldsymbol{y}_i$$

where $\mathcal{Y}_k$ is $\{\boldsymbol{y}_i | \boldsymbol{y}_i^T \boldsymbol{v}_k \leq b\}$. This has a single-pattern scheme

$$\boldsymbol{v}_{k+1} = \boldsymbol{v}_k + \rho_k \frac{b - \boldsymbol{v}_k^T \boldsymbol{y}_i}{|\boldsymbol{y}_i|^2} \boldsymbol{y}_i$$

where $\boldsymbol{v}_k^T \boldsymbol{y}_i \leq b$ (that is, the patterns $\boldsymbol{y}_i$ that cause the vector $\boldsymbol{v}$ to be corrected). This is the same as the fractional correction rule with a margin.

4.2.3 Fisher's criterion

The approach adopted by Fisher was to find a linear combination of the variables that separates the two classes as much as possible. That is, we seek the direction along which the two classes are best separated in some sense. The criterion proposed by Fisher is the ratio of between-class to within-class variances. Formally, we seek a direction $\boldsymbol{w}$ such that

$$J_F = \frac{\left|\boldsymbol{w}^T (\boldsymbol{m}_1 - \boldsymbol{m}_2)\right|^2}{\boldsymbol{w}^T \boldsymbol{S}_W \boldsymbol{w}} \tag{4.4}$$

is a maximum, where $\boldsymbol{m}_1$ and $\boldsymbol{m}_2$ are the group means and $\boldsymbol{S}_W$ is the pooled within-class sample covariance matrix, in its bias-corrected form given by

$$\frac{1}{n-2} \left(n_1 \hat{\boldsymbol{\Sigma}}_1 + n_2 \hat{\boldsymbol{\Sigma}}_2\right)$$

where $\hat{\boldsymbol{\Sigma}}_1$ and $\hat{\boldsymbol{\Sigma}}_2$ are the maximum likelihood estimates of the covariance matrices of classes ω_1 and ω_2 respectively and there are n_i samples in class ω_i $(n_1 + n + 2 = n)$. Maximising the above criterion gives a solution for the direction $\boldsymbol{w}$. The threshold weight w_0 is determined by an allocation rule. The solution for $\boldsymbol{w}$ that maximises J_F can be obtained by differentiating J_F with respect to $\boldsymbol{w}$ and equating to zero. This yields

$$\frac{\boldsymbol{w}^T(\boldsymbol{m}_1 - \boldsymbol{m}_2)}{\boldsymbol{w}^T \boldsymbol{S}_W \boldsymbol{w}} \left\{ 2(\boldsymbol{m}_1 - \boldsymbol{m}_2) + \left(\frac{\boldsymbol{w}^T(\boldsymbol{m}_1 - \boldsymbol{m}_2)}{\boldsymbol{w}^T \boldsymbol{S}_W \boldsymbol{w}} \right) \boldsymbol{S}_W \boldsymbol{w} \right\} = 0$$

Since we are interested in the direction of $\boldsymbol{w}$ (and noting that $\boldsymbol{w}^T(\boldsymbol{m}_1 - \boldsymbol{m}_2)/\boldsymbol{w}^T \boldsymbol{S}_W \boldsymbol{w}$ is a scalar), we must have

$$\boldsymbol{w} \propto \boldsymbol{S}_W^{-1}(\boldsymbol{m}_1 - \boldsymbol{m}_2) \tag{4.5}$$

We may take equality without loss of generality. The solution for $\boldsymbol{w}$ is a special case of the more general feature extraction criteria described in Chapter 9 that result in transformations that maximise a ratio of between-class to within-class variance. Therefore, it should be noted that Fisher's criterion does not provide us with an allocation rule, merely a mapping to a reduced dimension (actually one dimension in the two-class situation) in which discrimination is in some sense easiest. If we wish to determine an allocation rule, we must specify a threshold, w_0, so that we may assign $\boldsymbol{x}$ to class ω_1 if

$$\boldsymbol{w}^T \boldsymbol{x} + w_0 > 0$$

In Chapter 2 we have seen that if the data were normally distributed with equal covariance matrices, then the optimal decision rule is linear: assign $\boldsymbol{x}$ to ω_1 if $\boldsymbol{w}^T\boldsymbol{x} + w_0 > 0$ where (equations (2.4) and (2.5))

$$\begin{aligned} \boldsymbol{w} &= \boldsymbol{S}_W^{-1}(\boldsymbol{m}_1 - \boldsymbol{m}_2) \\ w_0 &= -\frac{1}{2}(\boldsymbol{m}_1 + \boldsymbol{m}_2)^T \boldsymbol{S}_W^{-1}(\boldsymbol{m}_1 - \boldsymbol{m}_2) - \log\left(\frac{p(\omega_2)}{p(\omega_1)}\right) \end{aligned}$$

Thus, the direction onto which $\boldsymbol{x}$ is projected is the same as that obtained through maximisation of (4.4) and given by (4.5). This suggests that if we take $\boldsymbol{w} = \boldsymbol{S}_W^{-1}(\boldsymbol{m}_1 - \boldsymbol{m}_2)$ (unit constant of proportionality giving equality in (4.5)), then we may choose a threshold to be given by w_0 above, although we note that it is optimal for normally distributed classes.

Note, however, that the discriminant direction (4.5) has been derived without any assumptions of normality. We have used normal assumptions to set a threshold for discrimination. In non-normal situations, a different threshold may be more appropriate. Nevertheless, we may still use the above rule in the more general non-normal case, giving: assign $\boldsymbol{x}$ to ω_1 if

$$\left\{ \boldsymbol{x} - \frac{1}{2}(\boldsymbol{m}_1 + \boldsymbol{m}_2) \right\}^T \boldsymbol{w} > \log\left(\frac{p(\omega_2)}{p(\omega_1)}\right) \tag{4.6}$$

but it will not necessarily be optimal. Note that the above rule is not guaranteed to give a separable solution even if the two groups are separable.

4.2.4 Least mean squared error procedures

The perceptron and related criteria in Section 4.2.2 are all defined in terms of misclassified samples. In this section, all the data samples are used and we attempt to find a solution vector for which the *equality* constraints

$$\boldsymbol{v}^T \boldsymbol{y}_i = t_i$$

are satisfied for *positive* constants t_i. (Recall that the vectors $\boldsymbol{y}_i$ are defined by $\boldsymbol{y}_i^T = (1, \boldsymbol{x}_i^T)$ for $\boldsymbol{x}_i \in \omega_1$ and $\boldsymbol{y}_i^T = (-1, -\boldsymbol{x}_i^T)$ for $\boldsymbol{x}_i \in \omega_2$, or the $(D+1)$-dimensional vectors $(1, \boldsymbol{\phi}_i^T)$ and $(-1, -\boldsymbol{\phi}_i^T)$ for transformations $\boldsymbol{\phi}$ of the data $\boldsymbol{\phi}_i = \boldsymbol{\phi}(\boldsymbol{x}_i)$). In general, it will not be possible to satisfy these constraints exactly and we seek a solution for $\boldsymbol{v}$ that minimises a cost function of the difference between $\boldsymbol{v}^T \boldsymbol{y}_i$ and t_i. The particular cost function we shall consider is the mean squared error.

Solution

Let $\boldsymbol{Y}$ be the $n \times (p+1)$ matrix of sample vectors, with the ith row $\boldsymbol{y}_i$, and $\boldsymbol{t} = (t_1, \ldots, t_n)^T$. Then the sum-squared error criterion is

$$J_S = \|\boldsymbol{Y}\boldsymbol{v} - \boldsymbol{t}\|^2 \tag{4.7}$$

The solution for $\boldsymbol{v}$ minimising J_S is (see Appendix C)

$$\boldsymbol{v} = \boldsymbol{Y}^{\dagger}\boldsymbol{t}$$

where $\boldsymbol{Y}^{\dagger}$ is the pseudo-inverse of $\boldsymbol{Y}$. If $\boldsymbol{Y}^T\boldsymbol{Y}$ is nonsingular, then another form is

$$\boldsymbol{v} = (\boldsymbol{Y}^T\boldsymbol{Y})^{-1}\boldsymbol{Y}^T\boldsymbol{t} \tag{4.8}$$

For the given solution for $\boldsymbol{v}$, the approximation to $\boldsymbol{t}$ is

$$\begin{aligned} \hat{\boldsymbol{t}} &= \boldsymbol{Y}\boldsymbol{v} \\ &= \boldsymbol{Y}(\boldsymbol{Y}^T\boldsymbol{Y})^{-1}\boldsymbol{Y}^T\boldsymbol{t} \end{aligned}$$

A measure of how well the linear approximation fits the data is provided by the absolute error in the approximation, or *error sum of squares*, which is

$$\|\hat{\boldsymbol{t}} - \boldsymbol{t}\|^2 = \|\{\boldsymbol{Y}(\boldsymbol{Y}^T\boldsymbol{Y})^{-1}\boldsymbol{Y}^T - \boldsymbol{I}\}\boldsymbol{t}\|^2$$

and we define the *normalised error* as

$$\epsilon = \left(\frac{\|\hat{\boldsymbol{t}} - \boldsymbol{t}\|^2}{\|\bar{\boldsymbol{t}} - \boldsymbol{t}\|^2}\right)^{\frac{1}{2}}$$

where $\bar{\boldsymbol{t}} = \bar{t}\mathbf{1}$, in which

$$\bar{t} = \frac{1}{n}\sum_{i=1}^{n} t_i$$

is the mean of the values t_i and $\mathbf{1}$ is a vector of 1s. The denominator, $\|\bar{\boldsymbol{t}} - \boldsymbol{t}\|^2$, is the *total sum of squares* or *total variation*.

Thus, a normalised error close to zero represents a good fit to the data and a normalised error close to one means that the model predicts the data in the mean and represents a poor fit. The normalised error can be expressed in terms of the *multiple coefficient of determination*, R^2, used in ordinary least squares regression as (Dillon and Goldstein, 1984)

$$R^2 = 1 - \epsilon^2$$

Relationship to Fisher's linear discriminant

We have still not said anything about the choice of the t_i. In this section we consider a specific choice which we write

$$t_i = \begin{cases} t_1 & \text{for all } \boldsymbol{x}_i \in \omega_1 \\ t_2 & \text{for all } \boldsymbol{x}_i \in \omega_2 \end{cases}$$

Order the rows of $\boldsymbol{Y}$ so that the first n_1 samples correspond to class ω_1 and the remaining n_2 samples correspond to class ω_2. Write the matrix $\boldsymbol{Y}$ as

$$\boldsymbol{Y} = \begin{bmatrix} \boldsymbol{u}_1 & \boldsymbol{X}_1 \\ -\boldsymbol{u}_2 & -\boldsymbol{X}_2 \end{bmatrix} \tag{4.9}$$

where $\boldsymbol{u}_i (i = 1, 2)$ is a vector of n_i 1s and there are n_i samples in class $\omega_i (i = 1, 2)$. The matrix $\boldsymbol{X}_i$ has n_i rows containing the training set patterns and p columns. Then (4.8) may be written

$$\boldsymbol{Y}^T \boldsymbol{Y} \boldsymbol{v} = \boldsymbol{Y}^T \boldsymbol{t}$$

and on substitution for $\boldsymbol{Y}$ from (4.9) and $\boldsymbol{v}$ as $(w_0, \boldsymbol{w})^T$ this may be rearranged to give

$$\begin{bmatrix} n & n_1\boldsymbol{m}_1^T + n_2\boldsymbol{m}_2^T \\ n_1\boldsymbol{m}_1 + n_2\boldsymbol{m}_2 & \boldsymbol{X}_1^T\boldsymbol{X}_1 + \boldsymbol{X}_2^T\boldsymbol{X}_2 \end{bmatrix} \begin{bmatrix} w_0 \\ \boldsymbol{w} \end{bmatrix} = \begin{bmatrix} n_1 t_1 - n_2 t_2 \\ t_1 n_1 \boldsymbol{m}_1 - t_2 n_2 \boldsymbol{m}_2 \end{bmatrix}$$

where $\boldsymbol{m}_i$ is the mean of the rows of $\boldsymbol{X}_i$. The top row of the matrix gives a solution for w_0 in terms of $\boldsymbol{w}$ as

$$w_0 = \frac{-1}{n}(n_1\boldsymbol{m}_1^T + n_2\boldsymbol{m}_2^T)\boldsymbol{w} + \frac{n_1}{n}t_1 - \frac{n_2}{n}t_2 \tag{4.10}$$

and the second row gives

$$(n_1\boldsymbol{m}_1 + n_2\boldsymbol{m}_2)w_0 + (\boldsymbol{X}_1^T\boldsymbol{X}_1 + \boldsymbol{X}_2^T\boldsymbol{X}_2)\boldsymbol{w} = t_1\boldsymbol{m}_1 n_1 - t_2\boldsymbol{m}_2 n_2$$

Substituting for w_0 from (4.10) and rearranging gives

$$\left\{ n\boldsymbol{S}_W + \frac{n_1 n_2}{n}(\boldsymbol{m}_1 - \boldsymbol{m}_2)(\boldsymbol{m}_1 - \boldsymbol{m}_2)^T \right\} \boldsymbol{w} = (\boldsymbol{m}_1 - \boldsymbol{m}_2)\frac{n_1 n_2}{n}(t_1 + t_2) \tag{4.11}$$

where $\boldsymbol{S}_W$ is the estimate of the assumed common covariance matrix, written in terms of $\boldsymbol{m}_i$ and $\boldsymbol{X}_i$, $i = 1, 2$, as

$$\boldsymbol{S}_W = \frac{1}{n}\left\{ \boldsymbol{X}_1^T\boldsymbol{X}_1 + \boldsymbol{X}_2^T\boldsymbol{X}_2 - n_1\boldsymbol{m}_1\boldsymbol{m}_1^T - n_2\boldsymbol{m}_2\boldsymbol{m}_2^T \right\}$$

Whatever the solution for $\boldsymbol{w}$, the term

$$\frac{n_1 n_2}{n}(\boldsymbol{m}_1 - \boldsymbol{m}_2)(\boldsymbol{m}_1 - \boldsymbol{m}_2)^T \boldsymbol{w}$$

in (4.11) is in the direction of $\boldsymbol{m}_1 - \boldsymbol{m}_2$. Thus, (4.11) may be written

$$n \boldsymbol{S}_W \boldsymbol{w} = \alpha(\boldsymbol{m}_1 - \boldsymbol{m}_2)$$

for some constant of proportionality, α, with solution

$$\boldsymbol{w} = \frac{\alpha}{n} \boldsymbol{S}_W^{-1}(\boldsymbol{m}_1 - \boldsymbol{m}_2)$$

the same solution obtained for Fisher's linear discriminant (4.5). Thus, provided that the value of t_i is the same for all members of the same class, we recover Fisher's linear discriminant. We require that $t_1 + t_2 \neq 0$ to prevent a trivial solution for $\boldsymbol{w}$.

In the spirit of this approach, discrimination may be performed according to whether $w_0 + \boldsymbol{w}^T\boldsymbol{x}$ is closer in the least squares sense to t_1 than $-w_0 - \boldsymbol{w}^T\boldsymbol{x}$ is to t_2. That is, assign $\boldsymbol{x}$ to ω_1 if $\|t_1 - (w_0 + \boldsymbol{w}^T\boldsymbol{x})\|^2 < \|t_2 + (w_0 + \boldsymbol{w}^T\boldsymbol{x})\|^2$. Substituting for w_0 and $\boldsymbol{w}$, this simplifies to (assuming $\alpha(t_1 + t_2) > 0$): assign $\boldsymbol{x}$ to ω_1 if

$$\left(\boldsymbol{S}_W^{-1}(\boldsymbol{m}_1 - \boldsymbol{m}_2)\right)^T (\boldsymbol{x} - \boldsymbol{m}) > \frac{t_1 + t_2}{2}\frac{n_2 - n_1}{\alpha} \tag{4.12}$$

where $\boldsymbol{m}$ is the sample mean, $(n_1\boldsymbol{m}_1 + n_2\boldsymbol{m}_2)/n$. The threshold on the right-hand side of the inequality above is independent of t_1 and t_2 – see the exercises at the end of the chapter.

Of course, other discrimination rules may be used, particularly in view of the fact that the least squares solution gives Fisher's linear discriminant, which we know is the optimal discriminant for two normally distributed classes with equal covariance matrices. Compare the one above with (4.6) that incorporates the numbers in each class in a different way.

Optimal discriminant

Another important property of the measured squared error solution is that it approaches the minimum mean squared error approximation to the Bayes discriminant function, $g(\boldsymbol{x})$, given by

$$g(\boldsymbol{x}) = p(\omega_1|\boldsymbol{x}) - p(\omega_2|\boldsymbol{x})$$

in the limit as the number of samples tends to infinity.

In order to understand what this statement means, consider J_S given by (4.7) where $t_i = 1$ for all $\boldsymbol{y}_i$, so that

$$J_S = \sum_{\boldsymbol{x} \in \omega_1} (w_0 + \boldsymbol{w}^T\boldsymbol{x} - 1)^2 + \sum_{\boldsymbol{x} \in \omega_2} (w_0 + \boldsymbol{w}^T\boldsymbol{x} + 1)^2 \tag{4.13}$$

where we have assumed linear dependence of the $\boldsymbol{y}_i$ on the $\boldsymbol{x}_i$. Figure 4.4 illustrates the minimisation process taking place. For illustration, five samples are drawn from each of

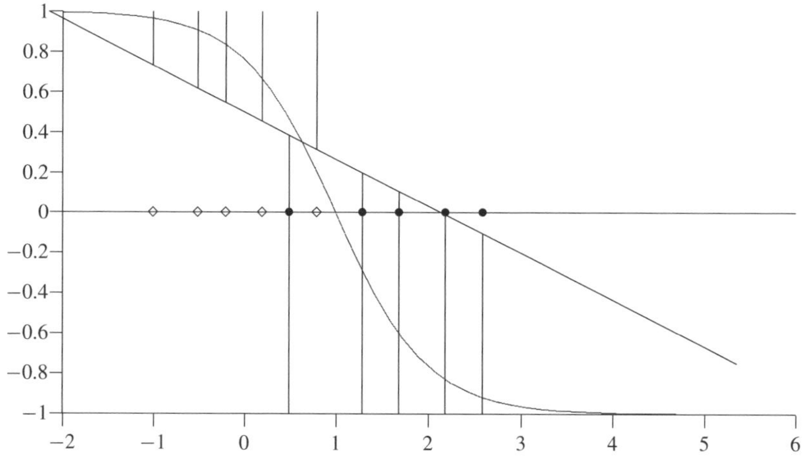

Figure 4.4 Optimality of least mean squared error rule – illustration of equation (4.13)

two univariate normal distributions of unit variance and means 0.0 and 2.0 and plotted on the x-axis of Figure 4.4, $\diamond$ for class ω_1 and $\bullet$ for class ω_2. Minimising J_S means that the sum of the squares of the distances from the straight line in Figure 4.4 to either $+1$ for class ω_1 or -1 for class ω_2 is minimised. Also plotted in Figure 4.4 is the optimal Bayes discriminant, $g(x)$, for the two normal distributions.

As the number of samples, n, becomes large, the expression J_S/n tends to

$$\frac{J_S}{n} = p(\omega_1)\int (w_0 + \boldsymbol{w}^T\boldsymbol{x} - 1)^2 p(\boldsymbol{x}|\omega_1)\,d\boldsymbol{x} + p(\omega_2)\int (w_0 + \boldsymbol{w}^T\boldsymbol{x} + 1)^2 p(\boldsymbol{x}|\omega_2)\,d\boldsymbol{x}$$

Expanding and simplifying, this gives

$$\begin{aligned}\frac{J_S}{n} &= \int (w_0 + \boldsymbol{w}^T\boldsymbol{x})^2 p(\boldsymbol{x})\,d\boldsymbol{x} + 1 - 2\int (w_0 + \boldsymbol{w}^T\boldsymbol{x}) g(\boldsymbol{x}) p(\boldsymbol{x})\,d\boldsymbol{x} \\ &= \int (w_0 + \boldsymbol{w}^T\boldsymbol{x} - g(\boldsymbol{x}))^2 p(\boldsymbol{x})\,d\boldsymbol{x} + 1 - \int g^2(\boldsymbol{x})\,d\boldsymbol{x}\end{aligned}$$

Since only the first integral in the above expression depends on w_0 and $\boldsymbol{w}$, we have the result that minimising (4.13) is equivalent, as the number of samples becomes large, to minimising

$$\int (w_0 + \boldsymbol{w}^T\boldsymbol{x} - g(\boldsymbol{x}))^2 p(\boldsymbol{x})\,d\boldsymbol{x} \qquad (4.14)$$

which is the minimum squared error approximation to the Bayes discriminant function. This is illustrated in Figure 4.5. The expression (4.14) above is the squared difference between the optimal Bayes discriminant and the straight line, integrated over the distribution, $p(\boldsymbol{x})$.

Note that if we were to choose a suitable basis $\phi_1, \ldots, \phi_D$, transform the feature vector $\boldsymbol{x}$ to $(\phi_1(\boldsymbol{x}), \ldots, \phi_D(\boldsymbol{x}))^T$ and then construct the linear discriminant function, we might get a closer approximation to the optimal discriminant, and the decision boundary

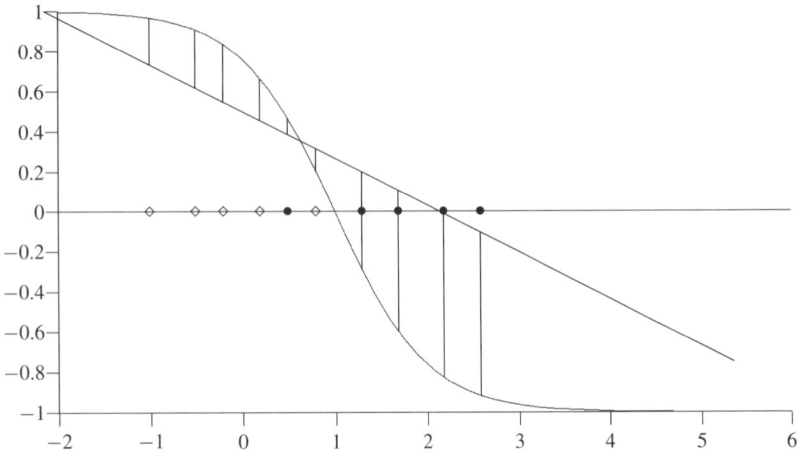

Figure 4.5 Least mean squared error approximation to the Bayes discriminant rule – illustration of equation (4.14)

would not necessarily be a straight line (or plane) in the original space of the variables, $\boldsymbol{x}$. Also, although asymptotically the solution gives the best approximation (in the least squares sense) to the Bayes discriminant function, it is influenced by regions of high density rather than samples close to the decision boundary. Although Bayesian heuristics motivate the use of a linear discriminant trained by least squares, it can give poor decision boundaries in some circumstances (Hastie *et al.*, 1994).

4.2.5 Support vector machines

As we stated in the introduction to this section, algorithms for linear discriminant functions may be applied to the original variables or in a transformed feature space defined by nonlinear transformations of the original variables. Support vector machines are no exception. They implement a very simple idea – they map pattern vectors to a high-dimensional feature space where a 'best' separating hyperplane (the *maximal margin* hyperplane) is constructed (see Figure 4.6).

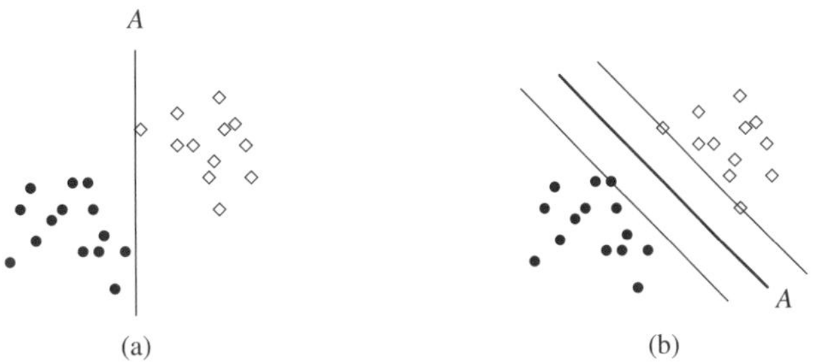

Figure 4.6 Two linearly separable sets of data with separating hyperplane. The separating hyperplane on the right (the thick line) leaves the closest points at maximum distance. The thin lines on the right identify the margin

In this section we introduce the basic ideas behind the support vector model and in Chapter 5 we develop the model further in the context of neural network classifiers. Much of the work on support vector classifiers relates to the binary classification problem, with the multiclass classifier constructed by combining several binary classifiers (see Section 4.3.7).

Linearly separable data

Consider the binary classification task in which we have a set of training patterns $\{\boldsymbol{x}_i, i = 1, \ldots, n\}$ assigned to one of two classes, ω_1 and ω_2, with corresponding labels $y_i = \pm 1$. Denote the linear discriminant function

$$g(\boldsymbol{x}) = \boldsymbol{w}^T \boldsymbol{x} + w_0$$

with decision rule

$$\boldsymbol{w}^T \boldsymbol{x} + w_0 \begin{cases} > & 0 \\ < & 0 \end{cases} \Rightarrow \boldsymbol{x} \in \begin{cases} \omega_1 \text{ with corresponding numeric value } y_i = +1 \\ \omega_2 \text{ with corresponding numeric value } y_i = -1 \end{cases}$$

Thus, all training points are correctly classified if

$$y_i(\boldsymbol{w}^T \boldsymbol{x}_i + w_0) > 0 \text{ for all } i$$

This is an alternative way of writing (4.1).

Figure 4.6a shows two separable sets of points with a separating hyperplane, A. Clearly, there are many possible separating hyperplanes. The maximal margin classifier determines the hyperplane for which the *margin* – the distance to two parallel hyperplanes on each side of the hyperplane A that separates the data – is the largest (Figure 4.6b). The assumption is that the larger the margin, the better the generalisation error of the linear classifier defined by the separating hyperplane.

In Section 4.2.2, we saw that a variant of the perceptron rule was to introduce a margin, $b > 0$, and seek a solution so that

$$y_i(\boldsymbol{w}^T \boldsymbol{x}_i + w_0) \geq b \tag{4.15}$$

The perceptron algorithm yields a solution for which all points $\boldsymbol{x}_i$ are at a distance greater than $b/|\boldsymbol{w}|$ from the separating hyperplane. A scaling of b, w_0 and $\boldsymbol{w}$ leaves this distance unaltered and the condition (4.15) still satisfied. Therefore, without loss of generality, a value $b = 1$ may be taken, defining what are termed the *canonical hyperplanes*, $H_1 : \boldsymbol{w}^T \boldsymbol{x} + w_0 = +1$ and $H_2 : \boldsymbol{w}^T \boldsymbol{x} + w_0 = -1$, and we have

$$\begin{aligned} \boldsymbol{w}^T \boldsymbol{x}_i + w_0 &\geq +1 \quad \text{for } y_i = +1 \\ \boldsymbol{w}^T \boldsymbol{x}_i + w_0 &\leq -1 \quad \text{for } y_i = -1 \end{aligned} \tag{4.16}$$

The distance between each of these two hyperplanes and the separating hyperplane, $g(\boldsymbol{x}) = 0$, is $1/|\boldsymbol{w}|$ and is termed the *margin*. Figure 4.7 shows the separating hyperplane and the canonical hyperplanes for two separable data sets. The points that lie on the canonical hyperplanes are called *support vectors* (circled in Figure 4.7).

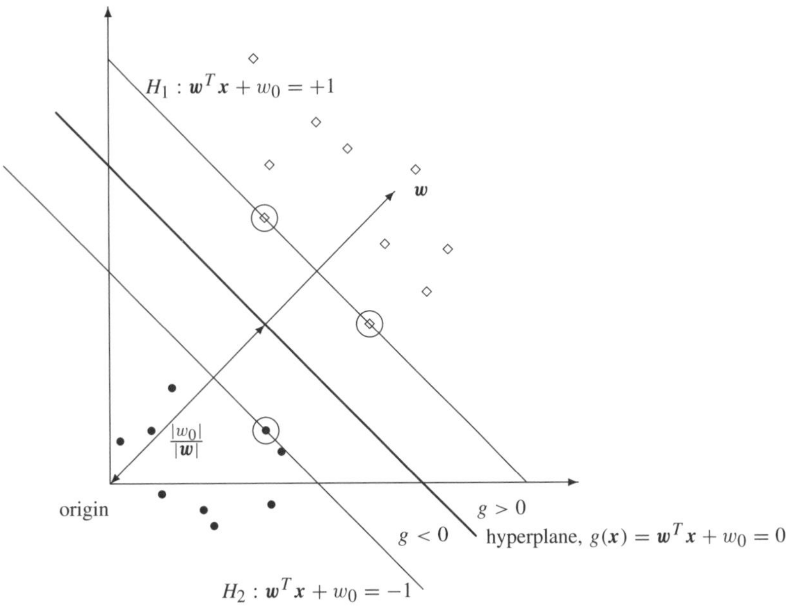

Figure 4.7 H_1 and H_2 are the canonical hyperplanes. The margin is the perpendicular distance between the separating hyperplane ($g(\boldsymbol{x}) = 0$) and a hyperplane through the closest points (marked by a ring around the data points). These are termed the *support vectors*

Therefore, maximising the margin means that we seek a solution that minimises $|\boldsymbol{w}|$ subject to the constraints

$$C1 : \quad y_i(\boldsymbol{w}^T\boldsymbol{x}_i + w_0) \geq 1 \quad i = 1, \ldots, n \tag{4.17}$$

A standard approach to optimisation problems with equality and inequality constraints is the Lagrange formalism (Fletcher, 1988) which leads to the *primal form* of the objective function, L_p, given by[1]

$$L_p = \frac{1}{2}\boldsymbol{w}^T\boldsymbol{w} - \sum_{i=1}^{n} \alpha_i(y_i(\boldsymbol{w}^T\boldsymbol{x}_i + w_0) - 1) \tag{4.18}$$

where $\{\alpha_i, i = 1, \ldots, n; \alpha_i \geq 0\}$ are the Lagrange multipliers. The *primal parameters* are $\boldsymbol{w}$ and w_0 and the number of parameters is $p + 1$, where p is the dimensionality of the feature space.

The solution to the problem of minimising $\boldsymbol{w}^T\boldsymbol{w}$ subject to constraints (4.17) is equivalent to determining the saddlepoint of the function L_p, at which L_p is minimised with respect to $\boldsymbol{w}$ and w_0 and maximised with respect to the α_i. Differentiating L_p with

[1]For inequality constraints of the form $c_i \geq 0$, the constraints equations are multiplied by positive Lagrange multipliers and subtracted from the objective function.

respect to $\boldsymbol{w}$ and w_0 and equating to zero yields

$$\begin{aligned}\sum_{i=1}^{n}\alpha_i y_i &= 0\\ \boldsymbol{w} &= \sum_{i=1}^{n}\alpha_i y_i \boldsymbol{x}_i\end{aligned} \tag{4.19}$$

Substituting into (4.18) gives the *dual form* of the Lagrangian

$$L_D = \sum_{i=1}^{n}\alpha_i - \frac{1}{2}\sum_{i=1}^{n}\sum_{j=1}^{n}\alpha_i\alpha_j y_i y_j \boldsymbol{x}_i^T\boldsymbol{x}_j \tag{4.20}$$

which is *maximised* with respect to the α_i subject to

$$\alpha_i \geq 0 \quad \sum_{i=1}^{n}\alpha_i y_i = 0 \tag{4.21}$$

The importance of the dual form is that it expresses the optimisation criterion as inner products of patterns, $\boldsymbol{x}_i$. This is a key concept and has important consequences for nonlinear support vector machines discussed in Chapter 5. The *dual variables* are the Lagrange multipliers, α_i, and so the number of parameters is n, the number of patterns.

Karush–Kuhn–Tucker conditions

In the above, we have reformulated the primal problem in an alternative dual form which is often easier to solve numerically. The *Kuhn–Tucker conditions* provide necessary and sufficient conditions to be satisfied when minimising an objective function subject to inequality and equality constraints. In the primal form of the objective function, these are

$$\begin{aligned}\frac{\partial L_p}{\partial \boldsymbol{w}} &= \boldsymbol{w} - \sum_{i=1}^{n}\alpha_i y_i \boldsymbol{x}_i = 0\\ \frac{\partial L_p}{\partial w_0} &= -\sum_{i=1}^{n}\alpha_i y_i = 0\\ y_i(\boldsymbol{x}_i^T\boldsymbol{w} + w_0) - 1 &\geq 0\\ \alpha_i &\geq 0\\ \alpha_i(y_i(\boldsymbol{x}_i^T\boldsymbol{w} + w_0) - 1) &= 0\end{aligned} \tag{4.22}$$

In particular, the condition $\alpha_i(y_i(\boldsymbol{x}_i^T\boldsymbol{w} + w_0) - 1) = 0$ (known as the Karush–Kuhn–Tucker complementarity condition–product of the Lagrange multiplier and the inequality constraint) implies that for *active* constraints (the solution satisfies $y_i(\boldsymbol{x}_i^T\boldsymbol{w} + w_0) - 1) = 0$) then $\alpha_i \geq 0$; otherwise, for inactive constraints $\alpha_i = 0$. For active constraints, the Lagrange multiplier represents the sensitivity of the optimal value of L_p to the particular constraint (Cristianini and Shawe-Taylor, 2000). These data points with non-zero Lagrange multiplier lie on the canonical hyperplanes. These are termed the *support vectors* and are the most informative points in the data set. If any of the other patterns (with $\alpha_i = 0$) were to be moved around (provided that they do not cross one of the outer–canonical–hyperplanes), they would not affect the solution for the separating hyperplane.

Classification

Recasting the constrained optimisation problem in its dual form enables numerical quadratic programming solvers to be employed. Once the Lagrange multipliers, α_i, have been obtained, the value of w_0 may be found from

$$\alpha_i(y_i(\boldsymbol{x}_i^T\boldsymbol{w} + w_0) - 1) = 0$$

using any of the support vectors (patterns for which $\alpha_i \neq 0$), or an average over all support vectors

$$n_{sv}w_0 + \boldsymbol{w}^T \sum_{i \in \mathcal{SV}} \boldsymbol{x}_i = \sum_{i \in \mathcal{SV}} y_i \tag{4.23}$$

where n_{sv} is the number of support vectors and the summations are over the set of support vectors, $\mathcal{SV}$. The solution for $\boldsymbol{w}$ used in the above is given by (4.19):

$$\boldsymbol{w} = \sum_{i \in \mathcal{SV}} \alpha_i y_i \boldsymbol{x}_i \tag{4.24}$$

since $\alpha_i = 0$ for other patterns. Thus, the support vectors define the separating hyperplane.

A new pattern, $\boldsymbol{x}$, is classified according to the sign of

$$\boldsymbol{w}^T\boldsymbol{x} + w_0$$

Substituting for $\boldsymbol{w}$ and w_0 gives the linear discriminant: assign $\boldsymbol{x}$ to ω_1 if

$$\sum_{i \in \mathcal{SV}} \alpha_i y_i \boldsymbol{x}_i^T\boldsymbol{x} - \frac{1}{n_{sv}} \sum_{i \in \mathcal{SV}} \sum_{j \in \mathcal{SV}} \alpha_i y_i \boldsymbol{x}_i^T \boldsymbol{x}_j + \frac{1}{n_{sv}} \sum_{i \in \mathcal{SV}} y_i > 0$$

Linearly non-separable data

In many real-world practical problems there will be no linear boundary separating the classes and the problem of searching for an optimal separating hyperplane is meaningless. Even if we were to use sophisticated feature vectors, $\boldsymbol{\phi}(\boldsymbol{x})$, to transform the data to a high-dimensional feature space in which classes are linearly separable, this would lead to an over-fitting of the data and hence poor generalisation ability. We shall return to nonlinear support vector machines in Chapter 5.

However, we can extend the above ideas to handle non-separable data by relaxing the constraints (4.16). We do this by introducing 'slack' variables $\xi_i, i = 1, \ldots, n$, into the constraints to give

$$\begin{array}{ll} \boldsymbol{w}^T\boldsymbol{x}_i + w_0 \geq +1 - \xi_i & \text{for } y_i = +1 \\ \boldsymbol{w}^T\boldsymbol{x}_i + w_0 \leq -1 + \xi_i & \text{for } y_i = -1 \\ \xi_i \geq 0 & i = 1, \ldots, n \end{array} \quad (4.25)$$

For a point to be misclassified by the separating hyperplane, we must have $\xi_i > 1$ (see Figure 4.8).

A convenient way to incorporate the additional cost due to non-separability is to introduce an extra cost term to the cost function by replacing $\boldsymbol{w}^T\boldsymbol{w}/2$ by $\boldsymbol{w}^T\boldsymbol{w}/2 + C\sum_i \xi_i$ where C is a 'regularisation' parameter. The term $C\sum_i \xi_i$ can be thought of as measuring some amount of misclassification – the lower the value of C, the smaller the penalty for 'outliers' and a 'softer' margin. Other penalty terms are possible, for example, $C\sum_i \xi_i^2$ (see Vapnik, 1998).

Thus, we minimise

$$\frac{1}{2}\boldsymbol{w}^T\boldsymbol{w} + C\sum_i \xi_i \quad (4.26)$$

subject to the constraints (4.25). The primal form of the Lagrangian (4.18) now becomes

$$L_p = \frac{1}{2}\boldsymbol{w}^T\boldsymbol{w} + C\sum_i \xi_i - \sum_{i=1}^{n} \alpha_i(y_i(\boldsymbol{w}^T\boldsymbol{x}_i + w_0) - 1 + \xi_i) - \sum_{i=1}^{n} r_i\xi_i \quad (4.27)$$

where $\alpha_i \geq 0$ and $r_i \geq 0$ are Lagrange multipliers; r_i are introduced to ensure positivity of ξ_i.

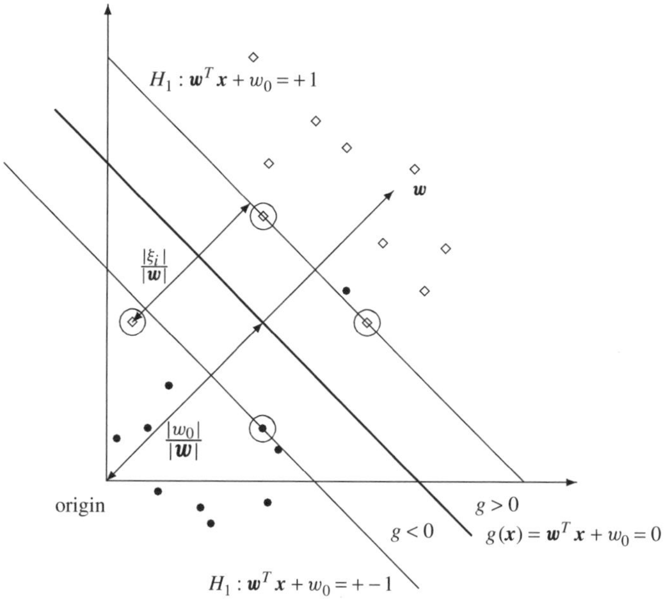

Figure 4.8 Linear separating hyperplane for non-separable data

Differentiating with respect to $\boldsymbol{w}$ and w_0 still results in (4.19)

$$\begin{aligned} \sum_{i=1}^{n} \alpha_i y_i &= 0 \\ \boldsymbol{w} &= \sum_{i=1}^{n} \alpha_i y_i \boldsymbol{x}_i \end{aligned} \tag{4.28}$$

and differentiating with respect to ξ_i yields

$$C - \alpha_i - r_i = 0 \tag{4.29}$$

Substituting the results (4.28) above into the primal form (4.27) and using (4.29) gives the dual form of the Lagrangian

$$L_D = \sum_{i=1}^{n} \alpha_i - \frac{1}{2} \sum_{i=1}^{n} \sum_{j=1}^{n} \alpha_i \alpha_j y_i y_j \boldsymbol{x}_i^T \boldsymbol{x}_j \tag{4.30}$$

which is the same form as the maximal margin classifier (4.20). This is maximised with respect to the α_i subject to

$$\begin{aligned} \sum_{i=1}^{n} \alpha_i y_i &= 0 \\ 0 \leq \alpha_i &\leq C \end{aligned}$$

The latter condition follows from (4.29) and $r_i \geq 0$. Thus, the only change to the maximisation problem is the upper bound on the α_i.

The Karush–Kuhn–Tucker complementarity conditions are

$$\begin{aligned} \alpha_i (y_i (\boldsymbol{x}_i^T \boldsymbol{w} + w_0) - 1 + \xi_i) &= 0 \\ r_i \xi_i = (C - \alpha_i)\xi_i &= 0 \end{aligned}$$

Patterns for which $\alpha_i > 0$ are termed the support vectors. Those satisfying $0 < \alpha_i < C$ must have $\xi_i = 0$ – that is, they lie on one of the canonical hyperplanes at a distance of $1/|\boldsymbol{w}|$ from the separating hyperplane (these support vectors are sometimes termed *margin vectors*). Non-zero slack variables can only occur when $\alpha_i = C$. In this case, the points $\boldsymbol{x}_i$ are misclassified if $\xi_i > 1$. If $\xi_i < 1$, they are classified correctly, but lie closer to the separating hyperplane than $1/|\boldsymbol{w}|$. As in the separable case, the value of w_0 is determined using the first condition above and any support vector or by summing over samples for which $0 < \alpha_i < C$ (for which $\xi_i = 0$) (equation (4.23)) and $\boldsymbol{w}$ is given by (4.24). This gives

$$w_0 = \frac{1}{N_{\widetilde{\mathcal{SV}}}} \left\{ \sum_{i \in \widetilde{\mathcal{SV}}} y_i - \sum_{i \in \mathcal{SV}, j \in \widetilde{\mathcal{SV}}} \alpha_i y_i \boldsymbol{x}_i^T \boldsymbol{x}_j \right\} \tag{4.31}$$

where $\mathcal{SV}$ is the set of support vectors with associated values of α_i satisfying $0 < \alpha_i \leq C$ and $\widetilde{\mathcal{SV}}$ is the set of $N_{\widetilde{\mathcal{SV}}}$ support vectors satisfying $0 < \alpha_i < C$ (those at the target distance of $1/|\boldsymbol{w}|$ from the separating hyperplane).

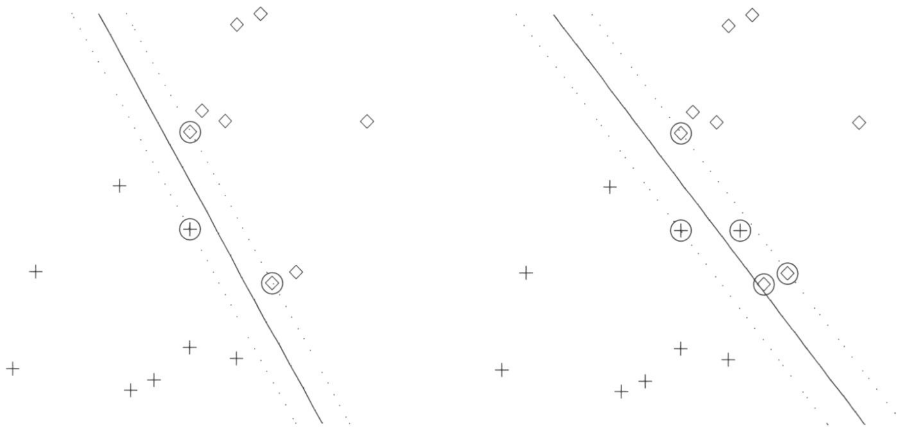

Figure 4.9 Linearly separable data (left) and non-separable data (right, $C = 20$)

Figure 4.9 shows the optimal separating hyperplane for linearly separable and non-separable data. The support vectors ($\alpha_i > 0$) are circled. All points that are not support vectors lie outside the margin strip ($\alpha_i = 0$, $\xi_i = 0$). In the right-hand figure, one of the support vectors (from class +) is incorrectly classified ($\xi_i > 1$).

The only free parameter is the regularisation parameter, C. A value may be chosen by varying C through a range of values and monitoring the performance of the classifier on a separate validation set, or by using cross-validation (see Chapter 11).

4.2.6 Example application study

The problem Classification of land cover using remote sensing satellite imagery (Brown *et al.*, 2000).

Summary A conventional classification technique developed in the remote sensing community (linear spectral mixture models) is compared, both theoretically and practically, to support vector machines. Under certain circumstances the methods are equivalent.

The data The data comprise measurements of two classes of land cover: developed and other (including slate, tarmac, concrete) and undeveloped and vegetation (including sand, water, soil, grass, shrubs, etc.). The measurements are Landsat images of the suburbs of Leicester, UK, in two frequency bands. Each image is 33×33 pixels in size and a pattern is a two-dimensional pair of measurements of corresponding pixels in each of the two bands. Training and test sets were constructed, the training set consisting of 'pure' pixels (those that relate to a region for which there is a single class).

The model A support vector machine model was adopted. Linear separable and linear non-separable support vector machines were trained (by maximising (4.18) and (4.27)) using patterns selected from the training set.

Training procedure The value of the regularisation parameter, C, was chosen to minimise a sum-squared error criterion evaluated on the test set.

4.2.7 Further developments

The main developments of the two-class linear algorithms are as follows.

1. Multiclass algorithms. These are discussed in the following section.
2. Nonlinear methods. Many classification methods that produce nonlinear decision boundaries are essentially *linear models*: they are linear combinations of nonlinear functions of the variables. Radial basis function networks are one example. Thus the machinery developed in this chapter is important. This is examined further in Chapter 5.
3. Regularisation – introducing a parameter that controls the sensitivity of the technique to small changes in the data or training procedure and improves generalisation. This includes combining multiple versions of the same classifier, trained under different conditions (see Chapter 8 and Skurichina, 2001).

4.2.8 Summary

In this section we have considered a range of techniques for performing linear discrimination in the two-class case. These are summarised in Table 4.1. They fall broadly into two groups: those techniques that minimise a criterion based on misclassified samples and those that use all samples, correctly classified or not. The former group includes the perceptron, relaxation and support vector machine algorithms. The latter group includes Fisher's criterion and criteria based on a least squares error measure, including the pseudo-inverse method.

Table 4.1 Summary of linear techniques

Procedure name	Criterion	Algorithm
Perceptron	$J_P(\boldsymbol{v}) = \sum_{\boldsymbol{y}_i \in \mathcal{Y}} (-\boldsymbol{v}^T \boldsymbol{y}_i)$	$\boldsymbol{v}_{k+1} = \boldsymbol{v}_k + \rho_k \sum_{\boldsymbol{y}_i \in \mathcal{Y}} \boldsymbol{y}_i$
Relaxation	$J_r = \frac{1}{2} \sum_{\boldsymbol{y}_i \in \mathcal{Y}} \frac{(\boldsymbol{v}^T \boldsymbol{y}_i - b)^2}{\lvert \boldsymbol{y}_i \rvert^2}$	$\boldsymbol{v}_{k+1} = \boldsymbol{v}_k + \rho_k \frac{b - \boldsymbol{v}_k^T \boldsymbol{y}_i}{\lvert \boldsymbol{y}_i \rvert^2} \boldsymbol{y}_i$
Fisher	$J_F = \frac{\lvert \boldsymbol{w}^T (\boldsymbol{m}_1 - \boldsymbol{m}_2) \rvert^2}{\boldsymbol{w}^T \boldsymbol{S}_W \boldsymbol{w}}$	$\boldsymbol{w} \propto \boldsymbol{S}_W^{-1} (\boldsymbol{m}_1 - \boldsymbol{m}_2)$
Least mean squared error–pseudo-inverse	$J_S = \lVert \boldsymbol{Y} \boldsymbol{v} - \boldsymbol{t} \rVert^2$	$\hat{\boldsymbol{v}} = \boldsymbol{Y}^\dagger \boldsymbol{t}$
Support vector machine	$\boldsymbol{w}^T \boldsymbol{w} + C \sum_i \xi_i$ subject to constraints (4.25)	quadratic programming

A perceptron is a trainable threshold logic unit. During training, weights are adjusted to minimise a specific criterion. For two separable classes, the basic error-correction procedure converges to a solution in which the classes are separated by a linear decision boundary. If the classes are not separable, the training procedure must be modified to ensure convergence. More complex decision surfaces can be implemented by using combinations and layers of perceptrons (Minsky and Papert, 1988; Nilsson, 1965). This we shall discuss further in Chapter 6.

Some of the techniques will find a solution that separates two classes if they are separable, others do not. A further dichotomy is between the algorithms that converge for non-separable classes and those that do not.

The least mean squared error design criterion is widely used in pattern recognition. It can be readily implemented in many of the standard computer programs for regression and we have shown how the discrimination problem may be viewed as an exercise in regression. The linear discriminant obtained by the procedure is optimal in the sense of providing a minimum mean squared error approximation to the Bayes discriminant function. The analysis of this section applies also to generalised linear discriminant functions (the variables $\boldsymbol{x}$ are replaced by $\boldsymbol{\phi}(\boldsymbol{x})$). Therefore, choosing a suitable basis for the $\phi_j(\boldsymbol{x})$ is important since a good set will lead to a good approximation to the Bayes discriminant function.

One problem with the least mean squared error procedure is that it is sensitive to outliers and does not necessarily produce a separable solution, even when the classes are separable by a linear discriminant. Modifications of the least mean square rule to ensure a solution for separable sets have been proposed (the Ho–Kashyap procedure which adjusts both the weight vector, $\boldsymbol{v}$, and the target vector, $\boldsymbol{t}$), but the optimal approximation to the Bayes discriminant function when the sets overlap is no longer achieved.

The least mean squared error criterion does possess some attractive theoretical properties that we can now quote without proof. Let E_1 denote the nearest-neighbour error rate, E_{mse} the least mean squared error rate and let $\boldsymbol{v}$ be the minimum error solution. Then (Devijver and Kittler, 1982)

$$\begin{aligned} \frac{J_S(\boldsymbol{v})/n}{1 - J_S(\boldsymbol{v})/n} &\geq E_{mse} \\ J_S(\boldsymbol{v})/n &\geq 2E_1 \geq 2E^* \\ J_S(\boldsymbol{v})/n &= 2E_1 \Rightarrow E_{mse} = E^* \end{aligned} \tag{4.32}$$

where E^* is the optimal Bayes error rate.

The first condition gives an upper bound on the error rate (and may easily be computed from the values of $J_S(\boldsymbol{v})$ delivered by the algorithm above). It seems sensible that if we have two possible sets of discriminant functions, ϕ_i and ν_i, then if $J_S^{\phi} < J_S^{\nu}$, then the set ϕ should be preferred since it gives a smaller upper bound for the error rate, E_{mse}. Of course, this is not sufficient but gives us a reasonable guideline.

The second two conditions show that the value of the criterion function J_S/n is bounded below by twice the nearest-neighbour error rate E_1, and $J_S/n = 2E_1$ if the linear discriminant function has the same sign as the Bayes discriminant function (crosses the x-axis at the same points).

Support vector machines have been receiving increasing research interest in recent years. They provide an optimally separating hyperplane in the sense that the margin between two groups is maximised. Development of this idea to the nonlinear classifier, discussed in Chapter 5, had led to classifiers with remarkable good generalisation ability.

4.3 Multiclass algorithms

4.3.1 General ideas

There are several ways of extending two-class procedures to the multiclass case.

One-against-all

For C classes, construct C binary classifiers. The kth classifier is trained to discriminate patterns in class ω_k from those in the remaining classes. Thus, determine the weight vector, $\boldsymbol{w}^k$, and the threshold, w_0^k, such that

$$(\boldsymbol{w}^k)^T\boldsymbol{x} + w_0^k \begin{Bmatrix} > & 0 \\ < & 0 \end{Bmatrix} \Rightarrow \boldsymbol{x} \in \begin{cases} \omega_k \\ \omega_1, \ldots, \omega_{k-1}, \omega_{k+1}, \ldots, \omega_C \end{cases}$$

Ideally, for a given pattern $\boldsymbol{x}$, the quantity $g_k(\boldsymbol{x}) = (\boldsymbol{w}^k)^T\boldsymbol{x} + w_0^k$ will be positive for one value of k and negative for the remainder, giving a clear indication of class. However, this procedure may results in a pattern $\boldsymbol{x}$ belonging to more than one class, or belonging to none.

If there is more than one class for which the quantity $g_k(\boldsymbol{x})$ is positive, $\boldsymbol{x}$ may be assigned to the class for which $((\boldsymbol{w}^k)^T\boldsymbol{x} + w_0^k)/|\boldsymbol{w}^k|$ (the distance to the hyperplane) is the largest. If all values of $g_k(\boldsymbol{x})$ are negative, then assign $\boldsymbol{x}$ to the class with smallest value of $|((\boldsymbol{w}^k)^T\boldsymbol{x} + w_0^k)|/|\boldsymbol{w}^k|$.

One-against-one

Construct $C(C-1)$ classifiers. Each classifier discriminates between two classes. A pattern $\boldsymbol{x}$ is assigned using each classifier in turn and a majority vote taken. This can lead to ambiguity, with no clear decision for some patterns.

Discriminant functions

A third approach is, for C classes, to define C linear discriminant functions $g_1(\boldsymbol{x}), \ldots, g_C(\boldsymbol{x})$ and assign $\boldsymbol{x}$ to class ω_i if

$$g_i(\boldsymbol{x}) = \max_j g_j(\boldsymbol{x})$$

that is, $\boldsymbol{x}$ is assigned to the class whose discriminant function is the largest value at $\boldsymbol{x}$. If

$$g_i(\boldsymbol{x}) = \max_j g_j(\boldsymbol{x}) \Leftrightarrow p(\omega_i|\boldsymbol{x}) = \max_j p(\omega_j|\boldsymbol{x})$$

then the decision boundaries obtained will be optimal in the sense of the Bayes minimum error.

The structure of this section follows that of Section 4.2, covering error-correction procedures, generalisations of Fisher's discriminant, minimum squared error procedures and support vector machines.

4.3.2 Error-correction procedure

A generalisation of the two-class error-correction procedure for $C > 2$ classes is to define C linear discriminants

$$g_i(\boldsymbol{x}) = \boldsymbol{v}_i^T \boldsymbol{z}$$

where $\boldsymbol{z}$ is the augmented data vector, $\boldsymbol{z}^T = (1, \boldsymbol{x}^T)$. The *generalised error-correction procedure* is used to train the classifier. Arbitrary initial values are assigned to the $\boldsymbol{v}_i$ and each pattern in the training set is considered one at a time. If a pattern belonging to class ω_i is presented and the maximum value of the discriminant functions is for the jth discriminant function (i.e. a pattern in class ω_i is classified as class ω_j) then the weight vectors $\boldsymbol{v}_i$ and $\boldsymbol{v}_j$ are modified according to

$$\begin{aligned} \boldsymbol{v}_i' &= \boldsymbol{v}_i + c\boldsymbol{z} \\ \boldsymbol{v}_j' &= \boldsymbol{v}_j - c\boldsymbol{z} \end{aligned}$$

where c is a positive correction increment. That is, the value of the ith discriminant function is increased for pattern $\boldsymbol{z}$ and the value of the jth discriminant is decreased. This procedure will converge in a finite number of steps if the classes are separable (see Nilsson, 1965). Convergence may require the data set to be cycled through several times (as in the two-class case).

Choosing c according to

$$c = (\boldsymbol{v}_j - \boldsymbol{v}_i)^T \frac{\boldsymbol{z}}{|\boldsymbol{z}|^2}$$

will ensure that after adjustment of the weight vectors, $\boldsymbol{z}$ will be correctly classified.

4.3.3 Fisher's criterion – linear discriminant analysis

The term *linear discriminant analysis* (LDA), although generically referring to techniques that produce discriminant functions that are linear in the input variables (and thus applying to the perceptron and all of the techniques of this chapter), is also used in a specific sense to refer to the technique of this subsection in which a transformation is sought that, in some sense, maximises between-class separability and minimises within-class variability. The characteristics of the method are:

1. A transformation is produced to a space of dimension at most $C - 1$, where C is the number of classes.

2. The transformation is distribution-free – for example, no assumption is made regarding normality of the data.

3. The axes of the transformed coordinate system can be ordered in terms of 'importance for discrimination'. Those most important can be used to obtain a graphical representation of the data by plotting the data in this coordinate system (usually two or three dimensions).

4. Discrimination may be performed in this reduced-dimensional space using any convenient classifier. Often improved performance is achieved over the application of the rule in the original data space. If a nearest class mean type rule is employed, the decision boundaries are linear (and equal to those obtained by a Gaussian classifier under the assumption of equal covariance matrices for the classes).

5. Linear discriminant analysis may be used as a post-processor for more complex, nonlinear classifiers.

There are several ways of generalising the criterion J_F (4.4) to the multiclass case. Optimisation of these criteria yields transformations that reduce to Fisher's linear discriminant in the two-class case and that, in some sense, maximise the between-class scatter and minimise the within-class scatter. We present one approach here.

We consider the criterion

$$J_F(\boldsymbol{a}) = \frac{\boldsymbol{a}^T \boldsymbol{S}_B \boldsymbol{a}}{\boldsymbol{a}^T \boldsymbol{S}_W \boldsymbol{a}} \tag{4.33}$$

where the sample-based estimates of $\boldsymbol{S}_B$ and $\boldsymbol{S}_W$ are given by

$$\boldsymbol{S}_B = \sum_{i=1}^{C} \frac{n_i}{n} (\boldsymbol{m}_i - \boldsymbol{m})(\boldsymbol{m}_i - \boldsymbol{m})^T$$

and

$$\boldsymbol{S}_W = \sum_{i=1}^{C} \frac{n_i}{n} \hat{\boldsymbol{\Sigma}}_i$$

where $\boldsymbol{m}_i$ and $\hat{\boldsymbol{\Sigma}}_i$, $i = 1, \ldots, C$, are the sample means and covariance matrices of each class (with n_i samples) and $\boldsymbol{m}$ is the sample mean. We seek a set of feature vectors $\boldsymbol{a}_i$ that maximise (4.33) subject to the normalisation constraint $\boldsymbol{a}_i^T \boldsymbol{S}_W \boldsymbol{a}_j = \delta_{ij}$ (class-centralised vectors in the transformed space are uncorrelated). This leads to the generalised symmetric eigenvector equation (Press *et al.*, 1992)

$$\boldsymbol{S}_B \boldsymbol{A} = \boldsymbol{S}_W \boldsymbol{A} \boldsymbol{\Lambda} \tag{4.34}$$

where $\boldsymbol{A}$ is the matrix whose columns are the $\boldsymbol{a}_i$ and $\boldsymbol{\Lambda}$ is the diagonal matrix of eigenvalues. If $\boldsymbol{S}_W^{-1}$ exists, this may be written

$$\boldsymbol{S}_W^{-1} \boldsymbol{S}_B \boldsymbol{A} = \boldsymbol{A} \boldsymbol{\Lambda} \tag{4.35}$$

The eigenvectors corresponding to the largest of the eigenvalues are used for feature extraction. The rank of $\boldsymbol{S}_B$ is at most $C - 1$; therefore the projection will be onto a space of dimension at most $C - 1$. The solution for $\boldsymbol{A}$ satisfying (4.34) satisfying the constraint also diagonalises the between-class covariance matrix, $\boldsymbol{A}^T \boldsymbol{S}_B \boldsymbol{A} = \boldsymbol{\Lambda}$, the diagonal matrix of eigenvalues.

When the matrix $\boldsymbol{S}_W$ is not ill-conditioned with respect to inversion, the eigenvectors of the generalised symmetric eigenvector equation can be determined by solving the equivalent equation

$$\boldsymbol{S}_W^{-1} \boldsymbol{S}_B \boldsymbol{a} = \lambda \boldsymbol{a} \tag{4.36}$$

though note that the matrix $\boldsymbol{S}_W^{-1}\boldsymbol{S}_B$ is not symmetric. However, the system may be reduced to a symmetric eigenvector problem using the Cholesky decomposition (Press *et al.*, 1992) of $\boldsymbol{S}_W$, which allows $\boldsymbol{S}_W$ to be written as the product $\boldsymbol{S}_W = \boldsymbol{L}\boldsymbol{L}^T$, for a lower triangular matrix $\boldsymbol{L}$. Then (4.36) is equivalent to

$$\boldsymbol{L}^{-1}\boldsymbol{S}_B(\boldsymbol{L}^{-1})^T\boldsymbol{y} = \lambda\boldsymbol{y}$$

where $\boldsymbol{y} = \boldsymbol{L}^T\boldsymbol{a}$. Efficient routines based on the QR algorithm (Stewart, 1973; Press *et al.*, 1992) may be used to solve the above eigenvector equation.

If $\boldsymbol{S}_W$ is close to singular, then $\boldsymbol{S}_W^{-1}\boldsymbol{S}_B$ cannot be computed accurately. One approach is to use the QZ (Stewart, 1973) algorithm, which reduces $\boldsymbol{S}_B$ and $\boldsymbol{S}_W$ to upper triangular form (with diagonal elements b_i and w_i respectively) and the eigenvalues are given by the ratios $\lambda_i = b_i/w_i$. If $\boldsymbol{S}_W$ is singular, the system will have 'infinite' eigenvalues, and the ratio cannot be formed. These 'infinite' eigenvalues correspond to eigenvectors in the null space of $\boldsymbol{S}_W$. L.-F. Chen *et al.* (2000) propose using these eigenvectors, ordered according to b_i, for the LDA feature space.

There are other approaches. Instead of solving (4.34) or (4.35), we may determine $\boldsymbol{A}$ by solving two symmetric eigenvector equations successively. The solution is given by

$$\boldsymbol{A} = \boldsymbol{U}_r\boldsymbol{\Lambda}_r^{-\frac{1}{2}}\boldsymbol{V}_\nu \tag{4.37}$$

where $\boldsymbol{U}_r = [\boldsymbol{u}_1, \dots, \boldsymbol{u}_r]$ are the eigenvectors of $\boldsymbol{S}_W$ with non-zero eigenvalues $\lambda_1, \dots, \lambda_r$; $\boldsymbol{\Lambda}_r = \text{diag}(\lambda_1, \dots, \lambda_r)$ and $\boldsymbol{V}_\nu$ is the matrix of eigenvectors of $\boldsymbol{S}'_B = \boldsymbol{\Lambda}_r^{-\frac{1}{2}}\boldsymbol{U}_r^T\boldsymbol{S}_B\boldsymbol{U}_r\boldsymbol{\Lambda}_r^{-\frac{1}{2}}$ and satisfies (4.34). (This is the Karhunen–Loève transformation proposed by Kittler and Young, 1973; see Chapter 9.)

Cheng *et al.* (1992) describe several methods for determining optimal discriminant transformations when $\boldsymbol{S}_W$ is ill-conditioned. These include:

1. *The pseudo-inverse method.* Replace $\boldsymbol{S}_W^{-1}$ by the pseudo-inverse, $\boldsymbol{S}_W^{\dagger}$ (Tian *et al.*, 1988).

2. *The perturbation method.* Stabilise the matrix $\boldsymbol{S}_W$ by adding a small perturbation matrix, $\boldsymbol{\Delta}$ (Hong and Yang, 1991). This amounts to replacing the singular values of $\boldsymbol{S}_W$, λ_r, by a small fixed positive value, δ, if $\lambda_r < \delta$.

3. *The rank decomposition method.* This is a two-stage process, similar to the one given above (4.37), with successive eigendecompositions of the total scatter matrix and between-class scatter matrix.

Discrimination

As in the two-class case, the transformation in itself does not provide us with a discrimination rule. The transformation is independent of the distributions of the classes and is defined in terms of matrices $\boldsymbol{S}_B$ and $\boldsymbol{S}_W$. However, if we were to assume that the data were normally distributed, with equal covariance matrices (equal to the within-class covariance matrix, $\boldsymbol{S}_W$) in each class and means $\boldsymbol{m}_i$, then the discrimination rule is: assign $\boldsymbol{x}$ to class ω_i if $g_i \geq g_j$ for all $j \neq i$, $j = 1, \dots, C$, where

$$g_i = \log(p(\omega_i)) - \frac{1}{2}(\boldsymbol{x} - \boldsymbol{m}_i)^T\boldsymbol{S}_W^{-1}(\boldsymbol{x} - \boldsymbol{m}_i)$$

or, neglecting the quadratic terms in $\boldsymbol{x}$,

$$g_i = \log(p(\omega_i)) - \tfrac{1}{2}\boldsymbol{m}_i^T \boldsymbol{S}_W^{-1} \boldsymbol{m}_i + \boldsymbol{x}^T \boldsymbol{S}_W^{-1} \boldsymbol{m}_i \tag{4.38}$$

the normal-based linear discriminant function (see Chapter 2). If $\boldsymbol{A}$ is the linear discriminant transformation, then $\boldsymbol{S}_W^{-1}$ may be written (see the exercises at the end of the chapter)

$$\boldsymbol{S}_W^{-1} = \boldsymbol{A}\boldsymbol{A}^T + \boldsymbol{A}_\perp \boldsymbol{A}_\perp^T$$

where $\boldsymbol{A}_\perp^T \boldsymbol{m}_j = 0$ for all j. Using the above expression for $\boldsymbol{S}_W^{-1}$ in (4.38) gives a discriminant function

$$g_i = \log(p(\omega_i)) - (\boldsymbol{y}(\boldsymbol{x}) - \boldsymbol{y}_i)^T (\boldsymbol{y}(\boldsymbol{x}) - \boldsymbol{y}_i) \tag{4.39}$$

and ignoring terms that are constant across classes, discrimination is based on

$$g_i = \log(p(\omega_i)) - \tfrac{1}{2}\boldsymbol{y}_i^T \boldsymbol{y}_i + \boldsymbol{y}^T(\boldsymbol{x})\boldsymbol{y}_i$$

a nearest class mean classifier in the transformed space, where $\boldsymbol{y}_i = \boldsymbol{A}^T \boldsymbol{m}_i$ and $\boldsymbol{y}(\boldsymbol{x}) = \boldsymbol{A}^T \boldsymbol{x}$.

This is simply the Gaussian classifier of Chapter 2 applied in the transformed space.

4.3.4 Least mean squared error procedures

Introduction

As in Section 4.2.4, we seek a linear transformation of the data $\boldsymbol{x}$ (or the transformed data $\boldsymbol{\phi}(\boldsymbol{x})$) that we can use to make a decision and which is obtained by minimising a squared error measure. Specifically, let the data be denoted by the $n \times p$ matrix $\boldsymbol{X} = [\boldsymbol{x}_1 | \ldots | \boldsymbol{x}_n]^T$ and consider the minimisation of the quantity

$$\begin{aligned} E &= \|\boldsymbol{W}\boldsymbol{X}^T + \boldsymbol{w}_0 \boldsymbol{1}^T - \boldsymbol{T}^T\|^2 \\ &= \sum_{i=1}^{n} (\boldsymbol{W}\boldsymbol{x}_i + \boldsymbol{w}_0 - \boldsymbol{t}_i)^T (\boldsymbol{W}\boldsymbol{x}_i + \boldsymbol{w}_0 - \boldsymbol{t}_i) \end{aligned} \tag{4.40}$$

where $\boldsymbol{W}$ is a $C \times p$ matrix of weights, $\boldsymbol{w}_0$ is a C-dimensional vector of biases and $\boldsymbol{1}$ is a vector with each component equal to unity. The $n \times C$ matrix of constants $\boldsymbol{T}$, sometimes termed the *target* matrix, is defined so that the ith row is

$$\boldsymbol{t}_i = \boldsymbol{\lambda}_j = \begin{pmatrix} \lambda_{j1} \\ \vdots \\ \lambda_{jC} \end{pmatrix} \text{ for } \boldsymbol{x}_i \text{ in class } \omega_j$$

that is, $\boldsymbol{t}_i$ has the same value for all patterns in the same class. Minimising (4.40) with respect to $\boldsymbol{w}_0$ gives

$$\boldsymbol{w}_0 = \bar{\boldsymbol{t}} - \boldsymbol{W}\boldsymbol{m} \tag{4.41}$$

where

$$\bar{\boldsymbol{t}} = \frac{1}{n}\sum_{j=1}^{C} n_j \boldsymbol{\lambda}_j$$

is the mean 'target' vector and

$$\boldsymbol{m} = \frac{1}{n}\sum_{i=1}^{n} \boldsymbol{x}_i$$

the mean data vector. Substituting for $\boldsymbol{w}_0$ from (4.41) into (4.40) allows us to express the error, E, as

$$E = \|\boldsymbol{W}\hat{\boldsymbol{X}}^T - \hat{\boldsymbol{T}}^T\|^2 \tag{4.42}$$

where $\hat{\boldsymbol{X}}$ and $\hat{\boldsymbol{T}}$ are defined as

$$\hat{\boldsymbol{X}} \triangleq \boldsymbol{X} - \boldsymbol{1}\boldsymbol{m}^T$$
$$\hat{\boldsymbol{T}} \triangleq \boldsymbol{T} - \boldsymbol{1}\bar{\boldsymbol{t}}^T$$

(data and target matrices with zero mean rows); $\boldsymbol{1}$ is an n-dimensional vector of 1s. The minimum (Frobenius) norm solution for $\boldsymbol{W}$ that minimises E is

$$\boldsymbol{W} = \hat{\boldsymbol{T}}^T(\hat{\boldsymbol{X}}^T)^{\dagger} \tag{4.43}$$

where $(\hat{\boldsymbol{X}}^T)^{\dagger}$ is the Moore–Penrose pseudo-inverse of $\hat{\boldsymbol{X}}^T$ ($\boldsymbol{X}^{\dagger} = (\boldsymbol{X}^T\boldsymbol{X})^{-1}\boldsymbol{X}^T$ if the inverse exists; see Appendix C), with matrix of fitted values

$$\tilde{\boldsymbol{T}} = \hat{\boldsymbol{X}}\hat{\boldsymbol{X}}^{\dagger}\hat{\boldsymbol{T}} + \boldsymbol{1}\bar{\boldsymbol{t}}^T \tag{4.44}$$

Thus, we can obtain a solution for the weights in terms of the data and the 'target matrix' $\boldsymbol{T}$, as yet unspecified.

Properties

Before we consider particular forms for $\boldsymbol{T}$, let us note one or two properties of the least mean squared error approximation. The large sample limit of (4.40) is

$$E/n \longrightarrow E_{\infty} = \sum_{j=1}^{C} p(\omega_j)\mathrm{E}[\|\boldsymbol{W}\boldsymbol{x} + \boldsymbol{w}_0 - \boldsymbol{\lambda}_j\|^2]_j \tag{4.45}$$

where $p(\omega_j)$ is the prior probability (the limit of n_j/n) and the expectation, $\mathrm{E}[.]_j$, is with respect to the conditional distribution of $\boldsymbol{x}$ on class ω_j, i.e. for any function z of $\boldsymbol{x}$

$$\mathrm{E}[z(\boldsymbol{x})]_j = \int z(\boldsymbol{x})p(\boldsymbol{x}|\omega_j)\,d\boldsymbol{x}$$

The solution for $\boldsymbol{W}$ and $\boldsymbol{w}_0$ that minimises (4.45) also minimises (Devijver, 1973; Wee, 1968; see also the exercises at the end of the chapter)

$$E' = \mathrm{E}[\|\boldsymbol{W}\boldsymbol{x} + \boldsymbol{w}_0 - \boldsymbol{\rho}(\boldsymbol{x})\|^2] \tag{4.46}$$

where the expectation is with respect to the unconditional distribution $p(\boldsymbol{x})$ of $\boldsymbol{x}$ and $\boldsymbol{\rho}(\boldsymbol{x})$ is defined as

$$\boldsymbol{\rho}(\boldsymbol{x}) = \sum_{j=1}^{C} \boldsymbol{\lambda}_j p(\omega_j|\boldsymbol{x}) \tag{4.47}$$

Thus, $\boldsymbol{\rho}(\boldsymbol{x})$ may be viewed as a 'conditional target' vector; it is the expected target vector given a pattern $\boldsymbol{x}$, with the property that

$$\mathrm{E}[\boldsymbol{\rho}(\boldsymbol{x})] = \int \boldsymbol{\rho}(\boldsymbol{x}) p(\boldsymbol{x}) d\boldsymbol{x} = \sum_{j=1}^{C} p(\omega_j)\boldsymbol{\lambda}_j$$

the mean target vector. From (4.45) and (4.46), the discriminant vector that minimises E_∞ has minimum variance from the discriminant vector $\boldsymbol{\rho}$.

Choice of targets

The particular interpretation of $\boldsymbol{\rho}$ depends on the choice we make for the target vectors for each class, $\boldsymbol{\lambda}_j$. If we interpret the prototype target matrix as

$$\lambda_{ji} = \text{loss in deciding } \omega_i \text{ when the true class is } \omega_j \tag{4.48}$$

then $\boldsymbol{\rho}(\boldsymbol{x})$ is the *conditional risk vector* (Devijver, 1973), where the conditional risk is the expected loss in making a decision, with the ith component of $\boldsymbol{\rho}(\boldsymbol{x})$ being the conditional risk of deciding in favour of ω_i. The Bayes decision rule for minimum conditional risk is

$$\text{assign } \boldsymbol{x} \text{ to } \omega_i \text{ if } \rho_i(\boldsymbol{x}) \leq \rho_j(\boldsymbol{x}), \quad j = 1, \dots, C$$

From (4.45) and (4.46), the discriminant rule that minimises the mean squared error E has minimum variance from the optimum Bayes discriminant function $\boldsymbol{\rho}$ as the number of samples tends to infinity.

For a coding scheme in which

$$\lambda_{ij} = \begin{cases} 1 & i = j \\ 0 & \text{otherwise} \end{cases} \tag{4.49}$$

the vector $\boldsymbol{\rho}(\boldsymbol{x})$ is equal to $\boldsymbol{p}(\boldsymbol{x})$, the vector of posterior probabilities. The Bayes discriminant rule for minimum error is

$$\text{assign } \boldsymbol{x} \text{ to } \omega_i \text{ if } \rho_i(\boldsymbol{x}) \geq \rho_j(\boldsymbol{x}), \quad j = 1, \dots, C$$

The change in the direction of the inequality results from the fact that the terms λ_{ij} are viewed as *gains*. For this coding scheme, the least mean squared error solution for $\boldsymbol{W}$ and $\boldsymbol{w}_0$ gives a vector discriminant function that asymptotically has minimum variance from the vector of *a posteriori* probabilities, shown for the two-class case in Section 4.2.4.

Figure 4.10 illustrates the least mean square procedure on some one-dimensional data. Data for three classes are positioned on the x-axis and the linear discriminant functions obtained by a least squares procedure are shown. These discriminant functions divide the data space into three regions, according to which linear discriminant function is the largest.

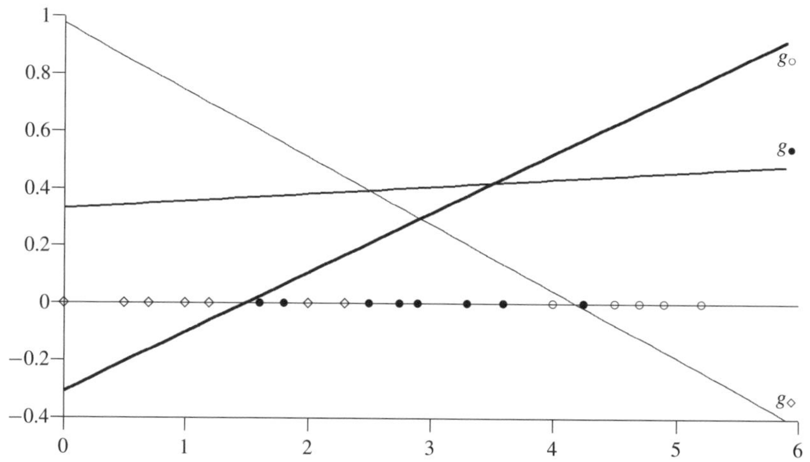

Figure 4.10 Linear discriminant functions

Decision rule

The above asymptotic results suggest that we use the same decision rules to assign a pattern $\boldsymbol{x}$ to a class assuming that the linear transformation $\boldsymbol{W}\boldsymbol{x}+\boldsymbol{w}_0$ had produced $\boldsymbol{\rho}(\boldsymbol{x})$. For example, with the coding scheme (4.49) for $\boldsymbol{\Lambda}$ that gives $\boldsymbol{\rho}(\boldsymbol{x}) = \boldsymbol{p}(\boldsymbol{x})$, we would assign $\boldsymbol{x}$ to the class corresponding to the largest component of the discriminant function $\boldsymbol{W}\boldsymbol{x} + \boldsymbol{w}_0$. Alternatively, in the spirit of the least squares approach, assign $\boldsymbol{x}$ to ω_i if

$$|\boldsymbol{W}\boldsymbol{x} + \boldsymbol{w}_0 - \boldsymbol{\lambda}_i|^2 < |\boldsymbol{W}\boldsymbol{x} + \boldsymbol{w}_0 - \boldsymbol{\lambda}_j|^2 \text{ for all } j \neq i \tag{4.50}$$

which leads to the linear discrimination rule: assign $\boldsymbol{x}$ to class ω_i if

$$\boldsymbol{d}_i^T\boldsymbol{x} + d_{0i} > \boldsymbol{d}_j^T\boldsymbol{x} + d_{0j} \quad \forall j \neq i$$

where

$$\boldsymbol{d}_i = \boldsymbol{\lambda}_i^T\boldsymbol{W}$$
$$d_{0i} = -|\boldsymbol{\lambda}_i|^2/2 + \boldsymbol{w}_0^T\boldsymbol{\lambda}_i$$

For $\boldsymbol{\lambda}_i$ given by (4.49), this decision rule is identical to the one that treats the linear discriminant function $\boldsymbol{W}\boldsymbol{x} + \boldsymbol{w}_0$ as $\boldsymbol{p}(\boldsymbol{x})$, but it is not so in general (Lowe and Webb, 1991).

We add a word of caution here. The result given above is an asymptotic result only. Even if we had a very large number of samples and a flexible set of basis functions $\phi(\boldsymbol{x})$ (replacing the measurements $\boldsymbol{x}$), then we do not necessarily achieve the Bayes optimal discriminant function. Our approximation may indeed become closer in the least squares sense, but this is weighted in favour of higher-density regions, not necessarily at class boundaries. In Section 4.3.5 we consider minimum-distance rules in the transformed space further.

A final result, which we shall quote without proof, is that for the 1-from-C coding scheme (4.49), the values of the vector $\boldsymbol{W}\boldsymbol{x} + \boldsymbol{w}_0$ do indeed sum to unity (Lowe and

Webb, 1991). That is, if we denote the linear discriminant vector z as

$$z = \boldsymbol{W}\boldsymbol{x} + \boldsymbol{w}_0$$

where $\boldsymbol{W}$ and $\boldsymbol{w}_0$ have been determined using the mean squared error procedure (equations (4.41) and (4.43)) with the columns of $\boldsymbol{\Lambda} = [\boldsymbol{\lambda}_1, \ldots, \boldsymbol{\lambda}_C]$ being the columns of the identity matrix, then

$$\sum_{i=1}^{C} z_i = 1$$

that is, the sum of the discriminant function values is unity. This does not mean that the components of z can necessarily be treated as probabilities since some may be negative.

4.3.5 Optimal scaling

Recall that for a C-class problem, the discriminant function approach constructs C discriminant functions $g_i(\boldsymbol{x})$ with decision rule: assign $\boldsymbol{x}$ to class ω_i if

$$g_i(\boldsymbol{x}) > g_j(\boldsymbol{x}) \qquad \text{for all } j \neq i$$

For a linear discriminant function,

$$g_i(\boldsymbol{x}) = \boldsymbol{w}_i^T \boldsymbol{x} + w_{i0}$$

and, for a least squares approach, we minimise an error of the form (4.40) with respect to the weight vectors and the offsets. This was illustrated in Figure 4.10 for three classes. Three linear discriminant functions are calculated for the data positioned on the x-axis. These linear discriminant functions can be used to partition the data space into three regions, according to which discriminant function is the largest.

An alternative to the zero–one coding for the response variable is to consider an approach where a single linear regression is determined, with different target values for each class, ω_i. For example, in Figure 4.11, the target for class $\diamond$ is -1, the target for class $\bullet$ is 0, and the target for class $\circ$ is taken to be +1. The linear regression $g(x)$ is shown. A pattern x' may be assigned to the class whose label is closest to $g(x')$. Thus, for $-\frac{1}{2} < g(x) < \frac{1}{2}$, $x \in$ class $\bullet$.

In practice, we want an automatic means of assigning values to class labels and, of course, the values -1, 0, $+1$ used above may not be the best ones to choose, in terms of minimising a least squares criterion.

Therefore we seek to minimise the squared error with respect to the class indicator values *and* the weight vectors $\boldsymbol{w}_i$. Define a set of *scorings* or *scalings* for the C classes by the vector $\boldsymbol{\theta} \in \mathbb{R}^C$. Generally we can find $K < C$ independent solutions for $\boldsymbol{\theta}$, namely $\boldsymbol{\theta}_1, \ldots, \boldsymbol{\theta}_K$, for which the squared error is minimised. Let $\boldsymbol{\Theta}$ be the $C \times K$ matrix $[\boldsymbol{\theta}_1, \ldots, \boldsymbol{\theta}_K]$ and $\boldsymbol{\Theta}^* = \boldsymbol{T}\boldsymbol{\Theta}$ be the $n \times K$ matrix of transformed values of the classes for each pattern, where we take the $n \times C$ matrix, $\boldsymbol{T}$, to be the class indicator matrix with $T_{ij} = 1$ if $\boldsymbol{x}_i \in$ class ω_j and 0 otherwise. Without loss of generality we shall

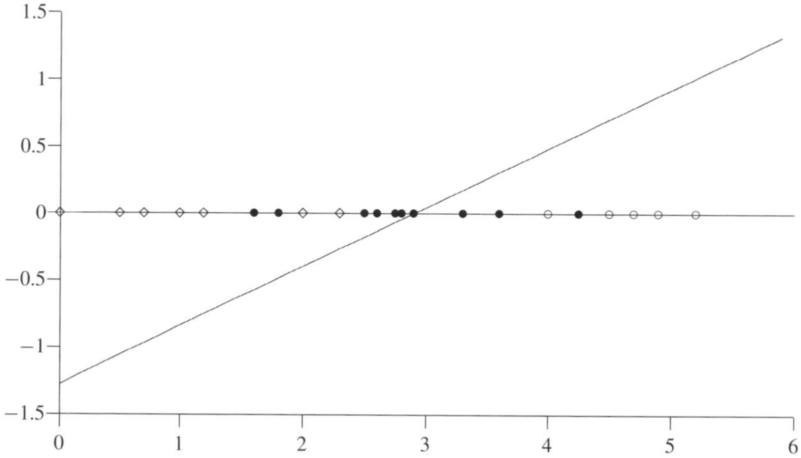

Figure 4.11 LDA regression for targets −1 and +1

assume that the data have zero mean and $\boldsymbol{X}$ is the $n \times p$ data matrix with ith row $\boldsymbol{x}_i^T$. Therefore, we seek to minimise the squared error

$$E = \|\boldsymbol{\Theta}^* - \boldsymbol{X}\boldsymbol{W}^T\|^2 \tag{4.51}$$

with respect to the scores $\boldsymbol{\Theta}$ *and* the $K \times p$ weight matrix $\boldsymbol{W}$. Compare the above with the standard least squares criterion (4.42). Note that an offset term $\mathbf{1}\boldsymbol{w}_0^T$ has not been included. The results of this section carry through with a slight modification with the inclusion of an offset.

The solution for $\boldsymbol{W}$ minimising E is given by

$$\boldsymbol{W}^T = \boldsymbol{X}^\dagger\boldsymbol{\Theta}^* \tag{4.52}$$

where $\boldsymbol{X}^\dagger$ is the pseudo-inverse of $\boldsymbol{X}$, with error, on substituting for $\boldsymbol{W}$, of

$$E = \mathrm{Tr}\{(\boldsymbol{T}\boldsymbol{\Theta})^T(\boldsymbol{I} - \boldsymbol{X}\boldsymbol{X}^\dagger)\boldsymbol{T}\boldsymbol{\Theta}\} \tag{4.53}$$

Minimising with respect to $\boldsymbol{\Theta}$ subject to the constraint $\boldsymbol{\Theta}^T\boldsymbol{D}_p\boldsymbol{\Theta} = \boldsymbol{I}_K$ (where $\boldsymbol{I}_K$ is the $K \times K$ identity matrix and $\boldsymbol{D}_p = \mathrm{diag}(n_1/n, \ldots, n_C/n)$), using the method of Lagrange multipliers, leads to the columns of $\boldsymbol{\Theta}$ satisfying the general symmetric eigenvector equation

$$\frac{1}{n}\boldsymbol{T}^T(\boldsymbol{X}\boldsymbol{X}^\dagger)\boldsymbol{T}\boldsymbol{\theta} = \lambda\boldsymbol{D}_p\boldsymbol{\theta} \tag{4.54}$$

Let the eigenvalues corresponding to solutions $\boldsymbol{\theta}_1, \ldots, \boldsymbol{\theta}_K$ be $\lambda_1, \ldots, \lambda_K$, in decreasing order. The matrix product $(\boldsymbol{X}\boldsymbol{X}^\dagger)\boldsymbol{T}$ in the above is the matrix of fitted values, $\tilde{\boldsymbol{T}}$, in a regression of $\boldsymbol{T}$ on $\boldsymbol{X}$ (see Section 4.3.4). Thus, the matrix on the left-hand side of (4.54) is simply the product of the targets and fitted values $\boldsymbol{T}^T\tilde{\boldsymbol{T}}$. This result is used in developments of this procedure to nonlinear discriminant analysis. The contribution to the average squared error of the lth scaling, $\boldsymbol{\theta}_l$, is given by

$$e_l^2 = 1 - \lambda_l$$

Relationship to the linear discriminant analysis solution

The between-class covariance matrix (assuming mean-centred data) is

$$\boldsymbol{S}_B = \sum_{i=1}^{C} \frac{n_i}{n} \boldsymbol{m}_i \boldsymbol{m}_i^T = \frac{1}{n^2} \boldsymbol{X}^T \boldsymbol{T} \left(\frac{\boldsymbol{T}^T \boldsymbol{T}}{n} \right)^{-1} \boldsymbol{T}^T \boldsymbol{X}$$

and using the solution (4.52) for $\boldsymbol{W}$, we may show that the between-class covariance matrix in the transformed space is

$$\boldsymbol{W} \boldsymbol{S}_B \boldsymbol{W}^T = \text{diag}(\lambda_1^2, \dots, \lambda_K^2), \tag{4.55}$$

a diagonal matrix. Similarly, we may show that the transformed within-class covariance matrix is given by

$$\boldsymbol{W} \boldsymbol{S}_W \boldsymbol{W}^T = \boldsymbol{D}_{\lambda(1-\lambda)} \stackrel{\triangle}{=} \text{diag}(\lambda_1(1-\lambda_1), \dots, \lambda_K(1-\lambda_K)) \tag{4.56}$$

Thus, the optimal scaling solution for $\boldsymbol{W}$ diagonalises both the between-class and within-class covariance matrices. In particular, the linear discriminant analysis transformation that transforms the within-class covariance matrix to the identity matrix and diagonalises the between-class covariance matrix is given by

$$\boldsymbol{W}_{LDA} = \boldsymbol{D}_{\lambda(1-\lambda)}^{\frac{1}{2}} \boldsymbol{W} \tag{4.57}$$

Discrimination

The optimal scaling solution may be used in discrimination in the same way as the linear discriminant solution in Section 4.3.3. Using the relationship between the linear discriminant solution and the optimal scaling solution (4.57), together with the LDA rule (4.39), discrimination is based on the rule: assign $\boldsymbol{x}$ to class ω_i if $g_i > g_j$ for all $j \neq i$, where

$$g_i = \log(p(\omega_i)) - \tfrac{1}{2}(\boldsymbol{y}_i - \boldsymbol{y}_{OS}(\boldsymbol{x}))^T \boldsymbol{D}_{\lambda(1-\lambda)}^{-1} (\boldsymbol{y}_i - \boldsymbol{y}_{OS}(\boldsymbol{x})) \tag{4.58}$$

where $\boldsymbol{y}_i$ is the transformation of the mean of the ith class and $\boldsymbol{y}_{OS}(\boldsymbol{x}) = \boldsymbol{W}\boldsymbol{x}$. An alternative form uses the solutions for the scalings and the transformed vector $y_{OS}(\boldsymbol{x})$ to give a discriminant function (Breiman and Ihaka, 1984; Hastie *et al.*, 1995)

$$g_i = \log(p(\omega_i)) - \tfrac{1}{2}(\boldsymbol{\theta}^i - \boldsymbol{y}_{OS}(\boldsymbol{x}))^T \boldsymbol{D}_{1-\lambda}^{-1} (\boldsymbol{\theta}^i - \boldsymbol{y}_{OS}(\boldsymbol{x})) - \|\boldsymbol{\theta}^i\|^2 \tag{4.59}$$

where $\boldsymbol{\theta}^i$ is the vector of scalings on class ω_i (the ith column of $\boldsymbol{\Theta}^T$). Equivalently,

$$g_i = \log(p(\omega_i)) - \tfrac{1}{2}(\boldsymbol{\theta}^i - \boldsymbol{y}_{OS}(\boldsymbol{x}))^T \boldsymbol{D}_{e^2}^{-1} (\boldsymbol{\theta}^i - \boldsymbol{y}_{OS}(\boldsymbol{x})) - \|\boldsymbol{\theta}^i\|^2 \tag{4.60}$$

where $\boldsymbol{D}_{e^2}$ is the diagonal matrix of contributions to the average squared error by the K solutions, $\text{diag}(e_1^2, \dots, e_K^2)$.

The main conclusion of this analysis is that a regression approach based on scalings and a linear transformation leads to a discriminant rule identical to that obtained from LDA. The discriminant function depends upon the optimal scalings, which in turn are eigensolutions of a matrix obtained from the fitted values in a linear regression and the targets. The advantage of this approach is that it provides a basis for developing nonlinear discriminant functions.

4.3.6 Regularisation

If the matrix $\boldsymbol{X}^T\boldsymbol{X}$ is close to singular, an alternative to the pseudo-inverse approach is to use a *regularised* estimator. The error, E (4.42), is modified by the addition of a regularisation term to give

$$E = \|\boldsymbol{W}\hat{\boldsymbol{X}}^T - \hat{\boldsymbol{T}}^T\|^2 + \alpha\|\boldsymbol{W}\|^2$$

where α is a *regularisation* parameter or *ridge parameter*. The solution for $\boldsymbol{W}$ that minimises E is

$$\boldsymbol{W} = \hat{\boldsymbol{T}}^T\hat{\boldsymbol{X}}(\hat{\boldsymbol{X}}\hat{\boldsymbol{X}}^T + \alpha\boldsymbol{I}_p)^{-1}$$

We still have to choose the ridge parameter, α, which may be different for each output dimension (corresponding to class in a discrimination problem). There are several possible choices (see Brown, 1993). The procedure of Golub *et al.* (1979) is to use a cross-validation estimate. The estimate $\hat{\alpha}$ of α is the value that minimises

$$\frac{\|(\boldsymbol{I} - \boldsymbol{A}(\alpha))\hat{\boldsymbol{t}}\|^2}{[\mathrm{Tr}(\boldsymbol{I} - \boldsymbol{A}(\alpha))]^2}$$

where

$$A(\alpha) = \hat{\boldsymbol{X}}(\boldsymbol{X}^T\boldsymbol{X} + \alpha\boldsymbol{I})^{-1}\hat{\boldsymbol{X}}^T$$

and $\hat{\boldsymbol{t}}$ is one of the columns of $\hat{\boldsymbol{T}}$, i.e. measurements on one of the output variables, that is being predicted.

4.3.7 Multiclass support vector machines

Support vector machines can be applied in multiclass problems either by using the binary classifier in a one-against-all or one-against-one situation, or by constructing C linear discriminant functions simultaneously (Vapnik, 1998).

Consider the linear discriminant functions

$$g_k(\boldsymbol{x}) = (\boldsymbol{w}^k)^T\boldsymbol{x} + w_0^k \quad k = 1, \dots, C$$

We seek a solution for $\{(\boldsymbol{w}^k, w_0^k), k = 1, \dots, C\}$ such that the decision rule: assign $\boldsymbol{x}$ to class ω_i if

$$g_i(\boldsymbol{x}) = \max_j g_j(\boldsymbol{x})$$

separates the training data without error. That is, there are solutions for $\{(\boldsymbol{w}^k, w_0^k), k = 1, \dots, C\}$ such that, for all $k = 1, \dots, C$,

$$(\boldsymbol{w}^k)^T\boldsymbol{x} + w_0^k - ((\boldsymbol{w}^j)^T\boldsymbol{x} + w_0^j) \geq 1$$

for all $\boldsymbol{x} \in \omega_k$ and for all $j \neq k$. This means that every pair of classes is separable. If a solution is possible, we seek a solution for which

$$\sum_{k=1}^{C}(\boldsymbol{w}^k)^T\boldsymbol{w}^k$$

is minimal. If the training data cannot be separated, slack variables are introduced and we minimise

$$L = \sum_{k=1}^{C} (\boldsymbol{w}^k)^T \boldsymbol{w}^k + C \sum_{i=1}^{n} \xi_i$$

subject to the constraints

$$(\boldsymbol{w}^k)^T \boldsymbol{x}_i + w_0^k - ((\boldsymbol{w}^j)^T \boldsymbol{x}_i + w_0^j) \geq 1 - \xi_i$$

for all $\boldsymbol{x}_i$ (where $\boldsymbol{x}_i \in \omega_k$), and for all $j \neq k$. The procedure for minimising the quantity L subject to inequality constraints is the same as that developed in the two-class case.

4.3.8 Example application study

The problem Face recognition using LDA (L.-F. Chen *et al.*, 2000).

Summary A technique, based on LDA, that is appropriate for the small-sample problem (when the within-class matrix, $\boldsymbol{S}_W$, is singular), is assessed in terms of error rate and computational requirements.

The data The data comprise 10 different facial images of 128 people (classes). The raw images are 155×175 pixels. These images are reduced to 60×60 and, after further processing and alignment, further dimension reduction, based on k-means clustering (see Chapter 10), to 32, 64, 128 and 256 values is performed.

The model The classifier is simple. The basic method is to project the data to a lower dimension and perform classification using a nearest-neighbour rule.

Training procedure The projection to the lower dimension is determined using the training data. Problems occur with the standard LDA approach when the $p \times p$ within-class scatter matrix, $\boldsymbol{S}_W$, is singular (of rank $s < p$). In this case, let $\boldsymbol{Q} = [\boldsymbol{q}_{s+1}, \ldots, \boldsymbol{q}_p]$ be the $p \times (p - s)$ matrix of eigenvectors of $\boldsymbol{S}_W$ with zero eigenvalue; that is, those that map into the null space of $\boldsymbol{S}_W$. The eigenvectors of $\tilde{\boldsymbol{S}}_B$, defined by $\tilde{\boldsymbol{S}}_B \triangleq \boldsymbol{Q}\boldsymbol{Q}^T \boldsymbol{S}_B (\boldsymbol{Q}\boldsymbol{Q}^T)^T$, are used to form the most discriminant set for LDA.

Results Recognition rates were calculated using a leave-one-out cross-validation strategy. Limited results on small sample size data sets show good performance compared to previously published techniques.

4.3.9 Further developments

The standard generalisation of Fisher's linear discriminant to the multiclass situation chooses as the columns of the feature extraction matrix $\boldsymbol{A}$, those vectors $\boldsymbol{a}_i$ that maximise

$$\frac{\boldsymbol{a}_i^T \boldsymbol{S}_B \boldsymbol{a}_i}{\boldsymbol{a}_i^T \boldsymbol{S}_W \boldsymbol{a}_i} \tag{4.61}$$

subject to the orthogonality constraint

$$\boldsymbol{a}_i^T \boldsymbol{S}_W \boldsymbol{a}_j = \delta_{ij}$$

i.e. the within-class covariance matrix in the transformed space is the identity matrix. For the two-class case ($C = 2$), only one discriminant vector is calculated. This is Fisher's linear discriminant.

An alternative approach, proposed by Foley and Sammon (1975) for the two-class case, and generalised to the multiclass case by Okada and Tomita (1985), is to seek the vector $\boldsymbol{a}_i$ that maximises (4.61) subject to the constraints

$$\boldsymbol{a}_i \boldsymbol{a}_j = \delta_{ij}$$

The first vector, $\boldsymbol{a}_1$, is Fisher's linear discriminant. The second vector, $\boldsymbol{a}_2$, maximises (4.61) subject to being orthogonal to $\boldsymbol{a}_1$ ($\boldsymbol{a}_2\boldsymbol{a}_1 = 0$), and so on. A direct analytic solution for the problem is given by Duchene and Leclercq (1988), and involves determining an eigenvector of a non-symmetric matrix. Okada and Tomita (1985) propose an iterative procedure for determining the eigenvectors. The transformation derived is not limited by the number of classes.

A development of this approach that uses error probability to select the vectors $\boldsymbol{a}_i$ is described by Hamamoto *et al.* (1991), and a comparison with the Fisher criterion on the basis of the discriminant score J is given by Hamamoto *et al.* (1993).

Another extension of LDA is to a transformation that is not limited by the number of classes. The set of discriminant vectors is augmented by an orthogonal set that maximises the projected variance. Thus the linear discriminant transformation is composed of two parts: a set of vectors determined, say, by the usual multiclass approach and a set orthogonal to the first set for which the projected variance is maximised. This combines linear discriminant analysis with a principal components analysis (Duchene and Leclercq, 1988).

There have been many developments of Fisher's linear discriminant both in the two-class case (for example, Aladjem, 1991) and in the multiclass situation (for example, Aladjem and Dinstein, 1992; Liu *et al.* 1993). The aims have been either to determine transformations that aid in exploratory data analysis or interactive pattern recognition or that may be used prior to a linear or quadratic discriminant rule (Schott, 1993). Several methods have been proposed for a small number of samples (Cheng *et al.*, 1992; Hong and Yang, 1991; Tian *et al.*, 1988) when the matrix $\boldsymbol{S}_W$ may be singular. These are compared by Liu *et al.* (1992).

A development of the least mean squared error approach which weights the error in favour of patterns near the class boundary has been proposed by Al-Alaoui (1977). Al-Alaoui describes a procedure, starting with the generalised inverse solution, that progressively replicates misclassified samples in the training set. The linear discriminant is iteratively updated and the procedure is shown to converge for the two-class separable case. Repeating the misclassified samples in the data set is equivalent to increasing the cost of misclassification of the misclassified samples or, alternatively, weighting the distribution of the training data in favour of samples near the class boundaries (see the description of boosting in Chapter 8).

The major developments of support vector machines presented in this chapter are to nonlinear models, to be discussed in Chapter 5.

4.3.10 Summary

In Chapter 2, we developed a multiclass linear discriminant rule based on normality assumptions for the class-conditional densities (with common covariance matrices). In this section, we have presented a different approach. We have started with the requirement of a linear discriminant rule, and sought ways of determining the weight vectors for each class. Several approaches have been considered, including perceptron schemes, Fisher's discriminant rule, least mean square procedures and support vector machines.

The least mean squared error approach has the following properties:

1. Asymptotically, the procedure produces discriminant functions that provide a minimum squared error approximation to the Bayes discriminant function.

2. By optimising with respect to a set of scores for each class, in addition to the transformation weights, a solution related to the linear discriminant transformation can be obtained and can be more accurate than a straight least squares solution (Hastie *et al.*, 1994).

3. For a '1-from-C' target coding ($T_{ij} = 0$ if $\boldsymbol{x}_i \in \omega_j$ and 0 otherwise), and minimising a squared error, the discriminant functions values sum to unity for any pattern $\boldsymbol{x}$.

One of the problems of the least mean squared error approach is that it places emphasis on regions of high density, which are not necessarily at class boundaries.

In many practical studies, when a linear discriminant rule is used, it is often the normal-based linear discriminant rule of Chapter 2. However, the least mean square rule, with binary-coded target vectors, is important since it forms a part of some nonlinear regression models. Also, there is a natural development of support vector machines to classification problems with nonlinear decision boundaries.

4.4 Logistic discrimination

In the previous sections, discrimination is performed using values of a linear function of the data sample $\boldsymbol{x}$ (or the transformed data samples $\boldsymbol{\phi}(\boldsymbol{x})$). We continue this theme here. We shall introduce logistic discrimination for the two-group case first, and then consider the multigroup situation.

4.4.1 Two-group case

The basic assumption is that the difference between the logarithms of the class-conditional density functions is linear in the variables $\boldsymbol{x}$:

$$\log\left(\frac{p(\boldsymbol{x}|\omega_1)}{p(\boldsymbol{x}|\omega_2)}\right) = \beta_0 + \boldsymbol{\beta}^T\boldsymbol{x} \tag{4.62}$$

This model is an exact description in a wide variety of situations including (Anderson, 1982):

1. when the class-conditional densities are multivariate normal with equal covariance matrices;
2. multivariate discrete distributions following a loglinear model with equal interaction terms between groups;
3. when situations 1 and 2 are combined: both continuous and categorical variables describe each sample.

Hence, the assumption is satisfied by many families of distributions and has been found to be applicable to a wide range of real data sets that depart from normality.

It is a simple matter to show from (4.62) that the assumption is equivalent to

$$\begin{aligned} p(\omega_2|\boldsymbol{x}) &= \frac{1}{1+\exp(\beta_0' + \boldsymbol{\beta}^T \boldsymbol{x})} \\ p(\omega_1|\boldsymbol{x}) &= \frac{\exp(\beta_0' + \boldsymbol{\beta}^T \boldsymbol{x})}{1+\exp(\beta_0' + \boldsymbol{\beta}^T \boldsymbol{x})} \end{aligned} \tag{4.63}$$

where $\beta_0' = \beta_0 + \log(p(\omega_1)/p(\omega_2))$.

Discrimination between two classes depends on the ratio $p(\omega_1|\boldsymbol{x})/p(\omega_2|\boldsymbol{x})$,

$$\text{assign } \boldsymbol{x} \text{ to } \begin{Bmatrix} \omega_1 \\ \omega_2 \end{Bmatrix} \text{ if } \frac{p(\omega_1|\boldsymbol{x})}{p(\omega_2|\boldsymbol{x})} \begin{Bmatrix} > \\ < \end{Bmatrix} 1$$

and substituting the expressions (4.63), we see that the decision about discrimination is determined solely by the linear function $\beta_0' + \beta^T \boldsymbol{x}$ and is given by

$$\text{assign } \boldsymbol{x} \text{ to } \begin{Bmatrix} \omega_1 \\ \omega_2 \end{Bmatrix} \text{ if } \beta_0' + \beta^T \boldsymbol{x} \begin{Bmatrix} > \\ < \end{Bmatrix} 0$$

This is an identical rule to that given in Section 4.2 on linear discrimination, and we gave several procedures for estimating the parameters. The only difference here is that we are assuming a specific model for the ratio of the class-conditional densities that leads to this discrimination rule, rather than specifying the rule *a priori*. Another difference is that we may use the models for the densities (4.63) to obtain maximum likelihood estimates for the parameters.

4.4.2 Maximum likelihood estimation

The parameters of the logistic discrimination model may be estimated using a maximum likelihood approach (Anderson, 1982; Day and Kerridge, 1967). An iterative nonlinear optimisation scheme may be employed using the likelihood function and its derivatives.

The estimation procedure depends on the sampling scheme used to generate the labelled training data (Anderson, 1982; McLachlan, 1992a), and three common sampling designs are considered by Anderson. These are: (i) sampling from the mixture distribution; (ii) sampling conditional on $\boldsymbol{x}$ in which $\boldsymbol{x}$ is fixed and one or more samples are taken (which may belong to ω_1 or ω_2); and (iii) separate sampling for each class in which the conditional distributions, $p(\boldsymbol{x}|\omega_i), i = 1, 2$, are sampled. Maximum likelihood estimates of $\boldsymbol{\beta}$ are independent of the sampling scheme, though one of the sampling designs considered (separate sampling from each group) derives estimates for β_0 rather than β_0' (which is the term required for discrimination). We assume in our derivation below a mixture sampling scheme, which arises when a random sample is drawn from a mixture of the groups. Each of the sampling schemes above is discussed in detail by McLachlan (1992a).

The likelihood of the observations is

$$L = \prod_{r=1}^{n_1} p(\boldsymbol{x}_{1r}|\omega_1) \prod_{r=1}^{n_2} p(\boldsymbol{x}_{2r}|\omega_2) \quad r = 1, \dots, n_s;\ s = 1, 2.$$

where $\boldsymbol{x}_{sr} (s = 1, 2;\ r = 1, \dots, n_s)$ are the observations in class ω_s. This may be rewritten as

$$\begin{aligned} L &= \prod_{r=1}^{n_1} p(\omega_1|\boldsymbol{x}_{1r}) \frac{p(\boldsymbol{x}_{1r})}{p(\omega_1)} \prod_{r=1}^{n_2} p(\omega_2|\boldsymbol{x}_{2r}) \frac{p(\boldsymbol{x}_{2r})}{p(\omega_2)} \\ &= \frac{1}{p(\omega_1)^{n_1} p(\omega_2)^{n_2}} \prod_{\text{all } \boldsymbol{x}} p(\boldsymbol{x}) \prod_{r=1}^{n_1} p(\omega_1|\boldsymbol{x}_{1r}) \prod_{r=1}^{n_2} p(\omega_2|\boldsymbol{x}_{2r}) \end{aligned}$$

The factor

$$\frac{1}{p(\omega_1)^{n_1} p(\omega_2)^{n_2}} \prod_{\text{all } \boldsymbol{x}} p(\boldsymbol{x})$$

is independent of the parameters of the model – the assumption in Anderson (1982) and Day and Kerridge (1967) is that we are free to choose $p(\boldsymbol{x})$; the only assumption we have made is on the log-likelihood ratio. Therefore, maximising the likelihood L is equivalent to maximising

$$L' = \prod_{r=1}^{n_1} p(\omega_1|\boldsymbol{x}_{1r}) \prod_{r=1}^{n_2} p(\omega_2|\boldsymbol{x}_{2r})$$

or

$$\log(L') = \sum_{r=1}^{n_1} \log(p(\omega_1|\boldsymbol{x}_{1r})) + \sum_{r=1}^{n_2} \log(p(\omega_2|\boldsymbol{x}_{2r}))$$

and using the functional forms (4.63)

$$\log(L') = \sum_{r=1}^{n_1} (\beta_0' + \boldsymbol{\beta}^T \boldsymbol{x}_{1r}) - \sum_{\text{all } \boldsymbol{x}} \log\{1 + \exp(\beta_0' + \boldsymbol{\beta}^T \boldsymbol{x})\}$$

The gradient of $\log(L')$ with respect to the parameters β_j is

$$\begin{aligned} \frac{\partial \log L'}{\partial \beta_0'} &= n_1 - \sum_{\text{all } \boldsymbol{x}} p(\omega_1|\boldsymbol{x}) \\ \frac{\partial \log L'}{\partial \beta_j} &= \sum_{r=1}^{n_1} (\boldsymbol{x}_{1r})_j - \sum_{\text{all } \boldsymbol{x}} p(\omega_1|\boldsymbol{x}) x_j, \quad j = 1, \dots, p \end{aligned}$$

Having written down an expression for the likelihood and its derivative, we may now use a nonlinear optimisation procedure to obtain a set of parameter values for which the function $\log(L')$ attains a local maximum. First of all, we need to specify initial starting values for the parameters. Anderson recommends taking zero as a starting value for all $p+1$ parameters, $\beta_0', \beta_1, \ldots, \beta_p$. Except in two special cases (see below), the likelihood has a unique maximum attained for finite $\boldsymbol{\beta}$ (Albert and Lesaffre, 1986; Anderson, 1982). Hence, the starting point is in fact immaterial.

If the two classes are separable, then there are non-unique maxima at infinity. At each stage of the optimisation procedure, it is easy to check whether $\beta_0' + \boldsymbol{\beta}^T \boldsymbol{x}$ gives complete separation. If it does, then the algorithm may be terminated. The second situation when L does not have a unique maximum at a finite value of $\boldsymbol{\beta}$ occurs with discrete data when the proportions for one of the variables are zero for one of the values. In this case, the maximum value of L is at infinity. Anderson (1974) suggests a procedure for overcoming this difficulty, based on the assumption that the variable is conditionally independent of the remaining variables in each group.

4.4.3 Multiclass logistic discrimination

In the multiclass discrimination problem, the basic assumption is that, for C classes,

$$\log\left(\frac{p(\boldsymbol{x}|\omega_s)}{p(\boldsymbol{x}|\omega_C)}\right) = \beta_{s0} + \boldsymbol{\beta}_s^T \boldsymbol{x}, \quad s = 1, \ldots, C-1$$

that is, the log-likelihood ratio is linear for any pair of likelihoods. Again, we may show that the posterior probabilities are of the form

$$p(\omega_s|\boldsymbol{x}) = \frac{\exp(\beta_{s0}' + \boldsymbol{\beta}_s^T \boldsymbol{x})}{1 + \sum_{s=1}^{C-1} \exp(\beta_{s0}' + \boldsymbol{\beta}_s^T \boldsymbol{x})}, \quad s = 1, \ldots, C-1$$

$$p(\omega_C|\boldsymbol{x}) = \frac{1}{1 + \sum_{s=1}^{C-1} \exp(\beta_{s0}' + \boldsymbol{\beta}_s^T \boldsymbol{x})}$$

where $\beta_{s0}' = \beta_{s0} + \log(p(\omega_s)/p(\omega_C))$. Also, the decision rule about discrimination depends solely on the linear functions $\beta_{s0}' + \boldsymbol{\beta}_j^T \boldsymbol{x}$ and the rule is: assign $\boldsymbol{x}$ to class ω_j if

$$\max\{\beta_{s0}' + \boldsymbol{\beta}_s^T \boldsymbol{x}\} = \beta_{j0}' + \boldsymbol{\beta}_j^T \boldsymbol{x} > 0, \quad s = 1, \ldots, C-1$$

otherwise assign $\boldsymbol{x}$ to class ω_C.

The likelihood of the observations is given by

$$L = \prod_{i=1}^{C} \prod_{r=1}^{n_i} p(\boldsymbol{x}_{ir}|\omega_i) \tag{4.64}$$

using the notation given previously. As in the two-class case, maximising L is equivalent to maximising

$$\log(L') = \sum_{s=1}^{C} \sum_{r=1}^{n_s} \log(p(\omega_s|\boldsymbol{x}_{sr})) \tag{4.65}$$

with derivatives

$$\frac{\partial \log(L')}{\partial \beta'_{j0}} = n_j - \sum_{\text{all } \boldsymbol{x}} p(\omega_j|\boldsymbol{x})$$

$$\frac{\partial \log(L')}{\partial (\beta_j)_l} = \sum_{r=1}^{n_j} (\mathrm{x}_{jr})_l - \sum_{\text{all } \boldsymbol{x}} p(\omega_j|\boldsymbol{x}) x_l$$

Again, for separable classes, the maximum of the likelihood is achieved at a point at infinity in the parameter space, but the algorithm may be terminated when complete separation occurs. Also, zero marginal sample proportions cause maxima at infinity and the procedure of Anderson may be employed.

4.4.4 Example application study

The problem To predict in the early stages of pregnancy the feeding method (bottle or breast) a woman would use after giving birth (Cox and Pearce, 1997).

Summary A 'robust' two-group logistic discriminant rule was developed and compared with the ordinary logistic discriminant. Both methods gave similar (good) performance.

The data Data were collected on 1200 pregnant women from two district general hospitals. Eight variables were identified as being important to the feeding method: presence of children under 16 years of age in the household, housing tenure, lessons at school on feeding babies, feeding intention, frequency of seeing own mother, feeding advice from relatives, how the woman was fed, previous experience of breast feeding.

Some patterns were excluded from the analysis for various reasons: incomplete information, miscarriage, termination, refusal and delivery elsewhere. This left 937 cases for parameter estimation.

The model Two models were assessed: the two-group ordinary logistic discrimination model (4.62) and a robust logistic discrimination model, designed to reduce the effect of outliers on the discriminant rule,

$$\frac{p(\boldsymbol{x}|\omega_1)}{p(\boldsymbol{x}|\omega_2)} = \frac{c_1 + c_2 \exp[\beta_0 + \boldsymbol{\beta}^T \boldsymbol{x}]}{1 + \exp[\beta_0 + \boldsymbol{\beta}^T \boldsymbol{x}]}$$

where c_1 and c_2 are fixed positive constants.

Training procedure The prior probabilities $p(\omega_1)$ and $p(\omega_2)$ are estimated from the data and c_1 and c_2 specified. This is required for the robust model, although it can be incorporated with β_0 in the standard model (see Section 4.4.1). Two sets of experiments were performed, both determining maximum likelihood estimates for the parameters: training on the full 937 cases and testing on the same data; training on 424 cases from one hospital and testing on 513 cases from the second hospital.

Results Both model gave similar performance, classifying around 85% cases correctly.

4.4.5 Further developments

Further developments of the basic logistic discriminant model have been to robust procedures (Cox and Ferry, 1991; Cox and Pearce, 1997) and to more general models. Several other discrimination methods are based on models of discriminant functions that are nonlinear functions of linear projections. These include the multilayer perceptron and projection pursuit discrimination (Chapter 6), in which the linear projection and the form of the nonlinear function are simultaneously determined.

4.4.6 Summary

Logistic discrimination makes assumptions about the log-likelihood ratios of one population relative to a reference population. As with the methods discussed in the previous sections, discrimination is made by considering a set of values formed from linear transformations of the explanatory variables. These linear transformations are determined by a maximum likelihood procedure. It is a technique which lies between the linear techniques of Chapters 1 and 4 and the nonlinear methods of Chapters 5 and 6 in that it requires a nonlinear optimisation scheme to estimate the parameters (and the posterior probabilities are nonlinear functions of the explanatory variables), but discrimination is made using a linear transformation.

One of the advantages of the maximum likelihood approach is that asymptotic results regarding the properties of the estimators may readily be derived. Logistic discrimination has further advantages, as itemised by Anderson (1982):

1. It is appropriate for both continuous and discrete-valued variables.
2. It is easy to use.
3. It is applicable over a wide range of distributions.
4. It has a relatively small number of parameters (unlike some of the nonlinear models discussed in Chapters 5 and 6).

4.5 Application studies

There have been several studies comparing logistic discrimination with linear discriminant analysis (for example, Bull and Donner, 1987; Press and Wilson, 1978). Logistic discrimination has been found to work well in practice, particularly for data that depart significantly from normality – something that occurs often in practice. The *Statlog* project (Michie *et al.*, 1994) compared a range of classification methods on various data sets. It reports that there is little practical difference between linear and logistic discrimination. Both methods were in the top five algorithms.

There are many applications of support vector machines, primarily concerned with the nonlinear variant, to be reported in Chapter 5. However, in a communications example, support vector machines have been used successfully to implement a decision feedback equaliser (to combat distortion and interference), giving superior performance to the conventional minimum mean squared error approach (S. Chen *et al.*, 2000).

4.6 Summary and discussion

The discriminant functions discussed in the previous section are all linear in the components of $\boldsymbol{x}$ or the transformed variables $\phi_i(\boldsymbol{x})$ (generalised linear discriminant functions). We have described several approaches for determining the parameters of the model (error-correction schemes, least squares optimisation, logistic model), but we have regarded the initial transformation, $\boldsymbol{\phi}$, as being prescribed. Some possible choices for the functions $\phi_i(\boldsymbol{x})$ were given in Chapter 1 and there are many other parametric and nonparametric forms that may be used, as we shall see in later chapters. But how should we choose the functions ϕ_i? A good choice for the ϕ_i will lead to better classification performance for a subsequent linear classifier than simply applying a linear classifier to the variables x_i. On the other hand, if we were to use a complex nonlinear classifier after the initial transformation of the variables, then the choice for the ϕ_i may not be too critical. Any inadequacies in a poor choice may be compensated for by the subsequent classification process. However, in general, if we have knowledge about which variables or transformations are useful for discrimination, then we should use it, rather than hope that our classifier will 'learn' the important relationships.

If we could choose the ϕ_i so that the conditional risk vector $\boldsymbol{\rho}(\boldsymbol{x})$ (4.47) has components $\rho_i(\boldsymbol{x})$ of the form

$$\rho_i(\boldsymbol{x}) = \sum_{j=1}^{D} a_{ij}\phi_j(\boldsymbol{x}) + a_{i0}$$

for weights a_{ij}, then a resulting linear classifier would be asymptotically optimum (Devijver, 1973). Unfortunately, this does not give us a prescription for the $\phi_i(\boldsymbol{x})$. Many sets of basis functions have been proposed, and the more basis functions we use in our classifier, the better we might expect our classifier to perform. This is not necessarily the case, since increasing the dimension of the vector $\boldsymbol{\phi}$, by the inclusion of more basis functions, leads to more parameters to estimate in the subsequent classification stage. Although error rate on the training set may in fact decrease, the true error rate may increase as generalisation performance deteriorates.

For polynomial basis functions, the number of terms increases rapidly with the order of the polynomial, restricting such an approach to polynomials of low order. However, an important recent development is that of support vector machines, replacing the need to calculate D-dimensional feature vectors $\boldsymbol{\phi}(\boldsymbol{x})$ with the evaluation of a kernel $K(\boldsymbol{x}, \boldsymbol{y})$ at points $\boldsymbol{x}$ and $\boldsymbol{y}$ in the training set.

In the following chapter, we turn to discriminant functions that are linear combinations of nonlinear functions $\boldsymbol{\phi}$ (i.e. generalised linear discriminant functions). The radial basis function network defines the nonlinear function *explicitly*. It is usually of a prescribed form with parameters set as part of the optimisation process. They are usually determined through a separate procedure and the linear parameters are obtained using the procedures of this chapter. The support vector machine defines the nonlinear function *implicitly*, through the specification of a kernel function.

4.7 Recommendations

1. Linear schemes provide a baseline from which more sophisticated methods may be judged. They are easy to implement and should be considered before more complex methods.

2. The error-correction scheme, or a support vector machine, can be used to test for separability. This might be important for high-dimensional data sets where classes may be separable due to the finite training set size. A classifier that achieves linear separability on a training set does not necessarily mean good generalisation performance, but it may indicate insufficient data to characterise distributions.

3. A regression on binary variables provides a least squares approach to the Bayes optimal discriminant. However, there is evidence to show that optimal scaling (linear discriminant analysis in the space of fitted values) provides a better classifier. The latter method (equivalent to the multiclass extension of Fisher's linear discriminant) is recommended.

4.8 Notes and references

The theory of algorithms for linear discrimination is well developed. The books by Nilsson (1965) and Duda *et al.* (2001) provide descriptions of the most commonly used algorithms (see also Ho and Agrawala, 1968; Kashyap, 1970). A more recent treatment of the perceptron can be found in the book by Minsky and Papert (1988).

Logistic discrimination is described in the survey article by Anderson (1982) and the book by McLachlan (1992a).

The development of optimal scaling and the relationship to linear discriminant analysis is described by Breiman and Ihaka (1984); the results in this chapter follow more closely the approach of Hastie *et al.* (1994).

Support vector machines were introduced by Vapnik and co-workers. The book by Vapnik (1998) provides a very good description with historical perspective. Cristianini and Shawe-Taylor (2000) present an introduction to support vector machines aimed at students and practitioners. Burges (1998) provides a very good tutorial on support vector machines for pattern recognition.

The website `www.statistical-pattern-recognition.net` contains references and links to further information on techniques and applications.

Exercises

1. Linear programming or linear optimisation techniques are procedures for maximising linear functions subject to equality and inequality constraints. Specifically, we find the vector $\boldsymbol{x}$ such that

$$z = \boldsymbol{a}_0^T \boldsymbol{x}$$

is minimised subject to the constraints

$$x_i \geq 0 \qquad i = 1, \ldots, n \tag{4.66}$$

and the additional constraints

$$\begin{array}{ll} \boldsymbol{a}_i^T \boldsymbol{x} \leq b_i & i = 1, \ldots, m_1 \\ \boldsymbol{a}_j^T \boldsymbol{x} \geq b_j \geq 0 & j = m_1 + 1, \ldots, m_1 + m_2 \\ \boldsymbol{a}_k^T \boldsymbol{x} = b_k \geq 0 & k = m_1 + m_2 + 1, \ldots, m_1 + m_2 + m_3 \end{array}$$

for given m_1, m_2 and m_3.

Consider optimising the perceptron criterion function as a problem in linear programming. The perceptron criterion function, with a positive margin vector $\boldsymbol{b}$, is given by

$$J_P = \sum_{\boldsymbol{y}_i \in \mathcal{Y}} (b_i - \boldsymbol{v}^T \boldsymbol{y}_i)$$

where now $\boldsymbol{y}_i$ are the vectors satisfying $\boldsymbol{v}^T \boldsymbol{y}_i \leq b_i$. A margin is introduced to prevent the trivial solution $\boldsymbol{v} = 0$. This can be reformulated as a linear programming problem as follows. We introduce the *artificial variables* a_i and consider the problem of minimising

$$\mathcal{Z} = \sum_{i=1}^{n} a_i$$

subject to

$$a_i \geq 0$$
$$a_i \geq b_i - \boldsymbol{v}^T \boldsymbol{y}_i$$

Show that minimising $\mathcal{Z}$ with respect to $\boldsymbol{a}$ and $\boldsymbol{v}$ will minimise the perceptron criterion.

2. Standard linear programming (see Exercise 1) requires all variables to be positive and the minimum of J_P will lead to a solution for $\boldsymbol{v}$ that may have negative components. Convert the $(p+1)$-dimensional vector $\boldsymbol{v}$ to a $2(p+1)$-dimensional vector

$$\begin{pmatrix} \boldsymbol{v}^+ \\ \boldsymbol{v}^- \end{pmatrix}$$

where $\boldsymbol{v}^+$ is the vector $\boldsymbol{v}$ with negative components set to zero, and $\boldsymbol{v}^-$ is the vector $\boldsymbol{v}$ with positive components set to zero and negative components multiplied by -1. For example, the vector $(1, -2, -3, 4, 5)^T$ is written as $(1, 0, 0, 4, 5, 0, 2, 3, 0, 0)^T$.

State the error-correction procedure as an exercise in linear programming.

3. Evaluate the constant of proportionality, α, in (4.12), and hence show that the offset on the right-hand side of (4.12) does not depend on the choice for t_1 and t_2 and is given by

$$\frac{p_2 - p_1}{2} \left(\frac{1 + p_1 p_2 d^2}{p_1 p_2} \right)$$

where $p_i = n_i/n$ and d^2 is given by

$$d^2 = (\boldsymbol{m}_1 - \boldsymbol{m}_2)^T \boldsymbol{S}_W^{-1} (\boldsymbol{m}_1 - \boldsymbol{m}_2)^T$$

the Mahalanobis distance between two normal distributions of equal covariance matrices (see Appendix A). Compare this with the optimal value for normal distributions, given by (4.6).

4. Show that the maximisation of

$$\mathrm{Tr}\{\boldsymbol{S}_W^{-1}\boldsymbol{S}_B\}$$

in the transformed space leads to the same feature space as the linear discriminant solution.

5. Show that the criterion

$$J_4 = \frac{\mathrm{Tr}\{\boldsymbol{A}^T \boldsymbol{S}_B \boldsymbol{A}\}}{\mathrm{Tr}\{\boldsymbol{A}^T \boldsymbol{S}_W \boldsymbol{A}\}}$$

is invariant to an orthogonal transformation of the matrix $\boldsymbol{A}$.

6. Consider the squared error,

$$\sum_{j=1}^{C} p(\omega_j)\mathrm{E}[\|\boldsymbol{W}\boldsymbol{x} + \boldsymbol{w}_0 - \boldsymbol{\lambda}_j\|^2]_j$$

where $\mathrm{E}[.]_j$ is with respect to the conditional distribution of $\boldsymbol{x}$ on class ω_j. By writing

$$\mathrm{E}[\|\boldsymbol{W}\boldsymbol{x} + \boldsymbol{w}_0 - \boldsymbol{\lambda}_j\|^2]_j = \mathrm{E}[\|(\boldsymbol{W}\boldsymbol{x} + \boldsymbol{w}_0 - \boldsymbol{\rho}(\boldsymbol{x})) + (\boldsymbol{\rho}(\boldsymbol{x}) - \boldsymbol{\lambda}_j)\|^2]_j$$

and expanding, show that the linear discriminant rule also minimises

$$E' = \mathrm{E}[\|\boldsymbol{W}\boldsymbol{x} + \boldsymbol{w}_0 - \boldsymbol{\rho}(\boldsymbol{x})\|^2]$$

where the expectation is with respect to the unconditional distribution $p(\boldsymbol{x})$ of $\boldsymbol{x}$ and $\boldsymbol{\rho}(\boldsymbol{x})$ is defined as

$$\boldsymbol{\rho}(\boldsymbol{x}) = \sum_{j=1}^{C} \boldsymbol{\lambda}_j p(\omega_j|\boldsymbol{x})$$

7. Show that if $\boldsymbol{A}$ is the linear discriminant transformation (transforming the within-class covariance matrix to the identity and diagonalising the between-class covariance matrix), then the inverse of the within-class covariance matrix may be written

$$\boldsymbol{S}_W^{-1} = \boldsymbol{A}\boldsymbol{A}^T + \boldsymbol{A}_\perp \boldsymbol{A}_\perp^T$$

where $\boldsymbol{A}_\perp^T \boldsymbol{m}_j = 0$ for all j.

8. Why is the normalisation criterion $\boldsymbol{\Theta}^T \boldsymbol{D}_p \boldsymbol{\Theta} = \boldsymbol{I}_K$ a convenient mathematical choice in optimal scaling? If an offset term $\boldsymbol{1}^T \boldsymbol{w}_0$ is introduced in (4.51), what would be the appropriate normalisation constraint?

9. Using the definition of $\boldsymbol{m}_j$,
$$\boldsymbol{m}_j = \frac{1}{n_j}[\boldsymbol{X}^T\boldsymbol{T}]_j$$
the solution for $\boldsymbol{W}$ (4.52) and the eigenvector equation (4.54), show that
$$\boldsymbol{y}_j^T \boldsymbol{D}_{\lambda(1-\lambda)}^{-1} \boldsymbol{y}_j = (\boldsymbol{\theta}^j)^T \boldsymbol{D}_{1-\lambda}^{-1} \boldsymbol{\theta}^j - |\boldsymbol{\theta}^j|^2$$
where $\boldsymbol{\theta}^j$ are the scorings on class ω_j. Hence derive the discriminant function (4.60) from (4.59).

10. Verify that the optimal scaling solution for $\boldsymbol{W}$ diagonalises the within- and between-class covariance matrices (equations (4.55) and (4.56)).

11. Show that the outputs of a linear discriminant function, trained using a least squares approach with 0–1 targets, sum to unity.

12. For normally distributed classes with equal covariance matrix, Fisher's linear discriminant, with a suitable choice of threshold, provides an optimal discriminant in the sense of obtaining the Bayes decision boundary. Is the converse true? That is, if Fisher's discriminant is identical to the Bayes decision boundary, are the classes normally distributed? Justify your answer.

13. Generate data from three bivariate normal distributions, with means $(-4, -4)$, $(0, 0)$, $(4, 4)$ and identity covariance matrices; 300 samples in train and test sets; equal priors. Train a least squares classifier (Section 4.3.4) and a linear discriminant function (Section 4.3.3) classifier. For each classifier, obtain the classification error and plot the data and the decision boundaries. What do you conclude from the results?

14. Derive the dual form of the Lagrangian for support vector machines applied to the multiclass case.

5

Nonlinear discriminant analysis – kernel methods

Overview

Developed primarily in the neural networks and machine learning literature, the radial basis function (RBF) network and the support vector machine (SVM) are flexible models for nonlinear discriminant analysis that give good performance on a wide range of problems. RBFs are sums of radially symmetric functions; SVMs define the basis functions implicitly through the specification of a kernel.

5.1 Introduction

In the previous chapter, classification of an object is achieved by a linear transformation whose parameters were determined as the result of some optimisation procedure. The linear transformation may be applied to the observed data, or some prescribed features of that data. Various optimisation schemes for the parameters were considered, including simple error-correction schemes (as in the case of the perceptron) and least squares error minimisation; in the logistic discrimination model, the parameters were obtained through a maximum likelihood approach using a nonlinear optimisation procedure.

In this chapter and the following one, we generalise the discriminant model still further by assuming parametric forms for the discriminant functions ϕ. Specifically, we assume a discriminant function of the form

$$g_j(\boldsymbol{x}) = \sum_{i=1}^{m} w_{ji}\phi_i(\boldsymbol{x}; \boldsymbol{\mu}_i) + w_{j0}, \quad j = 1, \dots, C \tag{5.1}$$

where there are m 'basis' functions, ϕ_i, each of which has n_m parameters $\boldsymbol{\mu}_i = \{\mu_{ik}, k = 1, \dots, n_m\}$ (the number of parameters may differ between the ϕ_i, but here we shall assume an equal number), and use the discriminant rule:

$$\text{assign } \boldsymbol{x} \text{ to class } \omega_i \text{ if } g_i(\boldsymbol{x}) = \max_j g_j(\boldsymbol{x})$$

that is, $\boldsymbol{x}$ is assigned to the class whose discriminant function is the largest. In (5.1) the parameters of the model are the values w_{ji} and μ_{ik} and the number of basis functions, m.

Equation (5.1) is exactly of the form of a generalised linear discriminant function, but we allow some flexibility in the nonlinear functions ϕ_i. There are several special cases of (5.1); these include

$\phi_i(\boldsymbol{x};\, \boldsymbol{\mu}_i) \equiv (\boldsymbol{x})_i$	linear discriminant function $m = p$, dimension of $\boldsymbol{x}$.
$\phi_i(\boldsymbol{x};\, \boldsymbol{\mu}_i) \equiv \phi_i(\boldsymbol{x})$	generalised linear discriminant function with fixed transformation; it can take any of the forms in Chapter 1, for example.

Equation (5.1) may be written

$$\boldsymbol{g}(\boldsymbol{x}) = \boldsymbol{W}\boldsymbol{\phi}(\boldsymbol{x}) + \boldsymbol{w}_0 \tag{5.2}$$

where $\boldsymbol{W}$ is the $C \times m$ matrix with (i, j)th component w_{ij}, $\boldsymbol{\phi}(\boldsymbol{x})$ is the m-dimensional vector with ith component $\phi_i(\boldsymbol{x}, \boldsymbol{\mu}_i)$ and $\boldsymbol{w}_0$ is the vector $(w_{10}, \ldots, w_{C0})^T$. Equation (5.2) may be regarded as providing a transformation of a data sample $\boldsymbol{x} \in \mathbb{R}^p$ to $\mathbb{R}^C$ through an intermediate space $\mathbb{R}^m$ defined by the nonlinear functions ϕ_i. This is a model of the *feed-forward type*. As we shall discuss later, models of this form have been widely used for functional approximation and (as with the linear and logistic models), they are not confined to problems in discrimination.

There are two problems to solve with the model (5.2). The first is to determine the complexity of the model or the model order. How many functions ϕ_i do we use (what is the value of m)? How complex should each function be (how many parameters do we allow)? The answers to these questions are data-dependent. There is an interplay between model order, training set size and the dimensionality of the data. Unfortunately there is no simple equation relating the three quantities – it is very much dependent on the data distribution. The problem of model order selection is non-trivial and is very much an active area of current research (see Chapter 11). The second problem is to determine the remaining parameters of the model ($\boldsymbol{W}$ and the μ_i), for a given model order. This is simpler and will involve some nonlinear optimisation procedure for minimising a cost function. We shall discuss several of the most commonly used forms.

In this chapter we introduce models that have been developed primarily in the neural network and machine learning literatures. The types of neural network model that we consider in this chapter are of the feed-forward type, and there is a very strong overlap between these models and those developed in the statistical literature, particularly kernel discrimination, logistic regression and projection pursuit (discussed in the next chapter). The radial basis function network and the multilayer perceptron (the latter also described in the next chapter) may be thought of as a natural progression of the generalised linear discriminant models described in the previous chapter.

Therefore, for the purpose of this chapter and the next, we consider neural network models to provide models of discriminant functions that are linear combinations of simple basis functions, usually of the same parametric form. The parameters of the basis functions, as well as the linear weights, are determined by a training procedure. Other models that we have described in earlier chapters could be termed neural network models (for example, linear discriminant analysis). Also, the development of classification and regression models in the neural network literature is no longer confined to such simple models.

The issues in neural network development are the common ones of pattern recognition: model specification, training and model selection for good generalisation performance. Section 5.2 discusses optimisation criteria for models of the form (5.2) above. Sections 5.3 and 5.4 then introduce two popular models, namely the radial basis function network and the support vector machine.

5.2 Optimisation criteria

All the optimisation criteria assume that we have a set of data samples $\{(\boldsymbol{x}_i, \boldsymbol{t}_i), i = 1, \ldots, n\}$ that we use to 'train' the model. In a regression problem, the $\boldsymbol{x}_i$ are measurements on the regressors and $\boldsymbol{t}_i$ are measurements on the dependent or response variables. In a classification problem, $\boldsymbol{t}_i$ are the class labels. In both cases, we wish to obtain an estimate of $\boldsymbol{t}$ given $\boldsymbol{x}$, a measurement. In the neural network literature, $\boldsymbol{t}_i$ are referred to as *targets*, the desired response of the model for measurements or *inputs*, $\boldsymbol{x}_i$; the actual responses of the model, $\boldsymbol{g}(\boldsymbol{x}_i)$, are referred to as *outputs*.

5.2.1 Least squares error measure

As in the linear case, we seek to minimise a squared error measure,

$$\begin{aligned} E &= \sum_{i=1}^{n} |\boldsymbol{t}_i - \boldsymbol{g}(\boldsymbol{x}_i)|^2 \\ &= \| -\boldsymbol{T}^T + \boldsymbol{W}\boldsymbol{\Phi}^T + \boldsymbol{w}_0\boldsymbol{1}^T\|^2 \end{aligned} \tag{5.3}$$

with respect to the parameters w_{ij} and μ_{jk}, where $\boldsymbol{T} = [\boldsymbol{t}_1, \ldots, \boldsymbol{t}_n]^T$ is an $n \times C$ target matrix whose ith row is the target for input $\boldsymbol{x}_i$; $\boldsymbol{\Phi} = [\boldsymbol{\phi}(\boldsymbol{x}_1), \ldots, \boldsymbol{\phi}(\boldsymbol{x}_n)]^T$ is an $n \times m$ matrix whose ith row is the set of basis function values evaluated at $\boldsymbol{x}_i$; $\|\boldsymbol{A}\|^2 = \text{Tr}\{\boldsymbol{A}\boldsymbol{A}^T\} = \sum_{ij} A_{ij}^2$; and $\boldsymbol{1}$ is a $n \times 1$ vector of 1s.

Properties

1. In a classification problem in which the target for pattern $\boldsymbol{x}_p \in \omega_j$ is the vector of losses $\boldsymbol{\lambda}_j$ with components λ_{ji} defined by the loss in deciding ω_i when the true class is ω_j, the solution for $\boldsymbol{g}$ that minimises (5.3) has minimum variance from the conditional risk vector, $\boldsymbol{\rho}$. In particular, for the 1-from-C coding ($\lambda_{ji} = 1$ for $i = j$; $\lambda_{ji} = 0$ for $i \neq j$), the vector $\boldsymbol{g}$ is a minimum square approximation to the vector of *a posteriori* probabilities asymptotically (see Chapter 4). This does not mean that the approximation $\boldsymbol{g}$ possesses the properties of the true *a posteriori* probability distribution (that its elements are positive and sum to unity) but it is an approximation to it (and the one with minimum variance from it).

2. There are two sets of parameters to determine – the linear weights $\boldsymbol{W}$ and the parameters of the nonlinear functions ϕ, namely $\{\mu\}$. For a given set of values of $\{\mu\}$, the solution for $\boldsymbol{W}$ with minimum norm that minimises (5.3) is

$$\boldsymbol{W} = \hat{\boldsymbol{T}}^T (\hat{\boldsymbol{\Phi}}^T)^{\dagger} \tag{5.4}$$

where † denotes the pseudo-inverse of a matrix (see Appendix C). The matrices $\hat{\boldsymbol{T}}$ and $\hat{\boldsymbol{\Phi}}$ are zero-mean matrices ($\hat{\boldsymbol{T}}^T \mathbf{1} = 0$; $\hat{\boldsymbol{\Phi}}^T \mathbf{1} = 0$) defined as

$$\hat{\boldsymbol{T}} \triangleq \boldsymbol{T} - \mathbf{1}\bar{\boldsymbol{t}}^T$$
$$\hat{\boldsymbol{\Phi}} \triangleq \boldsymbol{\Phi} - \mathbf{1}\bar{\boldsymbol{\phi}}^T$$

where

$$\bar{\boldsymbol{t}} = \frac{1}{n}\sum_{i=1}^{n} \boldsymbol{t}_i = \frac{1}{n}\boldsymbol{T}^T\mathbf{1}$$
$$\bar{\boldsymbol{\phi}} = \frac{1}{n}\sum_{i=1}^{n} \boldsymbol{\phi}(\boldsymbol{x}_i) = \frac{1}{n}\boldsymbol{\Phi}^T\mathbf{1}$$

are the mean values of the targets and the basis function outputs respectively. The solution for $\boldsymbol{w}_0$ is

$$\boldsymbol{w}_0 = \bar{\boldsymbol{t}} - \boldsymbol{W}\bar{\boldsymbol{\phi}}$$

Therefore, we may solve for the weights $\boldsymbol{W}$ using a linear method such as a singular value decomposition. However, we must use a nonlinear optimisation scheme to determine the parameters $\{\mu\}$.

3. In a classification problem, in which we have the 1-from-C coding for the class labels (that is, if $\boldsymbol{x}_i \in \omega_j$ then we have the target $\boldsymbol{t}_i = (0, 0, \dots, 0, 1, 0, \dots, 0)^T$, where the 1 is in the jth position), then using the pseudo-inverse solution (5.4) for the final layer weights gives the property that the components of $\boldsymbol{g}$ sum to unity for any data sample $\boldsymbol{x}$, i.e.

$$\sum_i g_i = 1$$

where

$$\boldsymbol{g} = \hat{\boldsymbol{T}}^T (\hat{\boldsymbol{\Phi}}^T)^{\dagger} \boldsymbol{\Phi}^T + \boldsymbol{w}_0 \tag{5.5}$$

The values of $\boldsymbol{g}$ are not constrained to be positive. One way to ensure positivity is to transform the values g_i by replacing g_i by $\exp(-g_i)/\sum_j \exp(-g_j)$. This forms the basis of the generalised logistic model (sometimes called *softmax*).

4. If the parameters $\{\boldsymbol{W}, \boldsymbol{w}_0\}$ are chosen to minimise the least squares error, then it may be shown that the parameters $\{\mu\}$ that minimise the least mean squared error maximise the feature extraction criterion (Lowe and Webb, 1991)

$$\mathrm{Tr}\{\boldsymbol{S}_B \boldsymbol{S}_T^{\dagger}\} \tag{5.6}$$

where $\boldsymbol{S}_B$ and $\boldsymbol{S}_T$ are defined as

$$\boldsymbol{S}_B \stackrel{\triangle}{=} \frac{1}{n^2}\hat{\boldsymbol{\Phi}}^T\hat{\boldsymbol{T}}\hat{\boldsymbol{T}}^T\hat{\boldsymbol{\Phi}}$$

$$\boldsymbol{S}_T \stackrel{\triangle}{=} \frac{1}{n}\hat{\boldsymbol{\Phi}}^T\hat{\boldsymbol{\Phi}}$$

The matrices $\boldsymbol{S}_T$ and $\boldsymbol{S}_B$ have the interpretation of being the total and between-class covariance matrices of the patterns $\boldsymbol{x}_i$ in the space spanned by the outputs of the nonlinear transformations, ϕ_i. Precise interpretations depend on the specific target coding schemes. Thus, the optimal method of solution of such a classifier is to find a nonlinear transformation into the space spanned by the nonlinear functions ϕ_i such that the patterns in different classes are somehow maximally separated (this information is contained in the between-class covariance matrix), while still maintaining an overall total normalisation (through the total covariance matrix). The form of the criterion (5.6) is independent of the transformation from the data space to the space of hidden unit outputs; i.e., it is not dependent on the form of the nonlinear functions ϕ, but is a property of the least mean square solution for the final layer. The expression (5.6) is also related to optimisation criteria used in clustering (see Chapter 10).

Incorporating priors and costs

A more general form of (5.3) is a weighted error function

$$\begin{aligned} E &= \sum_{i=1}^{n} d_i |\boldsymbol{t}_i - \boldsymbol{g}(\boldsymbol{x}_i)|^2 \\ &= \|(-\boldsymbol{T}^T + \boldsymbol{W}\boldsymbol{\Phi}^T + \boldsymbol{w}_0\boldsymbol{1}^T)\boldsymbol{D}\|^2 \end{aligned} \tag{5.7}$$

where the ith pattern is weighted by the real factor d_i, and $\boldsymbol{D}$ is diagonal with $\boldsymbol{D}_{ii} = \sqrt{d_i}$. Three different codings for the target matrix $\boldsymbol{T}$ and the weighting matrix $\boldsymbol{D}$ are described.

1. Cost-weighted target coding In a classification problem, with C classes, choose a uniform weighting for each pattern in the training set ($d_k = 1, k = 1, \ldots, n$) and employ a target coding scheme which, for a pattern in class ω_j, takes as the target vector the vector of losses $\boldsymbol{\lambda}_j$, whose ith component is the cost of assigning to class ω_i a pattern that belongs to class ω_j. The optimal discriminant vector is

$$\boldsymbol{\rho}(\boldsymbol{x}) = \sum_{j=1}^{C} \boldsymbol{\lambda}_j p(\omega_j|\boldsymbol{x})$$

which is the Bayes conditional risk vector (see Chapter 4).

If we make a decision by assigning a pattern to the class for which the classifier outputs are closest to the targets (minimum distance rule) we have: assign $\boldsymbol{x}$ to class ω_i if

$$\boldsymbol{\lambda}_i^T\boldsymbol{\lambda}_i - 2\boldsymbol{o}^T\boldsymbol{\lambda}_i \le \boldsymbol{\lambda}_j^T\boldsymbol{\lambda}_j - 2\boldsymbol{o}^T\boldsymbol{\lambda}_j, \quad j = 1, \ldots, C$$

where $\boldsymbol{o}$ is the output vector (4.50). Generally, this is not the same as the rule in which $\boldsymbol{x}$ is assigned to the class corresponding to the smallest value of $\boldsymbol{o}$ (treating $\boldsymbol{o}$ as an approximation to $\boldsymbol{\rho}$ and making a minimum-cost decision).

2. Prior-weighted patterns In this case, each pattern in the training set is weighted according to the *a priori* probabilities of class membership and the number in that class as

$$d_i = \frac{P_k}{n_k/n} \quad \text{for pattern } i \text{ in class } \omega_k$$

where P_k is the assumed known class probability (derived from knowledge regarding the relative expected class importance, or frequency of occurrence in operation) and n_k is the number of patterns in class ω_k in the training set.

The weighting above would be used in situations where the expected test conditions differ from the conditions described by the training data by the expected proportions in the classes. This may be a result of population drift (see Chapter 1), or when there are limited data available for training. The above weighting has been used by Munro *et al.* (1996) for learning low-probability events in order to reduce the size of the training set.

3. Cluster-weighted patterns The computation time for many neural network training schemes increases with the training set size. *Clustering* (see Chapter 10) is one means of finding a reduced set of prototypes that characterises the training data set. There are many ways in which clustering may be used to preprocess the data – it could be applied to classes separately, or to the whole training set. For example, when applied to each class separately, the patterns for that class are replaced by the cluster means, with d_i for the new patterns set proportional to the number in the cluster. When applied to the whole data set, if a cluster contains all members of the same class, those patterns are replaced in the data set by the cluster mean and d_i is set to the number of patterns in the cluster. If a cluster contains members of different classes, all patterns are retained.

Regularisation

Too many parameters in a model may lead to *over-fitting* of the data by the model and poor generalisation performance (see Chapter 1). One means of smoothing the model fit is to penalise the sum-squared error. Thus, we modify the squared error measure (5.3) and minimise

$$E = \sum_{i=1}^{n} |\boldsymbol{t}_i - \boldsymbol{g}(\boldsymbol{x}_i)|^2 + \alpha \int F(\boldsymbol{g}(\boldsymbol{x}))\, d\boldsymbol{x} \tag{5.8}$$

where α is a *regularisation* parameter and F is a function of the complexity of the model. For example, in a univariate curve-fitting problem, a popular choice for $F(\boldsymbol{g})$ is $\partial^2 \boldsymbol{g}/\partial x^2$, the second derivative of the fitting function $\boldsymbol{g}$. In this case, the solution for $\boldsymbol{g}$ that minimises (5.8) is a *cubic spline* (Green and Silverman, 1994).

In the neural network literature, a penalising term of the form

$$\alpha \sum_i \tilde{w}_i^2$$

is often used, where the summation is over all adjustable network parameters, $\tilde{w}$ (see Sections 5.3 and 6.2). This procedure is termed *weight decay*.

For the generalised linear model (5.2) we have seen that the penalised error is taken to be (see Chapter 4)

$$E = \| -\hat{\boldsymbol{T}}^T + \boldsymbol{W}\hat{\boldsymbol{\Phi}}^T \|^2 + \alpha \|\boldsymbol{W}\|^2 \tag{5.9}$$

expressed in terms of the zero-mean matrices $\hat{\boldsymbol{T}}$ and $\hat{\boldsymbol{\Phi}}$; and α is termed the *ridge parameter* (see Chapter 4). The solution for $\boldsymbol{W}$ that minimises E is

$$\boldsymbol{W} = \hat{\boldsymbol{T}}^T \hat{\boldsymbol{\Phi}} (\hat{\boldsymbol{\Phi}}^T \hat{\boldsymbol{\Phi}} + \alpha \boldsymbol{I}_m)^{-1}$$

This procedure has been used to regularise radial basis function networks (see Section 5.3).

5.2.2 Maximum likelihood

An alternative approach to the least squared error measure is to assume a parametric form for the class distributions and to use a maximum likelihood procedure. In the multiclass case, the basic assumption for the *generalised logistic discrimination* is

$$\log\left(\frac{p(\boldsymbol{x}|\omega_s)}{p(\boldsymbol{x}|\omega_C)}\right) = \beta_{s0} + \boldsymbol{\beta}_s^T \boldsymbol{\phi}(\boldsymbol{x}), \quad s = 1, \ldots, C-1 \tag{5.10}$$

where $\boldsymbol{\phi}(\boldsymbol{x})$ is a nonlinear function of the variables $\boldsymbol{x}$, with parameters $\{\mu\}$; that is, the log-likelihood ratio is a linear combination of the nonlinear functions, ϕ. The posterior probabilities are of the form (see Chapter 4)

$$\begin{aligned} p(\omega_s|\boldsymbol{x}) &= \frac{\exp(\beta'_{s0} + \boldsymbol{\beta}_s^T \boldsymbol{\phi}(\boldsymbol{x}))}{1 + \sum_{j=1}^{C-1} \exp(\beta'_{j0} + \boldsymbol{\beta}_j^T \boldsymbol{\phi}(\boldsymbol{x}))}, \quad s = 1, \ldots, C-1 \\ p(\omega_C|\boldsymbol{x}) &= \frac{1}{1 + \sum_{j=1}^{C-1} \exp(\beta'_{j0} + \boldsymbol{\beta}_j^T \boldsymbol{\phi}(\boldsymbol{x}))} \end{aligned} \tag{5.11}$$

where $\beta'_{s0} = \beta_{s0} + \log(p(\omega_s)/p(\omega_C))$. Discrimination depends solely on the $C-1$ functions $\beta'_{s0} + \boldsymbol{\beta}_s^T \phi(\boldsymbol{x})$ $(s = 1, \ldots, C-1)$ with the decision:

$$\text{assign } \boldsymbol{x} \text{ to class } \omega_j \text{ if } \max_{s=1,\ldots,C-1} \beta'_{s0} + \boldsymbol{\beta}_s^T \boldsymbol{\phi}(\boldsymbol{x}) = \beta'_{j0} + \boldsymbol{\beta}_j^T \boldsymbol{\phi}(\boldsymbol{x}) > 0$$

else assign $\boldsymbol{x}$ to class ω_C.

For a data set $\{\boldsymbol{x}_1, \ldots, \boldsymbol{x}_n\}$, the likelihood of the observations is given by:

$$L = \prod_{i=1}^{C} \prod_{\boldsymbol{x}_r \in \omega_i} p(\boldsymbol{x}_r|\omega_i) \tag{5.12}$$

where $\boldsymbol{x}_r$ is the rth pattern of class ω_i. The parameters of the model (in this case, the β terms and $\{\mu\}$, the set of parameters on which ϕ depends) may be determined using a maximum likelihood procedure (as in logistic discrimination; see Chapter 4). Equation (5.12) may be written

$$L = \left(\frac{1}{\prod_{i=1}^{C}[p(\omega_i)]^{n_i}}\right)\left(\prod_{\text{all } \boldsymbol{x}} p(\boldsymbol{x})\right)\prod_{i=1}^{C}\prod_{\boldsymbol{x}_r \in \omega_i} p(\omega_i|\boldsymbol{x}_r)$$

and assuming that the factor

$$\left(\frac{1}{\prod_{i=1}^{C}[p(\omega_i)]^{n_i}}\right)\prod_{\text{all } \boldsymbol{x}} p(\boldsymbol{x}) \tag{5.13}$$

is independent of the model parameters (the assumption that we are making is on the log-likelihood ratio – we are free to choose $p(\boldsymbol{x})$; see Chapter 4), then maximising L is equivalent to maximising L' given by

$$L' = \prod_{i=1}^{C}\prod_{\boldsymbol{x}_r \in \omega_i} p(\omega_i|\boldsymbol{x}_r) \tag{5.14}$$

Maximisation of the likelihood is achieved through the use of some form of numerical optimisation scheme, such as conjugate gradient methods or quasi-Newton procedures (see Press *et al.*, 1992).

The parameters of the generalised logistic model may also be found using a least squared error procedure by minimising

$$\sum_{i=1}^{n} |(\boldsymbol{t}_i - \boldsymbol{p}(\omega_{j(i)}|\boldsymbol{x}_i))|^2 \tag{5.15}$$

where $\boldsymbol{t}_i = (0, 0, \ldots, 0, 1, 0, \ldots, 0)^T$ (the 1 being in the jth position) for $\boldsymbol{x}_i \in \omega_j$ and $\boldsymbol{p}$ is the model for the vector of *a posteriori* probabilities given by (5.11). We know from Section 5.2.1 that $\boldsymbol{p}$ is asymptotically a minimum variance approximation to the Bayes discriminant function.

5.2.3 Entropy

Another optimisation criterion used for classification problems is the *relative* or *cross-entropy*. If $q(\omega|\boldsymbol{x})$ is the true posterior distribution and $p(\omega|\boldsymbol{x})$ is the approximation produced by a model, then the entropy measure to be minimised is

$$G = \sum_{i=1}^{C} \mathrm{E}\left[q(\omega_i|\boldsymbol{x})\log\left(\frac{q(\omega_i|\boldsymbol{x})}{p(\omega_i|\boldsymbol{x})}\right)\right] \tag{5.16}$$

where the expectation is with respect to the unconditional distribution of $\boldsymbol{x}$. If the training data are representative of this distribution, then we may approximate the expectation integral in (5.16) by a finite sum over the training set,

$$G = \sum_{i=1}^{C} \sum_{\boldsymbol{x}_r \in \omega_i} \log\left(\frac{q(\omega_i|\boldsymbol{x}_r)}{p(\omega_i|\boldsymbol{x}_r)}\right) \tag{5.17}$$

and minimising (5.17) is equivalent to maximising (cf. (5.12))

$$\hat{G} = \prod_{i=1}^{C} \prod_{\boldsymbol{x}_r \in \omega_i} p(\omega_i|\boldsymbol{x}_r)$$

(compare with (5.14); see the exercises at the end of the chapter). If our approximation for p is of the form (5.11), for example, then we have the generalised logistic discrimination model. Thus, minimising the cross-entropy is equivalent to maximum likelihood estimation.

5.3 Radial basis functions

5.3.1 Introduction

Radial basis functions (RBFs) were originally proposed in the functional interpolation literature (see the review by Powell, 1987; Lowe, 1995a) and first used for discrimination by Broomhead and Lowe (1988). However, RBFs have been around in one form or another for a very long time. They are very closely related to kernel methods for density estimation and regression developed in the statistics literature (see Chapter 3) and to normal mixture models (Chapter 2).

The RBF may be described mathematically as a linear combination of radially symmetric nonlinear basis functions. The RBF provides a transformation of a pattern $\boldsymbol{x} \in \mathbb{R}^p$ to an n'-dimensional output space according to[1]

$$g_j(\boldsymbol{x}) = \sum_{i=1}^{m} w_{ji}\phi_i(|\boldsymbol{x} - \boldsymbol{\mu}_i|) + w_{j0}, \quad j = 1, \ldots, n' \tag{5.18}$$

The parameters w_{ji} are often referred to as the *weights*; w_{j0} is the *bias* and the vectors $\boldsymbol{\mu}_i$ are the *centres*. The model (5.18) is very similar to the kernel density model described in Chapter 3 in which $n' = 1$, $w_{10} = 0$, and the number of centres m is taken to be equal to the number of data samples n, with $\boldsymbol{\mu}_i = \boldsymbol{x}_i$ (a centre at each data sample); $w_{ji} = 1/n$ and ϕ is one of the kernels given in Chapter 3, sometimes referred to as the *activation function* in the neural network literature.

In the case of exact interpolation, a basis function is also positioned at each data point ($m = n$). Suppose we seek a mapping g from $\mathbb{R}^p$ to $\mathbb{R}$ (taking $n' = 1$) through the points $(\boldsymbol{x}_i, t_i)$ which satisfies the condition that $g(\boldsymbol{x}_i) = t_i$; that is, under the assumed model (5.18) (and ignoring the bias), we seek a solution for $\boldsymbol{w} = (w_1, \ldots, w_n)^T$ that satisfies

$$\boldsymbol{t} = \boldsymbol{\Phi w}$$

[1]Here we use n' to denote the dimensionality of the output space. In a classification problem, we usually have $n' = C$, the number of classes.

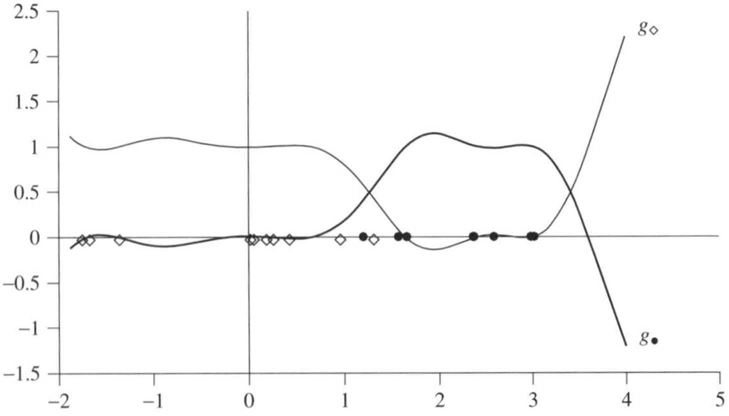

Figure 5.1 Discriminant functions constructed using a radial basis function network of normal kernels

where $\boldsymbol{t} = (t_1, \ldots, t_n)^T$ and $\boldsymbol{\Phi}$ is the $n \times n$ matrix with (i, j)th element $\phi(|\boldsymbol{x}_i - \boldsymbol{x}_j|)$, for a nonlinear function ϕ. For a large class of functions Micchelli (1986) has shown that the inverse of $\boldsymbol{\Phi}$ exists and the solution for $\boldsymbol{w}$ is given by

$$\boldsymbol{w} = \boldsymbol{\Phi}^{-1}\boldsymbol{t}$$

Exact interpolation is not a good thing to do in general. It leads to poor generalisation performance in many pattern recognition problems (see Chapter 1). The fitting function can be highly oscillatory. Therefore, we usually take $m < n$.

An often-cited advantage of the RBF model is its simplicity. Once the forms of the nonlinearity have been specified and the centres determined, we have a linear model whose parameters can be easily obtained by a least squares procedure, or indeed any appropriate optimisation procedure such as those described in Chapter 4 or Section 5.2.

For supervised classification, the RBF is used to construct a discriminant function for each class. Figure 5.1 illustrates a one-dimensional example. Data drawn from two univariate normal distributions of unit variance and means of 0.0 and 2.0 are plotted. Normal kernels are positioned over centres selected from these data. The weights w_{ji} are determined using a least squares procedure and the discriminant functions $g_\bullet$ and $g_\diamond$ plotted. Thus, a linear combination of 'blob'-shaped functions is used to produce two functions that can be used as the basis for discrimination.

5.3.2 Motivation

The RBF model may be motivated from several perspectives. We present the first two in a discrimination context and then one from a regression perspective.

Kernel discriminant analysis

Consider the multivariate kernel density estimate (see Chapter 3)

$$p(\boldsymbol{x}) = \frac{1}{nh^p}\sum_{i=1}^{n} K\left(\frac{1}{h}(\boldsymbol{x} - \boldsymbol{x}_i)\right)$$

where $K(\boldsymbol{x})$ is defined for p-dimensional $\boldsymbol{x}$ satisfying $\int_{\mathbb{R}^p} K(\boldsymbol{x})\,d\boldsymbol{x} = 1$. Suppose we have a set of samples $\boldsymbol{x}_i$ $(i = 1, \ldots, n)$ with n_j samples in class ω_j $(j = 1, \ldots, C)$. If we construct a density estimate for each class then the posterior probability of class membership can be written

$$p(\omega_j|\boldsymbol{x}) = \frac{p(\omega_j)}{p(\boldsymbol{x})}\frac{1}{n_j h^p}\sum_{i=1}^{n} z_{ji} K\left(\frac{1}{h}(\boldsymbol{x} - \boldsymbol{x}_i)\right) \tag{5.19}$$

where $p(\omega_j)$ is the prior probability of class ω_j and $z_{ji} = 1$ if $\boldsymbol{x}_i \in$ class ω_j, 0 otherwise. Thus, discrimination is based on a model of the form

$$\sum_{i=1}^{n} w_{ji}\phi_i(\boldsymbol{x} - \boldsymbol{x}_i) \tag{5.20}$$

where $\phi_i(\boldsymbol{x} - \boldsymbol{x}_i) = K((\boldsymbol{x} - \boldsymbol{x}_i)/h)$ and

$$w_{ji} = \frac{p(\omega_j)}{n_j} z_{ji} \tag{5.21}$$

(We neglect the term $p(\boldsymbol{x})h^p$ in the denominator of (5.19) since it is independent of j.) Equation (5.20) is of the form of an RBF with a centre at each data point and weights determined by class priors (equation (5.21)).

Mixture models

In discriminant analysis by Gaussian mixtures (Chapter 2), the class-conditional density for class ω_j is expressed as

$$p(\boldsymbol{x}|\omega_j) = \sum_{r=1}^{R_j} \pi_{jr} p(\boldsymbol{x}|\theta_{jr})$$

where class ω_j has R_j subgroups, π_{jr} are the mixing proportions ($\sum_{r=1}^{R_j} \pi_{jr} = 1$) and $p(\boldsymbol{x}|\theta_{jr})$ is the density of the rth subgroup of class ω_j evaluated at $\boldsymbol{x}$; (θ_{jr} denote the subgroup parameters: for normal mixtures, the mean and covariance matrix). The posterior probabilities of class membership are

$$p(\omega_j|\boldsymbol{x}) = \frac{p(\omega_j)}{p(\boldsymbol{x})}\sum_{r=1}^{R_j} \pi_{jr} p(\boldsymbol{x}|\theta_{jr})$$

where $p(\omega_j)$ is the prior probability of class ω_j.

For normal mixture components, $p(\boldsymbol{x}|\theta_{jr})$, with means $\boldsymbol{\mu}_{jr}$, $j = 1, \ldots, C$; $r = 1, \ldots, R_j$ and common diagonal covariance matrices, $\sigma^2 \boldsymbol{I}$, we have discriminant functions of the form (5.18) with – basis functions, where $m = \sum_{j=1}^{C} R_j$, weights set by the mixing proportions and class priors, and basis functions centred at the $\boldsymbol{\mu}_{jr}$.

Regularisation

Suppose that we have a data set $\{(\boldsymbol{x}_i, t_i), i = 1, \ldots, n\}$ where $\boldsymbol{x}_i \in \mathbb{R}^d$, and we seek a smooth surface g,

$$t_i = g(\boldsymbol{x}_i) + \text{error}$$

One such approach is to minimise the penalised sum of squares

$$S = \sum_{i=1}^{n} (t_i - g(\boldsymbol{x}_i))^2 + \alpha J(g)$$

where α is a regularisation or *roughness* (Green and Silverman, 1994) parameter and $J(g)$ is a penalty term that measures how 'rough' the fitted surface is and has the effect of penalising 'wiggly' surfaces (see Chapter 4 and Section 5.2.1).

A popular choice for J is one based on mth derivatives. Taking J as

$$J(g) = \int_{\mathbb{R}^d} \sum \frac{m!}{\nu_1! \dots \nu_d!} \left(\frac{\partial^m g}{\partial x_1^{\nu_1} \dots \partial x_d^{\nu_d}} \right)^2 dx_1 \dots dx_d$$

where the summation is over all non-negative integers $\nu_1, \nu_2, \dots, \nu_d$ such that $\nu_1 + \nu_2 + \cdots + \nu_d = m$, results in a penalty invariant under translations and rotations of the coordinate system (Green and Silverman, 1994).

Defining $\eta_{md}(r)$ by

$$\eta_{md}(r) = \begin{cases} \theta r^{2m-d} \log(r) & \text{if } d \text{ is even} \\ \theta r^{2m-d} & \text{if } d \text{ is odd} \end{cases}$$

and the constant of proportionality, θ, by

$$\theta = \begin{cases} (-1)^{m+1+d/2} 2^{1-2m} \pi^{-d/2} \dfrac{1}{(m-1)!} \dfrac{1}{(m-d/2)!} & \text{if } d \text{ is even} \\ \Gamma(d/2 - m) 2^{-2m} \pi^{-d/2} \dfrac{1}{(m-1)!} & \text{if } d \text{ is odd} \end{cases}$$

then (under certain conditions on the points $\boldsymbol{x}_i$ and m) the function g minimising $J(g)$ is a *natural thin-plate spline*. This is a function of the form

$$g(\boldsymbol{x}) = \sum_{i=1}^{n} b_i \eta_{md}(|\boldsymbol{x} - \boldsymbol{x}_i|) + \sum_{j=1}^{M} a_j \gamma_j(\boldsymbol{x})$$

where

$$M = \binom{m+d-1}{d}$$

and $\{\gamma_j, j = 1, \dots, M\}$ is a set of linearly independent polynomials spanning the M-dimensional space of polynomials in $\mathbb{R}^d$ of degree less than m. The coefficients $\{a_j, j = 1, \dots, M\}$, $\{b_i, i = 1, \dots, n\}$ satisfy certain constraints (see Green and Silverman, 1994, for further details). Thus, the minimising function contains radially symmetric terms, η, and polynomials.

An alternative derivation based on a different form for the penalty terms leading to Gaussian RBFs is provided by Bishop (1995).

5.3.3 Specifying the model

The basic RBF model is of the form

$$g_j(\boldsymbol{x}) = \sum_{i=1}^{m} w_{ji}\phi\left(\frac{|\boldsymbol{x} - \boldsymbol{\mu}_i|}{h}\right) + w_{j0}, \quad j = 1, \ldots, n'$$

that is, all the basis functions are of the same functional form ($\phi_i = \phi$) and a scaling parameter h has been introduced. In this model, there are five quantities to prescribe or to determine from the data:

the number of basis functions, m;
the form of the basis function, ϕ;
the smoothing parameter, h;
the positions of the centres, $\boldsymbol{\mu}_i$;
the weights, w_{ji}, and bias, w_{j0}.

There are three main stages in constructing an RBF model:

1. Specify the nonlinear functions, ϕ. To a large extent, this is independent of the data and the problem (though the parameters of these functions are not).

2. Determine the number and positions of the centres, and the kernel widths.

3. Determine the weights of the RBF. These values are data-dependent.

Stages 2 and 3 above are not necessarily carried out independently. Let us consider each in turn.

Specifying the functional form

The ideal choice of basis function is a matter for debate. However, although certain types of problem may be matched inappropriately to certain forms of nonlinearity, the actual form of the nonlinearity is relatively unimportant (as in kernel density estimation) compared to the number and the positions of the centres. Typical forms of nonlinearity are given in Table 5.1. Note that some RBF nonlinearities produce smooth approximations, in that the fitting function and its derivatives are continuous. Others (for example, $z\log(z)$ and $\exp(-z)$) have discontinuous gradients.

The two most popular forms are the thin-plate spline, $\phi(z) = z^2\log(z)$, and the normal or Gaussian form, $\phi(z) = \exp(-z^2)$. Each of these functions may be motivated from different perspectives (see Section 5.3.2): the normal form from a kernel regression and kernel density estimation point of view and the thin-plate spline from curve fitting (Lowe, 1995a). Indeed, each may be shown to be optimal under certain conditions: in fitting data in which there is normally distributed noise on the inputs, the normal form is the optimal basis function in a least squares sense (Webb, 1994); in fitting a surface through a set of points and using a roughness penalty, the *natural* thin-plate spline is the solution (Duchon, 1976; Meinguet, 1979).

Table 5.1 Radial basis function nonlinearities

Nonlinearity	Mathematical form $\phi(z)$, $z = \|\boldsymbol{x} - \boldsymbol{\mu}\|/h$
Gaussian	$\exp(-z^2)$
Exponential	$\exp(-z)$
Quadratic	$z^2 + \alpha z + \beta$
Inverse quadratic	$1/[1 + z^2]$
Thin-plate spline	$z^\alpha \log(z)$
Trigonometric	$\sin(z)$

These functions are very different: one is compact and positive, the second diverges at infinity and is negative over a region. However, in practice, this difference is to some extent superficial since, for training purposes, the function ϕ need only be defined in the feature space over the range $[s_{min}, s_{max}]$, where

$$\begin{aligned} s_{max} &= \max_{i,j} |\boldsymbol{x}_i - \boldsymbol{\mu}_j| \\ s_{min} &= \min_{i,j} |\boldsymbol{x}_i - \boldsymbol{\mu}_j| \end{aligned}$$

and therefore ϕ may be redefined over this region as $\hat{\phi}(s)$ given by

$$\hat{\phi} \longleftarrow \frac{\phi(s/h) - \phi_{min}}{\phi_{max} - \phi_{min}}$$

where ϕ_{max} and ϕ_{min} are the maximum and minimum values of ϕ over $[0, s_{max}]$ (taking $s_{min} = 0$) respectively ($0 \leq \hat{\phi} \leq 1$) and $s = |\boldsymbol{x} - \boldsymbol{\mu}_j|$ is the distance in the feature space. Scaling of ϕ may simply be compensated for by adjustment of weights $\{w_{ji}, i = 1, \ldots, m\}$ and bias w_{j0}, $j = 1, \ldots, n'$. The fitting function is unaltered.

Figure 5.2 illustrates the normalised form for the normal nonlinearity for several values of the smoothing parameter, h. For the Gaussian basis function, we see that there is little change in the normalised form for h/s_{max} greater than about 2. As $h \to \infty$, the normalised form for the nonlinearity tends to the quadratic

$$\hat{\phi}_\infty(s) \stackrel{\triangle}{=} 1 - \frac{s^2}{s_{max}^2} \qquad (5.22)$$

Thus, asymptotically, the normalised basis function is independent of h.

For large h/s_{max} (greater than about 2), changes in the value of h may be compensated for by adjustments of the weights, w_{ji}, and the radial basis function is a quadratic function of the input variables. For smaller values of h, the normalised function tends to the Gaussian form, thus allowing small-scale variation in the fitting function.

In some cases, particularly for the Gaussian kernel, it is important to choose an ‘appropriate’ value for the smoothing parameters in order to fit the structure in the data.

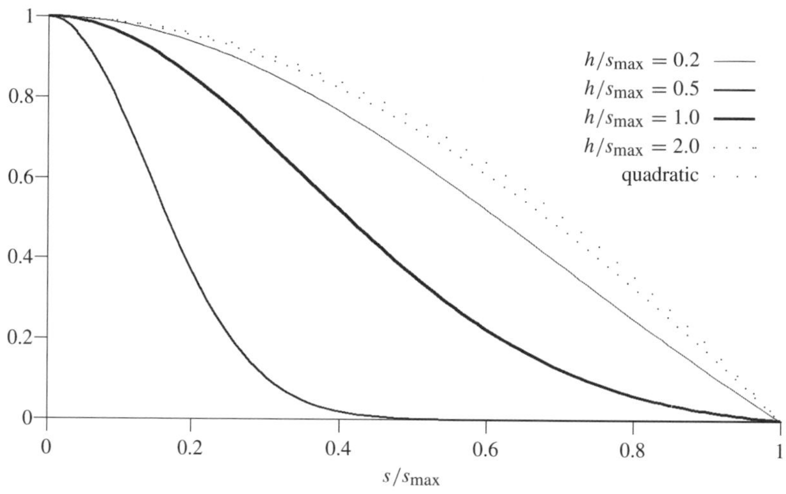

Figure 5.2 Normalised Gaussian basis functions, $\hat{\phi}(s)$, for $h/s_{max} = 0.2, 0.5, 1.0, 2.0$ and the limiting quadratic

This is a compromise between the one extreme of fitting noise in the data and the other of being unable to model the structure. As in kernel density estimation, it is more important to choose appropriate values for the smoothing parameters than the functional form. See Chapter 3 for a discussion on the choice of smoothing parameter in kernel density estimation. There is some limited empirical evidence to suggest that thin-plate splines fit data better in high-dimensional settings (Lowe, 1995b).

Centres and weights

The values of centres and weights may be found by minimising a suitable criterion (for example, least squares) using a nonlinear optimisation scheme. However, it is more usual to position the centres first, and then to calculate the weights using one of the optimisation schemes appropriate for linear models. Of course, this means that the optimisation criterion will not be at an extremum with respect to the positions of the centres, but in practice this does not matter.

Positioning of the centres can have a major effect on the performance of an RBF for discrimination and interpolation. In an interpolation problem, more centres should be positioned in regions of high curvature. In a discrimination problem, more centres should be positioned near class boundaries. There are several schemes commonly employed.

1. Select from data set–random selection Select randomly from the data set. We would expect there to be more centres in regions where the density is greater. A consequence of this is that sparse regions may not be 'covered' by the RBF unless the smoothing parameter is adjusted. Random selection is an approach that is commonly used. An advantage is that it is fast. A disadvantage is the failure to take into account the fitting function, or the class labels in a supervised classification problem; that is, it is an unsupervised placement scheme and may not provide the best solution for a mapping problem.

2. Clustering approach The values for the centres obtained by the previous approach could be used as seeds for a k-means clustering algorithm, thus giving centres for RBFs as cluster centres. The k-means algorithm seeks to partition the data into k groups or *clusters* so that the within-group sum of squares is minimised; that is, it seeks the cluster centres $\{\boldsymbol{\mu}_j, j = 1, \ldots, k\}$ that minimise

$$\sum_{j=1}^{k} S_j$$

where the within-group sum of squares for group j is

$$S_j = \sum_{i=1}^{n} z_{ji} |\boldsymbol{x}_i - \boldsymbol{\mu}_j|^2$$

in which $z_{ji} = 1$ if $\boldsymbol{x}_i$ is in group j (of size $n_j = \sum_{i=1}^{n} z_{ji}$) and zero otherwise; $\boldsymbol{\mu}_j$ is the mean of group j,

$$\boldsymbol{\mu}_j = \frac{1}{n_j} \sum_{i=1}^{n} z_{ji} \boldsymbol{x}_i$$

Algorithms for computing the cluster centres $\boldsymbol{\mu}_j$ using k-means are described in Chapter 10.

Alternatively, any other clustering approach could be used: either pattern based, or dissimilarity matrix based by first forming a dissimilarity matrix (see Chapter 10).

3. Normal mixture model If we are using normal nonlinearities, then it seems sensible to use as centres (and indeed widths) the parameters resulting from a normal mixture model of the underlying distribution $p(\boldsymbol{x})$ (see Chapter 2). We model the distribution as a mixture of normal models

$$p(\boldsymbol{x}) = \sum_{j=1}^{g} \pi_j p(\boldsymbol{x}|\boldsymbol{\mu}_j, h)$$

where

$$p(\boldsymbol{x}|\boldsymbol{\mu}_j, h) = \frac{1}{(2\pi)^{p/2} h^p} \exp\left\{ -\frac{1}{2h^2} (\boldsymbol{x} - \boldsymbol{\mu}_j)^T (\boldsymbol{x} - \boldsymbol{\mu}_j) \right\}$$

The values of h, π_j and $\boldsymbol{\mu}_j$ may be determined using the EM algorithm to maximise the likelihood (see Chapter 2). The weights π_j are ignored and the resulting normal basis functions, defined by h and $\boldsymbol{\mu}_i$, are used in the RBF model.

4. k-nearest-neighbour initialisation The approaches described above use the input data only to define the centres. Class labels, or the values of the dependent variables in a regression problem, are not used. Thus, unlabelled data from the same distribution as the training data may be used in the centre initialisation process. We now consider some supervised techniques. In Chapter 3 we found that in the k-nearest-neighbour classifier not all data samples are required to define the decision boundary. We may use an *editing* procedure to remove those prototypes that contribute to the error (with the hope of improving the generalisation performance) and a *condensing* procedure to reduce the number of samples needed to define the decision boundary. The prototypes remaining after editing and condensing may be retained as centres for an RBF classifier.

5. Orthogonal least squares The choice of RBF centres can be viewed as a problem in variable selection (see Chapter 9). Chen *et al.* (1991; see also Chen *et al.*, 1992) consider the complete set of data samples to be candidates for centres and construct a set of centres incrementally. Suppose that we have a set of $k-1$ centres, positioned over $k-1$ different data points. At the kth stage, the centre selected from the remaining $n-(k-1)$ data samples that reduces the prediction error the most is added to the set of centres. The number of centres is chosen as that for which an information criterion is minimised.

A naïve implementation of this method would solve for the network weights and evaluate equation (5.4) $n-(k-1)$ times at the kth stage. Chen *et al.* (1991) propose a scheme that reduces the computation based on an orthogonal least squares algorithm. We seek a solution for the $m \times n'$ matrix $\boldsymbol{W}$ that minimises[2]

$$\|\hat{\boldsymbol{T}} - \hat{\boldsymbol{\Phi}}\boldsymbol{W}^T\|^2$$

The matrix $\hat{\boldsymbol{\Phi}}$ is decomposed as $\hat{\boldsymbol{\Phi}} = \boldsymbol{V}\boldsymbol{A}$, where $\boldsymbol{A}$ is an $m \times m$ upper triangular matrix

$$\boldsymbol{A} = \begin{bmatrix} 1 & \alpha_{12} & \alpha_{23} & \dots & \alpha_{1m} \\ 0 & 1 & \alpha_{23} & \dots & \alpha_{2m} \\ 0 & 0 & \ddots & \ddots & \vdots \\ \vdots & \ddots & & & \vdots \\ 0 & \dots & & 0 & 1 \end{bmatrix}$$

and $\boldsymbol{V}$ is an $n \times m$ matrix $= [\boldsymbol{v}_1, \dots, \boldsymbol{v}_n]^T$ and $\boldsymbol{v}_i^T \boldsymbol{v}_j = 0$ if $i \neq j$. The error E is then given by

$$E = \|\hat{\boldsymbol{T}} - \boldsymbol{V}\boldsymbol{G}\|^2$$

where $\boldsymbol{G} = \boldsymbol{A}\boldsymbol{W}^T$. Using the minimum norm solution for $\boldsymbol{G}$ that minimises E, we may write (see the exercises)

$$\begin{aligned} E &= \text{Tr}\{\hat{\boldsymbol{T}}^T \hat{\boldsymbol{T}}\} - \text{Tr}\{\boldsymbol{G}^T \boldsymbol{V}^T \boldsymbol{V} \boldsymbol{G}\} \\ &= \text{Tr}\{\hat{\boldsymbol{T}}^T \hat{\boldsymbol{T}}\} - \sum_{j=1}^{m} |\boldsymbol{v}_j|^2 \left(\sum_{i=1}^{n'} g_{ji}^2 \right) \end{aligned} \tag{5.23}$$

The error reduction due to $\boldsymbol{v}_k$ is defined as

$$\text{err}_k \triangleq |\boldsymbol{v}_k|^2 \left(\sum_{i=1}^{n'} g_{ki}^2 \right)$$

[2]We work with the zero-mean matrices $\hat{\boldsymbol{T}}$ and $\hat{\boldsymbol{\Phi}}$.

where $\boldsymbol{G}_{ki} = g_{ki}$. Chen *et al.* (1991) propose an efficient algorithm for computing this error when an additional centre is introduced and terminate the algorithm using a criterion that balances the fitting error against complexity: specifically, the selection procedure is terminated when

$$n \log(\sigma_e^2) + k\chi$$

is a minimum, where χ is assigned a value 4 and σ_e^2 is the variance of the residuals for k centres. □

Having obtained centres, we now need to specify the smoothing parameters. These depend on the particular form we adopt for the nonlinearity. If it is normal, then the normal mixture approach will lead to values for the widths of the distributions naturally. The other approaches will necessitate a separate estimation procedure. Again, there are several heuristics which were discussed in Chapter 3 on kernel density estimation. Although they are suboptimal, they are fast to calculate. An alternative is to choose the smoothing parameter that minimises a cross-validation estimate of the sum-squared error.

We have not addressed the question of how to choose the *number* of centres (except as part of the orthogonal least squares centre selection procedure). This is very similar to many of the problems of model complexity discussed elsewhere in this book (see Chapter 11); for example, how many clusters are best, how many components in a normal mixture, how we determine intrinsic dimensionality, etc. It is not easy to answer. The number depends on several factors, including the amount and distribution of the data, the dimension and the form adopted for the nonlinearity. It is probably better to have many centres with limited complexity (single smoothing parameter) than an RBF with few centres and a complex form for the nonlinearity. There are several approaches to determining the number of centres. These include:

1. Using cross-validation. The cross-validation error (minimised over the smoothing parameter) is plotted as a function of the number of centres. The number of centres is chosen as that above which there is no appreciable decrease, or an increase in the cross-validation error (as used by Orr, 1995, in forward selection of centres).

2. Monitoring the performance on a separate test set. This is similar to the cross-validation procedure above, except that the error is evaluated on a separate test set.

3. Using an information complexity criterion. The sum-squared error is augmented by an additional term that penalises complexity (see for example, Chen *et al.*, 1991).

4. If we are using a normal mixture model to set the positions and widths of centres, we may use one of the methods for estimating the number of components in a normal mixture (see Chapter 2).

The final stage of determining the RBF is the calculation of the weights, either by optimising the squared error measure (5.3), the regularised error or the likelihood (5.12). Section 5.2 has described these procedures in some detail.

Most of the stages in the optimisation of an RBF use techniques described elsewhere in this book (for example, clustering/prototype selection, kernel methods, least squares or maximum likelihood optimisation). In many ways, a radial basis function is not new; all its constituent parts are widely used tools of pattern recognition. In Section 5.3.5 we put them together to derive a discrimination model based on normal nonlinearities.

5.3.4 Radial basis function properties

One of the properties of an RBF that has motivated its use in a wide range of applications both in functional approximation and in discrimination is that it is a universal approximator: it is possible (given certain conditions on the kernel function) to construct an RBF that approximates a given (integrable, bounded and continuous) function arbitrarily accurately (Park and Sandberg, 1993; Chen and Chen, 1995). This may require a very large number of centres. In most, if not all, practical applications the mapping we wish to approximate is defined by a finite set of data samples providing class-labelled data or, in an approximation problem, data samples and associated function values, and implemented in finite-precision arithmetic. Clearly, this limits the complexity of the model. We refer to Chapter 11 for a discussion of model order selection.

5.3.5 Simple radial basis function

We have now set up the machinery for implementing a simple RBF. The stages in a simple implementation are as follows.

1. Specify the functional form for the nonlinearity.
2. Prescribe the number of centres, m.
3. Determine the positions of the centres (for example, random selection or the k-means algorithm).
4. Determine the smoothing parameters (for example, simple heuristic or cross-validation).
5. Map the data to the space spanned by the outputs of the nonlinear functions; i.e. for a given data set $\boldsymbol{x}_i, i = 1, \ldots, n$, form the vectors $\boldsymbol{\phi}_i = \boldsymbol{\phi}(\boldsymbol{x}_i), i = 1, \ldots, n$.
6. Solve for the weights and the biases using (e.g.) least squares or maximum likelihood.
7. Calculate the final output on the train and test sets; classify the data if required.

The above is a simple prescription for an RBF network that uses unsupervised techniques for centre placement and width selection (and therefore is suboptimal). One of the often-quoted advantages of the RBF network is its simplicity – there is no nonlinear optimisation scheme required, in contrast to the multilayer perceptron classifier discussed in the following section. However, many of the sophisticated techniques for centre placement and width determination are more involved and increase the computational complexity of the model substantially. Nevertheless, the simple RBF can give acceptable performance for many applications.

5.3.6 Example application study

The problem The problem of source position estimation using measurements made on a radar focal-plane array using RBFs was treated by Webb and Garner (1999). This particular approach was motivated by a requirement for a compact integrated (hardware) implementation of a bearing estimator in a sensor focal plane (a solution was required that could readily be implemented in silicon on the same substrate as the focal-plane array).

Summary The problem is one of prediction rather than discrimination: given a set of training samples $\{(\boldsymbol{x}_i, \theta_i), i = 1, \ldots, n\}$, where $\boldsymbol{x}_i$ is a vector of measurements on the independent variables (array calibration measurements in this problem) and θ_i is the response variable (position), a predictor, f, was sought such that given a new measurement z, then $f(z)$ is a good estimate of the position of the source that gave rise to the measurement z. However, the problem differs from a standard regression problem in that there is noise on the measurement vector z, and this is similar to errors-in-variables models in statistics. Therefore, we seek a predictor that is robust to noise on the inputs.

The data The training data comprised detector outputs of an array of 12 detectors, positioned in the focal plane of a lens. Measurements were made on the detectors as the lens scanned across a microwave point source (thus providing measurements of the *point-spread function* of the lens. There were 3721 training samples measured at a signal-to-noise ratio (SNR) of about 45 dB. The test data were recorded for the source at specific positions over a range of lower SNR.

The model The model adopted is a standard RBF network with a Gaussian kernel with centres defined in the 12-dimensional space. The approach adopted to model parameter estimation was a standard least squares one. The problem may be regarded as one example from a wider class in discrimination and regression in which the expected operating conditions (the test conditions) differ from the training conditions in a known way. For example, in a discrimination problem, the class priors may differ considerably from the values estimated from the training data. In a least squares approach, this may be compensated for by modifying the sum-squared error criterion appropriately (see Section 5.2.1). Also, allowance for expected population drift may be made by modifying the error criterion. In the source position estimation problem, the training conditions are considered 'noiseless' (obtained through a calibration procedure) and the test conditions differ in that there is noise (of known variance) on the data. Again, this can be taken into account by modifying the sum-squared error criterion.

Training procedure An RBF predictor was designed with centres chosen using a k-means procedure. A ridge regression type solution was obtained for the weights (see Section 5.2.1 on regularisation), with the ridge parameter inversely proportional to an SNR term. Thus, there is no need to perform any search procedure to determine the ridge parameter. It can be set by measuring the SNR of the radar system. The theoretical development was validated by experimental results on a 12-element microwave focal-plane array in which two angle coordinates were estimated.

Results It was shown that it was possible to compensate for noisy test conditions by using a regularisation solution for the parameters, with regularisation parameter proportional to the inverse of the SNR.

5.3.7 Further developments

There have been many developments of the basic RBF model in the areas of RBF design, learning algorithms and Bayesian treatments.

Developments of the k-means approach to take account of the class labels of the data samples are described by Musavi *et al.* (1992), in which clusters contain samples from the same class, and Karayiannis and Wi (1997), in which a localised class-conditional variance is minimised as part of a network growing process.

Chang and Lippmann (1993) propose a supervised approach for allocating RBF centres near class boundaries. An alternative approach that chooses centres for a classification task in a supervised way based on the ideas of *support vectors* is described by Schölkopf *et al.* (1997) (see Chapter 4 and Section 5.4). In support vector learning of RBF networks with Gaussian basis functions, the separating surface obtained by a support vector machine (SVM) approach (that is, the decision boundary) is a linear combination of Gaussian functions centred at selected training points (the support vectors). The number and location of centres is automatically determined. In a comparative study reviewing several approaches to RBF training, Schwenker *et al.* (2001) find that the SVM learning approach is often superior on a classification task to the standard two-stage learning of RBFs (selecting or adapting the centres followed by calculating the weights).

The orthogonal least squares forward selection procedure has been developed to use a regularised error criterion by Chen *et al.* (1996) and Orr (1995), who uses a generalised cross-validation criterion as a stopping condition.

In the basic approach, all patterns are used at once in the calculation for the weights ('batch learning'). On-line learning methods have been developed (for example, Marinaro and Scarpetta, 2000). These enable the weights of the network to be updated sequentially according to the error computed on the last selected new example. This allows for possible temporal changes in the task being learned.

A Bayesian treatment that considers the number of basis functions and weights to be unknown has been developed by Holmes and Mallick (1998). A joint probability density function is defined over model dimension and model parameters. Using Markov chain Monte Carlo methods, inference is made by integrating over the model dimension and parameters.

5.3.8 Summary

Radial basis functions are simple to construct, easy to train and find a solution for the weights rapidly. They provide a very flexible model and give very good performance over a wide range of problems, both for discrimination and for functional approximation. The RBF model uses many of the standard pattern recognition building blocks (clustering and least squares optimisation, for example). There are many variants of the model (due to the choice of centre selection procedure, form of the nonlinear functions, procedure for determining the weights, and model selection method). This can make it difficult to draw meaningful conclusions about RBF performance over a range of studies on different applications, since the form of an RBF may vary from study to study.

The disadvantages of RBFs also apply to many, if not all, of the discrimination models covered in this book. That is, care must be taken not to construct a classifier that models noise in the data, or models the training set too well, which may give poor generalisation performance. Choosing a model of the appropriate complexity is very important, as we emphasise repeatedly in this book. Regularising the solution for the weights can improve generalisation performance. Model selection requirements add to the computational requirements of the model, and thus the often-claimed simplicity of RBF networks is perhaps overstated. Yet it should be said that a simple scheme can give good performance which, not unusually, exceeds that of many of the more 'traditional' statistical classifiers.

5.4 Nonlinear support vector machines

In Chapter 4 we introduced the support vector machine as a tool for finding the optimal separating hyperplane for linearly separable data and considered developments of the approach for situations when the data are not linearly separable. As we remarked in that chapter, the support vector algorithm may be applied in a transformed feature space, $\boldsymbol{\phi}(\boldsymbol{x})$, for some nonlinear function $\boldsymbol{\phi}$. Indeed, this is the principle behind many methods of pattern classification: transform the input features nonlinearly to a space in which linear methods may be applied (see also Chapter 1). We discuss this approach further in the context of SVMs.

For the binary classification problem, we seek a discriminant function of the form

$$g(\boldsymbol{x}) = \boldsymbol{w}^T \boldsymbol{\phi}(\boldsymbol{x}) + w_0$$

with decision rule

$$\boldsymbol{w}^T \boldsymbol{\phi}(\boldsymbol{x}) + w_0 \begin{cases} > & 0 \\ < & 0 \end{cases} \Rightarrow \boldsymbol{x} \in \begin{cases} \omega_1 \text{ with corresponding numeric value } y_i = +1 \\ \omega_2 \text{ with corresponding numeric value } y_i = -1 \end{cases}$$

The SVM procedure determines the maximum margin solution through the maximisation of a Lagrangian. The dual form of the Lagrangian (equation (4.30)) becomes

$$L_D = \sum_{i=1}^{n} \alpha_i - \frac{1}{2} \sum_{i=1}^{n} \sum_{j=1}^{n} \alpha_i \alpha_j y_i y_j \boldsymbol{\phi}^T(\boldsymbol{x}_i) \boldsymbol{\phi}(\boldsymbol{x}_j) \tag{5.24}$$

where $y_i = \pm 1, i = 1, \dots, n$, are class indicator values and $\alpha_i, i = 1, \dots, n$, are Lagrange multipliers satisfying

$$\begin{aligned} 0 \leq \alpha_i \leq C \\ \sum_{i=1}^{n} \alpha_i y_i = 0 \end{aligned} \tag{5.25}$$

for a 'regularisation' parameter, C. Maximising (5.24) subject to the constraints (5.25) leads to support vectors identified by non-zero values of α_i.

The solution for $\boldsymbol{w}$ (see Chapter 4) is

$$\boldsymbol{w} = \sum_{i \in \mathcal{SV}} \alpha_i y_i \boldsymbol{\phi}(\boldsymbol{x}_i)$$

and classification of a new data sample $\boldsymbol{x}$ is performed according to the sign of

$$g(\boldsymbol{x}) = \sum_{i \in \mathcal{SV}} \alpha_i y_i \boldsymbol{\phi}^T(\boldsymbol{x}_i)\boldsymbol{\phi}(\boldsymbol{x}) + w_0 \tag{5.26}$$

where

$$w_0 = \frac{1}{N_{\widetilde{\mathcal{SV}}}} \left\{ \sum_{i \in \widetilde{\mathcal{SV}}} y_i - \sum_{i \in \mathcal{SV}, j \in \widetilde{\mathcal{SV}}} \alpha_i y_i \boldsymbol{\phi}^T(\boldsymbol{x}_i)\boldsymbol{\phi}(\boldsymbol{x}_j) \right\} \tag{5.27}$$

in which $\mathcal{SV}$ is the set of support vectors with associated values of α_i satisfying $0 < \alpha_i \le C$ and $\widetilde{\mathcal{SV}}$ is the set of $N_{\widetilde{\mathcal{SV}}}$ support vectors satisfying $0 < \alpha_i < C$ (those at the target distance of $1/|\boldsymbol{w}|$ from the separating hyperplane).

Optimisation of L_D (5.24) and the subsequent classification of a sample ((5.26) and (5.27)) relies only on scalar products between transformed feature vectors, which can be replaced by a kernel function

$$K(\boldsymbol{x}, \boldsymbol{y}) = \boldsymbol{\phi}^T(\boldsymbol{x})\boldsymbol{\phi}(\boldsymbol{y})$$

Thus, we can avoid computing the transformation $\boldsymbol{\phi}(\boldsymbol{x})$ explicitly and replace the scalar product with $K(\boldsymbol{x}, \boldsymbol{y})$ instead. The discriminant function (5.26) becomes

$$g(\boldsymbol{x}) = \sum_{i \in \mathcal{SV}} \alpha_i y_i K(\boldsymbol{x}_i, \boldsymbol{x}) + w_0 \tag{5.28}$$

The advantage of the kernel representation is that we need only use K as the training algorithm and even do not need to know $\boldsymbol{\phi}$ explicitly, provided that the kernel can be written as an inner product. In some cases (for example, the exponential kernel), the feature space is infinite-dimensional and so it is more efficient to use a kernel.

5.4.1 Types of kernel

There are many types of kernel that may be used in an SVM. Table 5.2 lists some commonly used forms. Acceptable kernels must be expressible as an inner product in a feature space, which means that they must satisfy *Mercer's condition* (Courant and Hilbert, 1959; Vapnik, 1998): a kernel $K(\boldsymbol{x}, \boldsymbol{y}), \boldsymbol{x}, \boldsymbol{y} \in \mathbb{R}^p$, is an inner product in some feature space, or $K(\boldsymbol{x}, \boldsymbol{y}) = \boldsymbol{\phi}^T(\boldsymbol{x})\boldsymbol{\phi}(\boldsymbol{y})$, if and only if $K(\boldsymbol{x}, \boldsymbol{y}) = K(\boldsymbol{y}, \boldsymbol{x})$ and

$$\int K(\boldsymbol{x}, z) f(\boldsymbol{x}) f(z)\, d\boldsymbol{x}\, dz \ge 0$$

Table 5.2 Support vector machine kernels

Nonlinearity	Mathematical form $K(\boldsymbol{x}, \boldsymbol{y})$
Polynomial	$(1 + \boldsymbol{x}^T \boldsymbol{y})^d$
Gaussian	$\exp(-\lvert\boldsymbol{x} - \boldsymbol{y}\rvert^2/\sigma^2)$
Sigmoid	$\tanh(k\boldsymbol{x}^T \boldsymbol{y} - \delta)$

for all functions f satisfying

$$\int f^2(\boldsymbol{x})\,d\boldsymbol{x} < \infty$$

That is, $K(\boldsymbol{x}, \boldsymbol{y})$ may be expanded as

$$K(\boldsymbol{x}, \boldsymbol{y}) = \sum_{j=1}^{\infty} \lambda_j \hat{\phi}_j(\boldsymbol{x}) \hat{\phi}_j(\boldsymbol{y})$$

where λ_j and $\phi_j(\boldsymbol{x})$ are the eigenvalues and eigenfunctions satisfying

$$\int K(\boldsymbol{x}, \boldsymbol{y}) \phi_j(\boldsymbol{x})\,d\boldsymbol{x} = \lambda_j \phi_j(\boldsymbol{x})$$

and $\hat{\phi}_j$ is normalised so that $\int \hat{\phi}_j^2(\boldsymbol{x})\,d\boldsymbol{x} = 1$.

As an example, consider the kernel $K(\boldsymbol{x}, \boldsymbol{y}) = (1 + \boldsymbol{x}^T \boldsymbol{y})^d$ for $d = 2$ and $\boldsymbol{x}, \boldsymbol{y} \in \mathbb{R}^2$. This may be expanded as

$$\begin{aligned}(1 + x_1 y_1 + x_2 y_2)^2 &= 1 + 2x_1 y_1 + 2x_2 y_2 + 2x_1 x_2 y_1 y_2 + x_1^2 y_1^2 + x_2^2 y_2^2 \\ &= \boldsymbol{\phi}^T(\boldsymbol{x}) \boldsymbol{\phi}(\boldsymbol{y})\end{aligned}$$

where $\boldsymbol{\phi}(\boldsymbol{x}) = (1, \sqrt{2}x_1, \sqrt{2}x_2, \sqrt{2}x_1 x_2, x_1^2, x_2^2)$.

5.4.2 Model selection

The degrees of freedom of the SVM model are the choice of kernel, the parameters of the kernel and the choice of the regularisation parameter, C, which penalises the training errors. For most types of kernel, it is generally possible to find values for the kernel parameters for which the classes are separable. However, this is not a sensible strategy and leads to over-fitting of the training data and poor generalisation to unseen data.

The simplest approach to model selection is to reserve a validation set that is used to monitor performance as the model parameters are varied. More expensive alternatives that make better use of the data are data resampling methods such as cross-validation and bootstrapping (see Chapter 11).

5.4.3 Support vector machines for regression

Support vector machines may also be used for problems in regression. Suppose that we have a data set $\{(\boldsymbol{x}_i, y_i), i = 1, \ldots, n\}$ of measurements $\boldsymbol{x}_i$ on the independent variables and y_i on the response variables. Instead of the constraints (4.25) we have

$$\begin{aligned}&(\boldsymbol{w}^T \boldsymbol{x}_i + w_0) - y_i \le \epsilon + \xi_i & i = 1, \ldots, n \\ &y_i - (\boldsymbol{w}^T \boldsymbol{x}_i + w_0) \le \epsilon + \hat{\xi}_i & i = 1, \ldots, n \\ &\xi_i, \hat{\xi} \ge 0 & i = 1, \ldots, n\end{aligned} \tag{5.29}$$

This allows a deviation between the target values y_i and the function f,

$$f(\boldsymbol{x}) = \boldsymbol{w}^T\boldsymbol{x} + w_0$$

Two slack variables are introduced: one (ξ) for exceeding the target value by more than ϵ and $\hat{\xi}$ for being more than ϵ below the target value (see Figure 5.3). As in the classification case, a loss function is minimised subject to the constraints (5.29). For a linear ϵ-insensitive loss, we minimise

$$\frac{1}{2}\boldsymbol{w}^T\boldsymbol{w} + C\sum_{i=1}^{n}(\xi_i + \hat{\xi}_i)$$

The primal form of the Lagrangian is

$$\begin{aligned} L_p = \frac{1}{2}\boldsymbol{w}^T\boldsymbol{w} + C\sum_{i=1}^{n}(\xi_i + \hat{\xi}_i) - \sum_{i=1}^{n}\alpha_i(\xi_i + \epsilon - (\boldsymbol{w}^T\boldsymbol{x}_i + w_0 - y_i)) - \sum_{i=1}^{n} r_i\xi_i \\ - \sum_{i=1}^{n}\hat{\alpha}_i(\hat{\xi}_i + \epsilon - (y_i - \boldsymbol{w}^T\boldsymbol{x}_i - w_0)) - \sum_{i=1}^{n}\hat{r}_i\hat{\xi}_i \end{aligned} \quad (5.30)$$

where $\alpha_i, \hat{\alpha}_i \geq 0$ and $r_i, \hat{r}_i \geq 0$ are Lagrange multipliers. Differentiating with respect to $\boldsymbol{w}$, w_0, ξ_i and $\hat{\xi}_i$ gives

$$\begin{aligned} \boldsymbol{w} + \sum_{i=1}^{n}(\alpha_i - \hat{\alpha}_i)\boldsymbol{x}_i &= 0 \\ \sum_{i=1}^{n}(\alpha_i - \hat{\alpha}_i) &= 0 \\ C - \alpha_i - r_i &= 0 \\ C - \hat{\alpha}_i - \hat{r}_i &= 0 \end{aligned} \quad (5.31)$$

Substituting for $\boldsymbol{w}$ into (5.30) and using the relations above gives the dual form

$$L_D = \sum_{i=1}^{n}(\hat{\alpha}_i - \alpha_i)y_i - \epsilon\sum_{i=1}^{n}(\hat{\alpha}_i + \alpha_i) - \frac{1}{2}\sum_{i=1}^{n}\sum_{j=1}^{n}(\hat{\alpha}_i - \alpha_i)(\hat{\alpha}_j - \alpha_j)\boldsymbol{x}_i^T\boldsymbol{x}_j \quad (5.32)$$

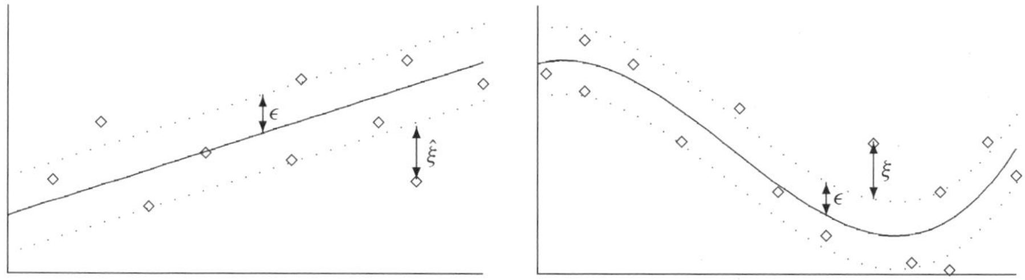

Figure 5.3 Linear (left) and nonlinear (right) SVM regression. The variables ξ and $\hat{\xi}$ measure the cost of lying outside the 'ϵ-insensitive band' around the regression function

This is maximised subject to

$$\begin{aligned} \sum_{i=1}^{n}(\alpha_i - \hat{\alpha}_i) = 0 \\ 0 \leq \alpha_i, \hat{\alpha}_i \leq C \end{aligned} \tag{5.33}$$

which follows from (5.31) and $r_i, \hat{r}_i \geq 0$. The Karush–Kuhn–Tucker complementarity conditions are

$$\begin{aligned} \alpha_i(\xi_i + \epsilon - (\boldsymbol{w}^T \boldsymbol{x}_i + w_0 - y_i)) &= 0 \\ \hat{\alpha}_i(\hat{\xi}_i + \epsilon - (y_i - \boldsymbol{w}^T \boldsymbol{x}_i - w_0)) &= 0 \\ r_i \xi_i = (\alpha_i - C)\xi_i &= 0 \\ \hat{r}_i \hat{\xi}_i = (\hat{\alpha}_i - C)\hat{\xi}_i &= 0 \end{aligned} \tag{5.34}$$

These imply $\alpha_i \hat{\alpha}_i = 0$ and $\xi_i \hat{\xi}_i = 0$. Those patterns $\boldsymbol{x}_i$ with $\alpha_i > 0$ or $\hat{\alpha}_i > 0$ are support vectors. If $0 < \alpha_i < C$ or $0 < \hat{\alpha}_i < C$ then $(\boldsymbol{x}_i, y_i)$ lies on the boundary of the tube surrounding the regression function at distance ϵ. If $\alpha_i = C$ or $\hat{\alpha}_i = C$, then the point lies outside the tube.

The solution for $f(\boldsymbol{x})$ is then

$$f(\boldsymbol{x}) = \sum_{i=1}^{n}(\hat{\alpha}_i - \alpha_i)\boldsymbol{x}_i^T \boldsymbol{x} + w_0 \tag{5.35}$$

using the expression for $\boldsymbol{w}$ in (5.31). The parameter w_0 is chosen so that

$$\begin{aligned} & f(\boldsymbol{x}_i) - y_i = \epsilon \quad \text{for any } i \text{ with } 0 < \alpha_i < C \\ \text{or} \quad & f(\boldsymbol{x}_i) - y_i = -\epsilon \quad \text{for any } i \text{ with } 0 < \hat{\alpha}_i < C \end{aligned}$$

by the Karush–Kuhn–Tucker complementarity conditions above.

Nonlinear regression

The above may also be generalised to a nonlinear regression function in a similar manner to the way in which the linear discriminant function, introduced in Chapter 4, was generalised to the nonlinear discriminant function. If the nonlinear function is given by

$$f(\boldsymbol{x}) = \boldsymbol{w}^T \boldsymbol{\phi}(\boldsymbol{x}) + w_0$$

then equation (5.32) is replaced by

$$L_D = \sum_{i=1}^{n}(\hat{\alpha}_i - \alpha_i)y_i - \epsilon \sum_{i=1}^{n}(\hat{\alpha}_i + \alpha_i) - \frac{1}{2}\sum_{i=1}^{n}\sum_{j=1}^{n}(\hat{\alpha}_i - \alpha_i)(\hat{\alpha}_j - \alpha_j)K(\boldsymbol{x}_i, \boldsymbol{x}_j) \tag{5.36}$$

where $K(\boldsymbol{x}, \boldsymbol{y})$ is a kernel satisfying Mercer's conditions. This is maximised subject to the constraints (5.33).

The solution for $f(\boldsymbol{x})$ is then (compare with (5.35))

$$f(\boldsymbol{x}) = \sum_{i=1}^{n}(\hat{\alpha}_i - \alpha_i)K(\boldsymbol{x}, \boldsymbol{x}_i) + w_0$$

The parameter w_0 is chosen so that

$$\begin{aligned} & f(\boldsymbol{x}_i) - y_i = \epsilon \quad \text{for any } i \text{ with } 0 < \alpha_i < C \\ \text{or} \quad & f(\boldsymbol{x}_i) - y_i = -\epsilon \quad \text{for any } i \text{ with } 0 < \hat{\alpha}_i < C \end{aligned}$$

by the Karush–Kuhn–Tucker complementarity conditions (5.34).

Implementation

There are many freely available and commercial software packages for solving quadratic programming optimisation problems, often based on standard numerical methods of nonlinear optimisation that iteratively hill-climb to find the maximum of the objective function. For very large data sets, however, they become impractical. Traditional quadratic programming algorithms require that the kernel be computed and stored in memory and can involve expensive matrix operations. There are many developments to handle large data sets. *Decomposition* methods (see Table 5.3) apply the standard optimisation package to a fixed subset of the data and revise the subset in the light of applying the classification/regression model learned to the training data not in the subset.

A special development of the decomposition algorithm is the sequential minimal optimisation algorithm, which optimises a subset of two points at a time, for which the optimisation admits an analytic solution. Pseudocode for the algorithm can be found in Cristianini and Shawe-Taylor (2000).

5.4.4 Example application study

The problem To predict protein secondary structure as a step towards the goal of predicting three-dimensional protein structures directly from protein sequences (Hua and Sun, 2001).

Table 5.3 The decomposition algorithm

1. Set b, the size of the subset ($b < n$, the total number of patterns). Set $\alpha_i = 0$ for all patterns.
2. Choose b patterns from the training data set to form a subset $\mathcal{B}$.
3. Solve the quadratic programming problem defined by the subset $\mathcal{B}$ using a standard routine.
4. Apply the model to all patterns in the training set.
5. If there are any patterns that do not satisfy the Karush–Kuhn–Tucker conditions, replace any patterns in $\mathcal{B}$ and the corresponding α_i with these patterns and their α_i values.
6. If not converged, go to step 3.

Summary As a result of the genome and other sequencing projects, the number of protein sequences is growing rapidly. However, this is much greater than the increasing number of known protein structures. The aim of this study is to classify the secondary structure (as helix (H), sheets (E), or coil (C)) based on sequence features. An SVM was compared with an algorithm based on a multilayer perceptron (see Chapter 6).

The data Two data sets were used to develop and test the algorithms. One is a data set of 126 protein chains. The second has 513 protein chains.

The model A standard SVM model was adopted, with a spherically symmetric Gaussian kernel (Table 5.2), with $\sigma^2 = 10.0$ and a regularisation parameter $C = 1.5$.

Training procedure Classifiers were constructed for different window lengths, l: the number of amino acids in the sequence. Each amino acid is encoded as a 21-dimensional binary vector corresponding to the 20 types of amino acid and the C or N terminus. Thus the pattern vector is of dimension $21l$. Three binary SVM classifiers were constructed: H versus (E, C); E versus (H, C); and C versus (H, E). These were combined in two main ways. The first method assigned a test pattern to the class with the largest positive distance to the optimal separating hyperplane. In the second, the binary classifiers were combined in a tree structure. A sevenfold cross-validation scheme was used to obtain results.

Results The performance of the SVM method matched or was significantly better than the neural network method.

5.4.5 Further developments

There are many developments of the basic SVM model for discrimination and regression presented in this chapter. Multiclass SVMs may be developed along the lines discussed in Chapter 4, whether by combining binary classifiers in a one-against-one or a one-against-all method, or by solving a single multiclass optimisation problem (the 'all-together' method). In an assessment of these methods, Hsu and Lin (2002) find that the all-together method yields fewer support vectors, but one-against-all is more suitable for practical use.

Incorporation of priors and costs into the SVM model (to allow for test conditions that differ from training conditions, as often happens in practice) is addressed by Lin *et al.* (2002).

The basic regression model has been extended to take account of different ϵ-insensitive loss functions and ridge regression solutions (Vapnik, 1998; Cristianini and Shawe-Taylor, 2000). The ν-support vector algorithm (Schölkopf *et al.*, 2000) introduces a parameter to control the number of support vectors and errors. The support vector method has also been applied to density estimation (Vapnik, 1998).

The relationship of the support vector method to other methods of classification is discussed by Guyon and Stork (1999) and Schölkopf *et al.* (1997).

5.4.6 Summary

Support vector machines comprise a class of algorithms that represent the decision boundary in a pattern recognition problem typically in terms of a small subset of the training samples. This generalises to problems in regression through the ϵ-insensitive loss function that does not penalise errors below $\epsilon > 0$.

The loss function that is minimised comprises two terms, a term $\boldsymbol{w}^T\boldsymbol{w}$ that characterises model complexity and a second term that measures training error. A single parameter, C, controls the trade-off between these two terms. The optimisation problem can be recast as a quadratic programming problem for both the classification and regression cases.

Several approaches for solving the multiclass classification problem have been proposed: combining binary classifiers and a one-step multiclass SVM approach.

SVMs have been applied to a wide range of applications, and demonstrated to be valuable for real-world problems. The generalisation performance often either matches or is significantly better that competing methods.

Once the kernel is fixed, SVMs have only one free parameter – the regularisation parameter that controls the balance between model complexity and training error. However, there are often parameters of the kernel that must be set and a poor choice can lead to poor generalisation. The choice of best kernel for a given problem is not resolved and special kernels have been derived for particular problems, for example document classification.

5.5 Application studies

There have been very many application studies involving neural networks (including the radial basis function network and the multilayer perceptron described in the next chapter). Examples are given here and in Chapter 6. Some review articles in specific application domains include the following:

- Face processing. Valentin *et al.* (1994) review connectionist models of face recognition (see also Samal and Iyengar, 1992).
- Speech recognition. Morgan and Bourlard (1995) review the use of artificial neural networks in automatic speech recognition, and describe hybrid hidden Markov models and artificial neural network models.
- Image compression. A summary of the use of neural network models as signal processing tools for image compression is provided by Dony and Haykin (1995).
- Fault diagnosis. Sorsa *et al.* (1991) consider several neural architectures, including the multilayer perceptron, for process fault diagnosis.
- Chemical science. Sumpter *et al.* (1994).
- Target recognition. Reviews of neural networks for automatic target recognition are provided by Roth (1990) and Rogers *et al.* (1995).
- Financial engineering. Refenes *et al.* (1997); Burrell and Folarin (1997).

There have been many special issues of journals focusing on different aspects of neural networks, including everyday applications (Dillon *et al.*, 1997), industrial electronics (Chow, 1993), general applications (Lowe, 1994), signal processing (Constantinides *et al.*, 1997; Unbehauen and Luo, 1998), target recognition, image processing (Chellappa *et al.*, 1998), machine vision (Dracopoulos and Rosin, 1998) and oceanic engineering (Simpson, 1992).

Other application domains include remote sensing, medical image analysis and character recognition. See also Chapter 6.

There have been many comparative studies assessing the performance of neural networks in terms of speed of training, memory requirements, and classification performance (or prediction error in regression problems), comparing the results with statistical classifiers. Probably the most comprehensive comparative study is that provided by the *Statlog* project (Michie *et al.*, 1994). A neural network method (an RBF) gave best performance on only one out of 22 data sets, but provided close to best performance in nearly all cases. Other comparative studies include, for example, assessments on character recognition (Logar *et al.*, 1994), fingerprint classification (Blue *et al.*, 1994) and remote sensing (Serpico *et al.*, 1996).

There is an increasing amount of application and comparative studies involving support vector machines. These include:

- Financial time series prediction. Cao and Tay (2001) investigate the feasibility of using SVMs in financial forecasting (see also van Gestel *et al.*, 2001). An SVM with Gaussian kernel is applied to multivariate data (five or eight variables) relating to the closing price of the S&P Daily Index in the Chicago Mercantile Exchange.
- Drug design. This is an application in structure-activity relationship analysis, a technique used to reduce the search for new drugs. Combinatorial chemistry enables the synthesis of millions of new molecular compounds at a time. Statistical techniques that direct the search for new drugs are required to provide an alternative to testing every molecular combination. Burbidge *et al.* (2001) compare SVMs with an RBF network and a classification tree (see Chapter 7). The training time for the classification tree was much smaller than for the other methods, but significantly better performance (measured in terms of error rate) was obtained with the SVM.
- Cancer diagnosis. There have been several applications of SVMs to disease diagnosis. Furey *et al.* (2000) address the problem of tissue sample labelling using measurements from DNA microarray experiments. The data sets comprise measurements on ovarian tissue; human tumour and normal colon tissues; and bone marrow and blood samples from patients with leukaemia. Similar performance to a linear perceptron was achieved. Further cancer studies are reported by Guyon *et al.* (2002) and Ramaswamy *et al.* (2001).
- Radar image analysis. Zhao *et al.* (2000) compare three classifiers, including an SVM, on an automatic target recognition task using synthetic aperture radar data. Experimental results show that the SVM and a multilayer perceptron (see Chapter 6) gave similar performance, but superior to nearest-neighbour.
- Text analysis. De Vel *et al.* (2001) use SVMs to analyse e-mail messages. In the growing field of *computer forensics*, of particular interest to investigators is the misuse of e-mail for the distribution of messages and documents that may be unsolicited,

inappropriate, unauthorised or offensive. The objectives of this study are to classify e-mails as belonging to a particular author.

5.6 Summary and discussion

In this chapter we have developed the basic linear discriminant model to one in which the model is essentially linear, but the decision boundaries are nonlinear. The radial basis function model is implemented in a straightforward manner and, in its simplest form, it requires little more than a matrix pseudo-inverse operation to determine the weights of the network. This hides the fact that optimum selection of the numbers and positions of centres is a more complicated process. Nevertheless, simple rules of thumb can result in acceptable values giving good performance.

Strongly related to kernel discriminant analysis and kernel regression, it possesses many of the asymptotic properties of those methods. Under appropriate conditions, the model provides a least squares approximation to the posterior probabilities of class membership. This enables changes in priors and costs to be readily incorporated into a trained model, without need for retraining.

The support vector machine defines the basis functions implicitly through the definition of a kernel function in the data space and has been found to give very good performance on many problems. There are few parameters to set: the kernel parameters and the regularisation parameter. These can be varied to give optimum performance on a validation set. The SVM focuses on the decision boundary and the standard model is not suitable for the non-standard situation where the operating conditions differ from the training conditions due to drifts in values for costs and priors. An SVM implementation of the RBF will automatically determine the number of centres, their positions and weights as part of the optimisation process (Vapnik, 1998).

5.7 Recommendations

The nonlinear discriminant methods described in this chapter are easy to implement and there are many sources of software for applying them to a data set. Before applying these techniques you should consider the reasons for doing so. Do you believe that the decision boundary is nonlinear? Is the performance provided by linear techniques below that desired or believed to be achievable? Moving to neural network techniques or support vector machines may be one way to achieve improved performance. This is not guaranteed. If the classes are not separable, a more complex model will not make them so. It may be necessary to make measurements on additional variables.

It is recommended that:

1. a simple pattern recognition technique (k-nearest-neighbour, linear discriminant analysis) is implemented as a baseline before considering neural network methods;
2. a simple RBF (unsupervised selection of centres, weights optimised using a squared error criterion) is tried to get a feel for whether nonlinear methods provide some gain for your problem;

3. a regularised solution for the weights of an RBF is used;
4. for a model that provides approximations to the posterior probabilities that enable changes of priors and costs to be incorporated into a trained model, an RBF is used;
5. knowledge of the data generation process is used, including noise on the data, for network design or data preprocessing;
6. for classification problems in high-dimensional spaces where training data are representative of test conditions and misclassification rate is an acceptable measure of classifier performance, support vector machines are implemented.

5.8 Notes and references

There are many developments of the techniques described in this chapter, in addition to other neural network methods. A description of these is beyond the scope of this book, but the use of neural techniques in pattern recognition and the relationship to statistical and structural pattern recognition can be found in the book by Schalkoff (1992).

A comprehensive account of feed-forward neural networks for pattern recognition is given in the book by Bishop (1995). Relationships to other statistical methods and other insights are described by Ripley (1996). Tarassenko (1998) provides a basic introduction to neural network methods (including radial basis functions, multilayer perceptrons, recurrent networks and unsupervised networks) with an emphasis on applications.

A good summary of Bayesian perspectives is given in the book by Ripley (1996); see also Bishop (1995); MacKay (1995); Buntine and Weigend (1991); Thodberg (1996).

Although many of the features of support vector machines can be found in the literature of the 1960s (large margin classifiers, optimisation techniques and sparseness, slack variables), the basic support vector machine for non-separable data was not introduced until 1995 (Cortes and Vapnik, 1995). The book by Vapnik (1998) provides an excellent description of SVMs. A very good self-contained introduction is provided by Cristianini and Shawe-Taylor (2000). The tutorial by Burges (1998) is an excellent concise account. A comprehensive treatment of this rapidly developing field is provided in the recent book by Schölkopf and Smola (2001). The decomposition algorithm was suggested by Osuna *et al.* (1997) and the sequential minimal optimisation algorithm proposed by Platt (1998).

The website `www.statistical-pattern-recognition.net` contains references and links to further information on techniques and applications.

Exercises

Data set 1: 500 samples in training, validation and test sets; p-dimensional; 3 classes; class $\omega_1 \sim N(\boldsymbol{\mu}_1, \boldsymbol{\Sigma}_1)$; class $\omega_2 \sim 0.5N(\boldsymbol{\mu}_2, \boldsymbol{\Sigma}_2) + 0.5N(\boldsymbol{\mu}_3, \boldsymbol{\Sigma}_3)$; class $\omega_3 \sim 0.2N(\boldsymbol{\mu}_4, \boldsymbol{\Sigma}_4) + 0.8N(\boldsymbol{\mu}_5, \boldsymbol{\Sigma}_5)$; $\boldsymbol{\mu}_1 = (-2, 2, \dots, 2)^T$, $\boldsymbol{\mu}_2 = (-4, -4, \dots, -4)^T$, $\boldsymbol{\mu}_3 = (4, 4, \dots, 4)^T$, $\boldsymbol{\mu}_4 = (0, 0, \dots, 0)^T$, $\boldsymbol{\mu}_5 = (-4, 4, \dots, 4)^T$ and $\boldsymbol{\Sigma}_i$ as the identity matrix; equal class priors.

Data set 2: Generate time series data according to the iterative scheme

$$u_{t+1} = \frac{4\left(1 - \frac{\Delta^2}{2}\right)u_t - (2 + \mu\Delta(1 - u_t^2))u_{t-1}}{2 - \mu\Delta(1 - u_t^2)}$$

initialised with $u_{-1} = u_0 = 2$. Plot the generated time series. Construct training and test sets of 500 patterns $(\boldsymbol{x}_i, \boldsymbol{t}_i)$ where $\boldsymbol{x}_i = (u_i, u_{i+1})^T$ and $\boldsymbol{t}_i = u_{i+2}$. Thus the problem is one of time series prediction: predict the next sample in the series given the previous two samples. Take $\mu = 4$, $\Delta = \pi/50$.

1. Compare a radial basis function classifier with a k-nearest-neighbour classifier. Take into account the type of classifier, computational requirements on training and test operations and the properties of the classifiers.

2. Compare and contrast a radial basis function classifier and a classifier based on kernel density estimation.

3. Implement a simple radial basis function classifier: m Gaussian basis functions of width h and centres selected randomly from the data. Using data from data set 1, evaluate performance as a function of dimensionality, p, and number of basis functions, where h is chosen based on the validation set.

4. Show that by appropriate adjustment of weights and biases in an RBF that the nonlinear function ϕ can be normalised to lie between 0 and 1 for a finite data set, without changing the fitting function. Verify for a Gaussian basis function that ϕ may be defined as (5.22) for large h.

5. For data set 2, train an RBF for an $\exp(-z^2)$ and a $z^2\log(z)$ nonlinearity for varying numbers of centres. Once trained, use the RBF in a generative mode: initialise u_{t-1}, u_t (as a sample from the training set), predict u_{t+1}; then predict u_{t+2} from u_t and u_{t+1} using the RBF. Continue for 500 samples. Plot the generated time series. Investigate the sensitivity of the final generated time series to starting values, number of basis functions and the form of the nonlinearity.

6. Consider the optimisation criterion (5.7)

$$E = \|(-\boldsymbol{T}^T + \boldsymbol{W}\boldsymbol{\Phi}^T + \boldsymbol{w}_0\boldsymbol{1}^T)\boldsymbol{D}\|^2$$

   By solving for $\boldsymbol{w}_0$, show that this may be written

$$E = \|(-\hat{\boldsymbol{T}}^T + \boldsymbol{W}\hat{\boldsymbol{\Phi}}^T)\boldsymbol{D}\|^2$$

   where $\hat{\boldsymbol{T}}$ and $\hat{\boldsymbol{\Phi}}$ are zero-mean matrices. By minimising with respect to $\boldsymbol{W}$ and substituting into the expression for E, show that minimising E is equivalent to maximising $\mathrm{Tr}(\boldsymbol{S}_B\boldsymbol{S}_T^{\dagger})$, where

$$\boldsymbol{S}_T = \frac{1}{n}\hat{\boldsymbol{\Phi}}^T\boldsymbol{D}^2\hat{\boldsymbol{\Phi}}$$
$$\boldsymbol{S}_B = \frac{1}{n^2}\hat{\boldsymbol{\Phi}}^T\boldsymbol{D}^2\hat{\boldsymbol{T}}\hat{\boldsymbol{T}}^T\boldsymbol{D}^2\hat{\boldsymbol{\Phi}}$$

7. For $\boldsymbol{D}$ equal to the identity matrix in the above and a target coding scheme

$$(\boldsymbol{t}_i)_k = \begin{cases} a_k & \boldsymbol{x}_i \in \omega_k \\ 0 & \text{otherwise} \end{cases}$$

determine the value of a_k for which $\boldsymbol{S}_B$ is the conventional between-class covariance matrix of the hidden output patterns.

8. Show that maximising L' given by (5.14) is equivalent to maximising (5.12) under the assumption that (5.13) does not depend on the model parameters.

9. Given a normal mixture model for each of C classes (see Chapter 2) with a common covariance matrix across classes, together with class priors $p(\omega_i)$, $i = 1, \ldots, C$, construct an RBF network whose outputs are proportional to the posterior probabilities of class membership. Write down the forms for the centres and weights.

10. Given

$$E = \|\hat{\boldsymbol{T}} - \boldsymbol{V}\boldsymbol{G}\|^2$$

in the orthogonal least squares training model, using the minimum norm solution for $\boldsymbol{V}$ and the properties of the pseudo-inverse (Appendix C) show that

$$E = \text{Tr}\{\hat{\boldsymbol{T}}^T \hat{\boldsymbol{T}}\} - \text{Tr}\{\boldsymbol{G}^T \boldsymbol{V}^T \boldsymbol{V}\boldsymbol{G}\}$$

11. Implement a support vector machine classifier and assess performance using data set 1. Use a Gaussian kernel and choose a kernel width and regularisation parameter using a validation set. Investigate performance as a function of dimensionality, p.

12. Let K_1 and K_2 be kernels defined on $\mathbb{R}^P \times \mathbb{R}^p$. Show that the following are also kernels:

$$K(\boldsymbol{x}, z) = K_1(\boldsymbol{x}, z) + K_2(\boldsymbol{x}, z)$$

$$K(\boldsymbol{x}, z) = aK_1(\boldsymbol{x}, z) \qquad \text{where } a \text{ is a positive real number}$$

$$K(\boldsymbol{x}, z) = f(\boldsymbol{x})f(z) \qquad \text{where } f(.) \text{ is a real-valued function on } \boldsymbol{x}$$

13. For the quadratic ϵ-insensitive loss, the primal form of the Lagrangian is written

$$L_p = \boldsymbol{w}^T\boldsymbol{w} + C\sum_{i=1}^{n}(\xi_i^2 + \hat{\xi}_i^2)$$

and is minimised subject to the constraints (5.29). Derive the dual form of the Lagrangian and state the constraints on the Lagrange multipliers.

14. Show that a support vector machine with a spherically symmetric kernel function satisfying Mercer's conditions implements an RBF classifier with numbers of centres and positions chosen automatically by the SVM algorithm.

6

Nonlinear discriminant analysis – projection methods

Overview

Nonlinear discriminant functions are constructed as sums of basis functions. These are nonlinear functions of linear projections of the data. Optimisation of an objective function is with respect to the projection directions and the basis function properties.

6.1 Introduction

In this chapter, further methods of nonlinear discrimination are explored. What makes these methods different from those in Chapter 5? Certainly, the approaches are closely related – discriminant functions are constructed through the linear combination of nonlinear basis functions and are generally of the form

$$g(\boldsymbol{x}) = \sum_{i=1}^{m} w_i \phi_i(\boldsymbol{x}, \boldsymbol{\mu}_i) \tag{6.1}$$

for weights w_i and parameters $\boldsymbol{\mu}_i$ of the nonlinear functions ϕ_i. For radial basis functions, the $\boldsymbol{\mu}_i$ represent the centres. In the support vector machine, the nonlinear functions are not defined explicitly, but implicitly through the specification of a kernel defined in the data space.

In this chapter, the discriminant functions are again of the form (6.1), where the combination of $\boldsymbol{x}$ and the parameter vector $\boldsymbol{\mu}$ is a scalar product, that is $\phi_i(\boldsymbol{x}, \boldsymbol{\mu}_i) = \phi_i(\boldsymbol{x}^T \boldsymbol{\mu}_i)$. Two models are described. In the first, the form of the function ϕ is prescribed and the optimisation procedure determines the parameters w_{ij} and $\boldsymbol{\mu}_i$ simultaneously. In the second, the form of ϕ is learned, in addition to the projection directions $\boldsymbol{\mu}_i$, and optimisation proceeds sequentially.

6.2 The multilayer perceptron

6.2.1 Introduction

The multilayer perceptron (MLP) is yet another tool in the pattern recognition toolbox, and one that has enjoyed considerable use in recent years. It was presented by Rumelhart *et al.* (1986) as a potential answer to the problems associated with the perceptron, and since that time it has been extensively employed in many pattern recognition tasks. In order to introduce some terminology, let us consider a simple model. We shall then consider some generalisations. The basic MLP produces a transformation of a pattern $\boldsymbol{x} \in \mathbb{R}^p$ to an n'-dimensional space according to

$$g_j(\boldsymbol{x}) = \sum_{i=1}^{m} w_{ji}\phi_i(\boldsymbol{\alpha}_i^T \boldsymbol{x} + \alpha_{i0}) + w_{j0}, \quad j = 1, \ldots, n' \tag{6.2}$$

The functions ϕ_i are fixed nonlinearities, usually identical and taken to be of the *logistic* form representing historically the mean firing rate of a neuron as a function of the input current,

$$\phi_i(z) = \phi(z) = \frac{1}{1 + \exp(-z)} \tag{6.3}$$

Thus, the transformation consists of projecting the data onto each of m directions described by the vectors $\boldsymbol{\alpha}_i$; then transforming the projected data (offset by a bias α_{i0}) by the nonlinear functions $\phi_i(z)$; and finally, forming a linear combination using the weights w_{ji}.

The MLP is often presented in diagrammatic form (see Figure 6.1). The *input nodes* accept the data vector or pattern. There are weights associated with the links between the input nodes and the *hidden nodes* that accept the weighted combination $z = \boldsymbol{\alpha}_i^T \boldsymbol{x} + \alpha_{i0}$ and perform the nonlinear transformation $\phi(z)$. The *output nodes* take a linear combination of the outputs of the hidden nodes and deliver these as *outputs*. In principle, there may

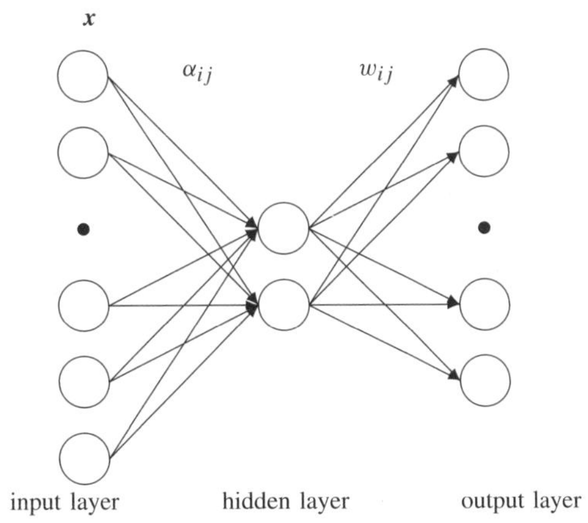

Figure 6.1 Multilayer perceptron

be many *hidden layers*, each one performing a transformation of a scalar product of the outputs of the previous layer and the vector of weights linking the previous layer to a given node. Also, there may be nonlinearities associated with the output nodes that, instead of producing a linear combination of weights w_{ij} and hidden unit outputs, perform nonlinear transformations of this value.

The emphasis of the treatment of MLPs in this section is on MLPs with a single hidden layer. Further, it is assumed either that the 'outputs' are a linear combination of the functions ϕ_i or at least that, in a discrimination problem, discrimination may be performed by taking a linear combination of the functions ϕ_i (a logistic discrimination model is of this type). There is some justification for using a *single* hidden layer model in that it has been shown that an MLP with a single hidden layer can approximate an arbitrary (continuous bounded integrable) function arbitrarily accurately (Hornik, 1993; see Section 6.2.4). Also, practical experience has shown that very good results can be achieved with a single hidden layer on many problems. There may be practical reasons for considering more than one hidden layer, and it is conceptually straightforward to extend the analysis presented here to do so.

The MLP is a nonlinear model: the output is a nonlinear function of its parameters and the inputs, and a nonlinear optimisation scheme must be employed to minimise the selected optimisation criterion. Therefore, all that can be hoped for is a local extremum of the criterion function that is being optimised. This may give satisfactory performance, but several solutions may have to be sought before an acceptable one is found.

6.2.2 Specifying the multilayer perceptron structure

To specify the network structure we must prescribe the number of hidden layers, the number of nonlinear functions within each layer and the form of the nonlinear functions. Most of the MLP networks to be found in the literature consist of layers of logistic processing units (6.3), with each unit connected to every unit in the previous layer (*fully layer connected*) and no connections between units in non-adjacent layers.

It should be noted, however, that many examples exist of MLP networks that are not fully layer connected. There has been some success with networks in which each processing unit is connected to only a small subset of the units in the previous layer. The units chosen are often part of a neighbourhood, especially if the input is some kind of image. In even more complex implementations the weights connected to units that examine similar neighbourhoods at different locations may be forced to be identical (shared weights). Such advanced MLP networks, although of great practical importance, are not universally applicable and are not considered further here.

6.2.3 Determining the multilayer perceptron weights

There are two stages to optimisation. The first is the initialisation of the parameter values; the second is the implementation of a nonlinear optimisation scheme.

Weight initialisation

There are several schemes for initialising the weights. The weights of the MLP are often started at small random values. In addition, some work has been carried out to investigate the benefits of starting the weights at values obtained from simple heuristics.

1. Random initialisation For random initialisation, the weights are initialised with values drawn from a uniform distribution over $[-\Delta, \Delta]$, where Δ depends on the scale of the values taken by the data. If all the variables are equally important and the sample variance is σ^2, then $\Delta = 1/(p\sigma)$ is a reasonable choice (p variables). Hush *et al.* (1992) assess several weight initialisation schemes and support initialisation to small random values.

2. Pattern classifier initialisation An alternative approach is to initialise the MLP to perform as a nearest class mean or nearest-neighbour classifier (Bedworth, 1988). As we saw in Chapter 3, for the nearest-neighbour classifier the decision surface is piecewise linear and is composed of segments of the perpendicular bisectors of the prototype vectors with dissimilar class labels. This decision surface can be implemented in an MLP with one hidden layer of scalar-product logistic processing units. If the decision regions are convex (as for a class mean classifier), then the structure of the network is particularly simple. There have been other studies that use various pattern recognition schemes for initialisation. For example, Weymaere and Martens (1994) propose a network initialisation procedure that is based on k-means clustering (see Chapter 10) and nearest-neighbour classification. Brent (1991) uses decision trees (see Chapter 7) for initialisation. Of course, given good starting values, training time is reduced, albeit at the expense of increased initialisation times.

3. Independence model initialisation An approach that initialises an MLP to deliver class-conditional probabilities under the assumption of independence of variables is described in Lowe and Webb (1990). This really only applies to categorical variables in which the data can be represented as binary patterns.

Optimisation

Many different optimisation criteria and many nonlinear optimisation schemes have been considered for the MLP. It would not be an exaggeration to say that out of all the classification techniques considered in this book, the MLP is the one which has been explored more than any other in recent years, particularly in the engineering literature, sometimes by researchers who do not have a real application, but whose imagination has been stimulated by the 'neural network' aspect of the MLP. It would be impossible to offer anything but a brief introduction to optimisation schemes for the MLP.

Most optimisation schemes involve the evaluation of a function and its derivatives with respect to a set of parameters. Here, the parameters are the multilayer perceptron weights and the function is a chosen error criterion. We shall consider two error criteria discussed earlier: one based on least squares minimisation and the other on a logistic discrimination model.

Least squares error minimisation The error to be minimised is the average squared distance between the approximation given by the model and the 'desired' value:

$$E = \sum_{i=1}^{n} |\boldsymbol{t}_i - \boldsymbol{g}(\boldsymbol{x}_i)|^2 \tag{6.4}$$

where $\boldsymbol{g}(\boldsymbol{x}_i)$ is the vector of 'outputs' and $\boldsymbol{t}_i$ is the desired pattern (sometimes termed the *target*) for data sample $\boldsymbol{x}_i$. In a regression problem, $\boldsymbol{t}_i$ are the dependent variables; in a discrimination problem, $\boldsymbol{t}_i = (t_{i1}, \dots, t_{iC})^T$ are the class labels usually coded as

$$t_{ij} = \begin{cases} 1 & \text{if } \boldsymbol{x}_i \text{ is in class } \omega_j \\ 0 & \text{otherwise} \end{cases}$$

Modifications of the error criterion to take into account the effect of priors and alternative target codings to incorporate costs of misclassification are described in Section 5.2.1.

Most of the nonlinear optimisation schemes used for the MLP require the error and its derivative to be evaluated for a given set of weight values.

The derivative of the error with respect to a weight v (which at the moment can represent either a weight α, between inputs and hidden units, or a weight w, between the hidden units and the outputs – see Figure 6.1) can be expressed as[1]

$$\frac{\partial E}{\partial v} = -2 \sum_{i=1}^{n} \sum_{l=1}^{n'} (\boldsymbol{t}_i - \boldsymbol{g}(\boldsymbol{x}_i))_l \frac{\partial g_l(\boldsymbol{x}_i)}{\partial v} \tag{6.5}$$

The derivative of g_l with respect to v, for v one of the weights w, $v = w_{jk}$ say, is

$$\frac{\partial g_l}{\partial w_{jk}} = \begin{cases} \delta_{lj} & k = 0 \\ \delta_{lj}\phi_k(\boldsymbol{\alpha}_k^T \boldsymbol{x}_i + \alpha_{k0}) & k \neq 0 \end{cases} \tag{6.6}$$

and for the weights α, $v = \alpha_{jk}$,

$$\frac{\partial g_l}{\partial \alpha_{jk}} = \begin{cases} w_{lj} \dfrac{\partial \phi_j}{\partial z} & k = 0 \\ w_{lj} \dfrac{\partial \phi_j}{\partial z} (\boldsymbol{x}_i)_k & k \neq 0 \end{cases} \tag{6.7}$$

where $\partial \phi_i / \partial z$ is the derivative of ϕ with respect to its argument, given by

$$\frac{\partial \phi}{\partial z} = \phi(z)(1 - \phi(z))$$

for the logistic form (6.3) and evaluated at $z = \boldsymbol{\alpha}_j^T \boldsymbol{x}_i + \alpha_{j0}$. Equations (6.5), (6.6) and (6.7) can be combined to give expressions for the derivatives of the squared error with respect to the weights w and α.

[1] Again, we use n' to denote the output dimensionality; in a classification problem, $n' = C$, the number of classes.

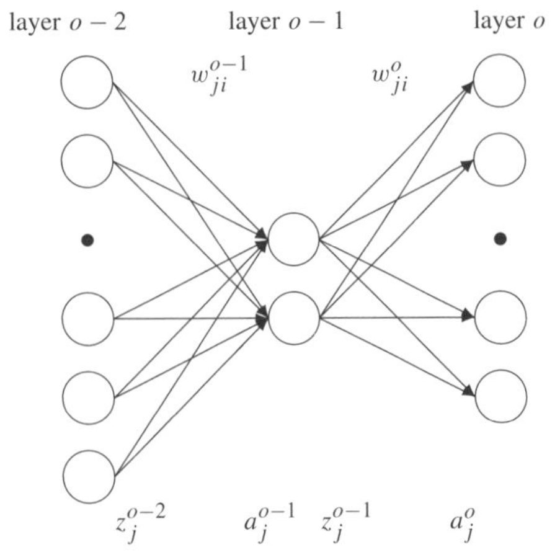

Figure 6.2 Notation used for back-propagation

Back-propagation In the above example, we explicitly calculated the derivatives for a single hidden layer network. Here we consider a more general treatment. *Back-propagation* is the term given to efficient calculation of the derivative of an error function in multilayer networks. We write the error, E, as

$$E = \sum_{k=1}^{n} E^k$$

where E^k is the contribution to the error from pattern k. For example,

$$E^k = |\boldsymbol{t}_k - \boldsymbol{g}(\boldsymbol{x}_k)|^2$$

in (6.4). We refer to Figure 6.2 for general notation: let the weights between layer $o-1$ and layer o be w^o_{ji} (the weight connecting the ith node of layer $o-1$ to the jth node of layer o); let a^o_j be the inputs to the nonlinearity at layer o and z^o_j be the outputs of the nonlinearity, so that

$$\begin{aligned} a^o_j &= \sum_{i=1}^{n_{o-1}} w^o_{ji} z^{o-1}_i \\ z^{o-1}_j &= \phi(a^{o-1}_j) \end{aligned} \tag{6.8}$$

where n_{o-1} is the number of nodes at layer $o-1$ and ϕ is the (assumed common) nonlinearity associated with a node.[2] These quantities are calculated during the process termed *forward propagation*: namely the calculation of the error, E, from its constituent parts, E^k.

Let layer o be the final layer of the network. The term E^k is a function of the inputs to the final layer,

$$E^k = E^k(a^o_1, \ldots, a^o_{n_o})$$

---

[2]Strictly, the terms a^o_j and z^{o-1}_j should be subscripted by k since they depend on the kth pattern.

For example, with a network with linear output units and using a squared error criterion

$$E^k = \sum_{j=1}^{n_o} ((\boldsymbol{t}_k)_j - a_j^o)^2$$

where $(\boldsymbol{t}_k)_j$ is the jth component of the kth target vector.

The derivatives of the error E are simply the sums of the derivatives of E^k, the contribution to the error by the kth pattern. The derivatives of E^k with respect to the final layer of weights, w_{ji}^o, are given by

$$\frac{\partial E^k}{\partial w_{ji}^o} = \frac{\partial E^k}{\partial a_j^o} \frac{\partial a_j^o}{\partial w_{ji}^o} = \delta_j^o z_i^{o-1} \tag{6.9}$$

where we use the convenient shorthand notation δ_j^o to denote $\partial E^k / \partial a_j^o$, the derivative with respect to the input of a particular node in the network.

The derivatives of E^k with respect to the weights w_{ji}^{o-1} are given by

$$\frac{\partial E^k}{\partial w_{ji}^{o-1}} = \frac{\partial E^k}{\partial a_j^{o-1}} \frac{\partial a_j^{o-1}}{\partial w_{ji}^{o-1}} = \delta_j^{o-1} z_i^{o-2} \tag{6.10}$$

where we have again used the notation δ_j^{o-1} for $\partial E^k / \partial a_j^{o-1}$. This may be expanded using the chain rule for differentiation as

$$\delta_j^{o-1} = \sum_{l=1}^{n_o} \frac{\partial E^k}{\partial a_l^o} \frac{\partial a_l^o}{\partial a_j^{o-1}} = \sum_{l=1}^{n_o} \delta_l^o \frac{\partial a_l^o}{\partial a_j^{o-1}}$$

and using the relationships (6.8), we have

$$\frac{\partial a_l^o}{\partial a_j^{o-1}} = w_{lj}^o \phi'(a_j^{o-1})$$

giving the *back-propagation formula*

$$\delta_j^{o-1} = \phi'(a_j^{o-1}) \sum_l \delta_l^o w_{lj}^o \tag{6.11}$$

The above result allows the derivatives of E^k with respect to the inputs to a particular node to be expressed in terms of derivatives with respect to the inputs to nodes higher up the network, that is, nodes in layers closer to the output layer. Once calculated, these are combined with node outputs in equations of the form (6.9) and (6.10) to give derivatives with respect to the weights.

Equation (6.11) requires the derivatives with respect to the nonlinearities ϕ to be specified. For ϕ given by (6.3),

$$\phi'(a) = \phi(a)(1 - \phi(a)) \tag{6.12}$$

The derivative calculation also requires the initialisation of δ_j^o, the derivative of the kth component of the error with respect to a_j^o. For the sum-squared error criterion, this is

$$\delta_j^o = -2((\boldsymbol{t}_k)_j - a_j^o) \tag{6.13}$$

Each layer in the network, up to the output layer, may have biases: the terms α_{i0} and w_{j0} in the single hidden layer model of (6.2). A bias node has $\phi(.) \equiv 1$ and in the general back-propagation scheme, and at a given layer (say $o-1$),

$$\frac{\partial a_j^{o-1}}{\partial w_{j0}^{o-1}} = 1$$

The above scheme is applied recursively to each layer to calculate derivatives. The term *back-propagation networks* is often used to describe multilayer perceptrons employing such a calculation, though strictly *back-propagation* refers to the method of derivative calculation, rather than the type of network.

Logistic classification We now consider an alternative error criterion based on the generalised logistic model. The basic assumption for the generalised logistic discrimination for an MLP model is

$$\log\left(\frac{p(\boldsymbol{x}|\omega_j)}{p(\boldsymbol{x}|\omega_C)}\right) = \sum_{i=1}^{m} w_{ji}\phi_i(a_i) + w_{j0}, \quad j = 1, \ldots, C-1$$

where ϕ_i is the ith nonlinearity (logistic function) and a_i represents the input to the ith nonlinearity. The terms a_i may be linear combinations of the outputs of a previous layer in a layered network system and depend on the parameters of a network. The posterior probabilities are given by (5.11)

$$\begin{aligned} p(\omega_j|\boldsymbol{x}) &= \frac{\exp\left(\sum_{i=1}^{m} w_{ji}\phi_i(a_i) + w'_{j0}\right)}{1 + \sum_{s=1}^{C-1}\exp\left(\sum_{i=1}^{m} w_{si}\phi_i(a_i) + w'_{s0}\right)}, \quad j = 1, \ldots, C-1 \\ p(\omega_C|\boldsymbol{x}) &= \frac{1}{1 + \sum_{s=1}^{C-1}\exp\left(\sum_{i=1}^{m} w_{si}\phi_i(a_i) + w'_{s0}\right)} \end{aligned} \tag{6.14}$$

where $w'_{j0} = w_{j0} + \log(p(\omega_j)/p(\omega_C))$.

Discrimination is based on the generalised linear model

$$g_j(\boldsymbol{x}) = \sum_{i=1}^{m} w_{ji}\phi_i(a_i) + w'_{j0}, \quad j = 1, \ldots, C-1$$

with decision rule:

$$\text{assign } \boldsymbol{x} \text{ to class } \omega_j \text{ if } \max_{s=1,\ldots,C-1} \sum_{i=1}^{m} w_{si}\phi_i(a_i) + w'_{s0} = \sum_{i=1}^{m} w_{ji}\phi_i(a_i) + w'_{j0} > 0$$

else assign $\boldsymbol{x}$ to class C.

In terms of the MLP model (Figure 6.2), equation (6.14) can be thought of as a final normalising layer in which w_{ji} are the final layer weights and the set of C inputs to the final layer (terms of the form $\sum_i w_{ji}\phi_i + w_{j0}$, $j = 1, \ldots, C$) is offset by prior terms $\log(p(\omega_j)/p(\omega_C))$ and normalised through (6.14) to give outputs $p(\omega_j|\boldsymbol{x})$, $j = 1, \ldots, C$.

Estimates of the parameters of the model may be achieved by maximising the log-likelihood, under a mixture sampling scheme given by (see Section 5.2.2)

$$\sum_{k=1}^{C} \sum_{\boldsymbol{x}_i \in \omega_k} \log(p(\omega_k|\boldsymbol{x}_i))$$

where the posterior probability estimates are of the form (6.14), or equivalently by minimising

$$-\sum_{i=1}^{n} \sum_{k=1}^{C} (\boldsymbol{t}_i)_k \log(p(\omega_k|\boldsymbol{x}_i))$$

where $\boldsymbol{t}_i$ is a target vector for $\boldsymbol{x}_i$: a vector of zeros, except that there is a 1 in the position corresponding to the class of $\boldsymbol{x}_i$. The above criterion is of the form $\sum_i E^i$, where

$$E^i = -\sum_{k=1}^{C} (\boldsymbol{t}_i)_k \log(p(\omega_k|\boldsymbol{x}_i)) \tag{6.15}$$

For the back-propagation algorithm, we require δ_j^o, $j = 1, \ldots, C-1$. This is given by

$$\delta_j^o = \frac{\partial E^i}{\partial a_j^o} = -(\boldsymbol{t}_i)_j + p(\omega_j|\boldsymbol{x}_i), \tag{6.16}$$

the difference between the jth component of the target vector and the jth output for pattern i (compare with (6.13); see the exercises).

Now that we can evaluate the error and its derivatives, we can, in principle, use one of many nonlinear optimisation schemes available (Press *et al.*, 1992). We have found that the *conjugate gradients algorithm* with the Polak–Ribière update scheme works well on many practical problems. For problems with a small number of parameters (less than about 250), the Broyden–Fletcher–Goldfarb–Shanno optimisation scheme is recommended (Webb *et al.*, 1988; Webb and Lowe, 1988). Storage of the inverse Hessian matrix ($n_p \times n_p$, where n_p is the number of parameters) limits its use in practice for large networks. For further details of optimisation algorithms see, for example, van der Smagt (1994) and Karayiannis and Venetsanopolous (1993).

Most algorithms for determining the weights of an MLP do not necessarily make direct use of the fact that the discriminant function is linear in the final layer of weights w and they solve for all the weights using a nonlinear optimisation scheme. Another approach is to alternate between solving for the weights w (using a linear pseudo-inverse method) and adjusting the parameters of the nonlinear functions ϕ, namely α. This is akin to the alternating least squares approach in projection pursuit and is equivalent to regarding the final layer weights as a function of the parameters α (Webb and Lowe, 1988; see also Stäger and Agarwal, 1997).

Stopping criterion The most common stopping criterion used in the nonlinear optimisation schemes is to terminate the algorithm when the relative change in the error is less than a specified amount (or the maximum number of allowed iterations has been exceeded).

An alternative that has been employed for classification problems is to cease training when the classification error (either on the training set or preferably a separate validation set) stops decreasing (see Chapter 11 for model selection). Another strategy is based on growing and pruning in which a network larger than necessary is trained and parts that are not needed are removed. See Reed (1993) for a survey of pruning algorithms.

Training strategies There are several training strategies that may be used to minimise the error. The one that we have described above uses the complete set of patterns to calculate the error and its derivatives with respect to the weights, w. These may then be used in a nonlinear optimisation scheme to find the values for which the error is a minimum.

A different approach that has been commonly used is to update the parameters using the gradients $\partial E_n/\partial w$. That is, calculate the contribution to the derivative of the error due to each pattern and use that in the optimisation algorithm. In practice, although the total error will decrease, it will not converge to a minimum, but tend to fluctuate around the local minimum. A stochastic update scheme that also uses a single pattern at a time will converge to a minimum, by ensuring that the influence of each gradient calculation decreases appropriately with iteration number.

Finally, incremental training is a heuristic technique for speeding up the overall learning time of neural networks whilst simultaneously improving the final classification performance. The method is simple: first the network is partially trained on a subset of the training data (there is no need to train to completion) and then the resulting network is further tuned using the entire training database. The motivation is that the subset training will perform a coarse search of weight space to find a region that 'solves' the initial problem. The hope is that this region will be a useful place to start for training on the full data set. The technique is often used with a small subset used for initial training and progressing through larger and larger training databases as the performance increases. The number of patterns used in each subset will vary according to the task although one should ensure that sufficient patterns representative of the data distribution are present to prevent the network over-fitting the subset data. Thus, in a discrimination problem, there should be samples from all classes in the initial training subset.

6.2.4 Properties

If we are free to choose the weights and nonlinear functions, then a single-layer MLP can approximate any continuous function to any degree of accuracy if and only if the nonlinear function is not a polynomial (Leshno *et al.*, 1993). Since such networks are simulated on a computer, the nonlinear function of the hidden nodes must be expressed as a finite polynomial. Thus, they are not capable of universal approximation (Wray and Green, 1995). However, for most practical purposes, the lack of a universal approximation property is irrelevant.

Classification properties of MLPs are addressed by Faragó and Lugosi (1993). Let L^* be the Bayes error rate (see Chapter 8); let g_{kn} be an MLP with one hidden layer of k nodes (with step function nonlinearities) trained on n samples to minimise the number of errors on the training data (with error probability $L(g_{kn})$); then, provided k is chosen so that

$$k \to \infty$$
$$k\frac{\log(n)}{n} \to 0$$

as $n \to \infty$, then

$$\lim_{n\to\infty} L(g_{kn}) = L^*$$

with probability 1. Thus, the classification error approaches the Bayes error as the number of training samples becomes large, provided k is chosen to satisfy the conditions above. However, although this result is attractive, the problem of choosing the parameters of g_{kn} to give minimum errors on a training set is computationally difficult.

6.2.5 Example application study

The problem The classification of radar clutter in a target recognition problem (Blacknell and White, 1994).

Summary A prerequisite for target detection in synthetic aperture radar imagery is the ability to classify background clutter in an optimal manner. Radar clutter is the unwanted return from the background (fields, woods, buildings), although in remote sensing land-use applications, the natural vegetation forms the wanted return.

The aim of the study is to investigate how well neural networks can approximate an optimum (Bayesian) classification. What form of preprocessing should be performed prior to network training? How should the network be constructed?

The data In terms of synthetic aperture radar imagery, a correlated K distribution provides a reasonable description of natural clutter textures arising from fields and woods. However, an analytical expression for correlated multivariate K distributions is not available. Therefore, the approach taken in this work is to develop methodology on simulated uncorrelated K-distributed data, before application to correlated data.

The form of the multivariate, uncorrelated K distribution is ($\boldsymbol{x} = (x_1, \ldots, x_p)^T$)

$$p(\boldsymbol{x}|\mu, \nu) = \prod_{i=1}^{p} \frac{2}{x_i \Gamma(\nu)} \left(\frac{\nu x_i}{\mu}\right)^{(\nu+1)/2} K_{\nu-1}\left\{2\sqrt{\left(\frac{\nu x_i}{\mu}\right)}\right\}$$

where μ is the mean, ν is the order parameter and $K_{\nu-1}$ is the modified Bessel function of order $\nu - 1$. Data for two 256×256 images were generated, each image with the same mean value but different order parameters. From each image 16-dimensional pattern vectors were extracted. These comprised measurements on 4×4 non-overlapping windows. (A second experiment used 2×1 windows.) Several discriminant functions were constructed, including an optimum Bayesian scheme, and evaluated on this two-class problem.

The model The neural network scheme was developed comprising an MLP with 100 hidden nodes, and a squared error objective function was used.

Training procedure Some simple preprocessing of the data (taking logarithms) was implemented to reduce network training times. Weights were optimised using gradient descent. The MLP error rate was 18.5% on the two-pixel data and 23.5% on the 16-pixel data (different order parameters), compared with optimum values of 16.9% and 20.9%.

Further improvements in MLP performance (reducing the two-pixel error rate to 17.0%) were reported through the use of a simple factorisation scheme. Separate MLPs were designed for different regions of the data space, these regions being defined by the original MLP classification.

Results

Preliminary results on correlated data showed that classifier combination (see Chapter 8) can improve performance.

Thus, careful network design and using knowledge of the data distribution to preprocess the data can lead to significant improvements in performance over the naïve implementation of an MLP.

6.2.6 Further developments

One important development is the application of Bayesian inference techniques to the fitting of neural network models, which has been shown to lead to practical benefits on real problems. In the *predictive approach* (Ripley, 1996), the posterior distribution of the observed target value ($\boldsymbol{t}$) given an input vector ($\boldsymbol{x}$) and the data set ($\mathcal{D}$), $p(\boldsymbol{t}|\boldsymbol{x}, \mathcal{D})$, is obtained by integrating over the posterior distribution of network weights, $\boldsymbol{w}$,

$$p(\boldsymbol{t}|\boldsymbol{x}, \mathcal{D}) = \int p(\boldsymbol{t}|\boldsymbol{x}, \boldsymbol{w})p(\boldsymbol{w}|\mathcal{D})d\boldsymbol{w} \tag{6.17}$$

where $p(\boldsymbol{w}|\mathcal{D})$ is the *posterior distribution* for $\boldsymbol{w}$ and $p(\boldsymbol{t}|\boldsymbol{x}, \boldsymbol{w})$ is the distribution of outputs for the given model, $\boldsymbol{w}$, and input, $\boldsymbol{x}$. For a data set $\mathcal{D} = \{(\boldsymbol{x}_i, \boldsymbol{t}_i), i = 1, \ldots, n\}$ of measurement vectors $\boldsymbol{x}_i$ with associated targets $\boldsymbol{t}_i$, the posterior of the weights $\boldsymbol{w}$ may be written

$$p(\boldsymbol{w}|\mathcal{D}) = \frac{p(\boldsymbol{w}, \mathcal{D})}{p(\mathcal{D})} = \frac{1}{p(\mathcal{D})}\prod_{i=1}^{n} p(\boldsymbol{t}_i|\boldsymbol{x}_i, \boldsymbol{w})p(\boldsymbol{x}_i|\boldsymbol{w})p(\boldsymbol{w})$$

assuming independent samples. Further, if we assume that $p(\boldsymbol{x}_i|\boldsymbol{w})$ does not depend on $\boldsymbol{w}$, then

$$p(\boldsymbol{w}|\mathcal{D}) \propto \prod_{i=1}^{n} p(\boldsymbol{t}_i|\boldsymbol{x}_i, \boldsymbol{w})p(\boldsymbol{w})$$

The *maximum a posteriori* (MAP) estimate of $\boldsymbol{w}$ is that for which $p(\boldsymbol{w}|\mathcal{D})$ is a maximum. Assuming a prior distribution,

$$p(\boldsymbol{w}) \propto \exp\left(-\frac{\alpha}{2}\|\boldsymbol{w}\|^2\right)$$

for parameter α, and a zero-mean Gaussian noise mode so that

$$p(\boldsymbol{t}|\boldsymbol{x}, \boldsymbol{w}) \propto \exp\left(-\frac{\beta}{2}|\boldsymbol{t} - \boldsymbol{g}(\boldsymbol{x}, \boldsymbol{w})|^2\right)$$

for the diagonal covariance matrix $(1/\beta)\boldsymbol{I}$ and network output $\boldsymbol{g}(\boldsymbol{x}; \boldsymbol{w})$, then

$$p(\boldsymbol{w}|\mathcal{D}) \propto \exp\left(-\frac{\beta}{2}\sum_i |\boldsymbol{t}_i - \boldsymbol{g}(\boldsymbol{x}_i; \boldsymbol{w})|^2 - \frac{\alpha}{2}\|\boldsymbol{w}\|^2\right)$$

and the MAP estimate is that for which

$$S(\boldsymbol{w}) \triangleq \sum_i |\boldsymbol{t}_i - \boldsymbol{g}(\boldsymbol{x}_i; \boldsymbol{w})|^2 + \frac{\alpha}{\beta}\|\boldsymbol{w}\|^2 \tag{6.18}$$

is a minimum. This is the regularised solution (equation (5.9)) derived as a solution for the MAP estimate for the parameters $\boldsymbol{w}$.

Equation (6.18) is not a simple function of $\boldsymbol{w}$ and may have many local minima (many local peaks of the posterior density), and the integral in (6.17) is computationally difficult to evaluate. Bishop (1995) approximates $S(\boldsymbol{w})$ using a Taylor expansion around its minimum value, $\boldsymbol{w}_{\text{MAP}}$ (although, as we have noted, there may be many local minima),

$$S(\boldsymbol{w}) = S(\boldsymbol{w}_{\text{MAP}}) + \tfrac{1}{2}(\boldsymbol{w} - \boldsymbol{w}_{\text{MAP}})^T \boldsymbol{A}(\boldsymbol{w} - \boldsymbol{w}_{\text{MAP}})$$

where $\boldsymbol{A}$ is the *Hessian* matrix

$$\boldsymbol{A}_{ij} = \frac{\partial}{\partial w_i}\frac{\partial}{\partial w_j} S(\boldsymbol{w})\bigg|_{\boldsymbol{w}=\boldsymbol{w}_{\text{MAP}}}$$

to give

$$p(\boldsymbol{w}|\mathcal{D}) \propto \exp\left\{-\frac{\beta}{2}(\boldsymbol{w} - \boldsymbol{w}_{\text{MAP}})^T \boldsymbol{A}(\boldsymbol{w} - \boldsymbol{w}_{\text{MAP}})\right\}$$

Also, expanding $g(\boldsymbol{x}; \boldsymbol{w})$ around $\boldsymbol{w}_{\text{MAP}}$ (assuming for simplicity a scalar quantity),

$$g(\boldsymbol{x}; \boldsymbol{w}) = g(\boldsymbol{x}; \boldsymbol{w}_{\text{MAP}}) + (\boldsymbol{w} - \boldsymbol{w}_{\text{MAP}})^T \boldsymbol{h}$$

where $\boldsymbol{h}$ is the gradient vector, evaluated at $\boldsymbol{w}_{\text{MAP}}$, gives

$$p(t|\boldsymbol{x}, \mathcal{D}) \propto \int \exp\left\{-\frac{\beta}{2}[t - g(\boldsymbol{x}; \boldsymbol{w}_{\text{MAP}}) - \Delta\boldsymbol{w}^T\boldsymbol{h}]^2 - \frac{\beta}{2}\Delta\boldsymbol{w}^T\boldsymbol{A}\Delta\boldsymbol{w}\right\} d\boldsymbol{w} \tag{6.19}$$

where $\Delta\boldsymbol{w} = (\boldsymbol{w} - \boldsymbol{w}_{\text{MAP}})$. This may be evaluated to give (Bishop, 1995)

$$p(t|\boldsymbol{x}, \mathcal{D}) = \frac{1}{(2\pi\sigma_t^2)^{\frac{1}{2}}} \exp\left\{-\frac{1}{2\sigma_t^2}(t - g(\boldsymbol{x}; \boldsymbol{w}_{\text{MAP}}))^2\right\} \tag{6.20}$$

where the variance σ_t^2 is given by

$$\sigma_t^2 = \frac{1}{\beta}(1 + \boldsymbol{h}^T\boldsymbol{A}^{-1}\boldsymbol{h}) \tag{6.21}$$

Equation (6.20) describes the distribution of output values for a given input value, $\boldsymbol{x}$, and given the data set, with the expression above providing error bars on the estimate. Further discussion of the Bayesian approach is given by Bishop (1993) and Ripley (1996).

6.2.7 Summary

The multilayer perceptron is a model that, in its simplest form, can be regarded as a generalised linear discriminant function in which the nonlinear functions ϕ are flexible and adapt to the data. This is the way that we have chosen to introduce it in this chapter as it forms a natural progression from linear discriminant analysis, through generalised linear discriminant functions with fixed nonlinearities to the MLP. It is related to the projection pursuit model described in the next section in that data are projected onto different axes, but unlike the projection pursuit model the nonlinear function does not adapt; it is usually a fixed logistic nonlinearity. The parameters of the model are determined through a nonlinear optimisation scheme, which is one of the drawbacks of the model. Computation time may be excessive.

There have been many assessments of gradient-based optimisation algorithms, variants and alternatives. Webb *et al.* (1988; see also Webb and Lowe, 1988) compared various gradient-based schemes on several problems. They found that the Levenberg–Marquardt optimisation scheme gave best overall performance for networks with a small number of parameters, but favoured conjugate gradient methods for networks with a large number of parameters. Further comparative studies include those of Karayiannis and Venetsanopolous (1993), van der Smagt (1994) and Stäger and Agarwal (1997). The addition of extra terms to the error involving derivatives of the error with respect to the input has been considered by Drucker and Le Cun (1992) as a means of improving generalisation (see also Bishop, 1993; Webb, 1994). The latter approach of Webb was motivated from an *error-in-variables* perspective (noise on the inputs). The addition of noise to the inputs as a means of improving generalisation has also been assessed by Holmström and Koistinen (1992) and Matsuoka (1992).

The MLP is a very flexible model, giving good performance on a wide range of problems in discrimination and regression. We have presented only a very basic model. There are many variants, some adapted to particular types of problem such as time series (for example, time-delay neural networks). Growing and pruning algorithms for MLP construction have also been considered in the literature, as well as the introduction of regularisation in the optimisation criteria in order to prevent over-fitting of the data. The implementation in hardware for some applications has also received attention. There are several commercial products available for MLP design and implementation.

6.3 Projection pursuit

6.3.1 Introduction

Projection pursuit is a term used to describe a technique for finding 'interesting' projections of data. It has been used for exploratory data analysis, density estimation and in multiple regression problems (Friedman and Tukey, 1974; Friedman and Stuetzle, 1981; Friedman *et al.*, 1984; Friedman, 1987; Huber, 1985; Jones and Sibson, 1987). It would therefore be at home in several of the chapters of this book.

One of the advantages of projection pursuit is that it is appropriate for sparse data sets in high-dimensional spaces. That is, in situations in which the sample size is too small to make kernel density estimates reliable. A disadvantage is that 'projection pursuit will uncover not only true but also spurious structure' (Huber, 1985).

The basic approach in projection pursuit regression is to model the regression surface as a sum of nonlinear functions of linear combinations of the variables

$$y = \sum_{j=1}^{m} \phi_j(\boldsymbol{\beta}_j^T \boldsymbol{x}) \tag{6.22}$$

where the parameters $\boldsymbol{\beta}_j$, $j = 1, \ldots, m$, and the function ϕ_j are determined from the data and y is the response variable. This is very similar to the multilayer perceptron model described in Section 6.2 where the ϕ_j are usually fixed as logistic functions.

Determination of the parameters $\boldsymbol{\beta}_j$ is achieved by optimising a *figure of merit* or *criterion of fit*, $I(\boldsymbol{\beta})$, in an iterative fashion. The steps in the optimisation procedure are as follows (Friedman and Stuetzle, 1981; Jones and Sibson, 1987).

1. Initialise $m = 1$; initialise residuals $r_i = y_i \triangleq y(\boldsymbol{x}_i)$.
2. Initialise $\boldsymbol{\beta}$ and project the independent variable $\boldsymbol{x}$ onto one dimension $z_i = \boldsymbol{\beta}^T \boldsymbol{x}_i$.
3. Calculate a smooth representation $\phi_{\beta}(z)$ of the current residuals (univariate nonparametric regression of current residuals on zs).
4. Calculate the figure of merit; for example, the fraction of unexplained variance for the linear combination $\boldsymbol{\beta}$,
$$I(\boldsymbol{\beta}) = 1 - \sum_{i=1}^{n} (r_i - \phi_{\beta}(\boldsymbol{\beta}^T \boldsymbol{x}_i))^2 \bigg/ \sum_{i=1}^{n} r_i^2$$
5. Find the vector $\boldsymbol{\beta}_m$ that maximises $I(\boldsymbol{\beta})$ and the corresponding $\phi_{\beta m}(\boldsymbol{\beta}_m^T \boldsymbol{x}_i)$.
6. Recalculate the figure of merit. Terminate if the figure of merit is less than a specified threshold (or the relative change in the figure of merit is less than a threshold) otherwise set $m = m + 1$, update the residuals
$$r_i = r_i - \phi_{\beta_m}(\boldsymbol{\beta}_m^T \boldsymbol{x}_i)$$
and go to step 2.

Step 3 requires a smoothing procedure to be implemented. There are many procedures for smoothing, including running means, kernel smoothers, median filters and regression splines (see also Hastie and Tibshirani, 1990). Green and Silverman (1994) describe efficient algorithms based on smoothing splines. Step 5 requires a search to find the minimum of the figure of merit. Jones and Sibson use an entropy projection index, with density calculated using a kernel density estimate. A steepest descent gradient method is used to determine $\boldsymbol{\beta}$. Alternatively, procedures that do not require gradient information may be employed, for example a multidimensional simplex method (Press *et al.*, 1992).

This procedure builds up a sequence of coefficient vectors $\boldsymbol{\beta}_i$. The projection pursuit procedure can also be implemented with readjustment of the previously determined coefficient vectors (termed *back-fitting* by Friedman and Stuetzle, 1981).

6.3.2 Projection pursuit for discrimination

The use of projection pursuit for discrimination has not been widely studied in the pattern recognition literature, but it is a straightforward development of the regression model given above.

We seek a set of functions $\boldsymbol{\phi}_j$ and a set of directions $\boldsymbol{\beta}_j$ such that

$$\sum_{i=1}^{n}\left|\boldsymbol{t}_i - \sum_{j=1}^{m}\boldsymbol{\phi}_j(\boldsymbol{\beta}_j^T\boldsymbol{x}_i)\right|^2 \tag{6.23}$$

is minimised, where $\boldsymbol{t}_i = (t_{i1}, \dots, t_{iC})^T$ are the class indicators ($t_{ij} = 1$ for $\boldsymbol{x}_i \in \omega_j$, $t_{ij} = 0$ otherwise) and we have taken m terms in the additive model (Hastie and Tibshirani, 1990). We know that the solution of (6.23) gives asymptotically a minimum variance approximation to the Bayes optimal discriminant function. Thus the function

$$g(\boldsymbol{x}) = \sum_{j=1}^{m}\boldsymbol{\phi}_j(\boldsymbol{\beta}_j^T\boldsymbol{x})$$

is our discriminant function.

In (6.23), there is a set of C nonlinear functions (the elements of the C-dimensional vector $\boldsymbol{\phi}_j$) to estimate for each projection direction $\boldsymbol{\beta}_j$, $j = 1, \dots, m$. An alternative approach is to use common basis functions for each class and write $\boldsymbol{\phi}_j = \boldsymbol{\gamma}_j\phi_j$, for a vector $\boldsymbol{\gamma}_j = (\gamma_{j1}, \dots, \gamma_{jC})^T$; that is, for a given component j, the form of the nonlinear functions is the same for each response variable, but it has different weights $\gamma_{jk}, k = 1, \dots, C$. Bias terms are also introduced and we minimise I defined by

$$I \triangleq \sum_{i=1}^{n}\left|\boldsymbol{t}_i - \boldsymbol{\gamma}_0 - \sum_{j=1}^{m}\boldsymbol{\gamma}_j\phi_j(\boldsymbol{\beta}_j^T\boldsymbol{x}_i)\right|^2$$

with respect to $\boldsymbol{\gamma}_0, \boldsymbol{\gamma}_j$, the projection directions $\boldsymbol{\beta}_j (j = 1, \dots, m)$ and the functions ϕ_j.

The basic strategy is to proceed incrementally, beginning with $m = 1$ and increasing the number of basis functions one at a time until I is less than some threshold. An alternating least squares procedure is adopted: alternately minimising with respect to $\boldsymbol{\beta}_m$ and ϕ_m for fixed $\boldsymbol{\gamma}_j$, $j = 0, \dots, m$ (and fixed $\boldsymbol{\beta}_j$, ϕ_j, $j = 1, \dots, m - 1$ – determined on previous iterations), and then minimising with respect to $\boldsymbol{\gamma}_j (j = 0, \dots, m)$ for fixed $\boldsymbol{\beta}_j$, ϕ_j, $j = 1, \dots, m$. Specifically:

1. Set $m = 1$. $\boldsymbol{\gamma}_0 = \frac{1}{n}\sum_{i=1}^{n}\boldsymbol{t}_i$.

2. At the mth stage, set

$$\boldsymbol{r}_i = \boldsymbol{t}_i - \boldsymbol{\gamma}_0 - \sum_{j=1}^{m-1}\boldsymbol{\gamma}_j\phi_j(\boldsymbol{\beta}_j^T\boldsymbol{x}_i)$$

and initialise ϕ_m and $\boldsymbol{\gamma}_m$.

3. Find the direction $\hat{\boldsymbol{\beta}}_m$ that minimises

$$I = \sum_{i=1}^{n} |\boldsymbol{r}_i - \boldsymbol{\gamma}_m \phi_m(\boldsymbol{\beta}_m^T \boldsymbol{x}_i)|^2 \tag{6.24}$$

   Set $\boldsymbol{\beta}_m = \hat{\boldsymbol{\beta}}_m$. Estimate $\hat{\phi}_m$ by smoothing $(\boldsymbol{\beta}_m^T \boldsymbol{x}_i, \boldsymbol{\gamma}_m^T \boldsymbol{r}_i / |\boldsymbol{\gamma}_m|^2)$. Set $\phi_m = \hat{\phi}_m$ and repeat this step several times.

4. Find the $\hat{\boldsymbol{\gamma}}_m$ that minimises (6.24). Set $\boldsymbol{\gamma}_m = \hat{\boldsymbol{\gamma}}_m$.

5. Repeat steps 3 and 4 until the loss function is minimised with respect to $\boldsymbol{\gamma}_0, \boldsymbol{\gamma}_j$ $\boldsymbol{\beta}_j, (j = 1, \dots, m)$ and the functions ϕ_j. If I is not less than some prespecified threshold, then set $m = m + 1$ and go to step 2.

Hwang *et al.* (1994b) discuss smoothers for determining ϕ_m in step 3 and implement a parametric smoother based on Hermite polynomials:

$$\begin{aligned} H_0(z) &= 1 \\ H_1(z) &= 2z \\ H_r(z) &= 2(zH_{r-1}(z) - (r-1)H_{r-2}(z)) \end{aligned}$$

The projection directions, $\boldsymbol{\beta}_j$, are determined through the use of a Newton–Raphson method (Press *et al.*, 1992) for solving systems of nonlinear equations, which in this case is

$$\frac{\partial I}{\partial \boldsymbol{\beta}_j} = 0$$

6.3.3 Example application study

The problem Data reduction of large volumes of hyperspectral imagery for remote sensing applications (Ifarraguerri and Chang, 2000).

Summary Hyperspectral imaging sensors are capable of generating large volumes of data and it is necessary to exploit techniques for reducing this volume while simultaneously preserving as much information as possible. A projection pursuit is applied to hyperspectral digital imagery and significant reduction demonstrated.

The data The data comprise measurements made using the hyperspectral digital imagery collection equipment (HYDICE), an imaging spectrometer operating in the visible to short-wave infrared wavebands. There are 224 spectral bands; 256 × 256 images of scenes including vehicles, trees, roads and other features were collected. Each pixel spectrum (224 elements) is a pattern vector.

The model A projection pursuit model is adopted with a projection index that is based on the information divergence of the projection's estimated probability distributions from the Gaussian distribution.

Training procedure Each pattern vector is considered as a candidate projection vector. The data are projected onto each candidate vector in turn and the projection index calculated. The pattern vector with the largest value of the projection index is selected as the first projection direction. The data are then projected onto the space orthogonal to this and the process repeated: the second projection direction is chosen from among the set of projected patterns. The process is repeated and the result is a set of orthogonal projection vectors.

The projection index is a measure of the difference of the distribution of projected values (normalised to zero mean, unit variance) from a standard normal distribution.

Results Most of the information was compressed into seven projections, with the components tending to correspond to different types of objects. For example, the first captured information relating to two objects in the scene, the second to roads and some of the vehicles and the third and fourth mainly to shadows and trees respectively.

6.3.4 Further developments

One of the main developments of projection pursuit has been in the area of neural network models and, in particular, the link with the multilayer perceptron model. In a comparative study on simulated data, Hwang *et al.* (1994a) report similar accuracy and similar computation times, but projection pursuit requires fewer functions. Zhao and Atkeson (1996) use a radial basis function smoother for the nonlinear functions, and solve for the weights as part of the optimisation process (see also Kwok and Yeung, 1996).

6.3.5 Summary

Projection pursuit may produce projections of the data that reveal structure that is not apparent by using the coordinates axes or simple projections such as principal components (see Chapter 9). As a classifier, there are obvious links to the multilayer perceptron of Section 6.2 and radial basis function models discussed in Chapter 5. If we take the vector of nonlinear functions of projections, $\boldsymbol{\phi}_j$, in (6.23) as

$$\boldsymbol{\phi}_j(y) = \boldsymbol{\lambda}_j f(\boldsymbol{\beta}_j^T \boldsymbol{x}) \tag{6.25}$$

for weights $\boldsymbol{\lambda}_j$ and a *fixed* nonlinear function f (a logistic function), then our discriminant model is identical to the MLP discussed in Section 6.2.

The optimisation procedures of the two models differ in that in the MLP all the projection directions are determined simultaneously as part of the optimisation procedure. In projection pursuit, they are usually calculated sequentially.

A major difference between the MLP and projection pursuit is one of application. Within recent years, the MLP has been extensively assessed and applied to a diversity of real-world applications. This is not true of projection pursuit.

6.4 Application studies

The term 'neural networks' is used to describe models that are combinations of many simple processing units – for example, the radial basis function is a linear combination of kernel basis functions and the multilayer perceptron is a weighted combination of logistic units – but the boundary between what does and what does not constitute a neural network is rather vague. In many respects, the distinction is irrelevant, but it still persists, largely through historical reasons relating to the application domain, with neural network methods tending to be developed more in the engineering and computer science literature.

Applications of multilayer perceptrons are widespread and many of the studies listed at the end of the previous chapter will include an assessment of an MLP model. Zhang (2000) surveys the area of classification from a neural network perspective. Further special issues of journals have been devoted to applications in process engineering (Fernández de Cañete and Bulsari, 2000), human–computer interaction (Yasdi, 2000), financial engineering (Abu-Mostafa *et al.*, 2001) and data mining and knowledge discovery (Bengio *et al.*, 2000). This latter area is one that is currently receiving considerable attention. Over the last decade there has been a huge increase in the amount of information available from the internet, business and other sources. One of the challenges in the area of *data mining and knowledge discovery* is to develop models that can handle large data sets, with a large number of variables (high-dimensional). Many standard data analysis techniques may not scale suitably.

Applications of MLPs, trained using a Bayesian approach, to image analysis are given by Vivarelli and Williams (2001) and Lampinen and Vehtari (2001).

Applications of projection pursuit include the following:

- Facial image recognition and document image analysis. Arimura and Hagita (1994) use projection pursuit to design screening filters for feature extraction in an image recognition application.
- Target detection. Chiang *et al.* (2001) use projection pursuit in a target detection application using data from hyperspectral imagery.

6.5 Summary and discussion

Two basic models have been considered, namely the multilayer perceptron and the projection pursuit model. Both models take a sum of univariate nonlinear functions ϕ of linear projections of the data, $\boldsymbol{x}$, onto a weight vector $\boldsymbol{\alpha}$, and use this for discrimination (in a classification problem). In the MLP, the nonlinear functions are of a prescribed form (usually logistic) and a weighted sum is formed. Optimisation of the objective function is performed with respect to the projection directions and the weights in the sum. Projection pursuit may be used in an unsupervised form, to obtain low-dimensional representations of the data, and also for regression and classification purposes. In the projection pursuit model, optimisation is performed with respect to the form of the nonlinear function and the projection directions and usually proceeds sequentially, finding the best projection direction first, then the second best, and so on. Backfitting allows readjustment of solutions. Both models have very strong links to other techniques for discrimination and regression. Consequently, algorithms to implement these models rely heavily on the techniques presented elsewhere in this book. Procedures for initialisation, optimisation of parameters, classification and performance estimation are common to other algorithms.

There have been many links between neural networks, such as the MLP, and established statistical techniques. The basic idea behind neural network methods in the combination of simple nonlinear processing in a hierarchical manner, together with efficient algorithms for their optimisation. In this chapter we have confined our attention primarily to single hidden layer networks, where the links to statistical techniques are more apparent. However, more complex networks can be built for specific tasks. Also, neural networks are not confined to the feed-forward types for supervised pattern classification or regression as presented here. Networks with feedback from one layer to the previous layer, and unsupervised networks have been developed for a range of applications (see Chapter 10 for an introduction to self-organising networks). Views of neural networks from statistical perspectives, and relationships to other pattern classification approaches, are provided by Barron and Barron (1988), Ripley (1994, 1996), Cheng and Titterington (1994) and Holmström *et al.* (1997).

6.6 Recommendations

1. Before assessing a nonlinear technique, obtain results on some standard linear techniques (Chapter 4) and simple classifiers (for example, Gaussian classifier, Chapter 2; nearest-neighbour, Chapter 3) for comparison.

2. Standardise the data (zero mean, unit variance) before training an MLP.

3. Take care in model selection so that over-training does not result – either use some form of regularisation in the training procedure or a separate validation set to determine model order.

4. Train the multilayer perceptron using a batch training method unless the data set is very large (in this case, divide the data set into subsets, randomly selected from the original data set).

5. Standardise the data before projection pursuit – zero mean, unit variance.

6. Run the projection pursuit algorithm several times to obtain suitable projection directions.

7. If a small number of projections is required (perhaps for visualisation purposes), projection pursuit is preferred since it gives similar performance to a multilayer perceptron with fewer projections.

6.7 Notes and references

Bishop (1995) provides a through account of feed-forward neural networks. An engineering perspective is provided by Haykin (1994). Tarassenko (1998) provides a basic introduction to neural network methods with an emphasis on applications.

Projection pursuit was originally proposed by Kruskal (1972) and developed by Friedman and Tukey (1974). Projection pursuit has been proposed in a regression context (Friedman and Stuetzle, 1981), density estimation (Friedman *et al.*, 1984) and exploratory data analysis (Friedman, 1987). Reviews are provided by Huber (1985) and Jones and Sibson (1987).

The website `www.statistical-pattern-recognition.net` contains references and links to further information on techniques and applications.

Exercises

1. Discuss the differences and similarities between a multilayer perceptron and a radial basis function classifier with reference to network structure, geometric interpretation, initialisation of parameters and algorithms for parameter optimisation.

2. For the logistic form of the nonlinearity in an MLP,

$$\phi(z) = \frac{1}{1 + \exp(-z)}$$

show that

$$\frac{\partial \phi}{\partial z} = \phi(z)(1 - \phi(z))$$

3. Consider an MLP with a final normalising layer ('softmax'). How would you initialise the MLP to give a nearest class mean classifier?

4. Describe two ways of estimating the posterior probabilities of class membership without modelling the class densities explicitly and using Bayes' theorem; state the assumptions and conditions of validity.

5. Logistic classification. Given (6.15) and (6.14), derive the expression for the derivative,

$$\frac{\delta E^i}{\delta a_j^0} = -(\boldsymbol{t}_i)_j + (\boldsymbol{o}_i)_j$$

where $\boldsymbol{o}_i$ is the output for input pattern $\boldsymbol{x}_i$. Show that for the squared error defined as

$$E = \frac{1}{2} \sum_{i=1}^{n} |\boldsymbol{t}_i - \boldsymbol{g}(\boldsymbol{x}_i)|^2$$

an identical result is obtained.

6. Write down a training strategy for a multilayer perceptron using the back-fitting procedure of projection pursuit. What are the potential advantages and disadvantages of such a training scheme compared with the conventional nonlinear optimisation scheme usually employed?

7. How could a multilayer perceptron be used as part of projection pursuit learning?

8. What are the main similarities and differences between a multilayer perceptron and a projection pursuit model? Take account of model structure, computational complexity, training strategy and optimisation procedures.

9. Using the results that for a nonsingular matrix $\boldsymbol{A}$ and vector $\boldsymbol{u}$,

$$(\boldsymbol{A} + \boldsymbol{u}\boldsymbol{u}^T)^{-1} = \boldsymbol{A}^{-1} - \boldsymbol{A}^{-1}\boldsymbol{u}\boldsymbol{u}^T\boldsymbol{A}^{-1}/(1 + \boldsymbol{u}^T\boldsymbol{A}^{-1}\boldsymbol{u})$$

derive the result (6.20) from (6.19) with σ_t^2 given by (6.21). What is the expression for the matrix $\boldsymbol{A}$ if an RBF model is used?

10. Consider a radial basis function network with a data-adaptive kernel (one that adapts to minimise the criterion optimised by training an RBF); that is, we are allowed to adjust the shape of the kernel to minimise the squared error, for example. Under what conditions will the radial basis function model approximate a projection pursuit model?

7

Tree-based methods

Overview

Classification or decision trees are capable of modelling complex nonlinear decision boundaries. An overly large tree is constructed and then pruned to minimise a cost–complexity criterion. The resulting tree is easily interpretable and can provide insight into the data structure. A procedure that can be regarded as a generalisation to provide continuous regression functions is described.

7.1 Introduction

Many discrimination methods are based on an expansion into sums of basis functions. The radial basis function uses a weighted sum of univariate functions of radial distances; the multilayer perceptron uses a weighted sum of sigmoidal functions of linear projections; projection pursuit uses a sum of smoothed functions of univariate regressions of projections. In this chapter, we consider two methods that construct the basis functions by recursively partitioning the data space. The classification and regression tree (CART) model uses an expansion into indicator functions of multidimensional rectangles. The multivariate adaptive regression spline (MARS) model is based on an expansion into sums of products of univariate linear spline functions.

7.2 Classification trees

7.2.1 Introduction

A *classification tree* or a decision tree is an example of a multistage decision process. Instead of using the complete set of features jointly to make a decision, different subsets of features are used at different levels of the tree. Let us illustrate with an

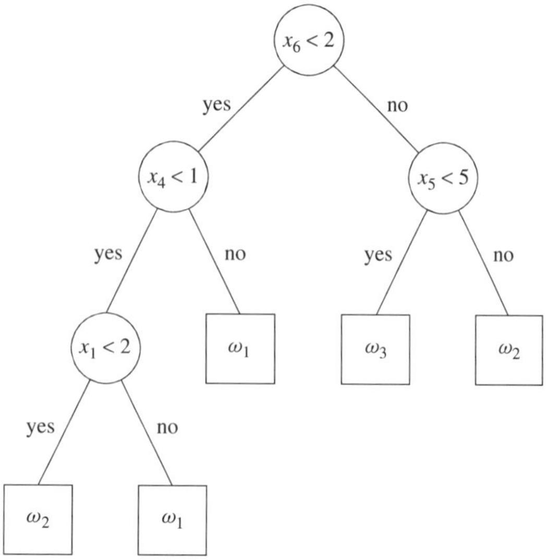

Figure 7.1 Classification tree

example. Figure 7.1 gives a classification tree solution to the head injury data problem[1] of Chapter 3. Associated with each *internal node* of the tree (denoted by a circle) are a variable and a threshold. Associated with each *leaf* or *terminal node* (denoted by a square) is a class label. The top node is the *root* of the tree. Now suppose we wish to classify the pattern $\boldsymbol{x} = (5, 4, 6, 2, 2, 3)$. Beginning at the root, we compare the value of x_6 with the threshold 2. Since the threshold is exceeded, we proceed to the *right child*. We then compare the value of the variable x_5 with the threshold 5. This is not exceeded, so we proceed to the *left child*. The decision at this node leads us to the terminal node with classification label ω_3. Thus, we assign the pattern $\boldsymbol{x}$ to class ω_3.

The classification tree of Figure 7.1 is a conceptually simple approximation to a complex procedure that breaks up the decision into a series of simpler decisions at each node. The number of decisions required to classify a pattern depends on the pattern. Figure 7.1 is an example of a *binary decision tree*. More generally, the outcome of a decision could be one of $m \geq 2$ possible categories. However, we restrict our treatment in this chapter to binary trees.

Binary trees successfully partition the feature space into two parts. In the example above, the partitions are hyperplanes parallel to the coordinate axes. Figure 7.2 illustrates this in two dimensions for a two-class problem and Figure 7.3 shows the corresponding binary tree. The tree gives 100% classification performance on the design set. The tree is not unique for the given partition. Other trees also giving 100% performance are possible.

The decisions at each node need not be thresholds on a single variable (giving hyperplanes parallel to coordinate axes as decision boundaries), but could involve a linear or nonlinear combination of variables. In fact, the data in Figure 7.2 are linearly separable

[1] Variables $x_1, \ldots, x_6$ are `age`, `EMV score`, `MRP`, `change`, `eye indicant` and `pupils` and classes ω_1, ω_2 and ω_3 dead/vegetative, severely disabled and moderate or good recovery.

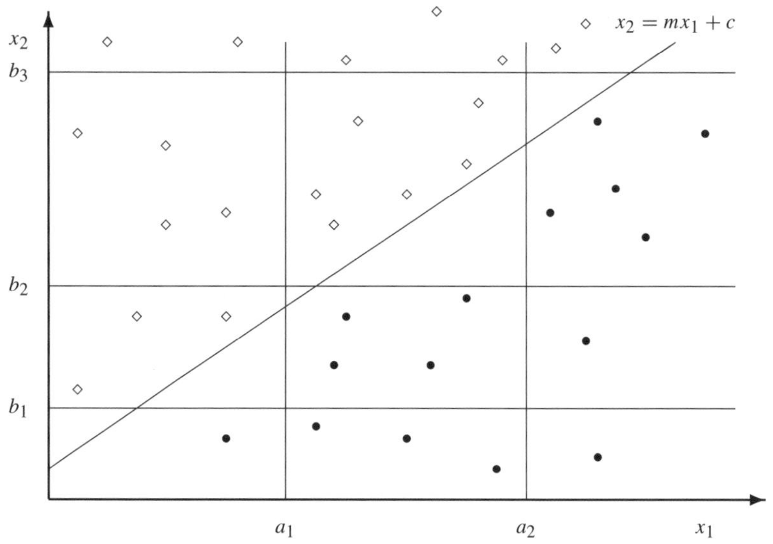

Figure 7.2 Boundaries on a two-class, two-dimension problem

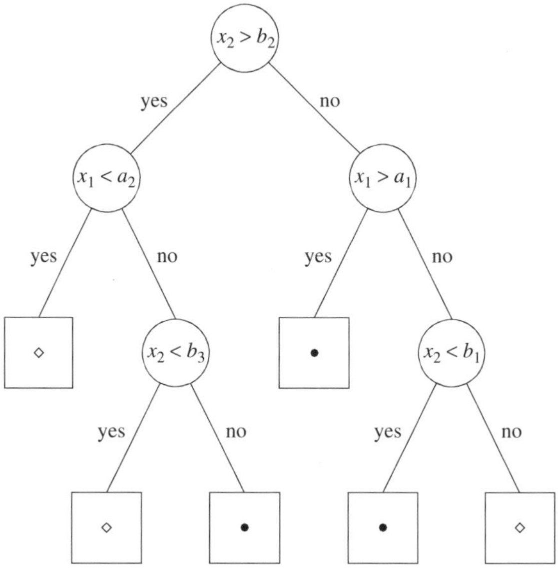

Figure 7.3 Binary decision tree for the two-class, two-dimension data of Figure 7.2

and the decision rule: assign $\boldsymbol{x}$ to class $\diamond$ if $x_2 - mx_1 - c > 0$ classifies all samples correctly.

Classification trees have been used on a wide range of problems. Their advantages are that they can be compactly stored; they efficiently classify new samples and have demonstrated good generalisation performance on a wide variety of problems. Possible disadvantages are the difficulty in designing an optimal tree, leading perhaps to large trees with poor error rates on certain problems, particularly if the separating boundary is complicated and a binary decision tree with decision boundaries parallel to the coordinate axes is used. Also, most approaches are non-adaptive – they use a fixed training set, and additional data may require redesign of the tree.

There are several heuristic methods for construction of decision-tree classifiers; see Safavian and Landgrebe (1991) for a survey of decision-tree classifier methodology. They are usually constructed top-down, beginning at the root node (a classification tree is usually drawn with its root at the top and its leaves at the bottom) and successively partitioning the feature space. The construction involves three steps:

1. Selecting a *splitting rule* for each internal node. This means determining the features, together with a threshold, that will be used to partition the data set at each node.

2. Determining which nodes are terminal nodes. This means that, for each node, we must decide whether to continue splitting or to make the node a terminal node and assign to it a class label. If we continue splitting until every terminal node has pure class membership (all samples in the design set that arrive at that node belong to the same class) then we are likely to end up with a large tree that over-fits the data and gives a poor error rate on an unseen test set. Alternatively, relatively impure terminal nodes (nodes for which the corresponding subset of the design set has mixed class membership) lead to small trees that may under-fit the data. Several *stopping rules* have been proposed in the literature, but the approach suggested by Breiman *et al.* (1984) is successively to grow and selectively prune the tree, using *cross-validation* to choose the subtree with the lowest estimated misclassification rate.

3. Assigning class labels to terminal nodes. This is relatively straightforward and labels can be assigned by minimising the estimated misclassification rate.

We shall now consider each of these stages in turn. The approach we present is based on the CART[2] description of Breiman *et al.* (1984).

7.2.2 Classifier tree construction

We begin by introducing some notation. A *tree* is defined to be a set T of positive integers together with two functions $l(.)$ and $r(.)$ from T to $T \cup \{0\}$. Each member of T corresponds to a node in the tree. Figure 7.4 shows a tree and the corresponding values of $l(t)$ and $r(t)$ (denoting the left and right nodes):

1. For each $t \in T$, either $l(t) = 0$ and $r(t) = 0$ (a terminal node) or $l(t) > 0$ and $r(t) > 0$ (a non-terminal node).

2. Apart from the root node (the smallest integer, $t = 1$ in Figure 7.4) there is a unique *parent* $s \in T$ of each node; that is, for $t \neq 1$, there is an s such that either $t = l(s)$ or $t = r(s)$.

[2]CART is a registered trademark of California Statistical Software, Inc.

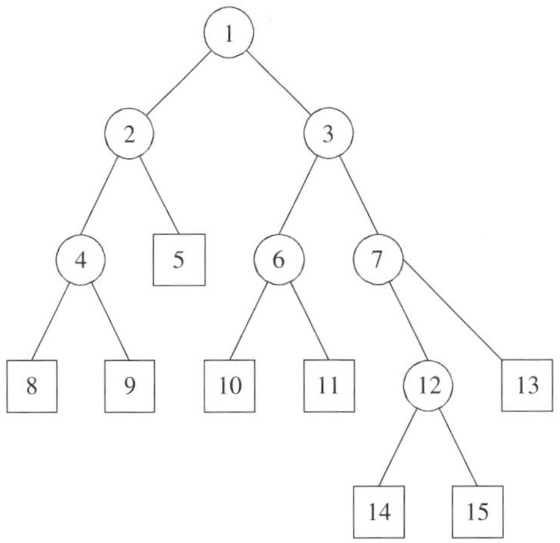

t	$l(t)$	$r(t)$
1	2	3
2	4	5
3	6	7
4	8	9
5	0	0
6	10	11
7	12	13
8	0	0
9	0	0
10	0	0
11	0	0
12	14	15
13	0	0
14	0	0
15	0	0

Figure 7.4 Classification tree and the values of $l(t)$ and $r(t)$

A *subtree* is a non-empty subset T_1 of T together with two functions l_1 and r_1 such that

$$l_1(t) = \begin{cases} l(t) & \text{if } l(t) \in T_1 \\ 0 & \text{otherwise} \end{cases}$$
$$r_1(t) = \begin{cases} r(t) & \text{if } r(t) \in T_1 \\ 0 & \text{otherwise} \end{cases} \tag{7.1}$$

and provided that $T_1, l_1(.)$ and $r_1(.)$ form a tree. For example, the set $\{3, 6, 7, 10, 11\}$ together with (7.1) forms a subtree, but the sets $\{2, 4, 5, 3, 6, 7\}$ and $\{1, 2, 4, 3, 6, 7\}$ do not; in the former case because there is no parent for both 2 and 3 and in the latter case because $l_1(2) > 0$ and $r_1(2) = 0$. A *pruned subtree* T_1 of T is a subtree of T that has the same root. This is denoted by $T_1 \leq T$. Thus, example (b) in Figure 7.5 is a *pruned* subtree, but example (a) is not (though it is a subtree).

Let $\tilde{T}$ denote the set of terminal nodes (the set $\{5, 8, 9, 10, 11, 13, 14, 15\}$ in Figure 7.4). Let $\{u(t), t \in \tilde{T}\}$ be a partition of the data space $\mathbb{R}^p$ (that is, $u(t)$ is a subspace of $\mathbb{R}^p$ associated with a terminal node such that $u(t) \cap u(s) = \oslash$ for $t \neq s, t, s \in \tilde{T}$; and $\cup_{t \in \tilde{T}} u(t) = \mathbb{R}^p$). Let $\omega_{j(t)} \in \{\omega_1, \ldots, \omega_C\}$ denote one of the class labels. Then a

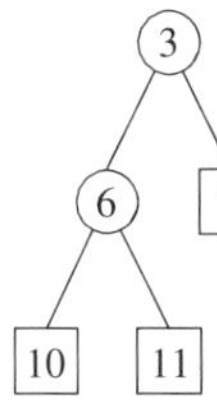

t	$l_1(t)$	$r_1(t)$
3	6	7
6	10	11
7	0	0
10	0	0
11	0	0

(a)

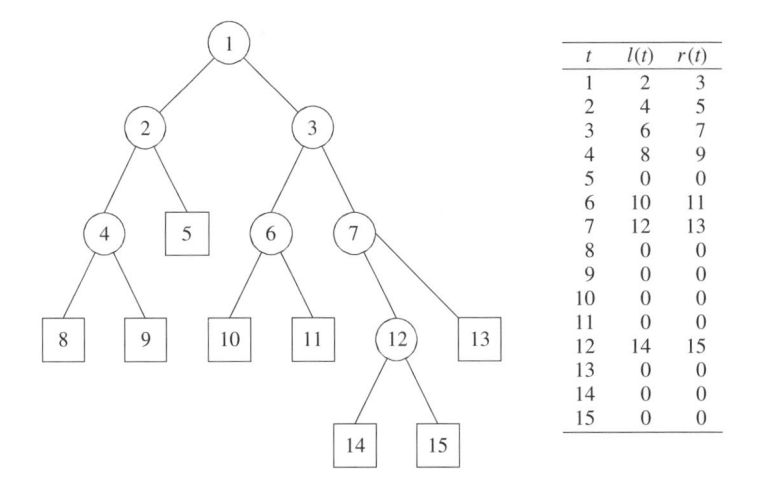

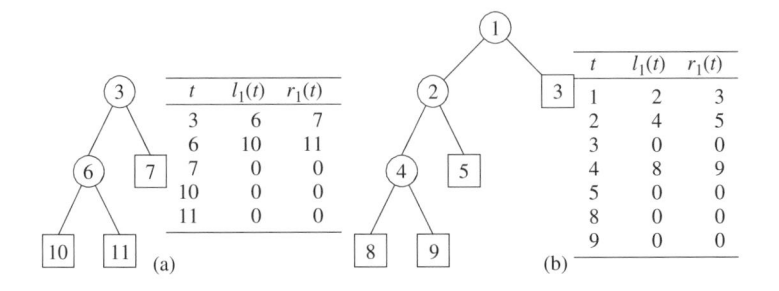

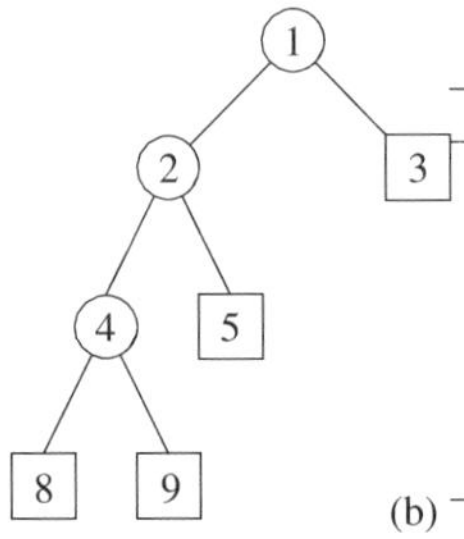

t	$l_1(t)$	$r_1(t)$
1	2	3
2	4	5
3	0	0
4	8	9
5	0	0
8	0	0
9	0	0

(b)

Figure 7.5 Possible subtrees of the tree in Figure 7.4; (a) is not a *pruned subtree*; (b) is a *pruned subtree*

classification tree consists of the tree T together with the class labels $\{\omega_{j(t)}, t \in \tilde{T}\}$ and the partition $\{u(t), t \in \tilde{T}\}$. All we are saying here is that associated with each terminal node is a region of the data space that we label as belonging to a particular class. There is also a subspace of $\mathbb{R}^p$, $u(t)$, associated with each *non-terminal* node, being the union of the subspaces of the terminal nodes that are its descendants.

A classification tree is constructed using a labelled data set, $\mathcal{L} = \{(\boldsymbol{x}_i, y_i), i = 1, \ldots, n\}$ where $\boldsymbol{x}_i$ are the data samples and y_i the corresponding class labels. If we let $N(t)$ denote the number of samples of $\mathcal{L}$ for which $\boldsymbol{x}_i \in u(t)$ and $N_j(t)$ be the number of samples for which $\boldsymbol{x}_i \in u(t)$ and $y_i = \omega_j$ ($\sum_j N_j(t) = N(t)$), then we may define

$$p(t) = \frac{N(t)}{n} \tag{7.2}$$

an estimate of $p(\boldsymbol{x} \in u(t))$ based on $\mathcal{L}$;

$$p(\omega_j|t) = \frac{N_j(t)}{N(t)} \tag{7.3}$$

an estimate of $p(y = \omega_j|\boldsymbol{x} \in u(t))$ based on $\mathcal{L}$; and

$$p_L = \frac{p(t_L)}{p(t)} \qquad p_R = \frac{p(t_R)}{p(t)}$$

where $t_L = l(t), t_R = r(t)$, as estimates of $p(\boldsymbol{x} \in u(t_L)|\boldsymbol{x} \in u(t))$ and $p(\boldsymbol{x} \in u(t_R)|\boldsymbol{x} \in u(t))$ based on $\mathcal{L}$ respectively.

We may assign a label to each node, t, according to the proportions of samples from each class in $u(t)$: assign label ω_j to node t if

$$p(\omega_j|t) = \max_i p(\omega_i|t)$$

We have now covered most of the terminology that we shall use. It is not difficult to understand, but it may be a bit much to take in all at once. Figure 7.6 and Table 7.1 illustrate some of these concepts using the data of Figure 7.2 and the tree of Figure 7.3.

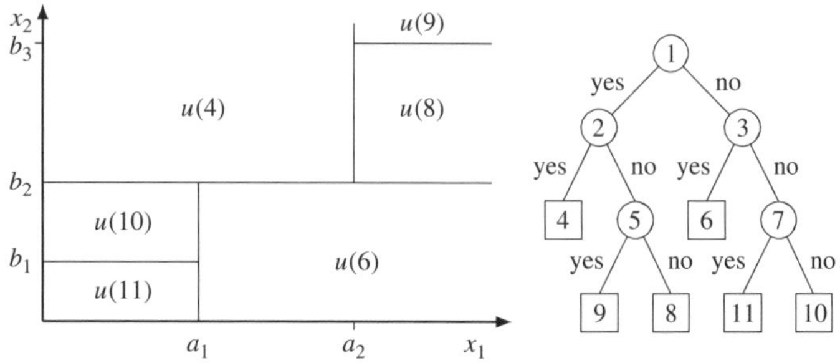

Figure 7.6 Classification tree and decision regions. A description of the nodes is given in Table 7.1

Table 7.1 Tree table – class $\omega_1 = \diamond$; class $\omega_2 = \bullet$

t	node	$l(t)$	$r(t)$	$N(t)$	$N_1(t)$	$N_2(t)$	$p(t)$	$p(1\|t)$	$p(2\|t)$	p_L	p_R
1	$x_2 > b_2$	2	3	35	20	15	1	$\frac{20}{35}$	$\frac{15}{35}$	$\frac{22}{35}$	$\frac{13}{35}$
2	$x_1 < a_2$	4	5	22	17	5	$\frac{22}{35}$	$\frac{17}{22}$	$\frac{5}{22}$	$\frac{15}{22}$	$\frac{7}{22}$
3	$x_1 > a_1$	6	7	13	4	9	$\frac{13}{35}$	$\frac{4}{13}$	$\frac{9}{13}$	$\frac{9}{13}$	$\frac{4}{13}$
4	$\diamond$	0	0	15	15	0	$\frac{15}{35}$	1	0		
5	$x_2 > b_3$	9	8	7	2	5	$\frac{7}{35}$	$\frac{2}{7}$	$\frac{5}{7}$	$\frac{5}{7}$	$\frac{2}{7}$
6	$\bullet$	0	0	9	0	9	$\frac{9}{35}$	0	1		
7	$x_2 < b_1$	11	10	4	3	1	$\frac{4}{35}$	$\frac{3}{4}$	$\frac{1}{4}$	$\frac{3}{4}$	$\frac{1}{4}$
8	$\bullet$	0	0	5	0	5	$\frac{5}{35}$	0	1		
9	$\diamond$	0	0	2	2	0	$\frac{2}{35}$	1	0		
10	$\diamond$	0	0	3	3	0	$\frac{3}{35}$	1	0		
11	$\bullet$	0	0	1	0	1	$\frac{1}{35}$	0	1		

Splitting rules

A splitting rule is a prescription for deciding which variable, or combination of variables, should be used at a node to divide the samples into subgroups, and for deciding what the threshold on that variable should be. A split consists of a condition on the coordinates of a vector $\boldsymbol{x} \in \mathbb{R}^p$. For example, we may define a split s_p to be

$$s_p = \{\boldsymbol{x} \in \mathbb{R}^p;\ x_4 \leq 8.2\}$$

a threshold on an individual feature, or

$$s_p = \{\boldsymbol{x} \in \mathbb{R}^p;\ x_2 + x_7 \leq 2.0\}$$

a threshold on a linear combination of features. Nonlinear functions have also been considered (Gelfand and Delp, 1991). Thus, at each non-terminal node, suppose $\boldsymbol{x} \in u(t)$. Then if $\boldsymbol{x} \in s_p$, the next step in the tree is to $l(t)$, the left branch, otherwise $r(t)$.

The question we now have to address is how to split the data that lie in the sub- $u(t)$ at node t. Following Breiman *et al.* (1984), we define the node impurity fu-

$\mathcal{I}(t)$, to be

$$\mathcal{I}(t) = \phi(p(\omega_1|t), \ldots, p(\omega_C|t))$$

where ϕ is a function defined on all C-tuples $(q_1, \ldots, q_C)$ satisfying $q_j \geq 0$ and $\sum_j q_j = 1$. It has the following properties:

1. ϕ is a maximum only when $q_j = 1/C$ for all j.

2. It is a minimum when $q_j = 1$, $q_i = 0, i \neq j$, for all j.

3. It is a symmetric function of $q_1, \ldots, q_C$.

One measure of the goodness of a split is the change in the impurity function. A split that maximises the decrease in the node impurity function when moving from one group to two

$$\Delta\mathcal{I}(s_p, t) \triangleq \mathcal{I}(t) - (\mathcal{I}(t_L)p_L + \mathcal{I}(t_R)p_R)$$

over all splits s_p is one choice. Several different forms for $\mathcal{I}(t)$ have been used. One suggestion is the *Gini criterion*

$$\mathcal{I}(t) = \sum_{i \neq j} p(\omega_i|t)p(\omega_j|t)$$

This is easily computed for a given split.

Different splits on the data at node t may now be evaluated using the above criterion, but what strategy do we use to search through the space of possible splits? First of all, we must confine our attention to splits of a given form. We shall choose the possible splits to consist of thresholds on individual variables

$$s_p = \{\boldsymbol{x};\ x_k \leq \tau\}$$

where $k = 1, \ldots, p$ and τ ranges over the real numbers. Clearly, we must restrict the number of splits we examine and so, for each variable x_k, τ is allowed to take one of a finite number of values within the range of possible values. Thus, we are dividing each variable into a number of categories, though this should be kept reasonably small to prevent excessive computation and need not be larger than the number of samples at each node.

There are many other approaches to splitting. Some are given in the survey by Safavian and Landgrebe (1991). The approach described above assumes ordinal variables. For a categorical variable with N unordered outcomes, partitioning on that variable means considering 2^N partitions. By exhaustive search, it is possible to find the optimal one, but this could result in excessive computation time if N is large. Techniques for nominal variables can be found in Breiman *et al.* (1984) and Chou (1991).

So now we can grow our tree by successively splitting nodes, but how do we stop? We could continue until each terminal node contained one observation only. This would lead to a very large tree, for a large data set, that would over-fit the data. We could implement a stopping rule: we do not split the node if the change in the impurity function is less than a prespecified threshold. Alternatively, we may grow a tree with terminal nodes that would have pure (or nearly pure) class membership and then prune it. This can lead to better performance than a stopping rule. We now discuss one pruning algorithm.

Pruning algorithm

We now consider one algorithm for pruning trees that will be required in the following subsection on classifier tree construction. The pruning algorithm is general in that it applies to trees that are not necessarily classification but regression trees. But first some more notation.

Let $R(t)$ be real numbers associated with each node t of a given tree T. If t is a terminal node, i.e. $t \in \tilde{T}$, then $R(t)$ could represent the proportion of misclassified samples – the number of samples in $u(t)$ that do not belong to the class associated with the terminal node, defined to be $M(t)$, divided by the total number of data points, n

$$R(t) = \frac{M(t)}{n} \qquad t \in \tilde{T}$$

Let $R_\alpha(t) = R(t) + \alpha$ for a real number α. Set[3]

$$R(T) = \sum_{t \in \tilde{T}} R(t)$$

$$R_\alpha(T) = \sum_{t \in \tilde{T}} R_\alpha(t) = R(T) + \alpha|\tilde{T}|$$

In a classification problem, $R(T)$ is the *estimated misclassification rate*; $|\tilde{T}|$ denotes the cardinality of the set $\tilde{T}$; $R_\alpha(T)$ is the *estimated complexity–misclassification rate* of a classification tree; and α is a constant that can be thought of as the cost of complexity per terminal node. If α is small, then there is a small penalty for having a large number of nodes. As α increases, the *minimising subtree* (the subtree $T' \leq T$ that minimises $R_\alpha(T')$) has fewer terminal nodes.

We shall describe the CART pruning algorithm (Breiman *et al.*, 1984) by means of an example, using the tree of Figure 7.7. Let the quantity $R(t)$ be given by

$$R(t) = r(t)p(t)$$

where $r(t)$ is the resubstitution estimate of the probability of misclassification (see Chapter 8) given that a case falls into node t,

$$r(t) = 1 - \max_{\omega_j} p(\omega_j|t)$$

and $p(t)$ and $p(\omega_j|t)$ are given by (7.2) and (7.3). Thus, if t is taken to be a terminal node, $R(t)$ is the contribution of that node to the total error.

Let T_t be the subtree with root t. If $R_\alpha(T_t) < R_\alpha(t)$, then the contribution to the cost of complexity of the subtree is less than that for the node t. This occurs for small α. As α increases, equality is achieved when

$$\alpha = \frac{R(t) - R(T_t)}{N_d(t) - 1}$$

[3]The argument of R may be a tree or a node; capital letters denote a tree.

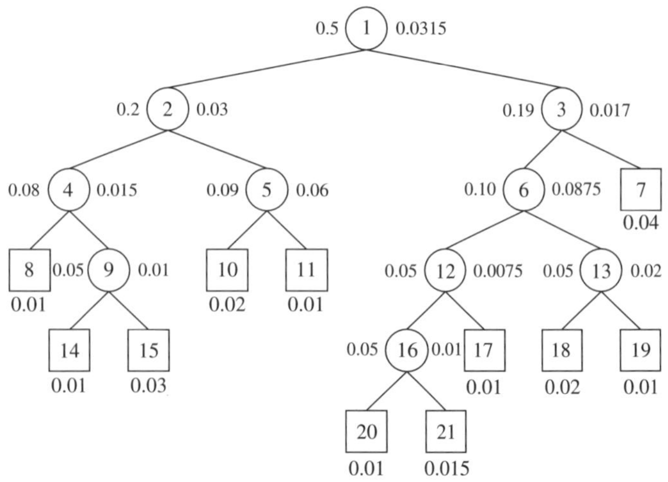

Figure 7.7 Pruning example – original tree

where $N_d(t)$ is the number of terminal nodes in T_t, i.e. $N_d(t) = |\tilde{T}_t|$ and termination of the tree at t is preferred. Therefore, we finally define

$$g(t) = \frac{R(t) - R(T_t)}{N_d(t) - 1} \tag{7.4}$$

as a measure of the strength of the link from node t.

In Figure 7.7, each terminal node has been labelled with a single number $R(t)$, the amount by which that node contributes to the error rate. Each non-terminal node has been labelled by two numbers. The number to the left of the node is the value of $R(t)$, the contribution to the error rate if that node were a terminal node. The number to the right is $g(t)$, calculated using (7.4). Thus, the value of $g(t)$ for node $t = 2$, say, is $0.03 = [0.2 - (0.01 + 0.01 + 0.03 + 0.02 + 0.01)]/4$.

The first stage of the algorithm searches for the node with the smallest value of $g(t)$. This is node 12, with a value of 0.0075. This is now made a terminal node and the value of $g(t)$ recalculated for all its ancestors. This is shown in Figure 7.8. The values of the nodes in the subtree beginning at node 2 are unaltered. The process is now repeated. The new tree is searched to find the node with the smallest value of $g(t)$. In this case, there are two nodes, 6 and 9, each with a value of 0.01. Both are made terminal nodes and again the values of $g(t)$ for the ancestors recalculated. Figure 7.9 gives the new tree (T^3). Node 4 now becomes the terminal node. This continues until all we are left with is the root node. Thus, the pruning algorithm generates a succession of trees. We denote the tree at the kth stage by T^k. Table 7.2 gives the value of the error rate for each successive tree, together with the number of terminal nodes in each tree and the value of $g(t)$ at each stage (denoted by α_k) that is used in the pruning to generate tree T^k. The tree T^k has all internal nodes with a value of $g(t) > \alpha_k$.

The results of the pruning algorithm are summarised in Figure 7.10. This figure shows the original tree together with the values $g_6(t)$ for the internal nodes, where g_k is defined

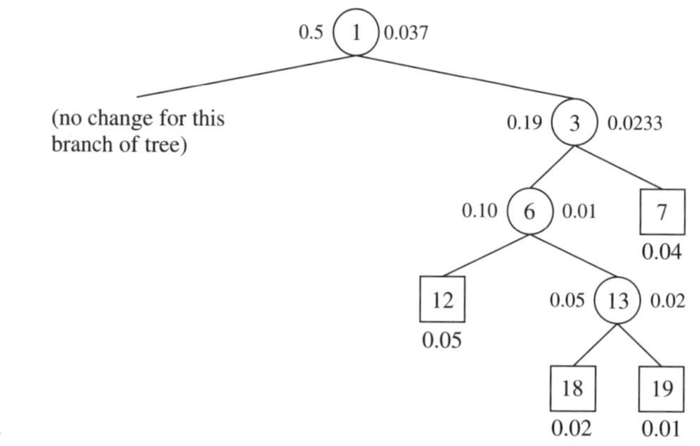

Figure 7.8 Pruning example – pruning at node 12

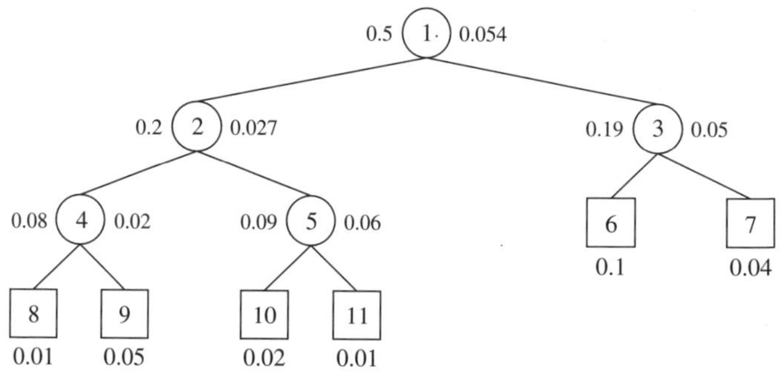

Figure 7.9 Pruning example – pruning at nodes 6 and 9

Table 7.2 Tree results

k	α_k	$\lvert\tilde{T}^k\rvert$	$R(T^k)$
1	0	11	0.185
2	0.0075	9	0.2
3	0.01	6	0.22
4	0.02	5	0.25
5	0.045	3	0.34
6	0.05	2	0.39
7	0.11	1	0.5

recursively ($0 \leq k \leq K - 1$; K = the number of pruning stages):

$$g_k = \begin{cases} g(t) & t \in T^k - \tilde{T}^k \quad (t \text{ an internal node of } T^k) \\ g_{k-1}(t) & \text{otherwise} \end{cases}$$

The values of $g_6(t)$ together with the tree T^1 are a useful means of summarising the pruning process. The smallest value is 0.0075 and shows that in going from T^1 to T^2

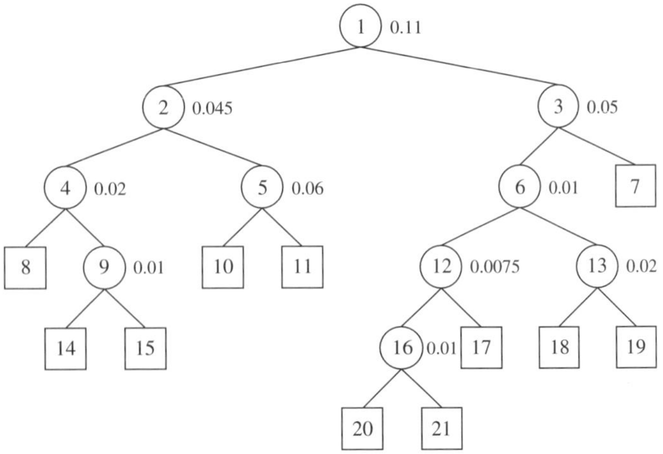

Figure 7.10 Pruning example – summary of pruning process

nodes 16, 17, 20 and 21 are removed. The smallest value of the pruned tree is 0.01 and therefore in going from T^2 to T^3 nodes 12, 13, 14, 15, 18 and 19 are removed; from T^3 to T^4 nodes 8 and 9 are removed; from T^4 to T^5 nodes 4, 5, 10 and 11; from T^5 to T^6 nodes 6 and 7; and finally, nodes 2 and 3 are removed to obtain T^7 the tree with a single node.

An explicit algorithm for the above process is given by Breiman *et al.* (1984).

Classification tree construction methods

Having considered approaches to growing and pruning classification trees and estimating their error rate, we are now in a position to present methods for tree construction using these features.

CART independent training and test set method The stages in the CART independent training and test set method are as follows. Assume that we are given a training set $\mathcal{L}_r$ and a test set (or, more correctly, a validation set) $\mathcal{L}_s$ of data samples. These are generated by partitioning the design set approximately into two groups.

The CART independent training and test set method is as follows.

1. Use the set $\mathcal{L}_r$ to generate a tree T by splitting all the nodes until all the terminal nodes are 'pure' – all samples at each terminal node belong to the same class. It may not be possible to achieve this with overlapping distributions, therefore an alternative is to stop when the number in each terminal node is less than a given threshold or the split of a node t results in a left son t_L or a right son t_R with $\min(N(t_L), N(t_R)) = 0$.
2. Use the CART pruning algorithm to generate a nested sequence of subtrees T^k using the set $\mathcal{L}_s$.
3. Select the smallest subtree for which $R(T^k)$ is a minimum.

CART cross-validation method For the cross-validation method, the training set $\mathcal{L}$ is divided into V subsets $\mathcal{L}_1, \ldots, \mathcal{L}_V$, with approximately equal numbers in each class.

Let $\mathcal{L}^v = \mathcal{L} - \mathcal{L}_v$, $v = 1, \ldots, V$. We shall denote by $T(\alpha)$ the pruned subtree with all internal nodes having a value of $g(t) > \alpha$. Thus, $T(\alpha)$ is equal to T^k, the pruned subtree at the kth stage, where k is chosen so that $\alpha_k \leq \alpha \leq \alpha_{k+1}$ $(\alpha_{K+1} = \infty)$.

The CART cross-validation method is as follows.

1. Use the set $\mathcal{L}$ to generate a tree T by splitting all the nodes until all the terminal nodes are 'pure' – all samples at each terminal node belong to the same class. It may not be possible to achieve this with overlapping distributions, therefore an alternative is to stop when the number in each terminal node is less than a given threshold or the split of a node t results in a left son t_L or a right son t_R with $\min(N(t_L), N(t_R)) = 0$. (This is step 1 of the CART independent training and test set method above.)

2. Using the CART pruning algorithm, generate a nested sequence of pruned subtrees $T = T^0 \geq T^1 \geq \cdots \geq T^K = \text{root}(T)$.

3. Use $\mathcal{L}_v$ to generate a tree T_v and assign class labels to the terminal nodes, for $v = 1, \ldots, V$.

4. Using the CART pruning algorithm, generate a nested sequence of pruned subtrees of T_v.

5. Calculate $R^{\text{cv}}(T^k)$ (the cross-validation estimate of the misclassification rate) given by

$$R^{\text{cv}}(T^k) = \frac{1}{V} \sum_{v=1}^{V} R_v(T_v(\sqrt{\alpha_k \alpha_{k+1}}))$$

   where R_v is the estimate of the misclassification rate based on the set $\mathcal{L}_v$ for the pruned subtree $T_v(\sqrt{\alpha_k \alpha_{k+1}})$.

6. Select the smallest $T^* \in \{T^0, \ldots, T^K\}$ such that

$$R^{\text{cv}}(T^*) = \min_k R^{\text{cv}}(T^k)$$

7. Estimate the misclassification rate by

$$\hat{R}(T^*) = R^{\text{cv}}(T^*)$$

The procedure presented in this section implements one of many approaches to recursive tree design which use a growing and pruning approach and that have emerged as reliable techniques for determining right-sized trees. The CART approach is appropriate for data sets of continuous or discrete variables with ordinal or nominal significance, including data sets of mixed variable types.

7.2.3 Other issues

Missing data

The procedure given above contains no mechanism for handling missing data. The CART algorithm deals with this problem through the use of *surrogate splits*. If the best split of a node is $\int$ on the variable x_m say, then the split $\int^*$ that predicts $\int$ most accurately, on a variable x_j other than x_m, is termed the *best surrogate for* $\int$. Similarly, a second best surrogate on a variable other than x_m and x_j can be found, and so on.

A tree is constructed in the usual manner, but at each node t, the best split $\int$ on a variable x_m is found by considering only those samples for which a value for x_m is available. Objects are assigned to the groups corresponding to t_L and t_R according to the value on x_m. If this is missing for a given test pattern, the split is made using the best surrogate for $\int$ (that is, the split is made on a different variable). If this value is also missing, the second best surrogate is used, and so on until the sample is split. Alternatively, we may use procedures that have been developed for use by conventional classifiers for dealing with missing data (see Chapter 11).

Priors and costs

The definitions (7.2) and (7.3) assume that the prior probabilities for each class, denoted by $\pi(i)$, are equal to N_j/n. If the distribution on the design set is not proportional to the expected occurrence of the classes, then the resubstitution estimates of the probabilities that a sample falls into node t, $p(t)$, and that it is in class ω_j given that it falls into node t, $p(j|t)$, are defined as

$$\begin{aligned} p(t) &= \sum_{j=1}^{C} \pi(j) \frac{N_j(t)}{N_j} \\ p(j|t) &= \frac{\pi(j)N_j(t)/N_j}{\sum_{j=1}^{C} \pi(j)N_j(t)/N_j} \end{aligned} \tag{7.5}$$

In the absence of costs (or assuming an equal cost loss matrix – see Chapter 1), the misclassification rate is given by

$$R(T) = \sum_{ij} q(i|j)\pi(j) \tag{7.6}$$

where $q(i|j)$ is the proportion of samples of class ω_j defined as class ω_i by the tree. If λ_{ji} is the cost of assigning an object of class ω_j to ω_i, then the misclassification cost is

$$R(T) = \sum_{ij} \lambda_{ji} q(i|j)\pi(j)$$

This may be written in the same form as (7.6), with redefined priors, provided that λ_{ji} is independent of i; thus, $\lambda_{ji} = \lambda_j$ and the priors are redefined as

$$\pi'(j) = \frac{\lambda_j \pi(j)}{\sum_j \lambda_j \pi(j)}$$

In general, if there is an asymmetric cost matrix with non-constant costs of misclassification for each class, then the costs cannot be incorporated into modified priors.

7.2.4 Example application study

The problem Identification of emitter type based on measurements of electronic characteristics (Brown *et al.*, 1993). This is important in military applications (to distinguish friend from foe) and civil applications such as communication spectrum management and air space management.

Summary A classification tree was compared with a multilayer perceptron on the task of identifying the emitter type given measurements made on a received waveform. Both methods produced comparable error rates.

The data The class distributions are typically multimodal since a given emitter can change its electronic signature by changing its settings. The data consisted of simulated measurements on three variables only (frequency, pulse repetition interval and pulse duration) of four radars, each of which had five operational settings.

The model A decision tree was constructed using the CART algorithm (Section 7.2.2).

Training procedure CART was trained employing the Gini criterion for impurity. The three features were also augmented by three additional features, being the sums of pairs of the original features. Thus some decision boundaries were not necessarily parallel to coordinate boundaries. This improved performance.

Results The performance of CART and the multilayer perceptron were similar, in terms of error rate, and the training time for the classification tree was considerably less than the MLP. Also, the classification tree had the advantage that it was easy to interpret, clearly identifying regions of feature space associated with each radar type.

7.2.5 Further developments

The splitting rules described have considered only a single variable. Some data sets may be naturally separated by hyperplanes that are not parallel to the coordinate axes, and Chapter 4 concentrated on means for finding linear discriminants. The basic CART algorithm attempts to approximate these surfaces by multidimensional rectangular regions, and this can result in very large trees. Extensions to this procedure that allow splits not orthogonal to axes are described by Loh and Vanichsetakul (1988) and Wu and Zhang (1991) (and indeed can be found in the CART book of Breiman *et al.*, 1984). Sankar and Mammone (1991) use a neural network to recursively partition the data space and allow general hyperplane partitions. Pruning of this *neural tree classifier* is also addressed (see also Sethi and Yoo, 1994, for multifeature splits methods using perceptron learning).

Speed-ups to the CART procedure are discussed by Mola and Siciliano (1997). Also, non-binary splits of the data may be considered (Loh and Vanichsetakul, 1988; Sturt, 1981; and in the vector quantisation context, Gersho and Gray, 1992).

There are many growing and pruning strategies. Quinlan (1987) describes and assesses four pruning approaches, the motivation behind the work being to simplify decision trees in order to use the knowledge in expert systems. The use of information-theoretic criteria in tree construction has been considered by several authors. Quinlan and Rivest (1989) describe an approach based on the *minimum description length principle* and Goodman and Smyth (1990) present a top-down mutual information algorithm for tree design. A comparative study of pruning methods for decision trees is provided by Esposito *et al.* (1997) (see also comments by Kay, 1997; Mingers, 1989). Averaging, as an alternative to pruning, is discussed by Oliver and Hand (1996).

Tree-based methods for vector quantisation are described by Gersho and Gray (1992). A tree-structured approach reduces the search time in the encoding process. Pruning the tree results in a variable-rate vector quantiser and the CART pruning algorithm may be used. Other approaches for growing a variable-length tree (in the vector quantisation context) without first growing a complete tree are described by Riskin and Gray (1991). Crawford (1989) describes some extensions to CART that improve upon the cross-validation estimate of the error rate and allow for incremental learning – updating an existing tree in the light of new data.

One of the problems with nominal variables with a large number of categories is that there may be many possible partitions to consider. Chou (1991) presents a clustering approach to finding a locally optimum partition without exhaustive search. Buntine (1992) develops a Bayesian statistics approach to tree construction that uses Bayesian techniques for node splitting, pruning and averaging of multiple trees (also assessed as part of the *Statlog* project). A Bayesian CART algorithm is described by Denison *et al.* (1998a).

7.2.6 Summary

One of the main attractions of CART is its simplicity: it performs binary splits on single variables in a recursive manner. Classifying a sample may require only a few simple tests. Yet despite its simplicity, it is able to give performance superior to many traditional methods on complex nonlinear data sets of many variables. Of course there are possible generalisations of the model to multiway splitting on linear (or even nonlinear) combinations of variables, but there is no strong evidence that these will lead to improved performance. In fact, the contrary has been reported by Breiman and Friedman (1988). Also, univariate splitting has the advantage that the models can be interpreted more easily. Lack of interpretability by many, if not most, of the other methods of discrimination described in this book can be a serious shortcoming in many applications.

Another advantage of CART is that it is a procedure that has been extensively evaluated and tested, both by the workers who developed it and numerous researchers who have implemented the software. In addition, the tree-structured approach may be used for regression using the same impurity function (for example, a least squares error measure) to grow the tree as to prune it.

A possible disadvantage of CART is that training can be time-consuming. An alternative is to use a parametric approach with the inherent underlying assumptions. The use of nonparametric methods in discrimination, regression and density estimation has been increasing with the continuing development of computing power, and, as Breiman and

Friedman point out, 'the cost for computation is decreasing roughly by a factor of two every year, whereas the price paid for incorrect assumptions is remaining the same'. It really depends on the problem. In most applications it is the cost of data collection that far exceeds any other cost and the nonparametric approach is appealing because it does not make the (often gross) assumptions regarding the underlying population distributions that other discrimination methods do.

7.3 Multivariate adaptive regression splines

7.3.1 Introduction

The multivariate adaptive regression spline (MARS) (Friedman, 1991) approach may be considered as a continuous generalisation of the regression tree methodology treated separately in Section 7.2, and the presentation here follows Friedman's approach.

Suppose that we have a data set of n measurements on p variables, $\{\boldsymbol{x}_i, i = 1, \ldots, n\}$, $\boldsymbol{x}_i \in \mathbb{R}^p$, and corresponding measurements on the response variable $\{y_i, i = 1, \ldots, n\}$. We shall assume that the data have been generated by a model

$$y_i = f(\boldsymbol{x}) + \epsilon$$

where ϵ denotes a residual term. Our aim is to construct an approximation, $\hat{f}$, to the function f.

7.3.2 Recursive partitioning model

The recursive partitioning model takes the form

$$\hat{f}(\boldsymbol{x}) = \sum_{m=1}^{M} a_m B_m(\boldsymbol{x})$$

The basis functions B_m are

$$B_m(\boldsymbol{x}) = I\{\boldsymbol{x} \in u_{i_m}\}$$

where I is an indicator function with value unity if the argument is true and zero otherwise. We have used the notation that $\{u_i, i = 1, \ldots, M\}$ is a partition of the data space $\mathbb{R}^p$ (that is, u_i is a subspace of $\mathbb{R}^p$ such that $u_i \cap u_j = \oslash$ for $i \neq j$, and $\cup_i u_i = \mathbb{R}^p$). The set $\{a_m, m = 1, \ldots, M\}$ are the coefficients in the expansion whose values are determined (often) by a least squares minimisation procedure for fitting the approximation to the data. In a classification problem, a regression for each class may be performed (using binary variables with value 1 for $\boldsymbol{x}_i \in$ class ω_j, and zero otherwise) giving C functions $\hat{f}_j$, on which discrimination is based.

The basis functions are produced by a recursive partitioning algorithm (Friedman, 1991) and can be represented as a product of step functions. Consider the partition

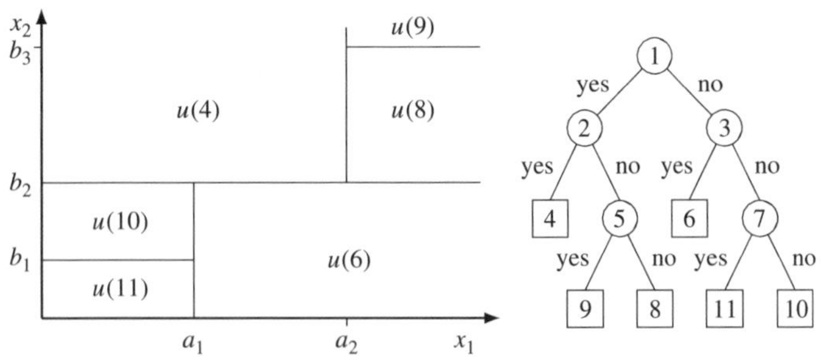

Figure 7.11 Classification tree and decision regions; the decision nodes (circular) are characterised as follows: node 1, $x_2 > b_2$; node 2, $x_1 < a_2$; node 5, $x_2 > b_3$; node 3, $x_1 > a_1$; node 7, $x_2 < b_1$. The square nodes correspond to regions in the feature space

produced by the tree given in Figure 7.11 . The first partition is on the variable x_2 and divides the plane into two regions, giving basis functions

$$H[(x_2 - b_2)] \text{ and } H[-(x_2 - b_2)]$$

where

$$H(x) = \begin{cases} 1 & x \geq 0 \\ 0 & \text{otherwise} \end{cases}$$

The region $x_2 < b_2$ is partitioned again, on variable x_1 with threshold a_1, giving the basis functions

$$H[-(x_2 - b_2)]H[+(x_1 - a_1)] \text{ and } H[-(x_2 - b_2)]H[-(x_1 - a_1)]$$

The final basis functions for the tree comprise the products

$$\begin{array}{l} H[-(x_2 - b_2)]H[-(x_1 - a_1)]H[+(x_2 - b_1)] \\ H[-(x_2 - b_2)]H[-(x_1 - a_1)]H[-(x_2 - b_1)] \\ H[-(x_2 - b_2)]H[+(x_1 - a_1)] \\ H[+(x_2 - b_2)]H[-(x_1 - a_2)] \\ H[+(x_2 - b_2)]H[+(x_1 - a_2)]H[+(x_2 - b_3)] \\ H[+(x_2 - b_2)]H[+(x_1 - a_2)]H[-(x_2 - b_3)] \end{array}$$

Thus, each basis function is a product of step functions, H.

In general, the basis functions of the recursive partitioning algorithm have the form

$$B_m(\boldsymbol{x}) = \prod_{k=1}^{K_m} H[s_{km}(x_{v(k,m)} - t_{km})] \tag{7.7}$$

where the s_{km} take the values ± 1 and K_m is the number of splits that give rise to $B_m(\boldsymbol{x})$; $x_{v(k,m)}$ is the variable split and the t_{km} are the thresholds on the variables.

The MARS procedure is a generalisation of this recursive partitioning procedure in the following ways.

Continuity

The recursive partitioning model described above is discontinuous at region boundaries. This is due to the use of the step function H. The MARS procedure replaces these step functions by spline functions. The two-sided truncated power basis functions for qth-order splines are

$$b_q^{\pm}(x - t) = [\pm(x - t)]_+^q$$

where $[.]_+$ denotes that the positive part of the argument is considered. The basis functions $b_q^+(x)$ are illustrated in Figure 7.12. The step function, H, is the special case of $q = 0$. The MARS algorithm employs $q = 1$. This leads to a continuous function approximation, but discontinuous first derivatives.

The basis functions now take the form

$$B_m^q(\boldsymbol{x}) = \prod_{k=1}^{K_m} [s_{km}(x_{v(k,m)} - t_{km})]_+^q \tag{7.8}$$

where the t_{km} are referred to as the *knot locations*.

Retention of parent functions

The basic recursive partitioning algorithm replaces an existing parent basis function by its product with a step function and the reflected step function. The number of basis functions therefore increases by one on each split. The MARS procedure retains the parent basis function. Thus the number of basis functions increases by two at each split. This provides a much more flexible model that is capable of modelling classes of functions that have no strong interaction effects, or strong interactions involving at most a few of the variables, as well as higher-order interactions. A consequence of the retention of the parent functions is that the regions corresponding to the basis functions may overlap.

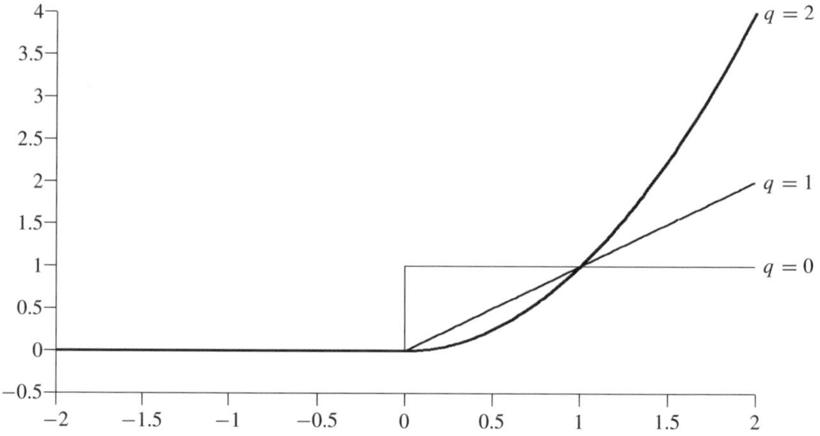

Figure 7.12 Spline functions, $b_q^+(x)$, for $q = 0, 1, 2$

Multiple splits

The basic recursive partitioning algorithm allows multiple splits on a single variable: as part of the modelling procedure, a given variable may be selected repeatedly and partitioned. Thus basis functions comprise products of repeated splits on a given variable (the basis functions for the tree in Figure 7.11 contain repeated splits). In the continuous generalisation, this leads to functional dependencies of higher order than q on individual variables. To take advantage of the properties of *tensor product spline basis functions*, whose factors involve a different variable, MARS restricts the basis functions to products involving single splits on a given variable. By reselecting the same parent for splitting (on the same variable), the MARS procedure is able to retain the flexibility of the repeated split model (it trades depth of tree for breadth of tree).

The second stage in the MARS strategy is a pruning procedure; basis functions are deleted one at a time – the basis function being removed being that which improves the fit the most (or degrades it the least). A separate validation set could be chosen to estimate the goodness of fit of the model. The lack-of-fit criterion proposed for the MARS algorithms is a modified form of the generalised cross-validation criterion of Craven and Wahba (1979). In a discrimination problem, a model could be chosen that minimises error rate on the validation set.

In applying MARS to discrimination problems, one approach is to use the usual binary coding for the response function, f_j, $j = 1, \ldots, C$, with $f_j(\boldsymbol{x}) = 1$ if $\boldsymbol{x}$ is in class ω_j and zero otherwise. Each class may have a separate MARS model; the more parsimonious model of having common basis functions for all classes will reduce the computation. The weights a_i in the MARS algorithms are replaced by vectors $\boldsymbol{a}_i$, determined to minimise a generalised cross-validation measure.

7.3.3 Example application study

The problem To understand the relationship between sea floor topography, ocean circulation and variations in the sea ice concentration in Antarctic regions (De Veaux *et al.*, 1993).

Summary The occurrence of *polynyas*, large areas of open water within an otherwise continuous cover of sea ice, is of interest to oceanographers. In particular, the Weddell Polynya was active for three years and is the subject of this investigation. A nonparametric regression model was used to quantify the effect of sea floor topography on sea ice surface characteristics.

The data Data were gathered from one ocean region (the Maud Rise). Ice concentration maps were derived from scanning multichannel microwave radiometer data. Bathymetry data were derived, primarily, from soundings taken from ships at approximately a 5 km interval along the ship tracks and at a vertical resolution of 1 m.

The model Five predictor variables were considered (bathymetry, its meridional and zonal derivatives and their derivatives). The MARS model was used with the interaction level restricted to two, resulting in 10 additional potential variables.

Training procedure The MARS procedure selected three two-variable interactions. Retaining only a few low-order interactions allowed the results of fitting the final model to be interpreted. This is important for other applications.

Results The results strongly support the hypothesis that there is a link between sea ice concentration and sea floor topography. In particular, the sea ice concentration can be predicted to some degree by the ocean depth and its first and second derivatives in both meridional and zonal directions.

7.3.4 Further developments

A development of the basic model to include procedures for mixed ordinal and categorical variables is provided by Friedman (1993). POLYMARS (Stone *et al.*, 1997) is a development to handle a categorical response variable, with application to classification problems.

Time series versions of MARS have been developed for forecasting applications (De Gooijer *et al.*, 1998). A Bayesian approach to MARS fitting, which averages over possible models (with a consequent loss of interpretability of the final model), is described by Denison *et al.* (1998b).

7.3.5 Summary

MARS is a method for modelling high-dimensional data in which the basis functions are a product of spline functions. MARS searches over threshold positions on the variables, or *knots*, across all variables and interactions. Once it has done this, it uses a least squares regression to provide estimates of the coefficients of the model.

MARS can be used to model data comprising measurements on variables of different type. The optimal model is achieved by growing a large model and then pruning the model by removing basis functions until a lack-of-fit criterion is minimised.

7.4 Application studies

Applications of decision-tree methodology are varied and include the following.

- Predicting stroke in patient rehabilitation outcome. Falconer *et al.* (1994) develop a classification tree model (CART) to predict rehabilitation outcomes for stroke patients. The data comprised measurements on 51 ordinal variables on 225 patients. A classification tree was used to identify those variables most informative for predicting favourable and unfavourable outcomes. The resulting tree used only 4 of the 51 variables measured on admission to a university-affiliated rehabilitation institute, improving the ability to predict rehabilitation outcomes, and correctly classifying 88% of the sample.

- Gait events. In a study into the classification of phases of the gait cycle (as part of the development of a control system for functional electrical stimulation of the lower limbs), Kirkwood *et al.* (1989) develop a decision-tree approach that enable redundant combinations of variables to be identified. Efficient rules are derived with high performance accuracy.
- Credit card applications. Carter and Catlett (1987) use Quinlan's (1986) *ID3* algorithm to assess credit card applications.
- Thyroid diseases. In a study of the use of a decision tree to synthesise medical knowledge, Quinlan (1986) employs *C4*, a descendant of *ID3* that implements a pruning algorithm, to generate a set of high-performance rules. The data consist of input from a referring doctor (patient's age, sex and 11 true–false indicators), a clinical laboratory (up to six assay results) and a diagnostician. Thus, variables are of mixed type, with missing values and some misclassified samples. The pruning algorithm leads to an improved simplicity and intelligibility of derived rules.

Other applications have been in the areas of telecommunications, marketing and industrial applications.

There have been several comparisons with neural networks and other discrimination methods:

- Digit recognition. Brown *et al.* (1993) compare a classification tree with a multilayer perceptron on a digit recognition (extracted from licence plate images) problem, with application to highway monitoring and tolling. All features were binary and the classification tree performance was poorer than the MLP, but performance improved when features that were a combination of the original variables were included.
- Various data sets. Curram and Mingers (1994) compare a multilayer perceptron with a decision tree on several data sets. The decision tree was susceptible to noisy data, but had the advantage of providing insight. Shavlik *et al.* (1991) compared *ID3* with a multilayer perceptron on four data sets, finding that the MLP handled noisy data and missing features slightly better than *ID3*, but took considerably longer to train.

Other comparative studies include speaker-independent vowel classification and load forecasting (Atlas *et al.*, 1989), finding the MLP superior to a classification tree; disk drive manufacture quality control and the prediction of chronic problems in large-scale communication networks (Apté *et al.*, 1994).

The MARS methodology has been applied to problems in classification and regression including the following.

- Economic time series. Sephton (1994) uses MARS to model three economic time series: annual US output, capital and labour inputs; interest rates and exchange rates using a generalised cross-validation score for model selection.
- Telecommunications. Duffy *et al.* (1994) compare neural networks with CART and MARS on two telecommunications problems: modelling switch processor memory, a regression problem; and characterising traffic data (speech and modem data at three different baud rates), a four-class discrimination problem.

- Particle detection. In a comparative study of four methods of discrimination, Holmström and Sain (1997) compare MARS with a quadratic classifier, a neural network and kernel discriminant analysis on a problem to detect a weak signal against a dominant background. Each event (in the two-class problem) was described by 14 variables and the training set comprised 5000 events. MARS appeared to give best performance.

7.5 Summary and discussion

Recursive partitioning methods have a long history and have been developed in many different fields of endeavour. Complex decision regions can be approximated by the union of simpler decision regions. A major step in this research was the development of CART, a simple nonparametric method of partitioning data. The approach described in this chapter for constructing a classification tree is based on that work.

There have been many comparative studies with neural networks, especially multilayer perceptrons. Both approaches are capable of modelling complex data. The MLP is usually longer to train and does not provide the insight of a tree, but has often shown better performance on the data sets used for the evaluation (which may favour an MLP anyway). Further work is required.

The multivariate adaptive regression spline approach is a recursive partitioning method that utilises products of spline functions as the basis functions. Like CART, it is also well suited to model mixed variable (discrete and continuous) data.

7.6 Recommendations

Although it cannot be said that classification trees perform substantially better than other methods (and, for a given problem, there may be a parametric method that will work better – but you do not know which one to choose), their simplicity and consistently good performance on a wide range of data sets have led to their widespread use in many disciplines. It is recommended that you try them for yourselves.

Specifically, classification tree approaches are recommended:

1. for complex data sets in which you believe decision boundaries are nonlinear and decision regions can be approximated by the sum of simpler regions;
2. for problems where it is important to gain an insight into the data structure and the classification rule, when the explanatory power of trees may lead to results that are easier to communicate than other techniques;
3. for problems with data consisting of measurements on variables of mixed type (continuous, ordinal, nominal);
4. where ease of implementation is required;
5. where speed of classification performance is important – the classifier performs simple tests on (single) variables.

MARS is simple to use and is recommended for high-dimensional regression problems, for problems involving variables of mixed type and for problems where some degree of interpretability of the final solution is required.

7.7 Notes and references

There is a very large literature on classification trees in the areas of pattern recognition, artificial intelligence, statistics and the engineering sciences, but by no means exclusively within these disciplines. Many developments have taken place by researchers working independently and there are many extensions of the approach described in this chapter as well as many alternatives. A survey is provided by Safavian and Landgrebe (1991); see also Feng and Michie (1994). Several tree-based approaches were assessed as part of the *Statlog* project (Michie *et al.*, 1994) including CART, which proved to be one of the better ones because it incorporates costs into the decision.

MARS was introduced by Friedman (1991). Software for CART, other decision-tree software and MARS is publicly available.

The website `www.statistical-pattern-recognition.net` contains references and links to further information on techniques and applications.

Exercises

1. In Figure 7.6, if regions $u(4)$ and $u(6)$ correspond to class ω_1 and $u(8)$, $u(9)$, $u(10)$ and $u(11)$ to class ω_2, construct a multilayer perceptron with the same decision boundaries.

2. A standard classification tree produces binary splits on a single variable at each node. For a two-class problem, using the results of Chapter 4, describe how to construct a tree that splits on a linear combination of variables at a given node.

3. The predictability index (relative decrease in the proportion of incorrect predictions for split s on node t) is written (using the notation of this chapter)

$$\tau(\omega|s) = \frac{\sum_{j=1}^{C} p^2(\omega_j|t_L)p_L + \sum_{j=1}^{C} p^2(\omega_j|t_R)p_R - \sum_{j=1}^{C} p^2(\omega_j|t)}{1 - \sum_{j=1}^{C} p^2(\omega_j|t)}$$

   Show that the decrease in impurity when passing from one group to two subgroups for the Gini criterion can be written

$$\Delta\mathcal{I}(s,t) = \sum_{j=1}^{C} p^2(\omega_j|t_L)p_L + \sum_{j=1}^{C} p^2(\omega_j|t_R)p_R - \sum_{j=1}^{C} p^2(\omega_j|t)$$

   and hence that maximising the predictability also maximises the decrease in impurity.

4. Consider the two-class problem with bivariate distributions characterised by x_1 and x_2 which can take three values. The training data are given by the following tables:

class ω_1

		x_1		
		1	2	3
x_2	1	3	0	0
	2	1	6	0
	3	4	1	2

class ω_2

		x_1		
		1	2	3
x_2	1	0	1	4
	2	0	5	7
	3	1	0	1

where, for example, there are four training samples in class ω_1 with $x_1 = 1$ and $x_2 = 3$. Using the Gini criterion, determine the split (variable and value) for the root node of a tree.

5. Construct a classification tree using data set 1 from the exercises in Chapter 5. Initially allow 10 splits per variable. Monitor the performance on the validation set as the tree is grown. Prune using the validation set to monitor performance. Investigate the performance of the approach for $p = 2, 5$ and 10. Describe the results and compare with a linear discriminant analysis.

6. Show that MARS can be recast in the form

$$a_0 + \sum_i f_i(x_i) + \sum_{ij} f_{ij}(x_i, x_j) + \sum_{ijk} f_{ijk}(x_i, x_j, x_k) + \cdots$$

7. Write the basis functions for the regions in Figure 7.11 as sums of products of splits on the variables x_1 and x_2, with at most one split on a given variable in the product.

8

Performance

Overview

Classifier performance assessment is an important aspect of the pattern recognition cycle. How good is the designed classifier and how well does it compare with competing techniques? Can improvements in performance be achieved with an ensemble of classifiers?

8.1 Introduction

The pattern recognition cycle (see Chapter 1) begins with the collection of data and initial data analysis followed, perhaps, by preprocessing of the data and then the design of the classification rule. Chapters 2 to 7 described approaches to classification rule design, beginning with density estimation methods and leading on to techniques that construct a discrimination rule directly. In this chapter we address two different aspects of performance: performance assessment and performance improvement.

Performance assessment, discussed in Section 8.2, should really be a part of classifier design and not an aspect that is considered separately, as it often is. A sophisticated design stage is often followed by a much less sophisticated evaluation stage, perhaps resulting in an inferior rule. The criterion used to design a classifier is often different from that used to assess it. For example, in constructing a discriminant rule, we may choose the parameters of the rule to optimise a squared error measure, yet assess the rule using a different measure of performance, such as error rate.

A related aspect of performance is that of comparing the performance of several classifiers trained on the same data set. For a practical application, we may implement several classifiers and want to choose the best, measured in terms of error rate or perhaps computational efficiency. In Section 8.3, comparing classifier performance is addressed.

Instead of choosing the best classifier from a set of classifiers, we ask the question, in Section 8.4, whether we can get improved performance by combining the outputs of several classifiers. This has been an area of growing research in recent years and is related to developments in the *data fusion* literature where, in particular, the problem of decision fusion (combining decisions from multiple target detectors) has been addressed extensively.

8.2 Performance assessment

Three aspects of the performance of a classification rule are addressed. The first is the *discriminability* of a rule (how well it classifies unseen data) and we focus on one particular method, namely the error rate. The second is the *reliability* of a rule. This is a measure of how well it estimates the posterior probabilities of class membership. Finally, the use of the *receiver operating characteristic* (ROC) as an indicator of performance for two-class rules is considered.

8.2.1 Discriminability

There are many measures of discriminability (Hand, 1997), the most common being the *misclassification rate* or the *error rate* of a classification rule. Generally, it is very difficult to obtain an analytic expression for the error rate and therefore it must be estimated from the available data. There is a vast literature on error rate estimation, but the error rate suffers from the disadvantage that it is only a single measure of performance, treating all correct classifications equally and all misclassifications with equal weight also (corresponding to a zero–one loss function – see Chapter 1). In addition to computing the error rate, we may also compute a *confusion* or *misclassification* matrix. The (i, j)th element of this matrix is the number of patterns of class ω_j that are classified as class ω_i by the rule. This is useful in identifying how the error rate is decomposed. A complete review of the literature on error rate estimation deserves a volume in itself, and is certainly beyond the scope of this book. Here, we limit ourselves to a discussion of the more popular types of error rate estimator.

Firstly, let us introduce some notation. Let the training data be denoted by $\boldsymbol{Y} = \{\boldsymbol{y}_i, i = 1, \ldots, n\}$, the pattern $\boldsymbol{y}_i$ consisting of two parts, $\boldsymbol{y}_i^T = (\boldsymbol{x}_i^T, \boldsymbol{z}_i^T)$, where $\{\boldsymbol{x}_i, i = 1, \ldots, n\}$ are the measurements and $\{\boldsymbol{z}_i, i = 1, \ldots, n\}$ are the corresponding class labels, now coded as a vector, $(\boldsymbol{z}_i)_j = 1$ if $\boldsymbol{x}_i \in$ class ω_j and zero otherwise. Let $\omega(\boldsymbol{z}_i)$ be the corresponding categorical class label. Let the decision rule designed using the training data be $\eta(\boldsymbol{x}; \boldsymbol{Y})$ (that is, η is the class to which $\boldsymbol{x}$ is assigned by the classifier designed using $\boldsymbol{Y}$) and let $Q(\omega(\boldsymbol{z}), \eta(\boldsymbol{x}; \boldsymbol{Y}))$ be the loss function

$$Q(\omega(\boldsymbol{z}), \eta(\boldsymbol{x}; \boldsymbol{Y})) = \begin{cases} 0 & \text{if } \omega(\boldsymbol{z}) = \eta(\boldsymbol{x}; \boldsymbol{Y}) \text{ (correct classification)} \\ 1 & \text{otherwise} \end{cases}$$

Apparent error rate The *apparent error rate*, e_A, or *resubstitution rate* is obtained by using the design set to estimate the error rate,

$$e_A = \frac{1}{n} \sum_{i=1}^{n} Q(\omega(\boldsymbol{z}_i), \eta(\boldsymbol{x}_i; \boldsymbol{Y}))$$

It can be severely optimistically biased, particularly for complex classifiers and a small data set when there is a danger of over-fitting the data – that is, the classifier models the noise on the data rather than its structure. Increasing the number of training samples reduces this bias.

True error rate The *true error rate* (or *actual error rate* or *conditional error rate*), e_T, of a classifier is the expected probability of misclassifying a randomly selected pattern. It is the error rate on an infinitely large test set drawn from the same distribution as the training data.

Expected error rate The *expected error rate*, e_E, is the expected value of the true error rate over training sets of a given size, $e_E = \mathrm{E}[e_T]$.

Bayes error rate The *Bayes error rate* or *optimal error rate*, e_B, is the theoretical minimum of the true error rate, the value of the true error rate if the classifier produced the true posterior probabilities of group membership, $p(\omega_i|\boldsymbol{x}), i = 1, \ldots, C$.

Holdout estimate

The holdout method splits the data into two mutually exclusive sets, sometimes referred to as the training and test sets. The classifier is designed using the training set and performance evaluated on the independent test set. The method makes inefficient use of the data (using only part of it to train the classifier) and gives a pessimistically biased error estimate (Devijver and Kittler, 1982). However, it is possible to obtain confidence limits on the true error rate given a set of n independent test samples, drawn from the same distribution as the training data. If the true error rate is e_T, and k of the samples are misclassified, then k is binomially distributed

$$p(k|e_T, n) = \mathrm{Bi}(k|e_T, n) \triangleq \binom{n}{k} e_T^k (1 - e_T)^{n-k} \tag{8.1}$$

The above expression gives the probability that k samples out of n of an independent test set are misclassified given that the true error rate is e_T. Using Bayes' theorem, we may write the conditional density of the true error rate, given the number of samples misclassified, as

$$p(e_T|k, n) = \frac{p(k|e_T, n)p(e_T, n)}{\int p(k|e_T, n)p(e_T, n)de_T}$$

Assuming $p(e_T, n)$ does not vary with e_T and $p(k|e_T, n)$ is the binomial distribution, we have a beta distribution for e_T,

$$p(e_T|k, n) = \mathrm{Be}(e_T|k + 1, n - k + 1) \triangleq \frac{e_T^k (1 - e_T)^{n-k}}{\int e_T^k (1 - e_T)^{n-k} de_T}$$

where $\mathrm{Be}(x|\alpha, \beta) = [\Gamma(\alpha+\beta)/(\Gamma(\alpha)\Gamma(\beta))]x^{\alpha-1}(1-x)^{\beta-1}$. The above posterior density provides a complete account of what can be learned given the test error. However, it may be summarised in several ways, one of which is to give an upper and lower bound (a percentage point) on the true error. For a given value of α (for example, 0.05), there are many intervals in which e_T lies with probability $1 - \alpha$. These are called $(1 - \alpha)$ credible regions, or Bayesian confidence intervals (O'Hagan, 1994). Among these intervals, the *highest posterior density* (HPD) credible region is the one with the additional property that every point within it has a higher probability than any point outside. It is also the

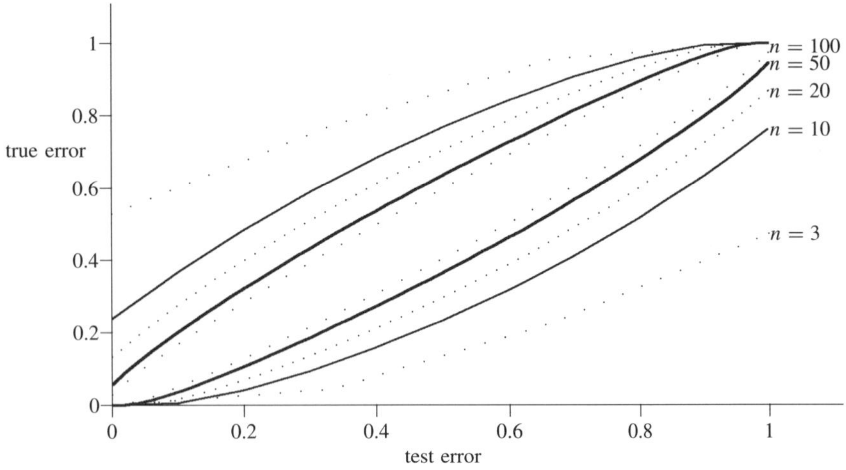

Figure 8.1 HPD credible region limits as a function of test error (number misclassified on test/size of test set) for several values of n, the number of test samples, and $\alpha = 0.05$ (i.e. the 95% credible region limits). From top to bottom, the limit lines correspond to $n = 3$, 10, 20, 50, 100, 100, 50, 20, 10, 3

shortest $(1 - \alpha)$ credible region. It is the interval E_α

$$E_\alpha = \{e_T : p(e_T|k, n) \geq c\}$$

where c is chosen such that

$$\int_{E_\alpha} p(e_T|k, n)de_T = 1 - \alpha \tag{8.2}$$

For multimodal densities, E_α may be discontinuous. However, for the beta distribution, E_α is a single region with lower and upper bounds $\epsilon_1(\alpha)$ and $\epsilon_2(\alpha)$ (both functions of k and n) satisfying

$$0 \leq \epsilon_1(\alpha) < \epsilon_2(\alpha) \leq 1$$

Turkkan and Pham-Gia (1993) provide a general-purpose Fortran subroutine for computing the HPD intervals of a given density function. Figure 8.1 displays the Bayesian confidence intervals as a function of test error for several values of n, the number of samples in the test set, and a value for α of 0.05, i.e. the bounds of the 95% credible region. For example, for 4 out of 20 test samples incorrectly classified, the $(1-\alpha)$ credible region (for $\alpha = 0.05$) is $[0.069, 0.399]$. Figure 8.2 plots the maximum length of the 95% HPD credible region (over the test error) as a function of the number of test samples. For example, we can see from the figure that, to be sure of having a HPD interval of less than 0.1, we must have more than 350 test samples.

Cross-validation

Cross-validation (also known as the *U-method*, the *leave-one-out estimate* or the *deleted estimate*) calculates the error by using $n - 1$ samples in the design set and testing on the remaining sample. This is repeated for all n subsets of size $n - 1$. For large n,

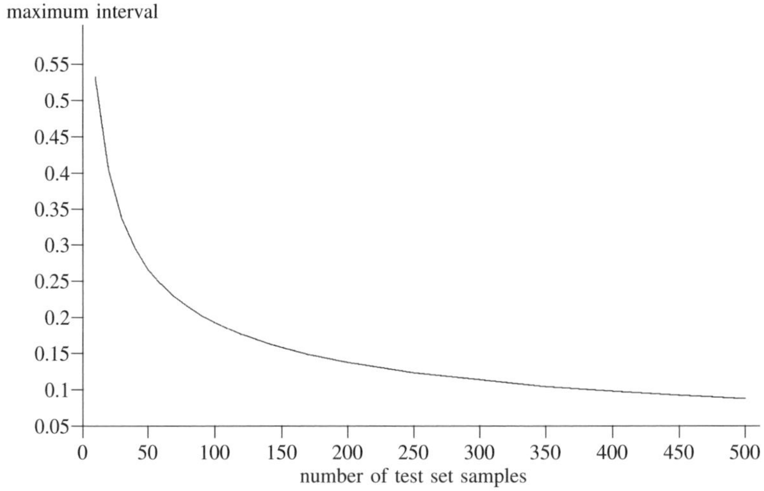

Figure 8.2 Maximum length of the 95% HPD credible region as a function of the number of test samples

it is computationally expensive, requiring the design of n classifiers. However, it is approximately unbiased, although at the expense of an increase in the variance of the estimator. Denoting by $\boldsymbol{Y}_j$ the training set with observation $\boldsymbol{x}_j$ deleted, then the cross-validation error is

$$e_{\text{CV}} = \frac{1}{n} \sum_{j=1}^{n} Q(\omega(z_j), \eta(\boldsymbol{x}_j, \boldsymbol{Y}_j))$$

One of the disadvantages of the cross-validation approach is that it may involve a considerable amount of computation. However, for discriminant rules based on multivariate normal assumptions, the additional computation can be considerably reduced through the application of the Sherman–Morisson formula (Fukunaga and Kessell, 1971; McLachlan, 1992a):

$$(\boldsymbol{A} + \boldsymbol{u}\boldsymbol{u}^T)^{-1} = \boldsymbol{A}^{-1} - \frac{\boldsymbol{A}^{-1}\boldsymbol{u}\boldsymbol{u}^T\boldsymbol{A}^{-1}}{1 + \boldsymbol{u}^T\boldsymbol{A}^{-1}\boldsymbol{u}} \tag{8.3}$$

for matrix $\boldsymbol{A}$ and vector $\boldsymbol{u}$.

The *rotation method* or *ν-fold cross-validation* partitions the training set into ν subsets, training on $\nu - 1$ and testing on the remaining set. This procedure is repeated as each subset is withheld in turn. If $\nu = n$ we have the standard cross-validation, and if $\nu = 2$ we have a variant of the holdout method in which the training set and test set are also interchanged. This method is a compromise between the holdout method and cross-validation, giving reduced bias compared to the holdout procedure, but less computation compared to cross-validation.

The jackknife

The *jackknife* is a procedure for reducing the bias of the apparent error rate. As an estimator of the true error rate, the apparent error rate bias is of order n^{-1}, for n samples. The jackknife estimate reduces the bias to the second order.

Let t_n denote a sample statistic based on n observations $\boldsymbol{x}_1, \ldots, \boldsymbol{x}_n$. We assume for large m that the expectation for sample size m takes the form

$$\mathrm{E}[t_m] = \theta + \frac{a_1(\theta)}{m} + \frac{a_2(\theta)}{m^2} + \mathcal{O}(m^{-3}) \tag{8.4}$$

where θ is the asymptotic value of the expectation and a_1 and a_2 do not depend on m. Let $t_n^{(j)}$ denote the statistic based on observations excluding $\boldsymbol{x}_j$. Finally, write $t_n^{(.)}$ for the average of the $t_n^{(j)}$ over $j = 1, \ldots, n$,

$$t_n^{(.)} = \frac{1}{n} \sum_{j=1}^{n} t_n^{(j)}$$

Then,

$$\begin{aligned} \mathrm{E}[t_n^{(.)}] &= \frac{1}{n} \sum_{j=1}^{n} \left(\theta + \frac{a_1(\theta)}{n-1} + \mathcal{O}(n^{-2}) \right) \\ &= \theta + \frac{a_1(\theta)}{n-1} + \mathcal{O}(n^{-2}) \end{aligned} \tag{8.5}$$

From (8.4) and (8.5), we may find a linear combination that has bias of order n^{-2},

$$t_J = n t_n - (n-1) t_n^{(.)}$$

t_J is termed the jackknifed estimate corresponding to t_n.

Applying this to error rate estimation, the jackknife version of the apparent error rate, e_J^0, is given by

$$\begin{aligned} e_J^0 &= n e_A - (n-1) e_A^{(.)} \\ &= e_A + (n-1)(e_A - e_A^{(.)}) \end{aligned}$$

where e_A is the apparent error rate; $e_A^{(.)}$ is given by

$$e_A^{(.)} = \frac{1}{n} \sum_{j=1}^{n} e_A^{(j)}$$

where $e_A^{(j)}$ is the apparent error rate when object j has been removed from the observations,

$$e_A^{(j)} = \frac{1}{n-1} \sum_{k=1, k \neq j}^{n} Q(\omega(z_k), \eta(\boldsymbol{x}_k; \boldsymbol{Y}_j))$$

As an estimator of the expected error rate, the bias of e_J^0 is of order n^{-2}. However, as an estimator of the true error rate, the bias is still of order n^{-1} (McLachlan, 1992a). To reduce the bias of e_J^0 as an estimator of the true error rate to second order, we use

$$e_J = e_A + (n-1)(\tilde{e}_A - e_A^{(.)}) \tag{8.6}$$

where $\tilde{e}_A$ is given by

$$\tilde{e}_A = \frac{1}{n^2} \sum_{j=1}^{n} \sum_{k=1}^{n} Q(\omega(z_k), \eta(\boldsymbol{x}_k; \boldsymbol{Y}_j))$$

The jackknife is closely related to the cross-validation method and both methods delete one observation successively to form bias-corrected estimates of the error rate. A difference is that in cross-validation, the contribution to the estimate is from the deleted sample only, classified using the classifier trained on the remaining set. In the jackknife, the error rate estimate is calculated from all samples, classified using the classifiers trained with each reduced sample set.

Bootstrap techniques

The term 'bootstrap' refers to a class of procedures that sample the observed distribution, with replacement, to generate sets of observations that may be used to correct for bias. Introduced by Efron (1979), it has received considerable attention in the literature during the past decade or so. It provides nonparametric estimates of the bias and variance of an estimator and, as a method of error rate estimation, it has proved superior to many other techniques. Although computationally intensive, it is a very attractive technique, and there have been many developments of the basic approach, largely by Efron himself; see Efron and Tibshirani (1986) and Hinkley (1988) for a survey of bootstrap methods.

The bootstrap procedure for estimating the bias correction of the apparent error rate is implemented as follows. Let the data be denoted by $\boldsymbol{Y} = \{(\boldsymbol{x}_i^T, \boldsymbol{z}_i^T)^T, i = 1, \ldots, n\}$. Let $\hat{F}$ be the *empirical distribution*. Under joint or mixture sampling it is the distribution with mass $1/n$ at each data point $\boldsymbol{x}_i, i = 1, \ldots, n$. Under separate sampling, $\hat{F}_i$ is the distribution with mass $1/n_i$ at point $\boldsymbol{x}_i$ in class ω_i (n_i patterns in class ω_i).

1. Generate a new set of data (the bootstrap sample) $\boldsymbol{Y}^b = \{(\tilde{\boldsymbol{x}}_i^T, \tilde{\boldsymbol{z}}_i^T)^T, i = 1, \ldots, n\}$ according to the empirical distribution.
2. Design the classifier using $\boldsymbol{Y}^b$.
3. Calculate the apparent error rate for this sample and denote it by $\tilde{e}_A$.
4. Calculate the actual error rate for this classifier (regarding the set $\boldsymbol{Y}$ as the entire population) and denote it by $\tilde{e}_c$.
5. Compute $w_b = \tilde{e}_A - \tilde{e}_c$.
6. Repeat steps 1–5 B times.
7. The bootstrap bias of the apparent error rate is
$$W_{\text{boot}} = \text{E}[\tilde{e}_A - \tilde{e}_c]$$
where the expectation is with respect to the sampling mechanism that generates the sets $\boldsymbol{Y}^b$, that is,
$$W_{\text{boot}} = \frac{1}{B}\sum_{b=1}^{B} w_b$$
8. The bias-corrected version of the apparent error rate is given by
$$e_A^{(B)} = e_A - W_{\text{boot}}$$

At step 1, under mixture sampling, n independent samples are generated from the distribution $\hat{F}$; some of these may be repeated in forming the set $\tilde{\boldsymbol{Y}}$ and it may happen that one or more classes are not represented in the bootstrap sample. Under separate sampling, n_i are generated using $\hat{F}_i, i = 1, \ldots, C$. Thus, all classes are represented in the bootstrap sample in the same proportions as the original data.

The number of bootstrap samples, B, used to estimate W_{boot} may be limited by computational considerations, but for error rate estimation it can be taken to be of the order of 25–100 (Efron, 1983, 1990; Efron and Tibshirani, 1986).

There are many variants of the basic approach described above (Efron, 1983; McLachlan, 1992a). These include the double bootstrap, the randomised bootstrap and the 0.632 estimator (Efron, 1983). The double bootstrap corrects for the bias of the ordinary bootstrap using a bootstrap to estimate the bias. The randomised bootstrap (for $C = 2$ classes) draws samples of size n from a data set of size $2n$, $\boldsymbol{Y}^{2n} = \{(\boldsymbol{x}_i^T, z_i^T)^T, (\boldsymbol{x}_i^T, \bar{z}_i^T)^T, i = 1, \ldots, n\}$, where $\bar{z}_i$ is the opposite class to z_i. Thus, the original data set is replicated with opposite class labels. The probability of choosing $\boldsymbol{x}_i$ is still $1/n$ (for mixture sampling), but the sample is taken as $(\boldsymbol{x}_i^T, z_i^T)^T$ or $(\boldsymbol{x}_i^T, \bar{z}_i^T)^T$ with probabilities π_i and $1-\pi_i$ respectively. Efron takes π_i equal to 0.9 for all i. This estimate was found to give a lower mean squared error over the ordinary bootstrap estimator.

The 0.632 estimator is a linear combination of the apparent error rate and another bootstrap error estimator, e_0,

$$e_{0.632} = 0.368 e_A + 0.632 e_0$$

where e_0 is an estimator that counts the number of training patterns misclassified that do not appear in the bootstrap sample. The number of misclassified samples is summed over all bootstrap samples and divided by the total number of patterns not in the bootstrap sample. If A_b is the set of patterns in $\boldsymbol{Y}$, but not in bootstrap sample $\boldsymbol{Y}^b$, then

$$e_0 = \frac{\sum_{b=1}^{B} \sum_{\boldsymbol{x} \in A_b} Q(\omega(z), \eta(\boldsymbol{x}, \boldsymbol{Y}^b))}{\sum_{b=1}^{B} |A_b|}$$

where $|A_b|$ is the cardinality of the set A_b. This estimator gave best performance in Efron's (1983) experiments.

The bootstrap may also be used for parametric distributions in which samples are generated according to the parametric form adopted, with the parameters replaced by their estimates. The procedure is not limited to estimates of the bias in the apparent error rate and has been applied to other measures of statistical accuracy, though bootstrap calculations for confidence limits require more bootstrap replications, typically 1000–2000 (Efron, 1990). More efficient computational methods aimed at reducing the number of bootstrap replications compared to the straightforward Monte Carlo approach above have been proposed by Davison *et al.* (1986) and Efron (1990). Application to classifiers such as neural networks and classification trees produces difficulties, however, due to multiple local optima of the error surface.

8.2.2 Reliability

The reliability (termed *imprecision* by Hand, 1997) of a discriminant rule is a measure of how well the posterior probabilities of group membership are estimated by the rule.

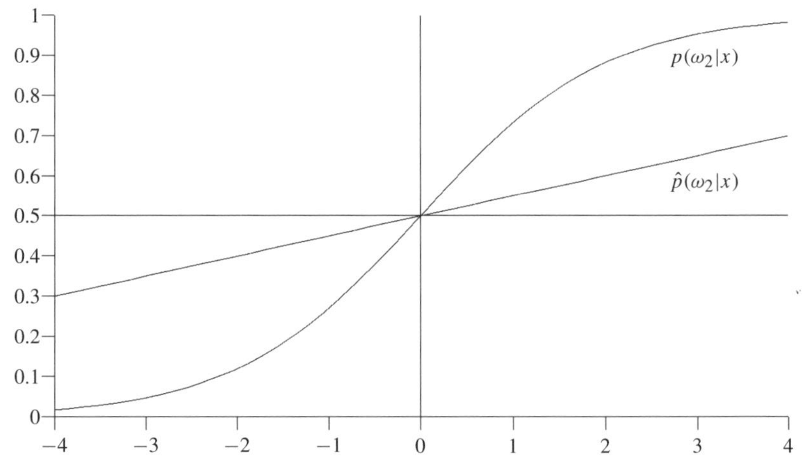

Figure 8.3 Good discriminability, poor reliability (following Hand, 1994a)

Thus, we are not simply interested in the class ω_i for which $p(\omega_j|\boldsymbol{x})$ is the greatest, but the value of $p(\omega_j|\boldsymbol{x})$ itself. Of course, we may not easily be able to estimate the reliability for two reasons. The first is that we do not know the true posterior probabilities. Secondly, some discriminant rules do not produce estimates of the posterior probabilities explicitly.

Figure 8.3 illustrates a rule with good discriminability but poor reliability for a two-class problem with equal priors. An object, $\boldsymbol{x}$, is assigned to class ω_2 if $p(\omega_2|\boldsymbol{x}) > p(\omega_1|\boldsymbol{x})$, or $p(\omega_2|\boldsymbol{x}) > 0.5$. The estimated posterior probabilities $\hat{p}(\omega_i|\boldsymbol{x})$ lead to good discriminability in the sense that the decision boundary is the same as a discriminant rule using the true posterior probabilities (i.e. $\hat{p}(\omega_2|\boldsymbol{x}) = 0.5$ at the same point as $p(\omega_2|\boldsymbol{x}) = 0.5$). However, the true and estimated posterior probabilities differ.

Why should we want good reliability? Is not good discriminability sufficient? In some cases, a rule with good discriminability may be all that is required. We may be satisfied with a rule that achieves the Bayes optimal error rate. On the other hand, if we wish to make a decision based on costs, or we are using the results of the classifier in a further stage of analysis, then good reliability is important. Hand (1997) proposes a measure of imprecision obtained by comparing an empirical sample statistic with an estimate of the same statistic computed using the classification function $\hat{p}(\omega_i|\boldsymbol{x})$,

$$R = \sum_{j=1}^{C} \frac{1}{n} \sum_{i=1}^{n} \left\{ \phi_j(\boldsymbol{x}_i)[z_{ji} - \hat{p}(\omega_j|\boldsymbol{x}_i)] \right\}$$

where $z_{ji} = 1$ if $\boldsymbol{x}_i \in$ class ω_j, 0 otherwise, and ϕ_j is a function that determines the test statistic (for example, $\phi_j(\boldsymbol{x}_i) = (1 - \hat{p}(\omega_j|\boldsymbol{x}_i))^2$).

Obtaining interval estimates for the posterior probabilities of class membership is another means of assessing the reliability of a rule and is discussed by McLachlan (1992a), both for the multivariate normal class-conditional distribution and in the case of arbitrary class-conditional probability density functions using a bootstrap procedure.

8.2.3 ROC curves for two-class rules

Introduction

The receiver operating characteristic (ROC) curve was introduced in Chapter 1, in the context of the Neyman–Pearson decision rule, as a means of characterising the performance of a two-class discrimination rule and provides a good means of visualising a classifier's performance in order to select a suitable decision threshold. The ROC curve is a plot of the true positive rate on the vertical axis against the false positive rate on the horizontal axis. In the terminology of signal detection theory, it is a plot of the probability of detection against the probability of false alarm, as the detection threshold is varied. Epidemiology has its own terminology: the ROC curve plots the *sensitivity* against $1 - S_e$, where S_e is the specificity.

In practice, the *optimal* ROC curve (the ROC curve obtained from the true class-conditional densities, $p(\boldsymbol{x}|\omega_i)$) is unknown, like error rate. It must be estimated using a trained classifier and an independent test set of patterns with known classes, although, in common with error rate estimation, a training set reuse method such as cross-validation or bootstrap methods may be used. Different classifiers will produce different ROC curves characterising performance of the classifiers.

Often however, we may want a single number as a performance indicator of a classifier, rather than a curve, so that we can compare the performance of competing classifier schemes.

In Chapter 1 it was shown that the minimum risk decision rule is defined on the basis of the likelihood ratio (see equation (1.15)); Assuming that there is no loss with correct classification, $\boldsymbol{x}$ is assigned to class ω_1 if

$$\frac{p(\boldsymbol{x}|\omega_1)}{p(\boldsymbol{x}|\omega_2)} > \frac{\lambda_{21}p(\omega_2)}{\lambda_{12}p(\omega_1)}, \tag{8.7}$$

where λ_{ji} is the cost of assigning a pattern $\boldsymbol{x}$ to ω_i when $\boldsymbol{x} \in \omega_j$, or alternatively

$$p(\omega_1|\boldsymbol{x}) > \frac{\lambda_{21}}{\lambda_{12} + \lambda_{21}} \tag{8.8}$$

and thus corresponds to a single point on the ROC curve determined by the relative costs and prior probabilities. The loss is given by (equation (1.11))

$$L = \lambda_{21}p(\omega_2)\epsilon_2 + \lambda_{12}p(\omega_1)\epsilon_1 \tag{8.9}$$

where $p(\omega_i)$ are the class priors and ϵ_i is the probability of misclassifying a class ω_i object. The ROC curve plots $1 - \epsilon_1$ against ϵ_2.

In the ROC curve plane (that is, the $(1 - \epsilon_1, \epsilon_2)$ plane), lines of constant loss (termed *iso-performance lines* by Provost and Fawcett, 2001) are straight lines at gradients of $\lambda_{21}p(\omega_2)/\lambda_{12}p(\omega_1)$ (see Figure 8.4), with loss increasing from top left to bottom right in the figure. Legitimate values for the loss are those for which the loss contours intercept the ROC curve (that is, a possible threshold on the likelihood ratio exists). The solution with minimum loss is that for which the loss contour is tangential with the ROC curve–the point where the ROC curve has gradient $\lambda_{21}p(\omega_2)/\lambda_{12}p(\omega_1)$. There are no other loss contours that intercept the ROC curve with lower loss.

For different values of the relative costs and priors, the loss contours are at different gradients, in general, and the minimum loss occurs at a different point on the ROC curve.

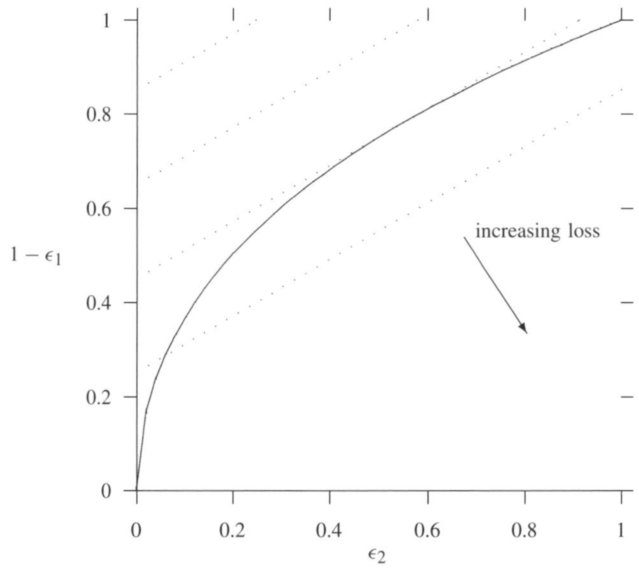

Figure 8.4 ROC curve with selected loss contours (straight lines) superimposed

Practical considerations

In many cases, the misclassification costs, λ_{12} and λ_{21}, are unknown and it unreasonable to assume equality (leading to the Bayes rule for minimum error). An alternative strategy is to compare the overall distribution of $\hat{p}(\boldsymbol{x}) \triangleq p(\omega_1|\boldsymbol{x})$ for samples from each of the classes ω_1 and ω_2. We would expect that the values of $p(\omega_1|\boldsymbol{x})$ are greater for samples $\boldsymbol{x}$ from class ω_1 than for samples $\boldsymbol{x}$ from class ω_2. Generally, the larger the difference between the two distributions, the better the classifier. A measure of the separation of these two distributions is the *area* under the ROC curve (AUC). This provides a single numeric value, based on the ROC curve, that ignores the costs, λ_{ij}. Thus, in contrast to the error rate, which assumes equal misclassification costs, it assumes nothing whatever is known about misclassification costs and thus is not influenced by factors that relate to the application of the rule. Both of these assumptions are unrealistic in practice since, usually, something will be known about likely values of the relative cost $\lambda_{12}/\lambda_{21}$. Also, the advantage of the AUC as a measure of performance (namely, that it is independent of the threshold applied to the likelihood ratio) can be a disadvantage when comparing rules. If two ROC curves cross each other, then in general one will be superior for some values of the threshold and the other superior for other values of the threshold. AUC fails to take this into account.

Interpretation

Let $\hat{p}(\boldsymbol{x}) = p(\omega_1|\boldsymbol{x})$, the estimated probability that an object $\boldsymbol{x}$ belongs to class ω_1. Let $f(\hat{p}) = f(\hat{p}(\boldsymbol{x})|\omega_1)$ be the probability density function for $\hat{p}$ values for patterns in class ω_1, and $g(\hat{p}) = g(\hat{p}(\boldsymbol{x})|\omega_2)$ be the probability density function for $\hat{p}$ values for patterns in class ω_2. If $F(\hat{p})$ and $G(\hat{p})$ are the cumulative distribution functions, then the ROC curve is a plot of $1 - F(\hat{p})$ against $1 - G(\hat{p})$ (see the exercise at the end of the chapter).

The area under the curve is given by

$$\int (1 - F(u))dG(u) = 1 - \int F(u)g(u)\,du \tag{8.10}$$

or alternatively

$$\int G(u)dF(u) = \int G(u)f(u)\,du \tag{8.11}$$

For an arbitrary point $\hat{p}(\boldsymbol{x}) = t \in [0, 1]$, the probability that a randomly chosen pattern $\boldsymbol{x}$ from class ω_2 will have a $\hat{p}(\boldsymbol{x})$ value smaller than t is $G(t)$. If t is chosen from the density f, then the probability that a randomly chosen class ω_2 pattern has a smaller value than a randomly chosen class ω_1 pattern is $\int G(u)f(u)\,du$. This is the same as the definition (8.11) for the area under the ROC curve.

A good classification rule (a rule for which the estimated values of $p(\omega_1|\boldsymbol{x})$ are very different for $\boldsymbol{x}$ from each of the two classes) lies in the upper left triangle. The closer that it gets to the upper corner the better.

A classification rule that is no better than chance produces an ROC curve that follows the diagonal from the bottom left to the top right.

Calculating the area under the ROC curve

The area under the ROC curve is easily calculated by applying the classification rule to a test set. For a classifier that produces estimates of $p(\omega_1|\boldsymbol{x})$ directly, we can obtain values $\{f_1, \ldots, f_{n_1}; f_i = p(\omega_1|\boldsymbol{x}_i), \boldsymbol{x}_i \in \omega_1\}$ and $\{g_1, \ldots, g_{n_2}; g_i = p(\omega_1|\boldsymbol{x}_i), \boldsymbol{x}_i \in \omega_2\}$ and use these to obtain a measure of how well separated are the distributions of $\hat{p}(\boldsymbol{x})$ for class ω_1 and class ω_2 patterns as follows (Hand and Till, 2001).

Rank the estimates $\{f_1, \ldots, f_{n_1}, g_1, \ldots, g_{n_2}\}$ in increasing order and let the rank of the ith pattern from class ω_1 be r_i. Then there are $r_i - i$ class ω_2 patterns with estimated value of $\hat{p}(\boldsymbol{x})$ less than that of the ith pattern of class ω_1. If we sum over class ω_1 test points, then we see that the number of pairs of points, one from class ω_1 and one from class ω_2, with $\hat{p}(\boldsymbol{x})$ smaller for class ω_2 than for class ω_1 is

$$\sum_{i=1}^{n_1}(r_i - i) = \sum_{i=1}^{n_1} r_i - \sum_{i=1}^{n_1} i = S_0 - \frac{1}{2}n_1(n_1 + 1)$$

where S_0 is the sum of the ranks of the class ω_1 test patterns. Since there are $n_1 n_2$ pairs, the estimate of the probability that a randomly chosen class ω_2 pattern has a lower estimated probability of belonging to class ω_1 than a randomly chosen class ω_1 pattern is

$$\hat{A} = \frac{1}{n_1 n_2}\left\{S_0 - \frac{1}{2}n_1(n_1 + 1)\right\}$$

This is equivalent to the area under the ROC curve and provides an estimate that has been obtained using the rankings alone and has not used threshold values to calculate it.

The standard deviation of the statistic $\hat{A}$ is (Hand and Till, 2001)

$$\sqrt{\frac{\hat{\theta}(1 - \hat{\theta}) + (n_1 - 1)(Q_0 - \hat{\theta}^2) + (n_2 - 1)(Q_1 - \hat{\theta}^2)}{n_1 n_2}}$$

where

$$\hat{\theta} = \frac{S_0}{n_1 n_2}$$
$$Q_0 = \frac{1}{6}(2n_1 + 2n_2 + 1)(n_1 + n_2) - Q_1$$
$$Q_1 = \sum_{j=1}^{n_1} (r_j - 1)^2$$

An alternative approach, considered by Bradley (1997), is to construct an estimate of the ROC curve directly for specific classifiers by varying a threshold and then to use an integration rule (for example, the trapezium rule) to obtain an estimate of the area beneath the curve.

8.2.4 Example application study

The problem This study (Bradley, 1997) comprises an assessment of AUC as a performance measure on six pattern recognition algorithms applied to data sets characterising medical diagnostic problems.

Summary The study estimates AUC through an integration of the ROC curve and its standard deviation is calculated using cross-validation.

The data There are six data sets comprising measurements on two classes:

1. Cervical cancer. Six features, 117 patterns; classes are normal and abnormal cervical cell nuclei.
2. Post-operative bleeding. Four features, 113 patterns (after removal of incomplete patterns); classes are normal blood loss and excessive bleeding.
3. Breast cancer. Nine features, 683 patterns; classes are benign and malignant.
4. Diabetes. Eight features, 768 patterns; classes are negative and positive test for diabetes.
5. Heart disease 1. Fourteen features, 297 patterns; classes are heart disease present and heart disease absent.
6. Heart disease 2. Eleven features, 261 patterns; classes are heart disease present and heart disease absent.

Incomplete patterns (patterns for which measurements on some features are missing) were removed from the data sets.

The models Six classifiers were trained on each data set:

1. quadratic discriminant function (Chapter 2);
2. k-nearest-neighbour (Chapter 3);

3. classification tree (Chapter 7);
4. multiscale classifier method (a development of classification trees);
5. perceptron (Chapter 4);
6. multilayer perceptron (Chapter 6).

The models were trained and classification performance monitored as a threshold was varied in order to estimate the ROC curves. For example, for the k-nearest-neighbour classifier, the five nearest neighbours in the training set to a test sample are calculated. If the number of neighbours belonging to class ω_1 is greater than L, where $L = [0, 1, 2, 3, 4, 5]$, then the test sample is assigned to class ω_1, otherwise it is assigned to the second class. This gives six points on the ROC curve.

For the multilayer perceptron, a network with a single output is trained and during testing, it is thresholded at values of $[0, 0.1, 0.2, \ldots, 1.0]$ to simulate different misclassification costs.

Training procedure A tenfold cross-validation scheme was used, with 90% of the samples used for training and 10% used in the test set, selected randomly. Thus, for each classifier on each data set, there are ten sets of results.

The ROC curve was calculated as a decision threshold was varied for each of the test set partitions and the area under the curve calculated using trapezoidal integration. The AUC for the rule is taken to be the average of the ten AUC values obtained from the ten partitions of the data set.

8.2.5 Further developments

The aspect of classifier performance that has received most attention in the literature is the subject of error rate estimation. Hand (1997) develops a much broader framework for the assessment of classification rules, defining four concepts.

Inaccuracy This is a measure of how (in)effective is a classification rule in assigning an object to the correct class. One example is error rate; another is the *Brier* or *quadratic score*, often used as an optimisation criterion for neural networks, defined as

$$\frac{1}{n}\sum_{i=1}^{n}\sum_{j=1}^{C}\left\{\delta(\omega_j|\boldsymbol{x}_i) - \hat{p}(\omega_j|\boldsymbol{x}_i)\right\}^2$$

where $\hat{p}(\omega_j|\boldsymbol{x}_i)$ is the estimated probability that pattern $\boldsymbol{x}_i$ belongs to class ω_j, and $\delta(\omega_j|\boldsymbol{x}_i) = 1$ if $\boldsymbol{x}_i$ is a member of class ω_j and zero otherwise.

Imprecision Equivalent to reliability defined in Section 8.2.2, this is a measure of the difference between the estimated probabilities of class membership, $\hat{p}(\omega_j|\boldsymbol{x})$, and the (unknown) true probabilities, $p(\omega_j|\boldsymbol{x})$.

Inseparability This measure is evaluated using the true probabilities of belonging to a class, and so it does not depend on a classifier. It measures the similarity of the true probabilities of class membership at a point $\boldsymbol{x}$, averaged over $\boldsymbol{x}$. If the probabilities at a point $\boldsymbol{x}$ are similar, then the classes are not separable.

Resemblance It measures the variation between the true probabilities, conditioned on the estimated ones. Does the predicted classification separate the true classes well? A low value of resemblance is to be hoped for.

In this section we have been unable to do full justice to the elegance of the bootstrap method, and we have simply presented the basic bootstrap approach for the bias correction of the apparent error rate, with some extensions. Further developments may be found in Efron (1983, 1990), Efron and Tibshirani (1986) and McLachlan (1992a). Further work on the AUC measure includes that of Hand and Till (2001) who develop it to the multiclass classification problem (see also Hajian-Tilaki *et al.*, 1997a, 1997b) and Adams and Hand (1999) who take account of some imprecisely known information on the relative misclassification costs (see also Section 8.3.3).

Provost and Fawcett (2001) propose an approach that uses the convex hull of ROC curves of different classifiers. Classifiers with ROC curves below the convex hull are never optimal (under any conditions on costs or priors) and can be ignored. Classifiers on the convex hull can be combined to produce a better classifier. This idea has been widely used in *data fusion* and is discussed in Sections 8.3.3 and 8.4.4.

8.2.6 Summary

In this section, we have given a rather brief treatment of classification rule performance assessment, covering three measures: discriminability, reliability (or imprecision) and the use of the ROC curve. In particular, we have given emphasis to the error rate of a classifier and schemes for reducing the bias of the apparent error rate, namely cross-validation, the jackknife and the bootstrap methods. These have the advantage over the holdout method in that they do not require a separate test set. Therefore, all the data may be used in classifier design.

The error rate estimators described in this chapter are all nonparametric estimators in that they do not assume a specific form for the probability density functions. Parametric forms of error rate estimators, for example based on a normal distribution model for the class-conditional densities, can also be derived. However, although parametric rules may be fairly robust to departures from the true model, parametric estimates of error rates may not be (Konishi and Honda, 1990). Hence our concentration in this chapter on nonparametric forms. For a further discussion of parametric error rate estimators we refer the reader to the book by McLachlan (1992a).

There are many other measures of discriminability, and limiting ourselves to a single measure, the error rate, may hide important information as to the behaviour of a rule. The error rate treats all misclassifications equally: misclassifying an object from class 0 as class 1 has the same severity as misclassifying an object from class 1 as class 0. Costs of misclassification may be very important in some applications but are rarely known precisely.

The reliability, or imprecision, of a rule tells us how well we can trust the rule – how close the estimated posterior densities are to the true posterior densities. Finally, the area under the ROC curve is a measure that summarises classifier performance over a range of relative costs.

8.3 Comparing classifier performance

8.3.1 Which technique is best?

Are neural network methods better than 'traditional' techniques? Is the classifier that you develop better than those previously published in the literature? There have been many comparative studies of classifiers and we have referenced these in previous chapters. Perhaps the most comprehensive study is the *Statlog* project (Michie *et al.*, 1994) which provides a study of more than 20 different classification procedures applied to about 20 data sets. Yet comparisons are not easy. Classifier performance varies with the data set, sample size, dimensionality of the data, and skill of the analyst. There are some important issues to be resolved, as outlined by Duin (1996):

1. An application domain must be defined. Although this is usually achieved by specifying a collection of data sets, these may not be representative of the problem domain that you wish to consider. Although a particular classifier may perform consistently badly on these data sets, it may be particularly suited to the one you have.

2. The skill of the analyst needs to be considered (and removed if possible). Whereas some techniques are fairly well defined (nearest-neighbour with a given metric), others require tuning. Can the results of classifications, using different techniques, on a given data set performed by separate analysts be sensibly compared? If one technique performs better than others, is it due to the superiority of the technique on that data set or the skill of the implementer in obtaining the best out of a favourite method? In fact, some classifiers are valuable because they have many free parameters and allow a trained analyst to incorporate knowledge into the training procedure. Others are valuable because they are largely automatic and do not require user input. The *Statlog* project was an attempt at developing automatic classification schemes, encouraging minimal tuning.

Related to the second issue above is that the main contribution to the final performance is the initial problem formulation (abstracting the problem to be solved from the customer, selecting variables and so on), again determined by the skill of the analyst. The classifier may only produce second-order improvements to performance.

In addition, what is the basis on which we make a comparison–error rate, reliability, speed of implementation, speed of testing, etc.?

There is no such thing as a best classifier, but there are several ways in which comparisons may be performed (Duin, 1996):

1. A comparison of experts. A collection of problems is sent to experts who may use whichever technique they feel is appropriate.
2. A comparison of toolsets by nonexperts. Here a collection of toolsets is provided to nonexperts for evaluation on several data sets.
3. A comparison of automatic classifiers (classifiers that require no tuning). This is performed by a single researcher on a benchmark set of problems. Although the results will be largely independent of the expert, they will probably be inferior to those obtained if the expert were allowed to choose the classifier.

8.3.2 Statistical tests

Bounds on the error rate are insufficient when comparing classifiers. Usually the test sets are not independent – they are common across all classifiers. There are several tests for determining whether one classification rule outperforms another on a particular dataset.

The question of measuring the accuracy of a classification rule using an independent training and test set was discussed in Section 8.2. This can be achieved by constructing a confidence interval or HPD region. Here we address the question: given two classifiers and sufficient data for a separate test set, which classifier will be more accurate on new test set examples?

Dietterich (1998) assesses five statistical tests, comparing them experimentally to determine the probability of incorrectly detecting a difference between classifier performance when no difference exists (Type I error).

Suppose that we have two classifiers, A and B. Let

$$n_{00} = \text{number of samples misclassified by both } A \text{ and } B$$
$$n_{01} = \text{number of samples misclassified by } A \text{ but not by } B$$
$$n_{10} = \text{number of samples misclassified by } B \text{ but not by } A$$
$$n_{11} = \text{number of samples misclassified by neither } A \text{ nor } B$$

Compute the z statistic

$$z = \frac{|n_{01} - n_{10}| - 1}{\sqrt{n_{10} + n_{01}}}$$

The quantity z^2 is distributed approximately as χ^2 with one degree of freedom. The null hypothesis (that the classifiers have the same error) can be rejected (with probability of incorrect rejection of 0.05) if $|z| > 1.96$. This is known as McNemar's test or the Gillick test.

8.3.3 Comparing rules when misclassification costs are uncertain

Introduction

Error rate or misclassification rate, discussed in Section 8.2, is often used as a criterion for comparing several classifiers. It requires no choice of costs, making the assumption that misclassification costs are all equal.

An alternative measure of performance is the area under the ROC curve (see also Section 8.2). This is a measure of the separability of the two distributions $f(\hat{p})$, the probability distribution of $\hat{p} = p(\omega_1|\boldsymbol{x})$ for patterns $\boldsymbol{x}$ in class ω_1, and $g(\hat{p})$, the probability distribution of $\hat{p}$ for patterns $\boldsymbol{x}$ in class ω_2. It has the advantage that it does not depend on the relative costs of misclassification.

There are difficulties with the assumptions behind both of these performance measures. In many, if not most, practical applications the assumption of equal costs is unrealistic. Also the minimum loss solution, which requires the specification of costs, is not sensible since rarely are costs and priors known precisely. In many real-world environments, misclassification costs and class priors are likely to change over time as the environment may change between design and test. Consequently, the point on the ROC curve corresponding to the minimum loss solution (where the threshold on the likelihood ratio is $\lambda_{21}p(\omega_2)/\lambda_{12}p(\omega_1)$ – equation (8.7)) changes. On the other hand, usually *something* is known about the relative costs and it is therefore inappropriate to summarise over all possible values.

ROC curves

Comparing classifiers on the basis of AUC is difficult when the ROC curves cross. Only in the case of one classifier dominating another will the AUC be a valid criterion for comparing different classifiers. If two ROC curves cross, then one curve will be superior for some values of the cost ratio and the other classifier will be superior for different values of the cost ratio. Two approaches for handling this situation are presented here.

LC index In this approach for comparing two classifiers A and B, the costs of misclassification, λ_{12} and λ_{21}, are rescaled so that $\lambda_{12}+\lambda_{21} = 1$ and the loss (8.9) calculated as a function of λ_{21} for each classifier. A function, $L(\lambda_{21})$, is defined to take the value $+1$ in regions of the [0, 1] interval for which classifier A is superior (it has a lower loss value than classifier B) and -1 in regions for which classifier B is superior. The confidence in any value of λ_{21} is the probability density function $D(\lambda_{21})$, defined later, and the LC index is defined as

$$\int_0^1 D(\lambda)L(\lambda)d\lambda$$

which ranges over ± 1, taking positive values when classifier A is more likely to lead to a smaller loss value than classifier B, and negative values when classifier B is more likely to lead to a smaller loss value than classifier A. A value of $+1$ means that A is certain to be a superior classifier since it is superior for all feasible values of λ_{21}.

How do we decide on the set of feasible values of λ_{21}? That is, what form do we choose for the distribution $D(\lambda_{21})$? One proposal is to specify an interval $[a, b]$ for the cost ratio, $\lambda_{12}/\lambda_{21}$, and a most likely value, m, and use this to define a unit-area triangle with base $[a, b]$ and apex at m. This is because, it is argued (Adams and Hand, 1999), that experts find it convenient to specify a cost ratio $\lambda_{12}/\lambda_{21}$ and an interval for the ratio.

The ROC convex hull method In this method a hybrid classification system is constructed from the set of available classifiers. For any value of the cost ratio, the combined

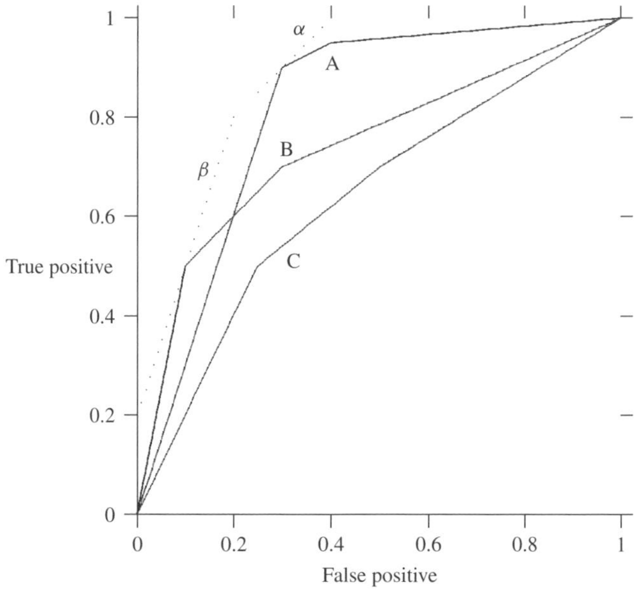

Figure 8.5 ROC convex hull method illustration

classifier will perform at least as well as the best classifier. The combined classifier is constructed to have an ROC curve that is the convex hull of the component classifiers.

Figure 8.5 illustrates the ROC convex hull method. For some values of costs and priors, the slope of the iso-performance lines is such that the optimal classifier (the point on the ROC curve lying to the top left) is classifier B. Line β is the iso-performance line with lowest loss that intercepts the ROC curve of classifier B. For much shallower gradients of the iso-performance lines (corresponding to different values of the priors or costs), the optimal classifier is classifier A. Here, line α is the lowest value iso-performance line (for a given value of priors and costs) that intercepts the ROC curve of classifier A. Classifier C is not optimal for any value of priors or costs. The points on the convex hull of the ROC curves define optimal classifiers for particular values of priors and costs. Provost and Fawcett (2001) present an algorithm for generating the ROC convex hull.

In practice, we need to store the range of threshold values $(\lambda_{21} p(\omega_2)/\lambda_{12} p(\omega_1))$ for which a particular classifier is optimal. Thus, the range of the threshold is partitioned into regions, each of which is assigned a classifier, the one that is optimal for that range of thresholds.

8.3.4 Example application study

The problem This study (Adams and Hand, 1999) develops an approach for comparing the performance of two classifiers when misclassification costs are uncertain, but not completely unknown. It is concerned with classifying customers according to their likely response to a promotional scheme.

Summary The LC index above is evaluated to compare a neural network classifier (Chapter 6) with quadratic discriminant analysis (Chapter 2).

The data The data comprise 8000 records (patterns) of measurements on 25 variables, mainly describing earlier credit card transaction behaviour. The classes are denoted class ω_1 and class ω_2, with class ω_2 thought likely to return a profit. The priors are set as $p(\omega_1) = 0.87$, $p(\omega_2) = 0.13$.

The model The multilayer perceptron had 25 input nodes and 13 hidden nodes, trained using 'weight decay' to avoid over-fitting, with the penalty term chosen by cross-validation.

Training procedure The LC index and AUC were computed. In order to obtain suitable values for the ratio of costs, banking experts were consulted and a model developed for the two types of misclassification based on factors such as cost of manufacture and distribution of marketing material, cost due to irritation caused by receiving junk mail and loss of potential profit by failing to mail a potential member of class ω_2.

An interval of possible values for the overall cost ratio $\lambda_{12}/\lambda_{21}$ was derived as [0.065, 0.15], with the most probable value at 0.095.

Results The AUC values for the neural network classifier and quadratic discriminant analysis were 0.7102 and 0.7244 respectively, suggesting that the quadratic discriminant is slightly preferable. The LC index was calculated to be −0.4, also suggesting that quadratic discriminant analysis is to be preferred.

8.3.5 Further developments

Adams and Hand (2000) present some guidelines for better methodology for comparing classifiers. They identify five common deficiencies in the practice of classifier performance assessment:

1. Assuming equal costs. In many practical applications, the two types of misclassification are not equal.

2. Integrating over costs. The AUC summarises performance over the entire range of costs. It is more likely that something will be known about costs and that a narrower range would be more appropriate.

3. Crossing ROC curves. The AUC measure is only appropriate if one ROC curve dominates over the entire range. If the ROC curves cross, then different classifiers will dominate for different ranges of the misclassification costs.

4. Fixing costs. It is improbable that exact costs can be given in many applications.

5. Variability. Error rate and the AUC measure are sample-based estimates. Standard errors should be given when reporting results.

8.3.6 Summary

There are several ways in which performance may be compared. It is important to use an assessment criterion appropriate to the real problem under investigation. Misclassification costs should be taken into account since they can influence the choice of method. Assuming equal misclassification costs is very rarely appropriate. Usually something can be said about costs, even if they are not known precisely.

8.4 Combining classifiers

8.4.1 Introduction

The approach to classifier design commonly taken is to identify a candidate set of plausible models, to train the classifiers using a training set of labelled patterns and to adopt the classifier that gives the best generalisation performance, estimated using an independent test set assumed representative of the true operating conditions. This results in a single 'best' classifier that may then be applied throughout the feature space. Earlier in this chapter we addressed the question of how we measure classifier performance to select a 'best' classifier.

We now consider the potential of combining classifiers for data sets with complex decision boundaries. It may happen that, out of our set of classifiers, no single classifier is clearly best (using some suitable performance measure, such as error rate). However, the set of misclassified samples may differ from one classifier to another. Thus, the classifiers may give complementary information and combination could prove useful. A simple example is illustrated in Figure 8.6. Two linear classifiers, Cl_1 and Cl_2, are defined on a univariate data space. Classifier Cl_1 predicts class $\bullet$ for data points to the left of B and class $\diamond$ for points to the right of B. Classifier Cl_2 predicts class $\bullet$ for points to the right of A and class $\diamond$ for points to the left. Neither classifier obtains 100% performance on the data set. However, 100% performance is achieved by combining them with the rule: assign x to $\bullet$ if Cl_1 and Cl_2 predict $\bullet$ else assign x to $\diamond$.

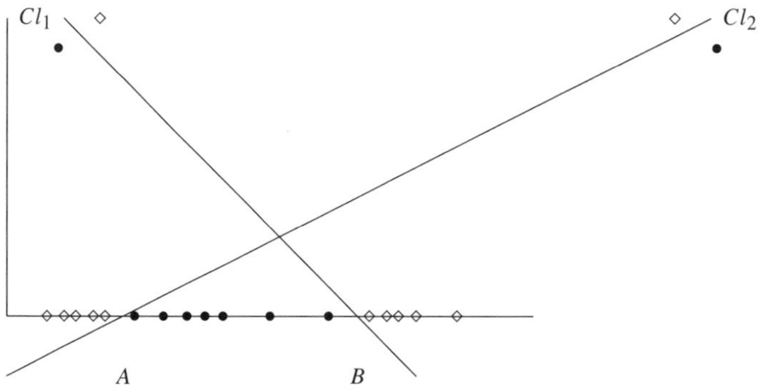

Figure 8.6 Two linear discriminants defined on a univariate data space

The idea of combining classifiers is not a new one, but one that has received increasing attention in recent years. Early work on multiclass discrimination developed techniques for combining the results of two-class discrimination rules (Devijver and Kittler, 1982). Also, recursive partitioning methods (for example, CART; see Chapter 7) lead to the idea of defining different rules for different parts of a feature space. The terms 'dynamic classifier selection' (Woods *et al.*, 1997) and 'classifier choice system' (Hand *et al.*, 2001) have been used for classifier systems that attempt to predict the best classifier for a given region of feature space. The term 'classifier fusion' or 'multiple classifier system' usually refers to the combination of predictions from multiple classifiers to yield a single class prediction.

8.4.2 Motivation

There are several ways of characterising multiple classifier systems, as indeed there are with basic component classifiers. We define three broad categories as follows.

C1. Different feature spaces

This describes the combination of a set of classifiers, each designed on different feature spaces (perhaps using data from different sensors). For example, in a person verification application, several classifier systems may be designed for use with different sensor data (for example, retina scan, facial image, handwritten signature) and we wish to combine the outputs of each system to improve performance. Figure 8.7 illustrates this situation. A set of L sensors $(S_1, S_2, \ldots, S_L)$ provides measurements, $x_1, x_2, \ldots, x_L$, on an object. Associated with each sensor is a classifier $(Cl_1, Cl_2, \ldots, Cl_L)$ providing, in this case, estimates of the posterior probabilities of class membership, $p(c|x_i)$ for sensor

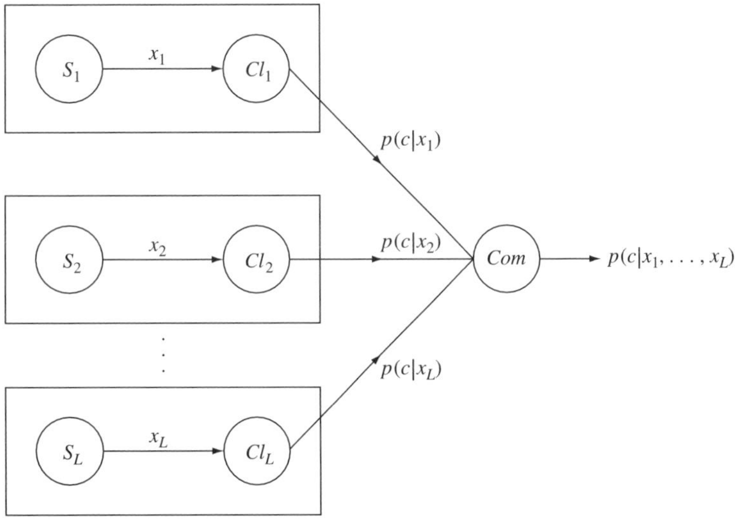

Figure 8.7 Classifier fusion architecture C1–component classifiers defined on different feature spaces

S_i. The combination rule (denoted Com in the figure), which is itself a classifier defined on a feature space of posterior probabilities, combines these posterior probabilities to provide an estimate of $p(c|x_1, \ldots, x_L)$. Thus, the individual classifiers can be thought of as performing a particular form of feature extraction prior to classification by the combiner.

The question usually addressed with this architecture is: given the component classifiers, what is the best combination rule? This is closely related to dynamic classifier selection and to architectures developed in the data fusion literature (see Section 8.4.4).

C2. Common feature space

In this case, we have a set of classifiers, $Cl_1, \ldots, Cl_L$, each defined on the same feature space and the combiner attempts to obtain a 'better' classifier through combination (see Figure 8.8). The classifiers can differ from each other in several ways.

1. The classifiers may be of different type, belonging to the set of favourite classifiers of a user: for example, nearest-neighbour, neural network, decision tree and linear discriminant.
2. They may be of similar type (for example, all linear discriminant functions or all neural network models) but trained using different training sets (or subsets of a larger training set), perhaps gathered at different times or with different noise realisations added to the input.
3. The classifiers may be of similar type, but with different random initialisation of the classifier parameters (for example, the weights in a neural network of a given architecture) in an optimisation procedure.

In contrast with category C1, a question that we might ask here is: given the combination rule, what is the best set of component classifiers? Equivalently, in case 2 above, for example, how do we train our neural network models to give the best performance on combination?

The issue of accuracy and diversity of component classifiers (Hansen and Salamon, 1990) is important for this multiple classifier architecture. Each component classifier is

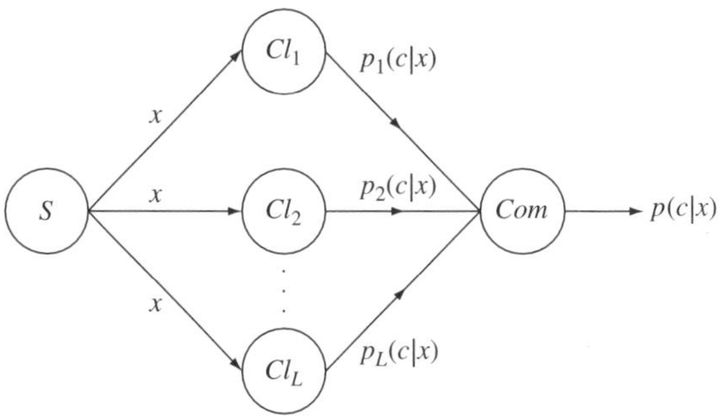

Figure 8.8 Classifier fusion architecture C2–component classifiers defined on a common feature space of measurements x made by sensor S

said to be *accurate* if it has an error rate lower than that obtained by random guessing on new patterns. Two classifiers are *diverse* if they make different errors in predicting the class of a pattern, $\boldsymbol{x}$. To see why both accuracy and diversity are important, consider an ensemble of three classifiers $h_1(\boldsymbol{x})$, $h_2(\boldsymbol{x})$ and $h_3(\boldsymbol{x})$, each predicting a class label as output. If all classifiers produce identical outputs, then there will be no improvement gained by combining the outputs: when h_1 is incorrect, h_2 and h_3 will also be incorrect. However, if the classifier outputs are uncorrelated, then when h_1 is incorrect, h_2 and h_3 may be correct and, if so, a majority vote will give the correct prediction. More specifically, consider L classifiers $h_1, \ldots, h_L$, each with an error rate of $p < \frac{1}{2}$. For the majority vote to be incorrect, we require that $L/2$ or more classifiers be incorrect. The probability that R classifiers are incorrect is

$$\frac{L!}{R!(L-R)!}p^R(1-p)^{L-R}$$

The probability that the majority vote is incorrect is therefore less than[1]

$$\sum_{R=\lfloor (L+1)/2 \rfloor}^{L} \frac{L!}{R!(L-R)!}p^R(1-p)^{L-R}$$

the area under the binomial distribution where at least $L/2$ are incorrect($\lfloor . \rfloor$ denotes the integer part). For example, with $L = 11$ classifiers, each with an error rate $p = 0.25$, the probability of six or more classifiers being incorrect is 0.034, which is much less than the individual error rate.

If the error rates of the individual classifiers exceed 0.5, then the error rate of the majority vote will increase, depending on the number of classes and the degree of correlation between the classifier outputs.

In both cases C1 and C2, each classifier is performing a particular form of feature extraction for inputs to the combiner classifier. If this combiner classifier is not fixed (that is, it is allowed to adapt to the forms of the input), then the general problem is one in which we seek the best forms of feature extractor matched to a combiner (see Figure 8.9). There is no longer a requirement that the outputs of the classifiers $Cl_1, \ldots, Cl_L$ are estimates of posterior probabilities (are positive and sum to unity over classes). Indeed, these may not be the best features. Thus the component classifiers are not classifiers at all and the procedure is essentially the same as many described elsewhere in this book – neural networks, projection pursuit, and so on. In this sense, there is no need for research on classifier combination methods since they impose an unnecessary restriction on the forms of the features input to the combiner.

C3. Repeated measurements

The final category of combination systems arise due to different classification of an object through repeated measurements. This may occur when we have a classifier designed on a feature space giving an estimate of the posterior probabilities of class membership, but in practice several (correlated) measurements may be made on the object (see Figure 8.10).

[1]For more than two classes, the majority vote could still produce a correct prediction even if more than $L/2$ classifiers are incorrect.

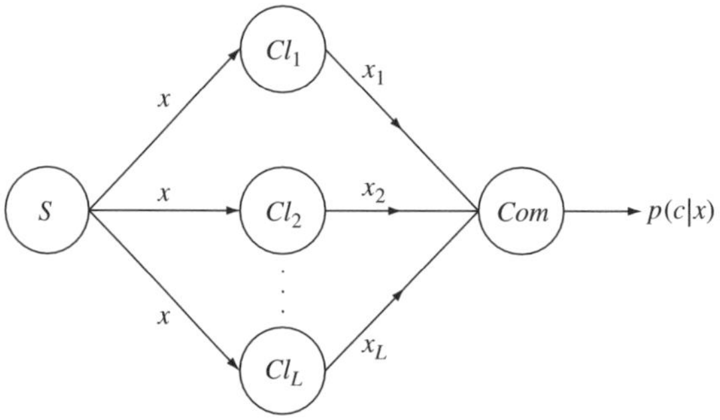

Figure 8.9 Classifier fusion architecture C2–component classifiers defined on a common feature space with L component classifiers delivering features x_i

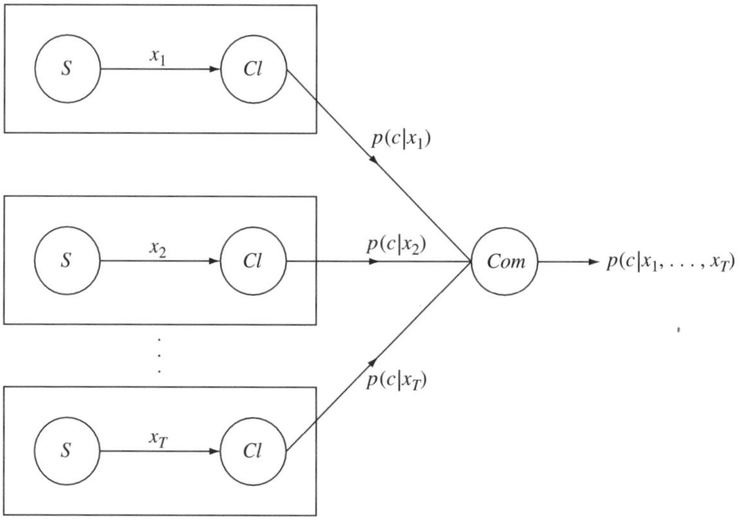

Figure 8.10 Classifier fusion architecture C3 repeated measurements on a common feature space. Sensor S produces a sequence of measurements x_i, $i = 1, \ldots, T$, which are input to classifier Cl

An example is that of recognition of aircraft from visual or infrared data. A probability density function of feature vectors may be constructed using a training data set. In the practical application, successive measurements are available from the sensor. How can we combine these predictions? This is sometimes described as *multiple observation fusion* or *temporal fusion*.

8.4.3 Characteristics of a combination scheme

Combination schemes may themselves be classified according to several characteristics, including their structure, input data type, form of component classifiers and training requirements. In this section, we summarise some of the main features of a combiner. We assume that there are measurements from C classes and there are L component classifiers.

Level

Combination may occur at different levels of component classifier output.

L1. Data level Raw sensor measurements are passed to a combiner that produces an estimate of the posterior probabilities of class membership. This simply amounts to defining a classifier on the augmented feature space comprising measurements on all sensor variables. That is, for sensors producing measurements $\boldsymbol{x}$, $\boldsymbol{y}$ and $\boldsymbol{z}$, a classifier is defined on the feature vector $(\boldsymbol{x}, \boldsymbol{y}, \boldsymbol{z})$. In the *data fusion* literature (see Section 8.4.4) this is referred to as a *centralised system*: all the information collected at distributed sensors is passed to a central processing unit that makes a final decision. A consequence of this procedure is that the combiner must be constructed using data of a high dimensionality. Consequently, it is usual to perform some feature selection or extraction prior to combination.

L2. Feature level Each constituent classifier (we shall use the term 'classifier' even though the output may not be an estimate of the posterior probabilities of class membership or a prediction of class) performs some local preprocessing, perhaps to reduce dimensionality. This could be important in some data fusion applications where the communication bandwidth between constituent classifiers and combiner is an important consideration. In data fusion, this is termed a *decentralised* system. The features derived by the constituent classifier for input (transmission) to the combiner can take several forms, including the following.

1. A reduced-dimensional representation, perhaps derived using principal components analysis.
2. An estimate of the posterior probabilities of class membership. Thus, each constituent processor is itself a classifier. This is a very specific form of preprocessing that may not be optimum, but it may be imposed by application constraints.
3. A coding of the constituent classifier's input. For an input, $\boldsymbol{x}$, the output of the constituent classifier is the index, y, in a codebook of vectors corresponding to the codeword, z, which is nearest to $\boldsymbol{x}$. At the combiner, the code index y is decoded to produce an approximation to $\boldsymbol{x}$ (namely, z). This is then used, together with approximations of the other constituent classifier inputs, to produce a classification. Such a procedure is important if there are bandwidth constraints on the links between constituent classifier and combiner. Procedures for vector quantisation (see Chapter 10) are relevant to this process.

L3. Decision level Each constituent classifier produces a unique class label. The combiner classifier is then defined on an L-dimensional space of categorical variables, each taking one of C values. Techniques for classifier construction on discrete variables (for example, histograms and generalisations – maximum weight dependence trees and Bayesian networks – see Chapter 3) are appropriate.

Degree of training

R1. Fixed classifiers In some cases, we may wish to use a fixed combination rule. This may occur if the training data used to design the constituent classifiers are unavailable, or different training sets have been used.

R2. Trainable classifiers Alternatively, the combination rule is adjusted based on knowledge of the training data used to define the constituent classifiers. This knowledge can take different forms:

1. *Probability density functions.* The joint probability density function of constituent classifier outputs is assumed known for each class through knowledge of the distribution of the inputs and the form of the classifiers. For example, in the two-class target detection problem, under the assumption of independent local decisions at each of the constituent classifiers, an optimal detection rule for the combiner can be derived, expressed in terms of the probability of false alarm and the probability of missed detection at each sensor (Chair and Varshney, 1986).

2. *Correlations.* Some knowledge concerning the correlations between constituent classifier outputs is assumed. Again, in the target detection problem under correlated local decisions (correlated outputs of constituent classifiers), Kam *et al.* (1992) expand the probability density function using the Bahadur–Lazarsfeld polynomials to rewrite the optimal combiner rule in terms of conditional correlation coefficients (see Section 8.4.4).

3. *Training data available.* It is assumed that the outputs of each of the individual constituent classifiers are known for a given input of known class. Thus, we have a set of labelled samples that may be used to train the combiner classifier.

Form of component classifiers

F1. Common form Classifiers may all be of the same form. For example, they all may be neural networks (multilayer perceptrons) of a given architecture, all linear discriminants or all decision trees. The particular form may be chosen for several reasons: *interpretability* (it is easy to interpret the classification process in terms of simple rules defined on the input space); *implementability* (the constituent classifiers are easy to implement and do not require excessive computation); *adaptability* (the constituent classifiers are flexible and it is easy to implement diverse classifiers whose combination leads to a lower error rate than any of the individuals).

F2. Dissimilar form The constituent classifiers may be a collection of neural networks, decision trees, nearest-neighbour methods and so on, the set perhaps arising through the analysis of a wide range of classifiers on different training sets. Thus, the classifiers have not necessarily been chosen so that their combination leads to the best improvement.

Structure

The structure of a multiple classifier system is often dictated by a practical application.

T1. Parallel The results from the constituent classifiers are passed to the combiner together before a decision is made by the combiner.

T2. Serial Each constituent classifier is invoked sequentially, with the results of one classifier being used by the next one in the sequence, perhaps to set a prior on the classes.

T3. Hierarchical The classifiers are combined in a hierarchy, with the outputs of one constituent classifier feeding as inputs to a parent node, in a similar manner to decision-tree classifiers (see Chapter 7). Thus the partition into a single combiner with several constituent classifiers is less apparent, with each classifier (apart from the leaf and root nodes) taking the output of a classifier as input and passing its own output as input to another classifier.

Optimisation

Different parts of the combining scheme may be optimised separately or simultaneously, depending on the motivating problem.

O1. Combiner Optimise the combiner alone. Thus, given a set of constituent classifiers, we determine the combining rule to give the greatest performance improvement.

O2. Constituent classifiers Optimise the constituent classifiers. For a fixed combiner rule, and the number and type of constituent classifiers, the parameters of these classifiers are determined to maximise performance.

O3. Combiner and constituent classifiers Optimise both the combiner rule and the parameters of the constituent classifiers. In this case, the constituent classifiers may not be classifiers in the strict sense, performing some form of feature extraction. Practical constraints such as limitations on the bandwidth between the constituent classifiers and the combiner may need to be considered.

O4. No optimisation We are provided with a fixed set of classifiers and use a standard combiner rule that requires no training (see Section 8.4.5).

8.4.4 Data fusion

Much of the work on multiple classifier systems reported in the pattern recognition, statistics and machine learning literature has strong parallels, and indeed overlaps substantially, with research on data fusion systems carried out largely within the engineering community (Dasarathy, 1994b; Varshney, 1997; Waltz and Llinas, 1990). In common with the research on classifier combination, different architectures may be considered (for example, serial or parallel) and different assumptions made concerning the joint distribution of classifier outputs, but the final architecture adopted and the constraints under which it is optimised are usually motivated by real problems. One application of special interest is that of distributed detection: detecting the presence of a target using a distributed array of sensors. In this section we review some of the work in this area.

Architectures

Figures 8.11 and 8.12 illustrate the two main architectures for a decentralised distributed detection system – the serial and parallel configurations. We assume that there are L sensors. The observations at each sensor are denoted by y_i and the decision at each sensor by $u_i, i = 1, \ldots, L$, where

$$u_i = \begin{cases} 1 & \text{if 'target present' declared} \\ 0 & \text{if 'target absent' declared} \end{cases}$$

and the final decision is denoted by u_0. Each sensor can be considered as a binary classifier and the problem is termed as one in *decision fusion*. This may be thought a very restrictive model, but it is one that may arise in some practical situations.

The main drawback with the serial network structure (Figure 8.11) is that it has a serious reliability problem. This problem arises because if there is a link failure between the $(i-1)$th sensor and the ith sensor then all the information used to make the previous decisions would be lost, resulting in the ith sensor becoming effectively the first sensor in the decision process.

The parallel system has been widely considered and is shown in Figure 8.12. Each sensor receives a local observation y_i, $i = 1, \ldots, L$, and produces a local decision u_i which is sent to the fusion centre. At the fusion centre all the local decisions u_i, $i = 1, \ldots, L$, are combined to obtain a global decision u_0. The parallel architecture is

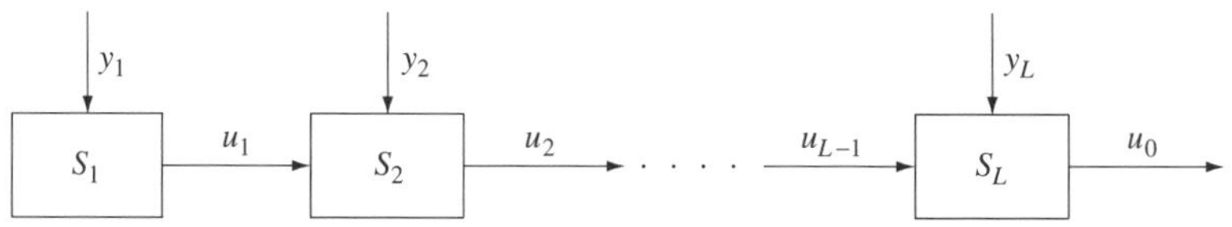

Figure 8.11 Sensors arranged in a serial configuration

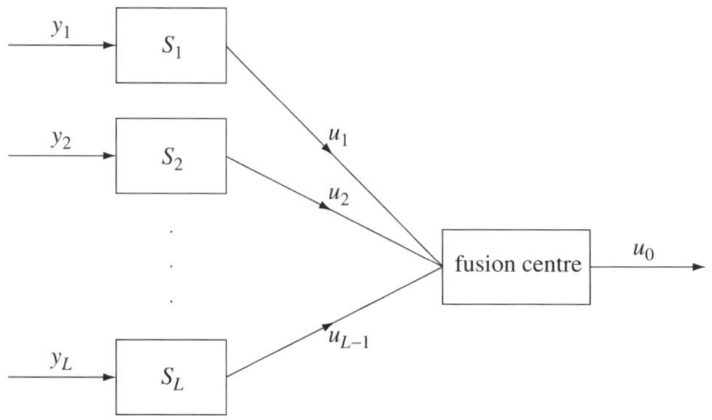

Figure 8.12 Sensors arranged in a parallel configuration

far more robust to link failure. A link failure between the ith sensor and the fusion centre does not seriously jeopardize the overall global decision, since it is only the decision of the ith sensor that is lost. The parallel distributed system has been considered under the assumptions of both correlated and independent local decisions, and methods have been proposed for solving for the optimal solution in both these cases.

The parallel decision system has also been extended to allow the local decisions to be passed to intermediate fusion centres, each processing all the L local decisions before passing their decisions on to another layer of intermediate fusion centres (Li and Sethi, 1993; Gini, 1997). After K such layers these intermediate fusion centre decisions are passed to the fusion centre and a global decision u_0 is made. This corresponds to the hierarchical model in multiple classifier systems (see Section 8.4.3).

The parallel architecture has also been used to handle repeated observations. One approach has been to use a memory term that corresponds to the decision made by the fusion centre using the last set of local decisions (Kam *et al.*, 1999). The memory term is used in conjunction with the next set of local decisions to make the next global decision. This memory term therefore allows the fusion centre to take into consideration the decision made on the last set of observations.

Bayesian approaches

We formulate the Bayes rule for minimum risk for the parallel configuration. Let class ω_2 be 'target absent' and class ω_1 be 'target present'. Then the Bayes rule for minimum risk is given by equation (1.12): declare a target present (class ω_1) if

$$\lambda_{11} p(\omega_1|\boldsymbol{u})p(\boldsymbol{u}) + \lambda_{21} p(\omega_2|\boldsymbol{u})p(\boldsymbol{u}) \leq \lambda_{12} p(\omega_1|\boldsymbol{u})p(\boldsymbol{u}) + \lambda_{22} p(\omega_2|\boldsymbol{u})p(\boldsymbol{u})$$

that is,

$$(\lambda_{21} - \lambda_{22})p(\boldsymbol{u}|\omega_2)p(\omega_2) \leq (\lambda_{12} - \lambda_{11})p(\boldsymbol{u}|\omega_1)p(\omega_1) \tag{8.12}$$

where $\boldsymbol{u} = (u_1, u_2, \dots, u_L)^T$ is the vector of local decisions; $p(\boldsymbol{u}|\omega_i), i = 1, 2$, are the class-conditional probability density functions; $p(\omega_i), i = 1, 2$, are the class priors and λ_{ji} are the costs of assigning a pattern $\boldsymbol{u}$ to ω_i when $\boldsymbol{u} \in \omega_j$. In order to evaluate the fusion rule (8.12), we require knowledge of the class-conditional densities and the costs. Several special cases have been considered. By taking the equal cost loss matrix (see Chapter 1), and assuming independence between local decisions, the fused decision (based on the evaluation of $p(\omega_i|\boldsymbol{u})$) can be expressed in terms of the probability of false alarm and the probability of missed detection at each sensor (see the exercises at the end of the chapter).

The problem of how to tackle the likelihood ratios when the local decisions are correlated has been addressed by Kam *et al.* (1992) who showed that by using the Bahadur–Lazarsfeld polynomials to form an expansion of the probability density functions, it is possible to rewrite the optimal data fusion rule in terms of the conditional correlation coefficients.

The Bahadur–Lazarsfeld expansion expresses the density $p(\boldsymbol{x})$ as

$$p(\boldsymbol{x}) = \prod_{j=1}^{L} (p_j^{x_j}(1 - p_j)^{1-x_j}) \times \left[1 + \sum_{i<j} \gamma_{ij} z_i z_j + \sum_{i<j<k} \gamma_{ijk} z_i z_j z_k + \dots \right]$$

where the γs are the correlation coefficients of the corresponding variables

$$\gamma_{ij} = \mathrm{E}[z_i z_j]$$
$$\gamma_{ijk} = \mathrm{E}[z_i z_j z_k]$$
$$\gamma_{ij\ldots L} = \mathrm{E}[z_i z_j \ldots z_L]$$

and

$$p_i = P(x_i = 1); \quad 1 - p_i = P(x_i = 0)$$

so that

$$\mathrm{E}[x_i] = 1 \times p_i + 0 \times (1 - p_i) = p_i$$
$$\mathrm{var}[x_i] = p_i(1 - p_i)$$

and

$$z_i = \frac{x_i - p_i}{\sqrt{\mathrm{var}(x_i)}}$$

Substituting each of the conditional densities in equation (8.12) by its Bahadur–Lazarsfeld expansion replaces the unknown densities by unknown correlation coefficients (see the exercises). However, this may simplify considerably under assumptions about the form of the individual detectors (see Kam *et al.*, 1992).

Neyman–Pearson formulation

In the Neyman–Pearson formulation, we seek a threshold on the likelihood ratio so that a specified false alarm rate is achieved (see Section 1.5.1). Since the data space is discrete (for an L-dimensional vector, $\boldsymbol{u}$, there are 2^L possible states) the decision rule of Chapter 1 is modified to become:

$$\text{if } \frac{p(\boldsymbol{u}|\omega_1)}{p(\boldsymbol{u}|\omega_2)} \begin{cases} > t & \text{then decide } u_0 = 1 \text{ (target present declared)} \\ = t & \text{then decide } u_0 = 1 \text{ with probability } \epsilon \\ < t & \text{then decide } u_0 = 0 \text{ (target absent declared)} \end{cases} \tag{8.13}$$

where ϵ and t are chosen to achieve the desired false alarm rate.

As an example, consider the case of two sensors, S_1 and S_2, operating with probabilities of false alarm pfa_1 and pfa_2 respectively, and probabilities of detection pd_1 and pd_2. Table 8.1 gives the probability density functions for $p(\boldsymbol{u}|\omega_1)$ and $p(\boldsymbol{u}|\omega_2)$ assuming independence.

There are four values for the likelihood ratio, $p(\boldsymbol{u}|\omega_1)/p(\boldsymbol{u}|\omega_2)$, corresponding to $\boldsymbol{u} = (0,0), (0,1), (1,0), (1,1)$. For $pfa_1 = 0.2$, $pfa_2 = 0.4$, $pd_1 = 0.6$ and $pd_2 = 0.7$, these values are 0.25, 0.875, 1.5 and 5.25. Figure 8.13 gives the ROC curve for the combiner, combined using rule (8.13). This is a piecewise linear curve, with four linear segments, each corresponding to one of the values of the likelihood ratio. For example, if we set $t = 0.875$ (one of the values of the likelihood ratio), then $u_0 = 1$ is decided if $\boldsymbol{u} = (1,0)$ and $\boldsymbol{u} = (1,1)$, and also for $\boldsymbol{u} = (0,1)$ with probability ϵ. This gives a probability of detection and a probability of false alarm of (using Table 8.1)

$$pd = pd_1 pd_2 + (1 - pd_2)pd_1 + \epsilon(1 - pd_1)pd_2 = 0.6 + 0.28\epsilon$$
$$pfa = pfa_1 pfa_2 + (1 - pfa_2)pfa_1 + \epsilon(1 - pfa_1)pfa_2 = 0.2 + 0.32\epsilon$$

a linear variation (as ϵ is varied) of (pfa, pd) values between (0.2, 0.6) and (0.52, 0.88).

Table 8.1 Probability density functions for $p(\boldsymbol{u}|\omega_1)$ (top) and $p(\boldsymbol{u}|\omega_2)$ (bottom)

		Sensor S_1	
		$u = 0$	$u = 1$
Sensor S_2	$u = 0$	$(1 - pd_1)(1 - pd_2)$	$(1 - pd_2)pd_1$
	$u = 1$	$(1 - pd_1)pd_2$	$pd_1 pd_2$

		Sensor S_1	
		$u = 0$	$u = 1$
Sensor S_2	$u = 0$	$(1 - pfa_1)(1 - pfa_2)$	$(1 - pfa_2)pfa_1$
	$u = 1$	$(1 - pfa_1)pfa_2$	$pfa_1 pfa_2$

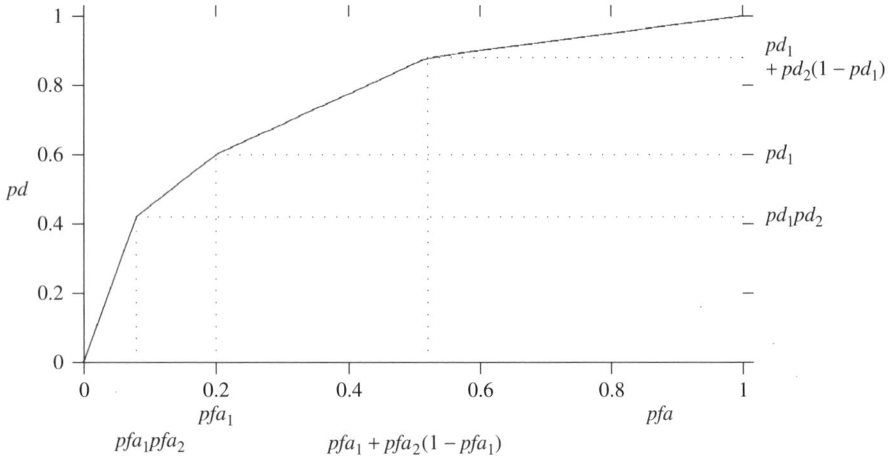

Figure 8.13 ROC curve for two sensors, assuming independence

It may be possible to achieve a better probability of detection of the combiner, for a given probability of false alarm, by operating the individual sensors at different local thresholds. For L sensors, this is an L-dimensional search problem and requires knowledge of the ROC of each sensor (Viswanathan and Varshney, 1997)

Approaches to the distributed detection problem using the Neyman–Pearson formulation have been proposed for correlated decisions. One approach is to expand the likelihood ratio using the Bahadur–Lazarsfeld polynomials, in a similar manner to the Bayesian formulation above. The Neyman–Pearson fusion rule can then be expressed as a function of the correlation coefficients.

If the independence assumption is not valid, and it is not possible to estimate the likelihood ratio through other means, then it may still be possible to achieve better performance than individual sensors through a 'random choice' fusion system. Consider two sensors S_1 and S_2, with ROC curves shown in Figure 8.14. For probabilities of false

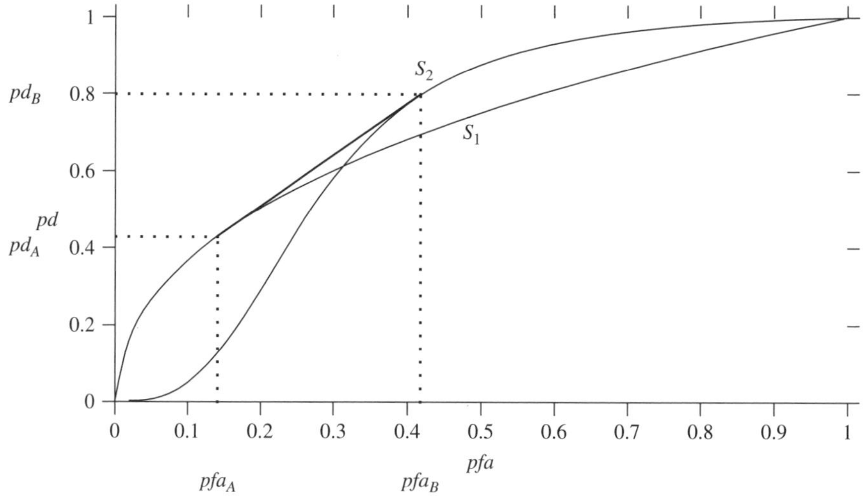

Figure 8.14 ROC curves for sensors S_1 and S_2 and points on the convex hull

alarm greater than pfa_B we operate sensor S_2, and for probabilities of false alarm less than pfa_A we operate sensor S_1. If, for probabilities of false alarm between pfa_A and pfa_B, we operate sensor S_1 with probability of false alarm pfa_A and sensor S_2 at the pfa_B point on its ROC curve, and randomly select sensor S_1 with probability ϵ and sensor S_2 with probability $1 - \epsilon$, then the probability of false alarm of the random choice fusion system is $\epsilon pfa_A + (1 - \epsilon) pfa_B$ and the probability of detection is $\epsilon pd_A + (1 - \epsilon) pd_B$. Thus, the best performance is achieved on the convex hull of the two ROC curves.

This differs from the example in Figure 8.13, where the combined output of both sensors is used, rather than basing a decision on a single sensor output.

Trainable rules

One of the difficulties with the Bayesian and the Neyman–Pearson formulations for a set of distributed sensors is that both methods require some knowledge of the probability density of sensor outputs. Often this information is not available and the densities must be estimated using a training set.

This is simply a problem of classifier design where the classifiers are defined on a feature space comprising the outputs of separate sensors (local decisions). Many of the techniques described elsewhere in this book, suitably adapted for binary variables, may be employed.

Fixed rules

There are a few 'fixed' rules for decision fusion that do not model the joint density of sensor predictions.

AND Class ω_1 (target present) is declared if all sensors predict class ω_1, otherwise class ω_2 is declared.

OR Class ω_1 (target present) is declared if at least one of the sensors predicts class ω_1, otherwise class ω_2 is declared.

Majority vote Class ω_1 (target present) is declared if a majority of the sensors predicts class ω_1, otherwise class ω_2 is declared.

k*-out-of-*N Class ω_1 (target present) is declared if at least k of the sensors predict class ω_1, otherwise class ω_2 is declared. All of the previous three rules are special cases of this rule.

It is difficult to draw general conclusions about the performance of these rules. For low false alarm rates, there is some evidence to show that the OR rule is inferior to the AND and majority-vote rules in a problem of signal detection in correlated noise. For similar local sensors, the optimal rule is the k-out-of-N decision rule, with k calculated from the prior probabilities and the sensor probability of false alarm and probability of detection.

8.4.5 Classifier combination methods

The characterising features of multiple classifier systems have been described in Section 8.4.3, and a practical motivating problem for the fusion of decisions from distributed sensors summarised in Section 8.4.4. We turn now to the methods of classifier fusion, many of which are multiclass generalisations of the binary classifiers employed for decision fusion.

We begin with the Bayesian decision rule and, following Kittler *et al.* (1998), make certain assumptions to derive combination schemes that are routinely used. Various developments of these methods are described.

We assume that we have an object Z that we wish to classify and that we have L classifiers with inputs $\boldsymbol{x}_1, \ldots, \boldsymbol{x}_L$ (as in Figure 8.7). The Bayes rule for minimum error (1.1) assigns Z to class ω_j if

$$p(\omega_j|\boldsymbol{x}_1, \ldots, \boldsymbol{x}_L) > p(\omega_k|\boldsymbol{x}_1, \ldots, \boldsymbol{x}_L) \quad k = 1, \ldots, C; k \neq j \tag{8.14}$$

or, equivalently (1.2), assigns Z to class ω_j if

$$p(\boldsymbol{x}_1, \ldots, \boldsymbol{x}_L|\omega_j)p(\omega_j) > p(\boldsymbol{x}_1, \ldots, \boldsymbol{x}_L|\omega_k)p(\omega_k) \quad k = 1, \ldots, C; k \neq j \tag{8.15}$$

This requires knowledge of the class-conditional joint probability densities $p(\boldsymbol{x}_1, \ldots, \boldsymbol{x}_L|\omega_j)$, $j = 1, \ldots, L$, which is assumed to be unavailable.

Product rule

If we assume conditional independence ($\boldsymbol{x}_1, \ldots, \boldsymbol{x}_L$ are conditionally independent given class), then the decision rule (8.15) becomes: assign Z to class ω_j if

$$\prod_{i=1}^{L}(p(\boldsymbol{x}_i|\omega_j))p(\omega_j) > \prod_{i=1}^{L}(p(\boldsymbol{x}_i|\omega_k))p(\omega_k) \quad k = 1, \ldots, C; k \neq j \tag{8.16}$$

or, in terms of the posterior probabilities of the individual classifiers: assign Z to class ω_j if

$$[p(\omega_j)]^{-(L-1)} \prod_{i=1}^{L} p(\omega_j|\boldsymbol{x}_i) > [p(\omega_k)]^{-(L-1)} \prod_{i=1}^{L} p(\omega_k|\boldsymbol{x}_i) \quad k = 1, \dots, C; k \neq j \tag{8.17}$$

This is the *product rule* and for equal priors simplifies to: assign Z to class ω_j if

$$\prod_{i=1}^{L} p(\omega_j|\boldsymbol{x}_i) > \prod_{i=1}^{L} p(\omega_k|\boldsymbol{x}_i) \quad k = 1, \dots, C; k \neq j \tag{8.18}$$

Both forms (8.17) and (8.18) have been used in studies. The independence assumption may seem rather severe, but it is one that has been successfully used in many practical problems (Hand and Yu, 2001). The rule requires the individual classifier posterior probabilities, $p(\omega_j|\boldsymbol{x})$, $j = 1, \dots, C$, to be calculated, and they are usually estimated from training data. The main problem with this method is that the product rule is sensitive to errors in the posterior probability estimates, and deteriorates more rapidly than the sum rule (see below) as the estimation errors increase. If one of the classifiers reports that the probability of a sample belonging to a particular class is zero, then the product rule will give a zero probability also, even if the remaining classifiers report that this is the most probable class.

The product rule would tend to be applied where each classifier receives input from different sensors.

Sum rule

Let us make the (rather strong) assumption that

$$p(\omega_k|\boldsymbol{x}_i) = p(\omega_k)(1 + \delta_{ki}) \tag{8.19}$$

where $\delta_{ki} \ll 1$, that is, the posterior probabilities $p(\omega_k|\boldsymbol{x}_i)$ used in the product rule (8.17) do not deviate substantially from the class priors $p(\omega_k)$. Then substituting for $p(\omega_k|\boldsymbol{x}_i)$ in the product rule (8.17), neglecting second-order and higher terms in δ_{ki}, and using (8.19) again leads to the sum rule (see the exercises at the end of the chapter): assign Z to class ω_j if

$$(1 - L)p(\omega_j) + \sum_{i=1}^{L} p(\omega_j|\boldsymbol{x}_i) > (1 - L)p(\omega_k) + \sum_{i=1}^{L} p(\omega_k|\boldsymbol{x}_i) \quad k = 1, \dots, C; k \neq j \tag{8.20}$$

This is the *sum rule* and for equal priors it simplifies to: assign Z to class ω_j if

$$\sum_{i=1}^{L} p(\omega_j|\boldsymbol{x}_i) > \sum_{i=1}^{L} p(\omega_k|\boldsymbol{x}_i) \quad k = 1, \dots, C; k \neq j \tag{8.21}$$

The assumption used to derive the sum rule approximation to the product rule, namely that the posterior probabilities are similar to the priors, will be unrealistic in many practical applications. However, it is a rule that is relatively insensitive to errors in the estimation of the joint densities and would be applied to classifiers used for a common input pattern (Figure 8.8).

In order to implement the above rule, each classifier must produce estimates of the posterior probabilities of class membership. In a comparison of the sum and product rules, Tax *et al.* (2000) concluded that the sum rule is more robust to errors in the estimated posterior probabilities (see also Kittler *et al.*, 1998). The averaging process reduces any effects of overtraining of the individual classifiers and may be thought of as a regularisation process.

It is also possible to apply a weighting to the sum rule to give: assign Z to class ω_j if

$$\sum_{i=1}^{L} w_i p(\omega_j|\boldsymbol{x}_i) > \sum_{i=1}^{L} w_i p(\omega_k|\boldsymbol{x}_i) \quad k = 1, \ldots, C; k \neq j \tag{8.22}$$

where $w_i, i = 1, \ldots, L$, are weights for the classifiers. A key question here is the choice of weights. These may be estimated using a training set to minimise the error rate of the combined classifier. In this case, the same weighting is applied throughout the data space. An alternative is to allow the weights to vary with the location of a given pattern in the data space. An extreme example of this is *dynamic classifier selection* where one weight is assigned the value unity and the remaining weights are zero. For a given pattern, dynamic feature selection attempts to select the best classifier. Thus, the feature space is partitioned into regions with a different classifier for each region.

Dynamic classifier selection has been addressed by Woods *et al.* (1997) who use local regions defined in terms of k-nearest-neighbour regions to select the most accurate classifier (based on the percentage of training samples correctly classified in the region); see also Huang and Suen (1995).

Min, max and median combiners

The max combiner may be derived by approximating the posterior probabilities in (8.20) by an upper bound, $L\max_i p(\omega_k|\boldsymbol{x}_i)$, to give the decision rule: assign Z to class ω_j if

$$(1-L)p(\omega_j) + L \max_i p(\omega_j|\boldsymbol{x}_i) > (1-L)p(\omega_k) + L \max_i p(\omega_k|\boldsymbol{x}_i) \quad k = 1, \ldots, C; \; k \neq j \tag{8.23}$$

This is the *max combiner* and for equal priors simplifies to

$$\max_i p(\omega_j|\boldsymbol{x}_i) > \max_i p(\omega_k|\boldsymbol{x}_i) \quad k = 1, \ldots, C, k \neq j \tag{8.24}$$

We can also approximate the product in (8.17) by an upper bound, $\min_i p(\omega_k|\boldsymbol{x}_i)$, to give the decision rule: assign Z to class ω_j if

$$[p(\omega_j)]^{-(L-1)} \min_i p(\omega_j|\boldsymbol{x}_i) > [p(\omega_k)]^{-(L-1)} \min_i p(\omega_k|\boldsymbol{x}_i) \quad k = 1, \ldots, C; k \neq j \tag{8.25}$$

This is the *min combiner* and for equal priors simplifies to: assign Z to class ω_j if

$$\min_i p(\omega_j|\boldsymbol{x}_i) > \min_i p(\omega_k|\boldsymbol{x}_i) \quad k = 1, \ldots, C; k \neq j \tag{8.26}$$

Finally, the *median combiner* is derived by noting that the sum rule calculates the mean of the classifier outputs and that a robust estimate of the mean is the median. Thus, under equal priors, the median combiner is: assign Z to class ω_j if

$$\underset{i}{\text{med}}\, p(\omega_j|\boldsymbol{x}_i) > \underset{i}{\text{med}}\, p(\omega_k|\boldsymbol{x}_i) \quad k = 1, \ldots, C;\ k \neq j \tag{8.27}$$

The min, max and median combiners are all easy to implement and require no training.

Majority vote

Among all the classifier combination methods described in this section, the majority vote is one of the easiest to implement. It is applied to classifiers that produce unique class labels as outputs (level L3) and requires no training. It may be considered as an application of the sum rule to classifier outputs where the posterior probabilities, $p(\omega_k|\boldsymbol{x}_i)$, have been 'hardened' (Kittler *et al.*, 1998); that is, $p(\omega_k|\boldsymbol{x}_i)$ is replaced by the binary-valued function, Δ_{ki}, where

$$\Delta_{ki} = \begin{cases} 1 & \text{if } p(\omega_k|\boldsymbol{x}_i) = \max_j p(\omega_j|\boldsymbol{x}_i) \\ 0 & \text{otherwise} \end{cases}$$

which produces decisions at the classifier outputs rather than posterior probabilities. A decision is made to classify a pattern to the class most often predicted by the constituent classifiers. In the event of a tie, a decision can be made according to the largest *prior* class probability (among the tying classes).

An extension to the method is the *weighted majority voting* technique in which classifiers are assigned unequal weights based on their performance. The weights for each classifier may be independent of predicted class, or they may vary across class depending on the performance of the classifier on each class. A key question is the choice of weights. The weighted majority vote combiner requires the results of the individual classifiers on a training set as training data for the allocation of the weights.

For weights that vary between classifiers but are independent of class, there are $L - 1$ parameters to estimate for L classifiers (we assume that the weights may be normalised to sum to unity). These may be determined by specifying some suitable objective function and an appropriate optimisation procedure. One approach is to define the objective function

$$F = R_e - \beta E$$

where R_e is the recognition rate and E is the error rate of the combiner (they do not sum to unity as the individual classifiers may reject patterns – see Chapter 1); β is a user-specified parameter that measures the relative importance of recognition and error rates and is problem-dependent (Lam and Suen, 1995). Rejection may be treated as an extra class by the component classifiers and thus the combiner will reject a pattern if the weighted majority of the classifiers also predicts a rejection. In a study of combination schemes applied to a problem in optical character recognition, Lam and Suen (1995) used a genetic optimisation scheme (a scheme that adjusts the weights using a learning method loosely motivated by an analogy to biological evolution) to maximise F and concluded that simple majority voting (all weights equal) gave the easiest and most reliable classification.

Borda count

The Borda count is a quantity defined on the ranked outputs of each classifier. If we define $B_i(j)$ as the number of classes ranked below class ω_j by classifier i, then the Borda count for class ω_j is B_j defined as

$$B_j = \sum_{i=1}^{L} B_i(j)$$

the sum of the number of classes ranked below ω_j by each classifier. A pattern is assigned to the class with the highest Borda count. This combiner requires no training, with the final decision being based on an average ranking of the classes.

Combiners trained on class predictions

The combiners described so far require no training, at least in their basic forms. General conclusions are that the sum rule and median rule can be expected to give better performance than other fixed combiners. We now turn to combiners that require some degree of training, and initially we consider combiners acting on discrete variables. Thus, the constituent classifiers deliver class labels and the combiner uses these class predictions to make an improved estimate of class (type L combination) – at least, that is what we hope.

'Bayesian combiner' This combiner simply uses the product rule with estimates of the posterior probabilities derived from the classifier predictions of each constituent classifier, together with a summary of their performance on a labelled training set.

Specifically, the Bayesian combination rule of Lam and Suen (1995) approximates the posterior probabilities by an estimate based on the results of a training procedure. Let $D^{(i)}$ denote the $C \times C$ confusion matrix (see Chapter 1) for the ith classifier based on the results of a classification of a training set by classifier i. The (j, k)th entry, $d_{jk}^{(i)}$, is the number of patterns with true class ω_k that are assigned to ω_j by classifier i. The total number of patterns in class ω_k is

$$n_k = \sum_{l=1}^{C} d_{lk}^{(i)}$$

for any i. The number of patterns assigned to class ω_l is

$$\sum_{k=1}^{C} d_{lk}^{(i)}$$

The conditional probability that a sample $\boldsymbol{x}$ assigned to class ω_l by classifier i actually belongs to ω_k is estimated as

$$p(\omega_k|\text{classifier } i \text{ predicts } \omega_l) = \frac{d_{lk}^{(i)}}{\sum_{k=1}^{C} d_{lk}^{(i)}}$$

Thus, for a given pattern, the posterior probability depends only on the predicted class: two distinct patterns $\boldsymbol{x}_i$ and $\boldsymbol{x}_j$, having the same predicted class, are estimated as having the same posterior probability. Substituting into the product rule (8.18), equal priors assumed, gives the decision rule: assign the pattern to class ω_j if

$$\prod_{i=1}^{L} \frac{d_{l_i j}^{(i)}}{\sum_{k=1}^{C} d_{l_i k}^{(i)}} > \prod_{i=1}^{L} \frac{d_{l_i m}^{(i)}}{\sum_{k=1}^{C} d_{l_i k}^{(i)}} \qquad m = 1, \ldots, C; m \neq j$$

where ω_{l_i} is the predicted class of pattern $\boldsymbol{x}_i$.

Density estimation in classifier output space An alternative approach is to regard the L class predictions from the constituent classifiers for an input, $\boldsymbol{x}$, as inputs to a classifier, the combiner, defined on an L-dimensional discrete-valued feature space (see Figure 8.15). Suppose that we have N training patterns $(\boldsymbol{x}_i, i = 1, \ldots, N)$ with associated class labels $(y_i, i = 1, \ldots, N)$; then the training data for the combiner comprises N L-dimensional vectors $(z_i, i = 1, \ldots, N)$ with associated class labels $(y_i, i = 1, \ldots, N)$. Each component of z_i is a discrete-valued variable taking 1 of C possible values corresponding to the class label from the component classifier (1 of $C + 1$ if a reject option is included in the constituent classifiers).

The combiner is trained using the training set $\{(z_i, y_i), i = 1, \ldots, N\}$ and an unknown pattern $\boldsymbol{x}$ classified by first applying each constituent classifier to obtain a vector of predictions z, which is then input to the combiner.

The most obvious approach to constructing the combiner is to estimate class-conditional probabilities, $p(z|\omega_i), i = 1, \ldots, C$, and to classify z to class ω_j if

$$p(z|\omega_j)p(\omega_j) > p(z|\omega_k)p(\omega_k) \qquad k = 1, \ldots, C; k \neq j$$

with the priors, $p(\omega_j)$, estimated from the training set (or perhaps using domain knowledge) and the densities, $p(z|\omega_j)$ estimated using a suitable nonparametric density estimation method appropriate for categorical variables.

Perhaps the simplest method is the histogram. This is the approach adopted by Huang and Suen (1995) and termed the *behaviour-knowledge space* method, and also investigated by Mojirsheibani (1999). However, this has the disadvantage of having to estimate

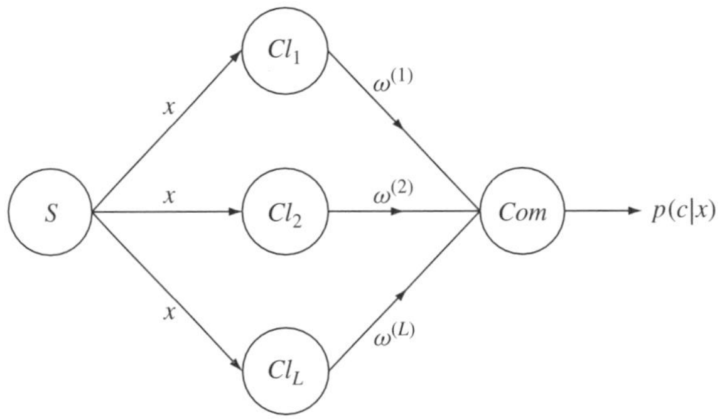

Figure 8.15 Classifier fusion architecture C2–component classifiers defined on a common feature space

and store high-order distributions which may be computationally expensive. The multidimensional histogram has C^L cells, which may be large, making reliable density estimation difficult in the absence of a large training data set. In Chapter 3 we described several ways around this difficulty:

1. Independence. Approximate the multivariate density as the product of the univariate estimates.

2. Lancaster models. Approximate the density using marginal distributions.

3. Maximum weight dependence trees. Approximate the density with a product of pairwise conditional densities.

4. Bayesian networks. Approximate the density with a product of more complex conditional densities.

Other approaches, based on constructing discriminant functions directly, rather than estimating the class-conditional densities and using Bayes' rule, are possible.

Stacked generalisation

Stacked generalisation, or simply *stacking*, constructs a generaliser using training data that consist of the 'guesses' of the component generalisers which are taught with different parts of the training set and try to predict the remainder, and whose output is an estimate of the correct class. Thus, in some ways it is similar to the models of the previous section – the combiner is a classifier (generaliser) defined on the outputs of the constituent classifiers – -but the training data used to construct the combiner comprise the prediction on held-out samples of the training set.

The basic idea is that the output of the constituent classifiers, termed level 1 data, $\mathcal{L}_1$ (level 0, $\mathcal{L}_0$, is the input level), has information that can be used to construct good combinations of the classifiers. We suppose that we have a set of constituent classifiers, $f_j, j = 1, \ldots, L$, and we seek a procedure for combining them. The level 1 data are constructed as follows.

1. Divide the $\mathcal{L}_0$ data (the training data, $\{(\boldsymbol{x}_i, y_i), i = 1, \ldots, n\}$) into V partitions.

2. For each partition, $v = 1, \ldots, V$, do the following.

   (a) Repeat the procedure for constructing the constituent classifiers using a subset of the data: train the constituent classifier j ($j = 1, \ldots, L$) on all the training data apart from partition v to give a classifier denoted, f_j^{-v}.

   (b) Test each classifier, f_j^{-v}, on all patterns in partition v.

This gives a data set of L predictions on each pattern in the training set. Together with the labels $\{y_i, i = 1, \ldots, n\}$, these comprise the training data for the combiner.

We must now construct a combiner for the outputs of the constituent classifiers. If the constituent classifiers produce class labels, then the training data for the combiner comprise L-dimensional measurements on categorical variables. Several methods are available to us, including those based on histogram estimates of the multivariate density, and variants, mentioned at the end of the previous section; for example, tree-based approaches and neural network methods. Merz (1999) compares an independence model and a multilayer perceptron model for the combiner with an approach based on a multivariate analysis method (*correspondence analysis*). For the multilayer perceptron, when the ith value of the variable occurs, each categorical variable is represented as C binary-valued inputs, with all inputs assigned the value of zero apart from the ith, which is assigned a value of one.

Mixture of experts

The *adaptive mixture of local experts* model (Jacobs *et al.*, 1991; Jordan and Jacobs, 1994) is a learning procedure that trains several component classifiers (the 'experts') and a combiner (the 'gating function') to achieve improved performance in certain problems. The experts each produce an output vector, $\boldsymbol{o}_i$ ($i = 1, \ldots, L$), for a given input vector, $\boldsymbol{x}$, and the gating network provides linear combination coefficients for the experts. The gating function may be regarded as assigning a probability to each of the experts, based on the current input (see Figure 8.16). The emphasis of the training procedure is to find the optimal gating function and, for a given gating function, to train each expert to give maximal performance.

In the basic approach, the output of the ith expert, $\boldsymbol{o}_i(\boldsymbol{x})$, is a *generalised linear function* of the input, $\boldsymbol{x}$,

$$\boldsymbol{o}_i(\boldsymbol{x}) = \boldsymbol{f}(\boldsymbol{w}_i^T \boldsymbol{x})$$

where $\boldsymbol{w}_i$ is a weight vector associated with the ith expert and $\boldsymbol{f}(.)$ is a fixed continuous nonlinear function. The gating network is also a generalised linear function, $\boldsymbol{g}$, of its input, with ith component

$$g_i(\boldsymbol{x}) = g(\boldsymbol{x}, \boldsymbol{v}_i) = \frac{\exp(\boldsymbol{v}_i^T \boldsymbol{x})}{\sum_{k=1}^{L} \exp(\boldsymbol{v}_i^T \boldsymbol{x})}$$

for weight vectors $\boldsymbol{v}_i, i = 1, \ldots, L$. These outputs of the gating network are used to weight the outputs of the experts to give the overall output, $\boldsymbol{o}(\boldsymbol{x})$, of the mixture of

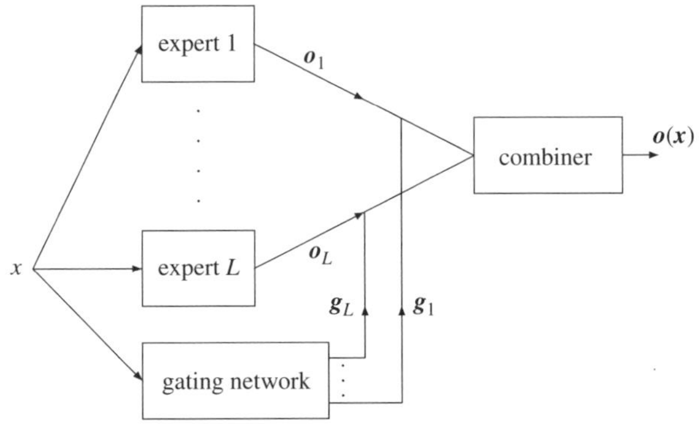

Figure 8.16 Mixture of experts architecture

experts architectures as

$$\boldsymbol{o}(\boldsymbol{x}) = \sum_{k=1}^{L} g_k(\boldsymbol{x}) \boldsymbol{o}_k(\boldsymbol{x}) \tag{8.28}$$

The above algorithm is very similar to (8.22) that giving rise to the weighted sum rule in that the model produces a linear combination of component classifiers. The key difference is that in the mixture of experts model, the combination model (the gating network) is trained simultaneously with the constituent classifiers (the experts). In many combination models, the basic models are trained first and then the combiner tuned to these trained models. A further difference is that the linear combination depends on the input pattern, $\boldsymbol{x}$. An interpretation is that the gating network provides a 'soft' partitioning of the input space, with experts providing local predictions (Jordan and Jacobs, 1994).

The mixture of experts model (8.28) is also similar to the multilayer perceptron[2] (Chapter 6) in that the output is a linear combination of nonlinear functions of projections of the data. The difference is that in (8.28) the linear combination depends on the input.

A probabilistic interpretation of the model is provided by Jordan and Jacobs (1994). We assume that we have an input variable $\boldsymbol{x}$ and a response variable $\boldsymbol{y}$ that depends probabilistically on $\boldsymbol{x}$. The mixing proportions, $g_i(\boldsymbol{x})$, are interpreted as multinomial probabilities associated with the process that maps $\boldsymbol{x}$ to an output $\boldsymbol{y}$. For a given $\boldsymbol{x}$, an output $\boldsymbol{y}$ is generated by selecting an expert according to the values of $g_k(\boldsymbol{x}), k = 1, \ldots, L$, say expert i, and then generating $\boldsymbol{y}$ according to a probability density $p(\boldsymbol{y}|\boldsymbol{x}, \boldsymbol{w}_i)$, where $\boldsymbol{w}_i$ denotes the set of parameters associated with expert i. Therefore, the total probability of generating $\boldsymbol{y}$ from $\boldsymbol{x}$ is the mixture of probabilities of generating $\boldsymbol{y}$ from each component density, where the mixing proportions are the multinomial probabilities $g_k(\boldsymbol{x})$, i.e.

$$p(\boldsymbol{y}|\boldsymbol{x}, \boldsymbol{\Phi}) = \sum_{k=1}^{L} g_k(\boldsymbol{x}) p(\boldsymbol{y}|\boldsymbol{x}, \boldsymbol{w}_k) \tag{8.29}$$

where $\boldsymbol{\Phi}$ is the set of all parameters, including both expert and gating network parameters.

The generating density $p(\boldsymbol{y}|\boldsymbol{x}, \boldsymbol{w}_k)$, can be taken to be one of several popular forms; for a problem in regression, a normal distribution with identical covariance matrices $\sigma^2 \boldsymbol{I}$ is often assumed,

$$p(\boldsymbol{y}|\boldsymbol{x}, \boldsymbol{w}_k) \sim \exp\left\{-\frac{1}{\sigma^2}(\boldsymbol{y} - \boldsymbol{o}_k(\boldsymbol{x}))^T (\boldsymbol{y} - \boldsymbol{o}_k(\boldsymbol{x}))\right\}$$

For binary classification, the Bernoulli distribution is generally assumed (single output, o_k; univariate binary response variable $y = 0, 1$)

$$p(\boldsymbol{y}|\boldsymbol{x}, \boldsymbol{w}_k) = o_k^y (1 - o_k)^{1-y}$$

and for multiclass problems, the multinomial distribution (L binary variables $y_i, i = 1, \ldots, L$, summing to unity)

$$p(\boldsymbol{y}|\boldsymbol{x}, \boldsymbol{w}_k) \sim \prod_{i=1}^{L} (o_k^i)^{y_i}$$

[2]The basic MLP, with single hidden layer and linear output layer. Further development of the mixture of experts model for hierarchical models is discussed by Jordan and Jacobs (1994).

Optimisation of the model (8.29) may be achieved via a maximum likelihood approach. Given a training set $\{(\boldsymbol{x}_i, y_i), i = 1, \ldots, n\}$ (in the classification case, y_i would be a C-dimensional vector coding the class – for class ω_j, all entries are zero apart from the jth, which is one), we seek a solution for $\boldsymbol{\Phi}$ for which the log-likelihood

$$\sum_t \log\left[\sum_{k=1}^{L} g_k(\boldsymbol{x}_t) p(\boldsymbol{y}_t|\boldsymbol{x}_t, \boldsymbol{w}_k)\right]$$

is a maximum. Jordan and Jacobs (1994) propose an approach based on the EM algorithm (see Chapter 2) for adjusting the parameters $\boldsymbol{w}_k$ and $\boldsymbol{v}_k$. At stage s of the iteration, the expectation and maximisation procedures are as follows:

1. E-step. Compute the probabilities

$$h_i^{(t)} = \frac{g(\boldsymbol{x}_t, \boldsymbol{v}_i^{(s)}) p(\boldsymbol{y}_t|\boldsymbol{x}_t, \boldsymbol{w}_i^{(s)})}{\sum_{k=1}^{L} g(\boldsymbol{x}_t, \boldsymbol{v}_k^{(s)}) p(\boldsymbol{y}_t|\boldsymbol{x}_t, \boldsymbol{w}_k^{(s)})}$$

for $t = 1, \ldots, n$; $i = 1, \ldots, L$.

2. M-step. For the parameters of the experts solve the maximisation problem

$$\boldsymbol{w}_i^{(s+1)} = \underset{\boldsymbol{w}_i}{\arg\max} \sum_{t=1}^{n} h_i^{(t)} \log[p(\boldsymbol{y}_t|\boldsymbol{x}_t, \boldsymbol{w}_i)]$$

and for the parameters of the gating network

$$V^{(s+1)} = \underset{V}{\arg\max} \sum_{t=1}^{n} \sum_{k=1}^{L} h_k^{(t)} \log[g(\boldsymbol{x}_t, \boldsymbol{v}_k)]$$

where $\boldsymbol{V}$ is the set of all $\boldsymbol{v}_i$.

Procedures for solving these maximisation problems are discussed by Chen *et al.* (1999), who propose a Newton–Raphson method, but other 'quasi-Newton' methods may be used (see Press *et al.*, 1992).

Bagging

Bagging and boosting (see the following section) are procedures for combining different classifiers generated using the same training set. Bagging or *bootstrap aggregating* (Breiman, 1996) produces replicates of the training set and trains a classifier on each replicate. Each classifier is applied to a test pattern $\boldsymbol{x}$ which is classified on a majority-vote basis, ties being resolved arbitrarily. Table 8.2 shows the algorithm. A *bootstrap sample* is generated by sampling n times from the training set with replacement. This provides a new training set, $\boldsymbol{Y}^b$, of size n. B bootstrap data sets, $\boldsymbol{Y}^b, b = 1, \ldots, B$, are generated and a classifier designed for each data set. The final classifier is that whose output is the class most often predicted by the subclassifiers.

A vital aspect of the bagging technique is that the procedure for producing the classifier is *unstable*. For a given bootstrap sample, a pattern in the training set has a probability of $1 - (1 - 1/n)^n$ of being selected at least once in the n times that patterns are randomly

Table 8.2 The bagging algorithm

Assume that we have a training set $(\boldsymbol{x}_i, z_i), i = 1, \ldots, n$, of patterns $\boldsymbol{x}_i$ and labels z_i.

1. For $b = 1, \ldots, B$, do the following.
   (a) Generate a bootstrap sample of size n by sampling with replacement from the training set; some patterns will be replicated, others will be omitted.
   (b) Design a classifier, $\eta_b(\boldsymbol{x})$.
2. Classify a test pattern $\boldsymbol{x}$ by recording the class predicted by $\eta_b(\boldsymbol{x})$, $b = 1, \ldots, B$, and assigning $\boldsymbol{x}$ to the class most represented.

selected from the training set. For large n, this is approximately $1 - 1/e = 0.63$, which means that each bootstrap sample contains only about 63% unique patterns from the training set. This causes different classifiers to be built. If the change in the classifiers is large (that is, small changes in a data set lead to large changes in the predictions), then the procedure is said to be unstable. Bagging of an unstable classifier should result in a better classifier and a lower error rate. However, averaging of a bad classifier can result in a poorer classifier. If the classifier is *stable* – that is, changes in the training data set lead to small changes in the classifier – then bagging will lead to little improvement.

The bagging procedure is particularly useful in classification problems using neural networks (Chapter 6) and classification trees (Chapter 7) since these are all unstable processes. For trees, a negative feature is that there is no longer the simple interpretation as there is with a single tree. Nearest-neighbour classifiers are stable and bagging offers little, if any, improvement.

In studies on *linear classifiers*, Skurichina (2001) reports that bagging may improve the performance on classifiers constructed on critical training sample sizes, but when the classifier is stable, bagging is usually useless. Also, for very large sample sizes, classifiers constructed on bootstrap replicates are similar and combination offers no benefit.

The procedure, as presented in Table 8.2, applies to classifiers whose outputs are class predictions. For classifier methods that produce estimates of the posterior probabilities, $\hat{p}(\omega_j|\boldsymbol{x})$, two approaches are possible. One is to make a decision for the class based on the maximum value of $\hat{p}(\omega_j|\boldsymbol{x})$ and then to use the voting procedure. Alternatively, the posterior probabilities can be averaged over all bootstrap replications, obtaining $\hat{p}_B(\omega_j|\boldsymbol{x})$, and then a decision based on the maximum value of $\hat{p}_B(\omega_j|\boldsymbol{x})$ is made. Breiman (1996) reports a virtually identical misclassification rate for the two approaches in a series of experiments on 11 data sets. However, bagged estimates of the posterior probabilities are likely to be more accurate than single estimates.

Boosting

Boosting is a procedure for combining or 'boosting' the performance of weak classifiers (classifiers whose parameter estimates are usually inaccurate and give poor performance) in order to achieve a better classifier. It differs from bagging in that it is a *deterministic* procedure and generates training sets and classifiers *sequentially*, based on the results of the previous iteration. In contrast, bagging generates the training sets *randomly* and can generate the classifiers *in parallel*.

Proposed by Freund and Schapire (1996), boosting assigns a weight to each pattern in the training set, reflecting its importance, and constructs a classifier using the training set and the set of weights. Thus, it requires a classifier that can handle weights on the training samples. Some classifiers may be unable to support weighted patterns. In this case, a subset of the training examples can be sampled according to the distribution of the weights and these examples used to train the classifier in the next stage of the iteration.

The basic boosting procedure is AdaBoost (Adaptive Boosting; Freund and Schapire, 1996). Table 8.3 presents the basic AdaBoost algorithm for the binary classification problem. Initially, all samples are assigned a weight $w_i = 1/n$. At each stage of the algorithm, a classifier $\eta_t(\boldsymbol{x})$ is constructed using the weights w_i (as though they reflect the probability of occurrence of the sample). The weight of misclassified patterns is increased and the weight of correctly classified patterns is decreased. The effect of this is that the higher-weight patterns influence the learning classifier more, and thus cause the classifier to focus more on the misclassifications, i.e. those patterns that are nearest the decision boundaries. There is a similarity with support vector machines in this respect (Chapters 4 and 5). The error e_t is calculated, corresponding to the sum of the weights of the misclassified samples. These get boosted by a factor $(1 - e_t)/e_t$, increasing the total weight on the misclassified samples (provided that $e_t < 1/2$). This process is repeated and a set of classifiers is generated. The classifiers are combined using a linear weighting whose coefficients are calculated as part of the training procedure.

There are several ways in which the AdaBoost algorithm has been generalised. One generalisation is for the classifiers to deliver a measure of confidence in the prediction. For example, in the two-class case, instead of the output being ± 1 corresponding to one of the two classes, the output is a number in the range $[-1, +1]$. The sign of the output is the

Table 8.3 The AdaBoost algorithm

1. Initialise the weights $w_i = 1/n, i = 1, \ldots, n$.
2. For $t = 1, \ldots, T$,
   (a) construct a classifier $\eta_t(\boldsymbol{x})$ from the training data with weights $w_i, i = 1, \ldots, n$;
   (b) calculate e_t as the sum of the weights w_i corresponding to misclassified patterns;
   (c) if $e_t > 0.5$ or $e_t = 0$ then terminate the procedure, otherwise set $w_i = w_i(1 - e_t)/e_t$ for the misclassified patterns and renormalise the weights so that they sum to unity.
3. For a two-class classifier, in which $\eta_t(\boldsymbol{x}) = 1$ implies $\boldsymbol{x} \in \omega_1$ and $\eta_t(\boldsymbol{x}) = -1$ implies $\boldsymbol{x} \in \omega_2$, form a weighted sum of the classifiers, η_t,

$$\hat{\eta} = \sum_{t=1}^{T} \log\left(\frac{1 - e_t}{e_t}\right) \eta_t(\boldsymbol{x})$$

   and assign $\boldsymbol{x}$ to ω_1 if $\hat{\eta} > 0$.

Table 8.4 The AdaBoost.MH algorithm converts the C-class problem into a two-class problem operating on the original training data with an additional 'feature'

---

1. Initialise the weights $w_{ij} = 1/(nC), i = 1, \ldots, n;\ j = 1, \ldots, C$.
2. For $t = 1, \ldots, T$,
   (a) construct a 'confidence-rated' classifier $\eta_t(\boldsymbol{x}, l)$ from the training data with weights $w_{ij}, i = 1, \ldots, n;\ j = 1, \ldots, C$;
   (b) calculate
   $$r_t = \sum_{i=1}^{n} \sum_{l=1}^{C} w_{il} y_{il} \eta_t(\boldsymbol{x}_i, l)$$
   and
   $$\alpha_t = \frac{1}{2} \log\left(\frac{1 + r_t}{1 - r_t}\right)$$
   (c) Set $w_{ij} = w_{ij} \exp(-\alpha_t y_{ij} \eta_t(\boldsymbol{x}_i, j))$ and renormalise the weights so that they sum to unity.
3. Set
   $$\hat{\eta}(\boldsymbol{x}, l) = \sum_{t=1}^{T} \alpha_t \eta_t(\boldsymbol{x}, l)$$
   and assign $\boldsymbol{x}$ to ω_j if
   $$\hat{\eta}(\boldsymbol{x}, j) \geq \hat{\eta}(\boldsymbol{x}, k) \quad k = 1, \ldots, C;\ k \neq j$$

---

predicted class label (-1 or $+1$) and the magnitude represents the degree of confidence: close to zero is interpreted as low confidence and close to unity as high confidence.

For the multiclass generalisation, Table 8.4 presents the AdaBoost.MH algorithm (Schapire and Singer, 1999). The basic idea is to expand the training set (of size n) to a training set of size $n \times C$ pairs,

$$((\boldsymbol{x}_i, 1), y_{i1}), ((\boldsymbol{x}_i, 2), y_{i2}), \ldots, ((\boldsymbol{x}_i, C), y_{iC}), \quad i = 1, \ldots, n$$

Thus, each training pattern is replicated C times and augmented with each of the class labels. The new labels for a pattern $(\boldsymbol{x}, l)$ take the values

$$y_{il} = \begin{cases} +1 & \text{if } \boldsymbol{x}_i \in \text{class } \omega_l \\ -1 & \text{if } \boldsymbol{x}_i \notin \text{class } \omega_l \end{cases}$$

A classifier, $\eta_t(\boldsymbol{x}, l)$, is trained and the final classifier, $\hat{\eta}(\boldsymbol{x}, l)$, is a weighted sum of the classifiers constructed at each stage of the iteration, with decision: assign $\boldsymbol{x}$ to class j if

$$\hat{\eta}(\boldsymbol{x}, j) \geq \hat{\eta}(\boldsymbol{x}, k) \quad k = 1, \ldots, C;\ k \neq j$$

Note that the classifier $\eta_t(\boldsymbol{x}, l)$ is defined on a data space of possible mixed variable type: real continuous variables $\boldsymbol{x}$ and categorical variable l. Care will be needed in classifier design (see Chapter 11).

In the study of linear classifiers by Skurichina (2001), it is reported that boosting is only useful for large sample sizes, when applied to classifiers that perform poorly. The performance of boosting depends on many factors, including training set size, choice of classifier, the way in which the classifier weights are incorporated and the data distribution.

8.4.6 Example application study

The problem This study (Sharkey *et al.*, 2000) addresses a problem in condition monitoring and fault detection – the early detection of faults in a mechanical system by continuous monitoring of sensor data.

Summary A neural network approach is developed and ensembles of networks, based on majority vote, assessed. A *multinet* system that selects the appropriate ensemble is evaluated.

The data The data source was a four-stroke, two-cylinder air-cooled diesel engine. The recorded data comprised a digitised signal representing cylinder pressure as a function of crank angle position. Data corresponding to normal operating conditions (N) and two fault conditions – leaky exhaust valve (E) and leaky fuel injector (F) – were acquired, each at 15 load levels ranging from no load to full load. Eighty samples representing the entire cycle were acquired for each operating condition and load (giving $1200 = 80 \times 15$ data samples for each condition, 3600 in total). Each sample comprised 7200 measurements (two cycles, sampled at intervals of 0.1 degrees). However, only the 200 in the neighbourhood of combustion were used (sometimes subsampled to 100 or 50 samples) to form the pattern vectors input to the classifiers.

The data were partitioned into a validation set (600 patterns), a test set (600 patterns), with 1200–2400 of the remainder used for training. The patterns were standardised in two ways: dividing the input pattern by its maximum value; and subtracting the mean of the pattern vector elements and dividing by the standard deviation.

The model The basic classifier was a multilayer perceptron with a single hidden layer comprising a logistic sigmoid nonlinearity and an output layer equal to the number of classes (either two or three, depending on the experiment).

Training procedure For each of four class groupings (NEF, NE, EF and NF), three networks were trained corresponding to different subsamplings of the training vector or different subsets of the training data. The results for the individual networks and for an ensemble based on error rate were recorded. The ensemble classifier was based on a majority vote.

A multinet system was also developed. Here, a pattern was first presented to the three-class (NEF) network and if the largest two network outputs were, say, classes N and E, then the pattern was presented to the NE network to make the final adjudication. Thus, this is a form of *dynamic classifier selection* in which a classifier is selected depending on the input pattern.

Results The ensemble classifier gave better results than the individual classifiers. The multinet system performance was superior to a single three-class network, indicating that partitioning a multiclass classification problem into a set of binary classification problems can lead to improved performance.

8.4.7 Further developments

Many of the methods of classifier combination, even the basic non-trainable methods, are active subjects of research and assessment. For example, further properties of fixed rules (sum, voting, ranking) are presented by Kittler and Alkoot (2001) and Saranli and Demirekler (2001).

Development of stacking to density estimation is reported by Smyth and Wolpert (1999). Further applications of stacking neural network models are given by Sridhar *et al.* (1999).

The basic mixture of experts model has been extended to a tree architecture by Jordan and Jacobs (1994). Termed 'hierarchical mixture of experts', non-overlapping sets of expert networks are combined using gating networks. Outputs of these gating networks are themselves grouped using a further gating network.

Boosting is classed by Breiman (1998) as an 'adaptive resampling and combining' or *arcing* algorithm. Definitions for the bias and variance of a classifier, C, are introduced and it is shown that

$$e_E = e_B + \text{bias}(C) + \text{var}(C)$$

where e_E and e_B are the expected and Bayes error rates respectively (see Section 8.2.1). Unstable classifiers can have low bias and high variance on a large range of data sets. Combining multiple versions can reduce variance significantly.

8.4.8 Summary

In this section, we have reviewed the characteristics of combination schemes, presented a motivating application (distributed sensor detection) and described the properties of some of the more popular methods of classifier combination. Combining the results of several classifiers can give improved performance over a single classifier. To some degree, research in this area has the flavour of a cottage industry, with many *ad hoc* techniques proposed and assessed. Motivation for some of the methodology is often very weak. On the other hand, some work is motivated by real-world practical applications such as the distributed detection problem and person verification using different identification systems. Often, in applications such as these, the constituent classifier is fixed and an optimal combination is sought. There is no universal best combiner, but simple methods such as the sum, product and median rules can work well.

Of more interest are procedures that simultaneously construct the component classifiers and the combination rule. Unstable classification methods (classification methods for which small perturbations in their training set or construction procedure may result in large changes in the predictor; for example, decision trees) can have their accuracy improved by combining multiple versions of the classifier. Bagging and boosting fall into this category. Bagging perturbs the training set repeatedly and combines by simple voting; boosting reweights misclassified samples and classifiers are combined by weighted voting. Unstable classifiers such as trees can have a high variance that is reduced by bagging and boosting. However, boosting may increase the variance of a stable classifier and be counter-productive.

8.5 Application studies

One of the main motivating applications for research on multiple classifier systems has been the detection and tracking of targets using a large number of different types of sensors. Much of the methodology developed applies to highly idealised scenarios, often failing to take into account practical considerations such as asynchronous measurements, data rates and bandwidth constraints on the communication channels between the sensors and the fusion centre. Nevertheless, methodology developed within the *data fusion* literature is relevant to other practical problems. Example applications include:

- Biomedical data fusion. Various applications include coronary care monitoring and ultrasound image segmentation for the detection of the oesophagus (Dawant and Garbay, 1999).
- Airborne target identification (Raju and Sarma, 1991).

Examples of the use of classifier fusion techniques described in this chapter include the following:

- Biometrics. Chatzis *et al.* (1999) combine the outputs of five methods for person verification, based on image and voice features, in a decision fusion application. Kittler *et al.* (1997) assess a multiple observation fusion (Figure 8.10) approach to person verification. In a writer verification application, Zois and Anastassopoulos (2001) use the Bahadur–Lazarsfeld expansion to model correlated decisions. Prabhakar and Jain (2002) use kernel-based density estimates (Chapter 3) to model the distributions of the component classifier outputs, each assumed to provide a measure of confidence in one of two classes, in a fingerprint verification application.
- Chemical process modelling. Sridhar *et al.* (1996) develop a methodology for stacking neural networks in plant-process modelling applications.
- Remote sensing. In a classification of land cover from remotely sensed data using decision trees, Friedl *et al.* (1999) assess a boosting procedure (see also Chan *et al.*, 2001). In a similar application, Giacinto *et al.* (2000) assess combination methods applied to five neural and statistical classifiers.

8.6 Summary and discussion

The most common measure of classifier performance assessment is misclassification rate or error rate. We have reviewed the different types of error rate and described procedures for error rate estimation. Other performance measures are reliability – how good is our classifier at estimating the true posterior probabilities – and the area under the receiver operating characteristic curve, AUC. Misclassification rate makes the rather strong assumption that misclassification costs are equal. In most practical applications, this is unrealistic. The AUC is a measure averaged over all relative costs, and it might be argued that this is equally inappropriate since usually something will be known about relative costs. The LC index was introduced as one attempt to make use of domain knowledge in performance assessment.

Combining the results of several classifiers, rather than selecting the best, may offer improved performance. There may be practical motivating problems for this – such as those in distributed data fusion – and many rules and techniques have been proposed and assessed. These procedures differ in several respects: they may be applied at different levels of processing (raw 'sensor' data, feature level, decision level); they may be trainable or fixed; the component classifiers may be similar (for example, all decision trees) or of different forms, developed independently; the structure may be serial or parallel; finally, the combiner may be optimised alone, or jointly with the component classifiers.

There is no universal best combination rule, but the choice of rule will depend on the data set and the training set size.

8.7 Recommendations

1. Use error rate with care. Are the assumptions of equal misclassification costs appropriate for your problem?

2. If you are combining prescribed classifiers, defined on the same inputs, the sum rule is a good start.

3. For classifiers defined on separate features, the product rule is a simple one to begin with.

4. Boosting and bagging are recommended to improve performance of unstable classifiers.

8.8 Notes and references

The subject of error rate estimation has received considerable attention. The literature up to 1973 is surveyed in the extensive bibliography of Toussaint (1974), and more recent advances by Hand (1986) and McLachlan (1987). The holdout method was considered by Highleyman (1962). The leave-one-out method for error estimation is usually attributed to Lachenbruch and Mickey (1968) and cross-validation in a wider context to Stone (1974).

The number of samples required to achieve good error rate estimates is discussed with application to a character recognition task by Guyon *et al.* (1998).

Quenouille (1949) proposed the method of sample splitting for overcoming bias, later termed the jackknife. The bootstrap procedure as a method of error rate estimation has been widely applied following the pioneering work of Efron (1979, 1982, 1983). Reviews of bootstrap methods are provided by Efron and Tibshirani (1986) and Hinkley (1988). There are several studies comparing the performance of the different bootstrap estimators (Efron, 1983; Fukunaga and Hayes, 1989b; Chernick *et al.*, 1985; Konishi and Honda, 1990). Davison and Hall (1992) compare the bias and variability of the bootstrap with cross-validation. They find that cross-validation gives estimators with higher variance but lower bias than the bootstrap. The main differences between the estimators are when there is large class overlap, when the bias of the bootstrap is an order of magnitude greater than that of cross-validation.

The 0.632 bootstrap for error rate estimation is investigated by Fitzmaurice *et al.* (1991) and the number of samples required for the double bootstrap by Booth and Hall (1994). The bootstrap has been used to compute other measures of statistical accuracy. The monograph by Hall (1992) provides a theoretical treatment of the bootstrap with some emphasis on curve estimation (including parametric and nonparametric regression and density estimation).

Reliability of posterior probabilities of group membership is discussed in the book by McLachlan (1992a). Hand (1997) also considers other measures of performance assessment.

The use of the ROC curves in pattern recognition for performance assessment and comparison is described by Bradley (1997), Hand and Till (2001), Adams and Hand (1999) and Provost and Fawcett (2001).

There is a large literature on combining classifiers. A good starting point is the statistical pattern recognition review by Jain *et al.* (2000). Kittler *et al.* (1998) describe a common theoretical framework for some of the fixed combination rules.

Within the defence and aerospace domain, data fusion has received considerable attention, particularly the detection and tracking of targets using multiple distributed sources (Dasarathy, 1994b; Varshney, 1997; Waltz and Llinas, 1990), with benefits in robust operational performance, reduced ambiguity, improved detection and improved system reliability (Harris *et al.*, 1997). Combining neural network models is reviewed by Sharkey (1999).

Stacking originated with Wolpert (1992) and the mixture of experts model with Jacobs *et al.* (1991; see also Jordan and Jacobs, 1994)

Bagging is presented by Breiman (1996). Comprehensive experiments on bagging and boosting for linear classifiers are described by Skurichina (2001). The first provable polynomial-time boosting algorithm was presented by Schapire (1990). The AdaBoost algorithm was introduced by Freund and Schapire (1996, 1999). Improvements to the basic algorithm are given by Schapire and Singer (1999). Empirical comparisons of bagging and boosting are given by Bauer and Kohavi (1999).

A statistical view of boosting is provided by Friedman *et al.* (1998).

The website `www.statistical-pattern-recognition.net` contains references and pointers to other websites for further information on techniques.

Exercises

1. Two hundred labelled samples are used to train two classifiers. In the first classifier, the data set is divided into training and test sets of 100 samples each and the classifier

designed using the training set. The performance on the test set is 80% correct. In the second classifier, the data set is divided into a training set of 190 samples and a test set of 10 samples. The performance on the test set is 90%.

Is the second classifier 'better' than the first? Justify your answer.

2. Verify the Sherman–Morisson formula (8.3). Describe how it may be used to estimate the error rate of a Gaussian classifier using cross-validation.

3. The ROC curve is a plot of $1 - \epsilon_1$, the 'true positive', against ϵ_2, the 'false positive' as the threshold on (see equation (8.8))

$$p(\omega_1|\boldsymbol{x})$$

is varied, where

$$\epsilon_1 = \int_{\Omega_2} p(\boldsymbol{x}|\omega_1)\,d\boldsymbol{x}$$

$$\epsilon_2 = \int_{\Omega_1} p(\boldsymbol{x}|\omega_2)\,d\boldsymbol{x}$$

and Ω_1 is the domain where $p(\omega_1|\boldsymbol{x})$ lies above the threshold.

Show, by conditioning on $p(\omega_1|\boldsymbol{x})$, that the true positive and false positive (for a threshold μ) may be written respectively as

$$1 - \epsilon_1 = \int_{\mu}^{1} dc \int_{p(\omega_1|\boldsymbol{x})=c} p(\boldsymbol{x}|\omega_1)\,d\boldsymbol{x}$$

and

$$\epsilon_2 = \int_{\mu}^{1} dc \int_{p(\omega_1|\boldsymbol{x})=c} p(\boldsymbol{x}|\omega_2)\,d\boldsymbol{x}$$

The term $\int_{p(\omega_1|\boldsymbol{x})=c} p(\boldsymbol{x}|\omega_1)\,d\boldsymbol{x}$ is the density of $p(\omega_1|\boldsymbol{x})$ values at c for class ω_1. Hence show that the ROC curve is defined as the cumulative density of $\hat{p} = p(\omega_1|\boldsymbol{x})$ for class ω_1 patterns plotted against the cumulative density for class ω_2 patterns.

4. Generate training data consisting of 25 samples from each of two bivariate normal distributions (means $(-d/2, 0)$ and $(d/2, 0)$ and identity covariance matrix). Compute the apparent error rate and a bias-corrected version using the bootstrap. Plot both error rates, together with an error rate computed on a test set (of appropriate size) as a function of separation, d. Describe the results.

5. What is the significance of the condition $e_t > 0.5$ in step 2 of the boosting algorithm in Section 8.4.4?

6. Design an experiment to evaluate the boosting procedure. Consider which classifier to use and data sets that may be used for assessment. How would weighted samples be incorporated into the classifier design? How will you estimate generalisation performance? Implement the experiment and describe the results.

7. Repeat Exercise 6, but assess the bagging procedure as a means of improving classifier performance.

8. Using expression (8.19) for the posterior probabilities, express the product rule in terms of the priors and δ_{ki}. Assuming $\delta_{ki} \ll 1$, show that the decision rule may be expressed (under certain assumptions) in terms of sums of the δ_{ki}. State your assumptions. Finally, derive (8.20) using (8.19).

9. Given measurements $\boldsymbol{u} = (u_1, \ldots, u_L)$ made by L detectors with probability of false alarm pfa_i and probability of detection pd_i, $(i = 1, \ldots, L)$, show (assuming independence and equal cost loss matrix)

$$\log\left(\frac{p(\omega_1|\boldsymbol{u})}{p(\omega_2|\boldsymbol{u})}\right) = \log\left(\frac{p(\omega_1)}{p(\omega_2)}\right) + \sum_{S_+}\log\left(\frac{pd_i}{pfa_i}\right) + \sum_{S_-}\log\left(\frac{1-pd_i}{1-pfa_i}\right)$$

   where S_+ is the set of all detectors such that $u_i = +1$ (target present declared – class ω_1) and S_- is the set of all detectors such that $u_i = 0$ (target absent declared – class ω_2).

   Therefore, express the data fusion rule as

$$u_0 = \begin{cases} 1 & \text{if } a_0 + \boldsymbol{a}^T\boldsymbol{u} > 0 \\ 0 & \text{otherwise} \end{cases}$$

   (see equation (8.12)) and determine a_0, $\boldsymbol{a}$.

10. Write a computer program to produce the ROC curve for the L-sensor fusion problem (L sensors with probability of false alarm pfa_i and probability of detection pd_i, $i = 1, \ldots, L$), using the decision rule (8.13).

11. Using the Bahadur–Lazarsfeld expansion, derive the Bayesian decision rule in terms of the conditional correlation coefficients,

$$\gamma^i_{ij\ldots L} = \mathrm{E}_i[z_i z_j \ldots z_L] = \int z_i z_j \ldots z_L \, p(\boldsymbol{u}|\omega_i)d\boldsymbol{u}$$

   for $i = 1, 2$.

9

Feature selection and extraction

Overview

Optimal and suboptimal search techniques for feature selection (selecting a subset of the original variables for classifier design) are considered. Feature extraction seeks a transformation (linear or nonlinear) of the original variables to a smaller set. The most widely used technique is principal components analysis.

9.1 Introduction

This chapter is concerned with representing data in a reduced number of dimensions. Reasons for doing this may be easier subsequent analysis, improved classification performance through a more stable representation, removal of redundant or irrelevant information or an attempt to discover underlying structure by obtaining a graphical representation. Techniques for representing data in a reduced dimension are termed *ordination methods* or *geometrical methods* in the multivariate analysis literature. They include such methods as principal components analysis and multidimensional scaling. In the pattern recognition literature they are termed *feature selection* and *feature extraction* methods and include linear discriminant analysis and methods based on the Karhunen–Loève expansion. Some of the methods are similar, if not identical, in certain circumstances and will be discussed in detail in the appropriate section of this chapter. Here, we approach the topic initially from a pattern recognition perspective and give a brief description of the terms *feature selection* and *feature extraction.*

Given a set of measurements, dimensionality reduction can be achieved in essentially two different ways. The first is to identify those variables that do not contribute to the classification task. In a discrimination problem, we would neglect those variables that do not contribute to class separability. Thus, the task is to seek d features out of the available p measurements (the number of features d must also be determined). This is termed *feature selection in the measurement space* or simply *feature selection* (see Figure 9.1a). There are situations other than for discrimination purposes in which it is desirable to select a subset from a larger number of features or variables. Miller (1990) discusses subset selection in the context of regression.

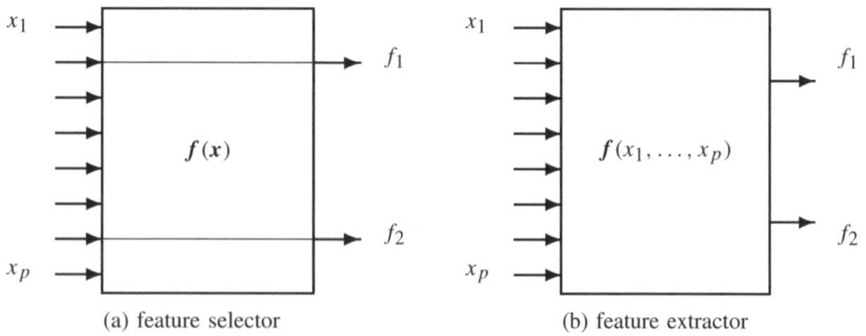

Figure 9.1 Dimensionality reduction by (a) feature selection and (b) feature extraction

The second approach is to find a transformation from the p measurements to a lower-dimensional feature space. This is termed *feature selection in the transformed space* or *feature extraction* (see Figure 9.1b). This transformation may be a linear or nonlinear combination of the original variables and may be supervised or unsupervised. In the supervised case, the task is to find the transformation for which a particular criterion of class separability is maximised.

Both of these approaches require the optimisation of some criterion function, J. For feature selection, the optimisation is over the set of all possible subsets of size d, $\mathcal{X}_d$, of the p possible measurements, $x_1, \ldots, x_p$. Thus we seek the subset $\tilde{X}_d$ for which

$$J(\tilde{X}_d) = \max_{X \in \mathcal{X}_d} J(X)$$

In feature extraction, the optimisation is performed over all possible transformations of the variables. The class of transformation is usually specified (for example, a linear transformation of the variable set) and we seek the transformation, $\tilde{A}$, for which

$$J(\tilde{A}) = \max_{A \in \mathcal{A}} J(A(\boldsymbol{x}))$$

where $\mathcal{A}$ is the set of allowable transformations. The feature vector is then $\boldsymbol{y} = \tilde{A}(\boldsymbol{x})$.

The above description is very much a simplification of dimensionality reduction techniques. The criterion function J is usually based on some measure of distance or dissimilarity between distributions, which in turn may require distances between objects to be defined. Distance measures, which are also important in some clustering schemes described in Chapter 10, are discussed in Appendix A.

We conclude this introductory section with a summary of notation. We shall denote the population covariance matrix by $\boldsymbol{\Sigma}$ and the covariance matrix of class ω_i by $\boldsymbol{\Sigma}_i$. We shall denote the maximum likelihood estimates of $\boldsymbol{\Sigma}$ and $\boldsymbol{\Sigma}_i$ by $\hat{\boldsymbol{\Sigma}}$ and $\hat{\boldsymbol{\Sigma}}_i$,

$$\hat{\boldsymbol{\Sigma}}_i = \frac{1}{n_i} \sum_{j=1}^{n} z_{ij} (\boldsymbol{x}_j - \boldsymbol{m}_i)(\boldsymbol{x}_j - \boldsymbol{m}_i)^T$$

$$\hat{\boldsymbol{\Sigma}} = \frac{1}{n} \sum_{j=1}^{n} (\boldsymbol{x}_j - \boldsymbol{m})(\boldsymbol{x}_j - \boldsymbol{m})^T$$

where

$$z_{ij} = \begin{cases} 1 & \text{if } \boldsymbol{x}_j \in \omega_i \\ 0 & \text{otherwise} \end{cases}$$

and $n_i = \sum_{j=1}^{n} z_{ij}$; $\boldsymbol{m}_i$ is the sample mean of class ω_i given by

$$\boldsymbol{m}_i = \frac{1}{n_i} \sum_{j=1}^{n} z_{ij} \boldsymbol{x}_j$$

and $\boldsymbol{m}$ the sample mean

$$\boldsymbol{m} = \sum_{i=1}^{C} \frac{n_i}{n} \boldsymbol{m}_i$$

The unbiased estimate of the covariance matrix is $n\hat{\boldsymbol{\Sigma}}/(n-1)$. We shall denote by $\boldsymbol{S}_W$ the within-class scatter matrix (or pooled within-class sample covariance matrix),

$$\boldsymbol{S}_W = \sum_{i=1}^{C} \frac{n_i}{n} \hat{\boldsymbol{\Sigma}}_i$$

with unbiased estimate $\boldsymbol{S} = n\boldsymbol{S}_W/(n-C)$. Finally, we denote by $\boldsymbol{S}_B$ the sample between-class covariance matrix

$$\boldsymbol{S}_B = \sum_{i=1}^{C} \frac{n_i}{n} (\boldsymbol{m}_i - \boldsymbol{m})(\boldsymbol{m}_i - \boldsymbol{m})^T$$

and note that $\boldsymbol{S}_W + \boldsymbol{S}_B = \hat{\boldsymbol{\Sigma}}$.

9.2 Feature selection

The problem Given a set of measurements on p variables, what is the best subset of size d? Thus, we are not considering a transformation of the measurements, merely selecting those d variables that contribute most to discrimination.

The solution Evaluate the optimality criterion for all possible combinations of d variables selected from p and select that combination for which this criterion is a maximum.

If it were quite so straightforward as this, then this section would not be so long as it is. The difficulty arises because the number of possible subsets is

$$n_d = \frac{p!}{(p-d)!d!}$$

which can be very large even for moderate values of p and d. For example, selecting the best 10 features out of 25 means that 3 268 760 feature sets must be considered, and evaluating the optimality criterion, J, for every feature set in an acceptable time may not be feasible. Therefore, we must consider ways of searching through the space of possible variable sets that reduce the amount of computation.

One reason for reducing the number of variables to measure is to eliminate redundancy. There is no need to waste effort (time and cost) making measurements on unnecessary variables. Also, reducing the number of variables may lead to a lower error rate since, as the number of variables increases, the complexity of the classifier, defined in terms of the number of parameters of the classifier to estimate, also increases. Given a finite design set, increasing the number of parameters of the classifier can lead to poor generalisation performance, even if the model is 'correct', that is, it is the model used to generate the data. The often-quoted example of this is the problem of discriminating between two classes of normally distributed data. The discrimination surface is quadratic, but the linear discriminant may give better performance on data sets of limited size.

There are two basic strategies for feature subset selection:

1. Optimal methods: these include exhaustive search methods which are feasible for only very small problems; accelerated search (we shall consider the branch and bound algorithm); and Monte Carlo methods (such as simulated annealing and genetic algorithms; Michalewicz, 1994) which can lead to a globally optimal solution, but are computationally expensive.

2. Suboptimal methods: the optimality of the above strategies is traded for computational efficiency.

The strategy adopted is independent of the optimality criterion, though the computational requirements do depend on the optimality criterion.

9.2.1 Feature selection criteria

In order to choose a good feature set, we require a means of measuring the ability of a feature set to discriminate accurately between two or more classes. This is achieved by defining a class separability measure that is optimised with respect to the possible subsets. We can choose the feature set in essentially two ways.

1. We can design a classifier on the reduced feature set and choose the feature sets for which the classifier performs well on a separate test/validation set. In this approach, the feature set is chosen to match the classifier. A different feature set may result with a different choice of classifier.

2. The second approach is to estimate the overlap between the distributions from which the data are drawn and favour those feature sets for which this overlap is minimal (that is, maximise *separability*). This is independent of the final classifier employed and it has the advantage that it is often fairly cheap to implement, but it has the disadvantage that the assumptions made in determining the overlap are often crude and may result in a poor estimate of the discriminability.

Error rate

A minimum expected classification error rate is often the main aim in classifier design. Error rate (or misclassification rate) is simply the proportion of samples incorrectly classified. Optimistic estimates of a classifier's performance will result if the data used to design the classifier are naïvely used to estimate the error rate. Such an estimate is termed the apparent error rate. Estimation should be based on a separate test set, but if data are limited in number, we may wish to use all available data in classifier design. Procedures such as the *jackknife* and the *bootstrap* have been developed to reduce the bias of the apparent error rate. Error rate estimates are discussed in Chapter 8.

Probabilistic distance

Probabilistic distance measures the distance between two distributions, $p(\boldsymbol{x}|\omega_1)$ and $p(\boldsymbol{x}|\omega_2)$, and can be used in feature selection. For example, the divergence is given by

$$J_D(\omega_1, \omega_2) = \int [p(\boldsymbol{x}|\omega_1) - p(\boldsymbol{x}|\omega_2)] \log \left\{ \frac{p(\boldsymbol{x}|\omega_1)}{p(\boldsymbol{x}|\omega_2)} \right\} d\boldsymbol{x}$$

A review of measures is given in Appendix A. All of the measures given in that appendix can be shown to give a bound on the error probability and have the property that they are maximised when classes are disjoint. In practice, it is not the tightness of the bound that is important but the computational requirements. Many of the commonly used distance measures, including those in Appendix A, simplify for normal distributions. The divergence becomes

$$J_D = \tfrac{1}{2}(\boldsymbol{\mu}_2 - \boldsymbol{\mu}_1)^T(\boldsymbol{\Sigma}_1^{-1} + \boldsymbol{\Sigma}_2^{-1})(\boldsymbol{\mu}_2 - \boldsymbol{\mu}_1) + \mathrm{Tr}\{\boldsymbol{\Sigma}_1^{-1}\boldsymbol{\Sigma}_2 + \boldsymbol{\Sigma}_1^{-1}\boldsymbol{\Sigma}_2 - 2\boldsymbol{I}\}$$

for normal distributions with means $\boldsymbol{\mu}_1$ and $\boldsymbol{\mu}_2$ and covariance matrices $\boldsymbol{\Sigma}_1$ and $\boldsymbol{\Sigma}_2$.

In a multiclass problem, the pairwise distance measures must be adapted. We may take as our cost function, J, the maximum overlap over all pairwise measures,

$$J = \max_{i,j(i \neq j)} J(\omega_i, \omega_j)$$

or the average of the pairwise measures,

$$J = \sum_{i<j} J(\omega_i, \omega_j) p(\omega_i) p(\omega_j)$$

Recursive calculation of separability measures

The search algorithms described later in this chapter, both optimal and suboptimal, construct the feature sets at the ith stage of the algorithm from that at the $(i-1)$th stage by the addition or subtraction of a small number of features from the current optimal set. For the parametric forms of the probabilistic distance criteria, the value of the criterion function at stage i can be evaluated by updating its value already calculated for stage $i-1$ instead of computing the criterion functions from their definitions. This can result in substantial computational savings.

All the parametric measures given in Appendix A, namely the Chernoff, Bhattacharyya, divergence, Patrick–Fischer and Mahalanobis distances, are functions of three basic building blocks of the form

$$\boldsymbol{x}^T \boldsymbol{S}^{-1} \boldsymbol{x}, \quad \mathrm{Tr}\{\boldsymbol{T}\boldsymbol{S}^{-1}\}, \quad |\boldsymbol{S}| \tag{9.1}$$

where $\boldsymbol{S}$ and $\boldsymbol{T}$ are positive definite symmetric matrices and $\boldsymbol{x}$ is a vector of parameters. Thus, to calculate the criteria recursively, we only need to consider each of the building blocks. For a $k \times k$ positive definite matrix $\boldsymbol{S}$, let $\tilde{\boldsymbol{S}}$ be the matrix with the kth element of the feature vector removed. Then $\boldsymbol{S}$ may be written

$$\boldsymbol{S} = \begin{bmatrix} \tilde{\boldsymbol{S}} & \boldsymbol{y} \\ \boldsymbol{y}^T & s_{kk} \end{bmatrix}$$

Assuming that $\tilde{\boldsymbol{S}}^{-1}$ is known, then $\boldsymbol{S}^{-1}$ may be written as

$$\boldsymbol{S}^{-1} = \begin{bmatrix} \tilde{\boldsymbol{S}}^{-1} + \frac{1}{d}\tilde{\boldsymbol{S}}^{-1}\boldsymbol{y}\boldsymbol{y}^T\tilde{\boldsymbol{S}}^{-1} & -\frac{1}{d}\tilde{\boldsymbol{S}}^{-1}\boldsymbol{y} \\ -\frac{1}{d}\boldsymbol{y}^T\tilde{\boldsymbol{S}}^{-1} & \frac{1}{d} \end{bmatrix} \tag{9.2}$$

where $d = s_{kk} - \boldsymbol{y}^T\tilde{\boldsymbol{S}}^{-1}\boldsymbol{y}$. Alternatively, if we know $\boldsymbol{S}^{-1}$, which may be written as

$$\boldsymbol{S}^{-1} = \begin{bmatrix} \boldsymbol{A} & \boldsymbol{c} \\ \boldsymbol{c}^T & b \end{bmatrix}$$

then we may calculate $\tilde{\boldsymbol{S}}^{-1}$ as

$$\tilde{\boldsymbol{S}}^{-1} = \boldsymbol{A} - \frac{1}{b}\boldsymbol{c}\boldsymbol{c}^T$$

Thus, we can compute the inverse of a matrix if we know the inverse before a feature is added to or deleted from the feature set.

In some cases it may not be necessary to calculate the inverse $\tilde{\boldsymbol{S}}^{-1}$ from $\boldsymbol{S}^{-1}$. Consider the quadratic form $\boldsymbol{x}^T\boldsymbol{S}^{-1}\boldsymbol{x}$, where $\boldsymbol{x}$ is a k-dimensional vector and $\tilde{\boldsymbol{x}}$ denotes the vector with the kth value removed. This can be expressed in terms of the quadratic form before the kth feature is removed as

$$\tilde{\boldsymbol{x}}^T\tilde{\boldsymbol{S}}^{-1}\tilde{\boldsymbol{x}} = \boldsymbol{x}^T\boldsymbol{S}^{-1}\boldsymbol{x} - \frac{1}{b}[(\boldsymbol{c}^T : b)\boldsymbol{x}]^2$$

where $[\boldsymbol{c}^T : b]$ is the row of $\boldsymbol{S}^{-1}$ corresponding to the feature that is removed. Thus, the calculation of $\tilde{\boldsymbol{S}}^{-1}$ can be deferred until it is confirmed that this feature is to be permanently removed from the candidate feature set.

The second term to consider in (9.1) is $\mathrm{Tr}\{\boldsymbol{T}\boldsymbol{S}^{-1}\}$. We may use the relationship

$$\mathrm{Tr}\{\tilde{\boldsymbol{T}}\tilde{\boldsymbol{S}}^{-1}\} = \mathrm{Tr}\{\boldsymbol{T}\boldsymbol{S}^{-1}\} - \frac{1}{b}(\boldsymbol{c}^T : b)\boldsymbol{T}\begin{pmatrix}\boldsymbol{c} \\ b\end{pmatrix} \tag{9.3}$$

Finally, the determinants satisfy

$$|\boldsymbol{S}| = (s_{kk} - \boldsymbol{y}^T\tilde{\boldsymbol{S}}\boldsymbol{y})|\tilde{\boldsymbol{S}}|$$

Criteria based on scatter matrices

The probabilistic distance measures require knowledge, or estimation, of a multivariate probability density function followed by numerical integration, except in the case of a parametric density function. This is clearly computationally very expensive. Other measures have been developed based on the between-class and within-class scatter matrices. Those given in this chapter are discussed in more detail by Devijver and Kittler (1982).

We define a measure of the separation between two data sets, ω_1 and ω_2, as

$$J_{as} = \frac{1}{n_1 n_2} \sum_{i=1}^{n_1} \sum_{j=1}^{n_2} d(\boldsymbol{x}_i, \boldsymbol{y}_j)$$

for $\boldsymbol{x}_i \in \omega_1$, $\boldsymbol{y}_j \in \omega_2$ and $d(\boldsymbol{x}, \boldsymbol{y})$ a measure of distance between samples $\boldsymbol{x}$ and $\boldsymbol{y}$. This is the average separation. Defining the average distance between classes as

$$J = \frac{1}{2} \sum_{i=1}^{C} p(\omega_i) \sum_{j=1}^{C} p(\omega_j) J_{as}(\omega_i, \omega_j)$$

where $p(\omega_i)$ is the prior probability of class ω_i (estimated as $p_i = n_i/n$), the measure J may be written, using a Euclidean distance squared for $d(\boldsymbol{x}, \boldsymbol{y})$,

$$J = J_1 = \text{Tr}\{\boldsymbol{S}_W + \boldsymbol{S}_B\} = \text{Tr}\{\hat{\boldsymbol{\Sigma}}\}$$

The criterion J_1 is not very satisfactory as a feature selection criterion: it is simply the total variance, which does not depend on class information.

Our aim is to find a set of variables for which the within-class spread is small and the between-class spread is large in some sense. Several criteria have been proposed for achieving this. One popular measure is

$$J_2 = \text{Tr}\{\boldsymbol{S}_W^{-1} \boldsymbol{S}_B\} \tag{9.4}$$

Another is the ratio of the determinants

$$J_3 = \frac{|\hat{\boldsymbol{\Sigma}}|}{|\boldsymbol{S}_W|} \tag{9.5}$$

the ratio of the total scatter to the within-class scatter. A further measure used is

$$J_4 = \frac{\text{Tr}\{\boldsymbol{S}_B\}}{\text{Tr}\{\boldsymbol{S}_W\}} \tag{9.6}$$

As with the probabilistic distance measures, each of these distance measures may be calculated recursively.

9.2.2 Search algorithms for feature selection

The problem of feature selection is to choose the 'best' possible subset of size d from a set of p features. In this section we consider strategies for doing that – both optimal and suboptimal. The basic approach is to build up a set of d features incrementally, starting with the empty set (a 'bottom-up' method) or to start with the full set of measurements and remove redundant features successively (a 'top-down' approach).

If X_k represents a set of k features or variables then, in a bottom-up approach, the best set at a given iteration, $\tilde{X}_k$, is the set for which the feature (extraction) selection criterion has its maximum value

$$J(\tilde{X}_k) = \max_{X \in \mathcal{X}_k} J(X)$$

The set $\mathcal{X}_k$ of all sets of features at a given step is determined from the set at the previous iteration. This means that the set of measurements at one stage of an iterative procedure is used as a starting point to find the set at the next stage. This does not imply that the sets are necessarily nested ($\tilde{X}_k \subset \tilde{X}_{k+1}$), though they may be.

Branch and bound procedure

This is an optimal search procedure that does not involve exhaustive search. It is a top-down procedure, beginning with the set of p variables and constructing a tree by deleting variables successively. It relies on one very important property of the feature selection criterion, namely that for two subsets of the variables, X and Y,

$$X \subset Y \Rightarrow J(X) < J(Y) \tag{9.7}$$

That is, evaluating the feature selection criterion on a subset of variables of a given set yields a smaller value of the feature selection criterion. This is termed the *monotonicity property*.

We shall describe the method by way of example. Let us assume that we wish to find the best three variables out of a set of five. A tree is constructed whose nodes represent all possible subsets of cardinality 3, 4 and 5 of the total set as follows. Level 0 in the tree contains a single node representing the total set. Level 1 contains subsets of the total set with one variable removed, and level 2 contains subsets of the total set with two variables removed. The numbers to the right of each node in the tree represent a subset of variables. The number to the left represents the variable that has been removed from the subset of the parent node in order to arrive at a subset for the child node. Level 2 contains all possible subsets of five variables of size three. Note that the tree is not symmetrical. This is because removing variables 4 then 5 from the original set (to give the subset (123)) has the same result as removing variable 5 then variable 4. Therefore, in order for the subsets not to be replicated, we have only allowed variables to be removed in increasing order. This removes unnecessary repetitions in the calculation.

Now we have obtained our tree structure, how are we going to use it? The tree is searched from the least dense part to the part with the most branches (right to left in Figure 9.2). Figure 9.3 gives a tree structure with values of the criterion J printed at the nodes. Starting at the rightmost set (the set (123) with a value of $J = 77.2$), the search backtracks to the nearest branching node and proceeds down the rightmost branch evaluating $J(\{1, 2, 4, 5\})$, then $J(\{1, 2, 4\})$ which gives a lower value than the current maximum value of J, J^*, and so is discarded. The set $J(\{1, 2, 5\})$ is next evaluated and retained as the current best value (largest on a subset of three variables), $J^* = 80.1$. $J(\{1, 3, 4, 5\})$ is evaluated next, and since this is less than the current best, the search of

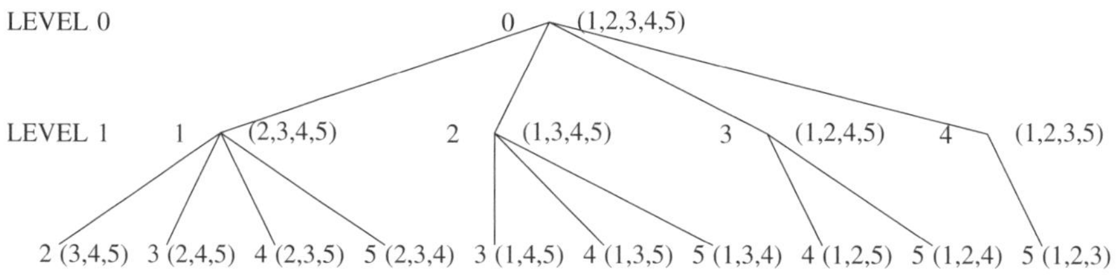

Figure 9.2 Tree figure for branch and bound method

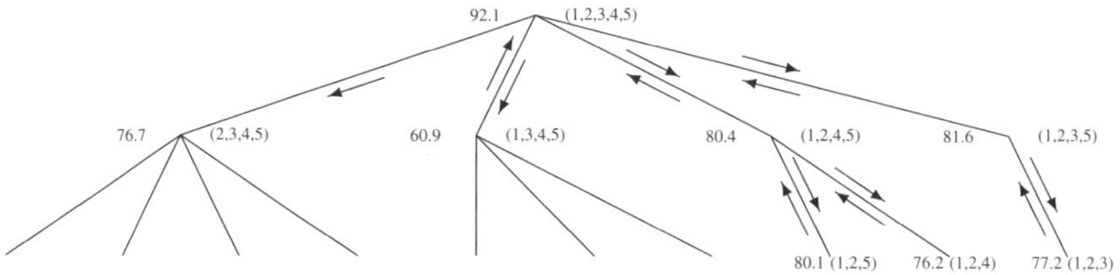

Figure 9.3 Tree figure for branch and bound method with feature selection criterion value at each node

the section of the tree originating from this node is not performed. This is because we know from the monotonicity property that all subsets of this set will yield a lower value of the criterion function. The algorithm then backtracks to the nearest branching node and proceeds down the next rightmost branch (in this case, the final branch). $J(\{2, 3, 4, 5\})$ is evaluated, and again since this is lower than the current best value on a subset of three variables, the remaining part of the tree is not evaluated.

Thus, although not all subsets of size 3 are evaluated, the algorithm is optimal since we know by condition (9.7) that those not evaluated will yield a lower value of J.

From the specific example above, we can see a more general strategy: start at the top level and proceed down the rightmost branch, evaluating the cost function J at each node. If the value of J is less than the current threshold then abandon the search down that particular branch and backtrack to the previous branching node. Continue the search down the next rightmost branch. If, on the search of any branch the bottom level is reached (as is bound to happen on the initial branch), then if the value of J for this level is larger than the current threshold, the threshold is updated and backtracking begins. Note from Figure 9.2 (and you can verify that this is true in general), that the candidates for removal at level i, given that variable n_{i-1} has been removed at the previous level, are

$$n_{i-1} + 1, \ldots, i + m$$

where m is the size of the final subset.

Note that if the criterion function is evaluated for the successor of a given node, i.e. for a node which is one level below and connected by a single link to a given node, then during the branch and bound procedure it will be evaluated for all 'brothers and sisters' of that node – i.e. for all other direct successors of the given node. Now since a node with a low value of J is more likely to be discarded than a node with a high value of J, it is sensible to order these sibling nodes so that those that have lower values have more branches. This is the case in Figure 9.3 where the nodes at level 2 are ordered so that those yielding a smaller value of J have the larger number of branches. Since all the sibling feature sets will be evaluated anyway, this results in no extra computation. This scheme is due to Narendra and Fukunaga (1977).

The feature selection criterion most commonly used in the literature for feature selection both in regression (Miller, 1990) and in classification (Narendra and Fukunaga, 1977) is the quadratic form

$$\boldsymbol{x}_k^T \boldsymbol{S}_k^{-1} \boldsymbol{x}_k$$

where $\boldsymbol{x}_k$ is a k-dimensional vector and $\boldsymbol{S}_k$ is a $k \times k$ positive definite matrix when k features are used. For example, in a two-class problem, the Mahalanobis distance (see Appendix A) between two groups with means $\boldsymbol{\mu}_i$ $(i = 1, 2)$ and covariance matrices $\boldsymbol{\Sigma}_i (i = 1, 2)$

$$J = (\boldsymbol{\mu}_1 - \boldsymbol{\mu}_2)^T \left(\frac{\boldsymbol{\Sigma}_1 + \boldsymbol{\Sigma}_2}{2} \right)^{-1} (\boldsymbol{\mu}_1 - \boldsymbol{\mu}_2) \tag{9.8}$$

satisfies (9.7). In a multiclass problem, we may take the sum over all pairs of classes. This will also satisfy (9.7) since each component of the sum does. There are many other feature selection criteria satisfying the monotonicity criterion, including probabilistic distance measures (for example, Bhattacharyya distance, divergence (Fukunaga, 1990)) and measures based on the scatter matrices (for example, J_2 (9.4), but not J_3 (9.5) or J_4 (9.6)).

9.2.3 Suboptimal search algorithms

There are many problems where suboptimal methods must be used. The branch and bound algorithm may not be computationally feasible (the growth in the number of possibilities that must be examined is still an exponential function of the number of variables) or may not be appropriate if the monotonicity property (9.7) does not hold. Suboptimal algorithms, although not capable of examining every feature combination, will assess a set of potentially useful feature combinations. We consider several techniques, varying in complexity.

Best individual N

The simplest method, and perhaps the one giving the poorest performance, for choosing the best N features is to assign a discrimination power estimate to each of the features in the original set, $\mathcal{X}$, individually. Thus, the features are ordered so that

$$J(x_1) \geq J(x_2) \geq \cdots \geq J(x_p)$$

and we select as our best set of N features the N features with the best individual scores:

$$\{x_i | i \leq N\}$$

In some cases this method can produce reasonable feature sets, especially if the features in the original set are uncorrelated, since the method ignores multivariate relationships.

However, if the features of the original set are highly correlated, the chosen feature set will be suboptimal as some of the features will be adding little discriminatory power. There are cases when the N best features are not the best N features even when the variables are independent (Hand, 1981a).

Sequential forward selection

Sequential forward selection (SFS, or the method of set addition) is a bottom-up search procedure that adds new features to a feature set one at a time until the final feature set is reached. Suppose we have a set of d_1 features, X_{d_1}. For each of the features ξ_j not yet selected (i.e. in $\mathcal{X} - X_{d_1}$) the criterion function $J_j = J(X_{d_1} + \xi_j)$ is evaluated. The feature that yields the maximum value of J_j is chosen as the one that is added to the set X_{d_1}. Thus, at each stage, the variable is chosen that, when added to the current set, maximises the selection criterion. The feature set is initialised to the null set. When the best improvement makes the feature set worse, or when the maximum allowable number of features is reached, the algorithm terminates. The main disadvantage of the method is that it does not include a mechanism for deleting features from the feature set once they have been added should further additions render them unnecessary.

Generalised sequential forward selection

Instead of adding a single feature at a time to a set of measurements, in the generalised sequential forward selection (GSFS) algorithm r features are added as follows. Suppose we have a set of d_1 measurements, X_{d_1}. All possible sets of size r are generated from the remaining $n - d_1$ features – this gives

$$\binom{n - d_1}{r}$$

sets. For each set of r features, Y_r, the cost function is evaluated for $X_{d_1} + Y_r$ and the set that maximises the cost function is used to increment the feature set. This is more costly than SFS, but has the advantage that at each stage it is possible to take into account to some degree the statistical relationship between the available measurements.

Sequential backward selection

Sequential backward selection (SBS), or sequential backward elimination, is the top-down analogy to SFS. Variables are deleted one at a time until d measurements remain. Starting with the complete set, the variable ξ_j is chosen for which $J(\mathcal{X} - \xi_j)$ is the largest (i.e. ξ_j decreases J the least). The new set is $\{\mathcal{X} - \xi_j\}$. This process is repeated until a set of the required cardinality remains. The procedure has the disadvantage over SFS that it is computationally more demanding since the criterion function J is evaluated over larger sets of variables.

Generalised sequential backward selection

If you have read the previous sections, you will not be surprised to learn that generalised sequential backward selection (GSBS) decreases the current set of variables by several variables at a time.

Plus l – take away r selection

This is a procedure that allows some backtracking in the feature selection process. If $l > r$, it is a bottom-up procedure. l features are added to the current set using SFS and then the worst r features are removed using SBS. This algorithm removes the problem of nesting since the set of features obtained at a given stage is not necessarily a subset of the features at the next stage of the procedure. If $l < r$ then the procedure is top-down, starting with the complete set of features, removing r, then adding l successively until the required number is achieved.

Generalised plus l – take away r selection

The generalised version of the l–r algorithm uses the GSFS and the GSBS algorithms at each stage rather than the SFS and SBS procedures. Kittler (1978a) generalises the procedure further by allowing the integers l and r to be composed of several components $l_i, i = 1, \ldots, n_l$, and $r_j, j = 1, \ldots, n_r$ (where n_l and n_r are the number of components), satisfying

$$0 \le l_i \le l \qquad 0 \le r_j \le r$$

$$\sum_{i=1}^{n_l} l_i = l \qquad \sum_{j=1}^{n_r} r_j = r$$

In this generalisation, instead of applying the generalised sequential forward selection in one step of l variables (denoted GSFS(l)), the feature set is incremented in n_l steps by adding l_i features ($i = 1, \ldots, n_l$) at each increment; i.e. apply GSFS(l_i) successively for $i = 1, \ldots, n_l$. This reduces the computational complexity. Similarly, GSBS(r) is replaced by applying GSBS(r_j), $j = 1, \ldots, n_r$, successively. The algorithm is referred to as the (Z_l, Z_r) algorithm, where Z_l and Z_r denote the sequence of integers l_i and l_j,

$$Z_l = (l_1, l_2, \ldots, l_{n_l})$$
$$Z_r = (r_1, r_2, \ldots, r_{n_r})$$

The suboptimal search algorithms discussed in this subsection and the exhaustive search strategy may be considered as special cases of the (Z_l, Z_r) algorithm (Devijver and Kittler, 1982).

Floating search methods

Floating search methods, sequential forward floating selection (SFFS) and sequential backward floating selection (SBFS), may be regarded as a development of the l–r algorithm above in which the values of l and r are allowed to 'float' – that is, they may change at different stages of the selection procedure.

Suppose that at stage k we have a set of subsets $X_1, \ldots, X_k$ of sizes 1 to k respectively. Let the corresponding values of the feature selection criteria be J_1 to J_k, where $J_i = J(X_i)$, for the feature selection criterion, $J(.)$. Let the total set of features be $\mathcal{X}$. At the kth stage of the SFFS procedure, do the following.

1. Select the feature x_j from $Y - X_k$ that increases the value of J the greatest and add it to the current set, $X_{k+1} = X_k + x_j$.

2. Find the feature, x_r, in the current set, X_{k+1}, that reduces the value of J the least; if this feature is the same as x_j then set $J_{k+1} = J(X_{k+1})$; increment k; go to step 1; otherwise remove it from the set to form $X'_k = X_{k+1} - x_r$.

3. Continue removing features from the set X'_k to form reduced sets X'_{k-1} while $J(X'_{k-1}) > J_{k-1}$; $k = k - 1$; or $k = 2$; then continue with step 1.

The algorithm is initialised by setting $k = 0$ and $X_0 = \oslash$ (the empty set) and using the SFS method until a set of size 2 is obtained.

9.2.4 Example application study

The problem Classification of land use using synthetic aperture radar images (Zongker and Jain, 1996).

Summary As part of an evaluation of feature selection algorithms on several data sets (real and simulated), classification performance using features produced by the SFS algorithm is compared with that using the SFFS algorithm.

The data The data comprised synthetic aperture radar images (approximately 22 000 pixels). A total of 18 features were computed from four different texture models: local statistics (five features), grey level co-occurrence matrices (six features), fractal features (two features) and a log-normal random field model (five features). One of the goals is to see how measures from different models may be utilised to provide better performance.

The model The best performance was assessed in terms of recognition rate of a 3-nearest-neighbour classifier.

Training procedure The data were divided into independent train and test sets and the performance of the classifier evaluated as a function of the number of features produced by the SFS and SFFS algorithms.

Results The recognition rate is not a monotonic function of the number of features. The best performance was achieved by SFFS using an 11-feature subset.

9.2.5 Further developments

There are other approaches to feature selection. Chang (1973) considers a dynamic programming approach. Monte Carlo methods based on simulated annealing and *genetic algorithms* (see, for example, Mitchell, 1997) are described by Siedlecki and Sklansky (1988), Brill *et al.* (1992) and Chang and Lippmann (1991).

Developments of the floating search methods to adaptive floating search algorithms that determine the number of additions or deletions dynamically as the algorithm proceeds are proposed by Somol *et al.* (1999). Other approaches include node pruning in neural networks and methods based on modelling the class densities as finite mixtures of a special type (Pudil *et al.*, 1995).

9.2.6 Summary

Feature selection is the process of selecting from the original features (or variables) those features that are important for classification. A feature selection criterion, J, is defined on subsets of the features and we seek that combination of features for which J is maximised.

In this section we have described some statistical pattern recognition approaches to feature selection, both optimal and suboptimal techniques for maximising J. Some of the techniques are dependent on a specific classifier through error rate. This may provide computational problems if the classifier is complex. Different feature sets may be obtained for different classifiers.

Generally there is a trade-off between algorithms that are computationally feasible, but not optimal, and those that are optimal or close to optimal but are computationally complex even for moderate feature set sizes. Studies of the floating methods suggest that these offer close to optimal performance at an acceptable cost.

9.3 Linear feature extraction

Feature extraction is the transformation of the original data (using all variables) to a data set with a reduced number of variables.

In the problem of feature selection covered in the previous section, the aim is to select those variables that contain the most discriminatory information. Alternatively, we may wish to limit the number of measurements we make, perhaps on grounds of cost, or we may want to remove redundant or irrelevant information to obtain a less complex classifier.

In feature extraction, all available variables are used and the data are transformed (using a linear or nonlinear transformation) to a reduced dimension space. Thus, the aim is to replace the original variables by a smaller set of underlying variables. There are several reasons for performing feature extraction:

1. to reduce the bandwidth of the input data (with the resulting improvements in speed and reductions in data requirements);

2. to provide a relevant set of features for a classifier, resulting in improved performance, particularly from simple classifiers;

3. to reduce redundancy;

4. to recover new meaningful underlying variables or features that describe the data, leading to greater understanding of the data generation process;

5. to produce a low-dimensional representation (ideally in two dimensions) with minimum loss of information so that the data may easily be viewed and relationships and structure in the data identified.

The techniques covered in this section are to be found in the literature on a diverse range of topics. Many are techniques of *exploratory data analysis* described in textbooks on multivariate analysis. Sometimes referred to as *geometric methods* or methods of *ordination*, they make no assumption about the existence of groups or clusters in the data. They have found application in a wide range of subjects including ecology, agricultural science, biology and psychology. Geometric methods are sometimes further categorised as being *variable-directed* when they are primarily concerned with relationships between variables, or *individual-directed* when they are primarily concerned with relationships between individuals.

In the pattern recognition literature, the data transformation techniques are termed *feature selection in the transformed space* or *feature extraction* and they be supervised (make use of class label information) or unsupervised. They may be based on the optimisation of a class separability measure, such as those described in the previous section.

9.3.1 Principal components analysis

Introduction

Principal components analysis originated in work by Pearson (1901). It is the purpose of principal components analysis to derive new variables (in decreasing order of importance) that are linear combinations of the original variables and are uncorrelated. Geometrically, principal components analysis can be thought of as a rotation of the axes of the original coordinate system to a new set of orthogonal axes that are ordered in terms of the amount of variation of the original data they account for.

One of the reasons for performing a principal components analysis is to find a smaller group of underlying variables that describe the data. In order to do this, we hope that the first few components will account for most of the variation in the original data. A representation in fewer dimensions may aid the user for the reasons given in the introduction to this chapter. Even if we are able to characterise the data by a few variables, it does not follow that we will be able to assign an interpretation to these new variables.

Principal components analysis is a variable-directed technique. It makes no assumptions about the existence or otherwise of groupings within the data and so is described as an unsupervised feature extraction technique.

So far, our discussion has been purely descriptive. We have introduced many terms without proper definition. There are several ways in which principal components analysis can be described mathematically, but let us leave aside the mathematics for the time being and continue with a geometrical derivation. We must necessarily confine our illustrations to two dimensions, but nevertheless we shall be able to define most of the attendant terminology and consider some of the problems of a principal components analysis.

In Figure 9.4 are plotted a dozen objects, with the x and y values for each point in the figure representing measurements on each of the two variables. They could represent

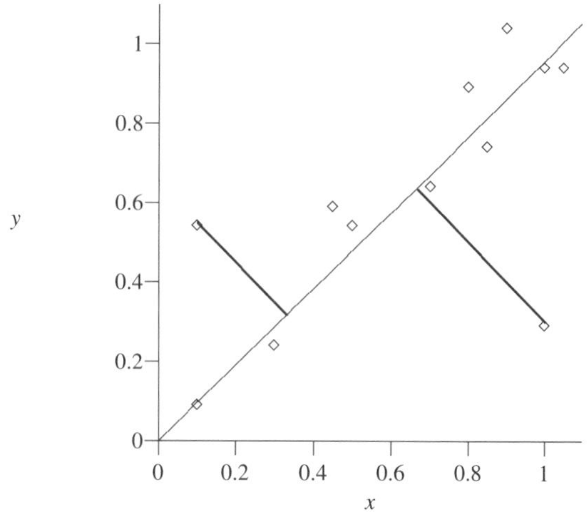

Figure 9.4 Principal components line of best fit

the height and weight of a group of individuals, for example, in which case one variable would be measured in metres or centimetres or inches and the other variable in grams or pounds. So the units of measurement may differ.

The problem we want to solve is: what is the best straight line through this set of points? Before we can answer, we must clarify what we mean by 'best'. If we consider the variable x to be an input variable and y a dependent variable so that we wish to calculate the expected value of y given x, $E[y|x]$, then the best (in a least squares sense) regression line of y on x

$$y = mx + c$$

is the line for which the sum of the squared distances of points from the line is a minimum, and the distance of a point from the line is the vertical distance.

If y were the *regressor* and x the dependent variable, then the linear regression line is the line for which the sum of squares of horizontal distances of points from the line is a minimum. Of course, this gives a different solution. (A good illustration of the two linear regressions on a bivariate distribution is given in Stuart and Ord, 1991.)

So we have two lines of best fit, and a point to note is that changing the scale of the variables does not alter the predicted values. If the scale of x is compressed or expanded, the slope of the line changes but the predicted value of y does not alter. Principal components analysis produces a single best line and the constraint that it satisfies is that the sum of the squares of the *perpendicular* distances from the sample points to the line is a minimum (see Figure 9.4). A standardisation procedure that is often carried out (and almost certainly if the variables are measured in different units) is to make the variance of each variable unity. Thus the data are transformed to new axes, centred at the centroid of the data sample and in coordinates defined in terms of units of standard deviation. The principal components line of best fit is not invariant to changes of scale.

The variable defined by the line of best fit is the first principal component. The second principal component is the variable defined by the line that is orthogonal with the first and so it is uniquely defined in our two-dimensional example. In a problem with higher-dimensional data, it is the variable defined by the vector orthogonal to the line of best fit of the first principal component that, together with the line of best fit, defines a plane of best fit, i.e. the plane for which the sum of squares of perpendicular distances of points from the plane is a minimum. Successive principal components are defined in a similar way.

Another way of looking at principal components (which we shall derive more formally in Section 9.3.1) is in terms of the variance of the data. If we were to project the data in Figure 9.4 onto the first principal axis (that is, the vector defining the first principal component), then the variation in the direction of the first principal component is proportional to the sum of the squares of the distances from the second principal axis (the constant of proportionality depending on the number of samples, $1/(n-1)$). Similarly, the variance along the second principal axis is proportional to the sum of the squares of the perpendicular distances from the first principal axis. Now, since the total sum of squares is a constant, minimising the sum of squared distances from a given line is the same as maximising the sum of squares from its perpendicular or, by the above, maximising the variance in the direction of the line. This is another way of deriving principal components: find the direction that accounts for as much of the variance as possible (the direction along which the variance is a maximum); the second principal component is defined by the direction orthogonal to the first for which the variance is a maximum, and so on. The variances are the *principal values*.

Principal components analysis produces an orthogonal coordinate system in which the axes are ordered in terms of the amount of variance in the original data for which the corresponding principal components account. If the first few principal components account for most of the variation, then these may be used to describe the data, thus leading to a reduced-dimension representation. We might also like to know if the new components can be interpreted as something meaningful in terms of the original variables. This wish is not often granted, and in practice the new components will be difficult to interpret.

Derivation of principal components

There are at least three ways in which we can approach the problem of deriving a set of principal components. Let $x_1, \ldots, x_p$ be our set of original variables and let $\xi_i, i = 1, \ldots, p$, be linear combinations of these variables

$$\xi_i = \sum_{j=1}^{p} a_{ij} x_j$$

or

$$\boldsymbol{\xi} = \boldsymbol{A}^T \boldsymbol{x}$$

where $\boldsymbol{\xi}$ and $\boldsymbol{x}$ are vectors of random variables and $\boldsymbol{A}$ is the matrix of coefficients. Then we can proceed as follows:

1. We may seek the orthogonal transformation $\boldsymbol{A}$ yielding new variables ξ_j that have stationary values of their variance. This approach, due to Hotelling (1933), is the one that we choose to present in more detail below.

2. We may seek the orthogonal transformation that gives uncorrelated variables ξ_j.
3. We may consider the problem geometrically and find the line for which the sum of squares of perpendicular distances is a minimum, then the plane of best fit and so on. We used this geometric approach in our two-dimensional illustration above (Pearson, 1901).

Consider the first variable ξ_1:

$$\xi_1 = \sum_{j=1}^{p} a_{1j} x_j$$

We choose $\boldsymbol{a}_1 = (a_{11}, a_{12}, \ldots, a_{1p})^T$ to maximise the variance of ξ_1, subject to the constraint $\boldsymbol{a}_1^T \boldsymbol{a}_1 = |\boldsymbol{a}_1|^2 = 1$. The variance of ξ_1 is

$$\begin{aligned} \text{var}(\xi_1) &= \text{E}[\xi_1^2] - \text{E}[\xi_1]^2 \\ &= \text{E}[\boldsymbol{a}_1^T \boldsymbol{x}\boldsymbol{x}^T \boldsymbol{a}_1] - \text{E}[\boldsymbol{a}_1^T \boldsymbol{x}]\text{E}[\boldsymbol{x}^T \boldsymbol{a}_1] \\ &= \boldsymbol{a}_1^T (\text{E}[\boldsymbol{x}\boldsymbol{x}^T] - \text{E}[\boldsymbol{x}]\text{E}[\boldsymbol{x}^T])\boldsymbol{a}_1 \\ &= \boldsymbol{a}_1^T \boldsymbol{\Sigma} \boldsymbol{a}_1 \end{aligned}$$

where $\boldsymbol{\Sigma}$ is the covariance matrix of $\boldsymbol{x}$ and E[.] denotes expectation. Finding the stationary value of $\boldsymbol{a}_1^T \boldsymbol{\Sigma} \boldsymbol{a}_1$ subject to the constraint $\boldsymbol{a}_1^T \boldsymbol{a}_1 = 1$ is equivalent to finding the unconditional stationary value of

$$f(\boldsymbol{a}_1) = \boldsymbol{a}_1^T \boldsymbol{\Sigma} \boldsymbol{a}_1 - \nu \boldsymbol{a}_1^T \boldsymbol{a}_1$$

where ν is a Lagrange multiplier. (The method of Lagrange multipliers can be found in most textbooks on mathematical methods; for example, Wylie and Barrett, 1995.) Differentiating with respect to each of the components of $\boldsymbol{a}_1$ in turn and equating to zero gives

$$\boldsymbol{\Sigma} \boldsymbol{a}_1 - \nu \boldsymbol{a}_1 = 0$$

For a non-trivial solution for $\boldsymbol{a}_1$ (that is, a solution other than the null vector), $\boldsymbol{a}_1$ must be an eigenvector of $\boldsymbol{\Sigma}$ with ν an eigenvalue. Now $\boldsymbol{\Sigma}$ has p eigenvalues $\lambda_1, \ldots, \lambda_p$, not all necessarily distinct and not all non-zero, but they can be ordered so that $\lambda_1 \geq \lambda_2 \geq \cdots \geq \lambda_p \geq 0$. We must chose one of these for the value of ν. Now, since the variance of $\boldsymbol{\xi}_1$ is

$$\begin{aligned} \boldsymbol{a}_1^T \boldsymbol{\Sigma} \boldsymbol{a}_1 &= \nu \boldsymbol{a}_1^T \boldsymbol{a}_1 \\ &= \nu \end{aligned}$$

and we wish to maximise this variance, then we choose ν to be the largest eigenvalue λ_1, and $\boldsymbol{a}_1$ is the corresponding eigenvector. This eigenvector will not be unique if the value of ν is a repeated root of the *characteristic equation*

$$|\boldsymbol{\Sigma} - \nu \boldsymbol{I}| = 0$$

The variable ξ_1 is the first *principal component* and has the largest variance of any linear function of the original variables $x_1, \ldots, x_p$.

The second principal component, $\xi_2 = \boldsymbol{a}_2^T \boldsymbol{x}$, is obtained by choosing the coefficients $a_{2i}, i = 1, \dots, p$, so that the variance of ξ_2 is maximised subject to the constraint $|\boldsymbol{a}_2| = 1$ *and* that ξ_2 is uncorrelated with the first principal component ξ_1. This second constraint implies

$$E[\xi_2 \xi_1] - E[\xi_2]E[\xi_1] = 0$$

or

$$\boldsymbol{a}_2^T \boldsymbol{\Sigma} \boldsymbol{a}_1 = 0 \tag{9.9}$$

and since $\boldsymbol{a}_1$ is an eigenvector of $\boldsymbol{\Sigma}$, this is equivalent to $\boldsymbol{a}_2^T \boldsymbol{a}_1 = 0$, i.e. $\boldsymbol{a}_2$ is orthogonal to $\boldsymbol{a}_1$.

Using the method of Lagrange's undetermined multipliers again, we seek the unconstrained maximisation of

$$\boldsymbol{a}_2^T \boldsymbol{\Sigma} \boldsymbol{a}_2 - \mu \boldsymbol{a}_2^T \boldsymbol{a}_2 - \eta \boldsymbol{a}_2^T \boldsymbol{a}_1$$

Differentiating with respect to the components of $\boldsymbol{a}_2$ and equating to zero gives

$$2\boldsymbol{\Sigma} \boldsymbol{a}_2 - 2\mu \boldsymbol{a}_2 - \eta \boldsymbol{a}_1 = 0 \tag{9.10}$$

Multiplying by $\boldsymbol{a}_1^T$ gives

$$2\boldsymbol{a}_1 \boldsymbol{\Sigma} \boldsymbol{a}_2 - \eta = 0$$

since $\boldsymbol{a}_1^T \boldsymbol{a}_2 = 0$. Also, by (9.9), $\boldsymbol{a}_2^T \boldsymbol{\Sigma} \boldsymbol{a}_1 = \boldsymbol{a}_1^T \boldsymbol{\Sigma} \boldsymbol{a}_2 = 0$, therefore $\eta = 0$. Equation (9.10) becomes

$$\boldsymbol{\Sigma} \boldsymbol{a}_2 = \mu \boldsymbol{a}_2$$

Thus, $\boldsymbol{a}_2$ is also an eigenvector of $\boldsymbol{\Sigma}$, orthogonal to $\boldsymbol{a}_1$. Since we are seeking to maximise the variance, it must be the eigenvector corresponding to the largest of the remaining eigenvalues, that is, the second largest eigenvalue overall.

We may continue this argument, with the kth principal component $\xi_k = \boldsymbol{a}_k^T \boldsymbol{x}$, where $\boldsymbol{a}_k$ is the eigenvector corresponding to the kth largest eigenvalue of $\boldsymbol{\Sigma}$ and with variance equal to the kth largest eigenvalue.

If some eigenvalues are equal, the solution for the eigenvectors is not unique, but it is always possible to find an orthonormal set of eigenvectors for a real symmetric matrix with non-negative eigenvalues.

In matrix notation,

$$\boldsymbol{\xi} = \boldsymbol{A}^T \boldsymbol{x} \tag{9.11}$$

$\boldsymbol{A} = [\boldsymbol{a}_1, \dots, \boldsymbol{a}_p]$, the matrix whose columns are the eigenvectors of $\boldsymbol{\Sigma}$.

So now we know how to determine the principal components – by performing an eigenvector decomposition of the symmetric positive definite matrix $\boldsymbol{\Sigma}$, and using the eigenvectors as coefficients in the linear combination of the original variables. But how do we determine a reduced-dimension representation of some given data? Let us consider the variance.

The sum of the variances of the principal components is given by

$$\sum_{i=1}^{p} \text{var}(\xi_i) = \sum_{i=1}^{p} \lambda_i$$

the sum of the eigenvalues of the covariance matrix $\mathbf{\Sigma}$, equal to the total variance of the original variables. We can then say that the first k principal components account for

$$\sum_{i=1}^{k} \lambda_i \Big/ \sum_{i=1}^{p} \lambda_i$$

of the total variance.

We can now consider a mapping to a reduced dimension by specifying that the new components must account for at least a fraction d of the total variance. The value of d would be specified by the user. We then choose k so that

$$\sum_{i=1}^{k} \lambda_i \geq d \sum_{i=1}^{p} \lambda_i \geq \sum_{i=1}^{k-1} \lambda_i$$

and transform the data to

$$\boldsymbol{\xi}_k = \boldsymbol{A}_k^T \boldsymbol{x}$$

where $\boldsymbol{\xi}_k = (\xi_1, \ldots, \xi_k)^T$ and $\boldsymbol{A}_k = [\boldsymbol{a}_1, \ldots, \boldsymbol{a}_k]$ is a $p \times k$ matrix. Choosing a value of d between 70% and 90% preserves most of the information in $\boldsymbol{x}$ (Jolliffe, 1986). Jackson (1991) advises against the use of this procedure: it is difficult to choose an appropriate value for d – it is very much problem-specific.

An alternative approach is to examine the eigenvalue spectrum and see if there is a point where the values fall sharply before levelling off at small values (the 'scree' test). We retain those principal components corresponding to the eigenvalues before the cut-off point or 'elbow' (see Figure 9.5). However, on occasion the eigenvalues drift downwards with no obvious cutting point and the first few eigenvalues account for only a small proportion of the variance.

It is very difficult to determine the 'right' number of components and most tests are for limited special cases and assume multivariate normality. Jackson (1991) describes a range of procedures and reports the results of several comparative studies.

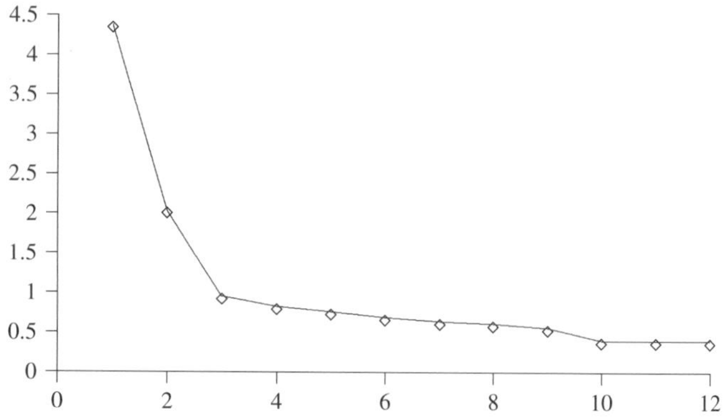

Figure 9.5 Eigenvalue spectrum

Remarks

Sampling The derivation of principal components above has assumed that the covariance matrix $\boldsymbol{\Sigma}$ is known. In most practical problems, we shall have an estimate of the sample covariance matrix from a set of sample vectors. We can use the sample covariance matrix to calculate principal components and take these as estimates of the eigenvectors of the covariance matrix $\boldsymbol{\Sigma}$. Note also, as far as deriving a reduced-dimension representation is concerned, the process is *distribution-free* – a mathematical method with no underlying statistical model. Therefore, unless we are prepared to assume some model for the data, it is difficult to obtain results on how good the estimates of the principal components are.

Standardisation The principal components are dependent on the scales used to measure the original variables. Even if the units used to measure the variables are the same, if one of the variables has a range of values that greatly exceeds the others, then we expect the first principal component to lie in the direction of this axis. If the units of each variable differ (for example, height, weight), then the principal components will depend on whether the height is measured in feet, inches or centimetres, etc. This does not occur in regression (which is independent of scale) but it does in principal components analysis in which we are minimising a perpendicular distance from a point to a line, plane, etc. and right angles do not transform to right angles under changes of scale. The practical solution to this problem is to standardise the data so that the variables have equal range. A common form of standardisation is to transform the data to have zero mean and unit variance, so that we find the principal components from the correlation matrix. This gives equal importance to the original variables. We recommend that all data are standardised to zero mean and unit variance. Other forms of standardisation are possible; for example, the data may be transformed logarithmically before a principal components analysis – Baxter (1995) compares several approaches.

Mean correction Equation (9.11) relates the principal components $\boldsymbol{\xi}$ to the observed random vector $\boldsymbol{x}$. In general, $\boldsymbol{\xi}$ will not have zero mean. In order for the principal components to have zero mean, they should be defined as

$$\boldsymbol{\xi} = \boldsymbol{A}^T(\boldsymbol{x} - \boldsymbol{\mu}) \tag{9.12}$$

for mean $\boldsymbol{\mu}$. In practice $\boldsymbol{\mu}$ is the sample mean, $\boldsymbol{m}$.

Approximation of data samples We have seen that in order to represent data in a reduced dimension, we retain only the first few principal components (the number is usually determined by the data). Thus, a data vector $\boldsymbol{x}$ is projected onto the first r (say) eigenvectors of the sample covariance matrix, giving

$$\boldsymbol{\xi}_r = \boldsymbol{A}_r^T(\boldsymbol{x} - \boldsymbol{\mu}) \tag{9.13}$$

where $\boldsymbol{A}_r = [\boldsymbol{a}_1, \ldots, \boldsymbol{a}_r]$ is the $p \times r$ matrix whose columns are the first r eigenvectors of the sample covariance matrix and $\boldsymbol{\xi}_r$ is used to denote the measurements on variables $\xi_1, \ldots, \xi_r$ (the first r principal components).

In representing a data point as a point in a reduced dimension, there is usually an error involved and it is of interest to know what the point $\boldsymbol{\xi}_r$ corresponds to in the original space.

The variable $\boldsymbol{\xi}$ is related to $\boldsymbol{x}$ by equation (9.12) ($\boldsymbol{A}^T = \boldsymbol{A}^{-1}$), giving

$$\boldsymbol{x} = \boldsymbol{A}\boldsymbol{\xi} + \boldsymbol{\mu}$$

If $\boldsymbol{\xi} = (\boldsymbol{\xi}_r, 0)^T$, a vector with its first r components equal to $\boldsymbol{\xi}_r$ and the remaining ones equal to zero, then the point corresponding to $\boldsymbol{\xi}_r$, namely $\boldsymbol{x}_r$, is

$$\begin{aligned}\boldsymbol{x}_r &= \boldsymbol{A}\begin{pmatrix}\boldsymbol{\xi}_r \\ 0\end{pmatrix} + \boldsymbol{\mu} \\ &= \boldsymbol{A}_r\boldsymbol{\xi}_r + \boldsymbol{\mu}\end{aligned}$$

and by (9.13),

$$\boldsymbol{x}_r = \boldsymbol{A}_r\boldsymbol{A}_r^T(\boldsymbol{x} - \boldsymbol{\mu}) + \boldsymbol{\mu}$$

The transformation $\boldsymbol{A}_r\boldsymbol{A}_r^T$ is of rank r and maps the original data distribution to a distribution that lies in an r-dimensional subspace (or on a *manifold* of dimension r) in $\mathbb{R}^p$. The vector $\boldsymbol{x}_r$ is the position the point $\boldsymbol{x}$ maps down to, given in the original coordinate system (the projection of $\boldsymbol{x}$ onto the space defined by the first r principal components); see Figure 9.6.

Singular value decomposition In Appendix C, the result is given that the right singular vectors of a matrix $\boldsymbol{Z}$ are the eigenvectors of $\boldsymbol{Z}^T\boldsymbol{Z}$. The sample covariance matrix (unbiased estimate) can be written in such a form

$$\frac{1}{n-1}\sum_{i=1}^{n}(\boldsymbol{x}_i - \boldsymbol{m})(\boldsymbol{x}_i - \boldsymbol{m})^T = \frac{1}{n-1}\tilde{\boldsymbol{X}}^T\tilde{\boldsymbol{X}}$$

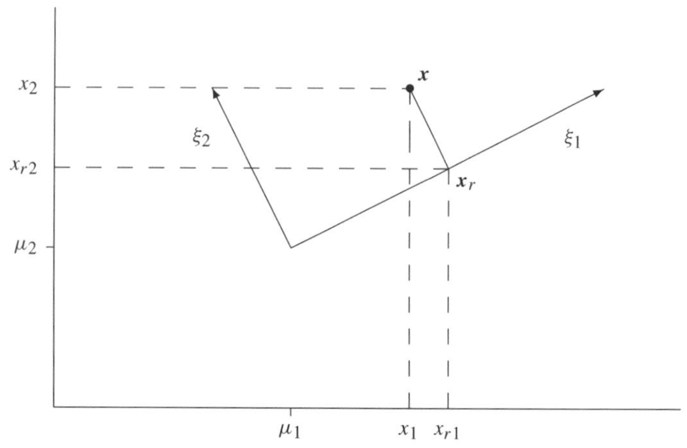

Figure 9.6 Reconstruction from projections: $\boldsymbol{x}$ is approximated by $\boldsymbol{x}_r$ using the first principal component

where $\tilde{X} = X - \mathbf{1}m^T$, X is the $n \times p$ data matrix, m is the sample mean and $\mathbf{1}$ is an n-dimensional vector of ones. Therefore, if we define

$$Z = \frac{1}{\sqrt{n-1}}\tilde{X} = \frac{1}{\sqrt{n-1}}(X - \mathbf{1}m^T)$$

then the right singular vectors of Z are the eigenvectors of the covariance matrix and the singular values are standard deviations of the principal components. Furthermore, setting

$$Z = \frac{1}{\sqrt{n-1}}\tilde{X}D^{-1} = \frac{1}{\sqrt{n-1}}(X - \mathbf{1}m^T)D^{-1}$$

where D is a $p \times p$ diagonal matrix with D_{ii} equal to the square root of the variance of the original variables, then the right singular vectors of Z are the eigenvectors of the *correlation* matrix. Thus, given a data matrix, it is not necessary to form the sample covariance matrix in order to determine the principal components.

If the singular value decomposition of $X - \mathbf{1}m^T$ is USV^T, where $U = [u_1, \ldots, u_p]$, $S = \text{diag}(\sigma_1, \ldots, \sigma_p)$ and $V = [v_1, \ldots, v_p]$, then the $n \times r$ matrix Z_r defined by

$$Z_r = U_r \Sigma_r V_r^T + \mathbf{1}m^T$$

where $U_r = [u_1, \ldots, u_r]$, $S_r = \text{diag}(\sigma_1, \ldots, \sigma_r)$ and $V_r = [v_1, \ldots, v_r]$, is the projection of the original data points onto the hyperplane spanned by the first r principal components and passing through the mean.

Selection of components There have been many methods proposed for the selection of components in a principal components analysis. There is no single best method as the strategy to adopt will depend on the objectives of the analysis: the set of components that gives a good fit to the data (in a predictive analysis) will differ from that which provides good discrimination (predictive analysis). Ferré (1995) and Jackson (1993) provide comparative studies. Jackson finds the *broken stick method* one of the most promising, and simple to implement. Observed eigenvectors are considered interpretable if their eigenvalues exceed a threshold based on random data: the kth eigenvector is retained if its eigenvalue λ_k exceeds $\sum_{i=k}^{p}(1/i)$. Prakash and Murty (1995) consider a genetic algorithm approach to the selection of components for discrimination. However, probably the most common approach is the percentage of variance method, retaining eigenvalues that account for approximately 90% of the variance. Another rule of thumb is to retain those eigenvectors with eigenvalues greater than the average (greater than unity for a correlation matrix). For a descriptive analysis, Ferré recommends a rule of thumb method. However, if we do use another approach then we must ask ourselves why we are performing a principal components analysis in the first place. There is no guarantee that a subset of the variables derived will be better for discrimination than a subset of the original variables.

Interpretation The first few principal components are the most important ones, but it may be very difficult to ascribe meaning to the components. One way this may be done is to consider the eigenvector corresponding to a particular component and select those variables for which the coefficients in the eigenvector are relatively large in magnitude. Then a purely subjective analysis takes place in which the user tries to see what these variables have in common.

An alternative approach is to use an algorithm for orthogonal rotation of axes such as the *varimax* algorithm (Kaiser, 1958, 1959). This rotates the given set of principal axes so that the variation of the squared loadings for a given component is large. This is achieved by making the loadings large on some variables and small on others, though unfortunately it does not necessarily lead to more easily interpretable principal components. Jackson (1991) considers techniques for the interpretation of principal components, including rotation methods.

Discussion

Principal components analysis is often the first stage in a data analysis and is often used to reduce the dimensionality of the data while retaining as much as possible of the variation present in the original dataset.

Principal components analysis takes no account of groups within the data (i.e. it is unsupervised). Although separate groups may emerge as a result of projecting data to a reduced dimension, this is not always the case and dimension reduction may obscure the existence of separate groups. Figure 9.7 illustrates a data set in two dimensions with two separate groups and the principal component directions. Projection onto the first eigenvector will remove group isolation, while projection onto the second retains group separation. Therefore, although dimension reduction may be necessary, the space spanned by the vectors associated with the first few principal components will not necessarily be the best for discrimination.

Summary

The stages in performing a principal components analysis are:

1. form the sample covariance matrix or standardise the data by forming the correlation matrix;

2. perform an eigendecomposition of the correlation matrix.

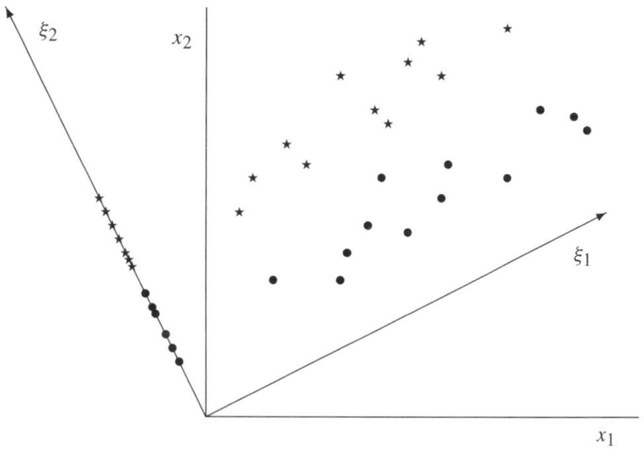

Figure 9.7 Two-group data and the principal axes

Alternatively:

1. standardise the data matrix;
2. perform a singular value decomposition of the standardised data matrix.

For a reduced-dimension representation of the data, project the data onto the first m eigenvectors, where, for example, m is chosen using a criterion based on the proportion of variance accounted for.

9.3.2 Karhunen–Loève transformation

A separate section on the Karhunen–Loève transformation, in addition to that on principal components analysis, may seem superfluous at first sight. After all, the Karhunen–Loève transformation is, in one of its most basic forms, identical to principal components analysis. It is included here because there are variants of the method, occurring in the pattern recognition literature under the general heading of Karhunen–Loève expansion, that incorporate class information in a way in which principal components analysis does not.

The Karhunen–Loève expansion was originally developed for representing a non-periodic random process as a series of orthogonal functions with uncorrelated coefficients. If $x(t)$ is a random process on $[0, T]$, then $x(t)$ may be expanded as

$$x(t) = \sum_{n=1}^{\infty} x_n \phi_n(t) \tag{9.14}$$

where the x_n are random variables and the basis functions ϕ are deterministic functions of time satisfying

$$\int_0^T \phi_n(t)\phi_m^*(t) = \delta_{mn}$$

where ϕ_m^* is the complex conjugate of ϕ_m. Define a correlation function

$$\begin{aligned} R(t,s) &= \mathrm{E}[x(t)x^*(s)] \\ &= \mathrm{E}\left[\sum_n \sum_m x_n x_m^* \phi_n(t)\phi_m^*(s)\right] \\ &= \sum_n \sum_m \phi_n(t)\phi_m^*(s)\mathrm{E}[x_n x_m^*] \end{aligned}$$

If the coefficients are uncorrelated ($\mathrm{E}[x_n x_m^*] = \sigma_n^2 \delta_{mn}$) then

$$R(t,s) = \sum_n \sigma_n^2 \phi_n(t)\phi_n^*(s)$$

and multiplying by $\phi_n(s)$ and integrating gives

$$\int R(t,s)\phi_n(s)ds = \sigma_n^2 \phi_n(t)$$

Thus, the functions ϕ_n are the eigenfunctions of the integral equation, with kernel $R(t, s)$ and eigenvalues σ_n^2. We shall not develop the continuous Karhunen–Loève expansion here but proceed straight away to the discrete case.

If the functions are uniformly sampled, with p samples, then (9.14) becomes

$$\boldsymbol{x} = \sum_{n=1}^{\infty} x_n \boldsymbol{\phi}_n \tag{9.15}$$

and the integral equation becomes

$$\boldsymbol{R}\boldsymbol{\phi}_k = \sigma_k^2 \boldsymbol{\phi}_k$$

where $\boldsymbol{R}$ is now the $p \times p$ matrix with (i, j)th element $R_{ij} = \mathrm{E}[x_i x_j^*]$. The above equation has only p distinct solutions for $\boldsymbol{\phi}$ and so the summation in (9.15) must be truncated to p. The eigenvectors of $\boldsymbol{R}$ are termed the *Karhunen–Loève coordinate axes* (Devijver and Kittler, 1982).

Apart from the fact that we have assumed zero mean for the random variable $\boldsymbol{x}$, this derivation is identical to that for principal components. Other ways of deriving the Karhunen–Loève coordinate axes are given in the literature, but these correspond to other views of principal components analysis, and so the end result is the same.

So where does the Karhunen–Loève transformation differ from principal coordinates analysis in a way which warrants its inclusion in this book? Strictly it does not, except that in the pattern recognition literature various methods for linearly transforming data to a reduced-dimension space defined by eigenvectors of a matrix of second-order statistical moments have been proposed under the umbrella term 'Karhunen–Loève expansion'. These methods could equally be referred to using the term 'generalised principal components' or something similar.

The properties of the Karhunen–Loève expansion are identical to those of the principal components analysis. It produces a set of mutually uncorrelated components, and dimensionality reduction can be achieved by selecting those components with the largest variances. There are many variants of the basic method that incorporate class information or that use different criteria for selecting features.

KL1: SELFIC – Self-featuring information-compression

In this procedure, class labels are not used and the Karhunen–Loève feature transformation matrix is $\boldsymbol{A} = [\boldsymbol{a}_1, \ldots, \boldsymbol{a}_p]$, where $\boldsymbol{a}_j$ are the eigenvectors of the sample covariance matrix, $\hat{\boldsymbol{\Sigma}}$, associated with the largest eigenvalues (Watanabe, 1985),

$$\hat{\boldsymbol{\Sigma}}\boldsymbol{a}_i = \lambda_i \boldsymbol{a}_i$$

and $\lambda_1 \geq \cdots \geq \lambda_p$.

This is identical to principal components analysis and is appropriate when class labels are unavailable (unsupervised learning).

KL2: Within-class information

If class information is available for the data, then second-order statistical moments can be calculated in a number of different ways. This leads to different Karhunen–Loève coordinate systems. Chien and Fu (1967) propose using the average within-class covariance matrix, $\boldsymbol{S}_W$, as the basis of the transformation. The feature transformation matrix, $\boldsymbol{A}$, is again the matrix of eigenvectors of $\boldsymbol{S}_W$ associated with the largest eigenvalues.

KL3: Discrimination information contained in the means

Again, the feature space is constructed from eigenvectors of the averaged within-class covariance matrix, but discriminatory information contained in the class means is used to select the subset of features that will be used in further classification studies (Devijver and Kittler, 1982). For each feature (with eigenvector $\boldsymbol{a}_j$ of $\boldsymbol{S}_W$ and corresponding eigenvalue λ_j) the quantity

$$J_j = \frac{\boldsymbol{a}_j^T \boldsymbol{S}_B \boldsymbol{a}_j}{\lambda_j}$$

where $\boldsymbol{S}_B$ is the between-class scatter matrix, is evaluated and the coordinate axes arranged in descending order of J_j.

KL4: Discrimination information contained in the variances

Another means of ordering the feature vectors (eigenvectors $\boldsymbol{a}_j$ of $\boldsymbol{S}_W$) is to use the discriminatory information contained in class variances (Kittler and Young, 1973). There are situations where class mean information is not sufficient to separate the classes and the measure given here uses the dispersion of class-conditional variances. The variance of feature j in the ith class weighted by the prior probability of class ω_i is given by

$$\lambda_{ij} = p(\omega_i)\boldsymbol{a}_j^T \hat{\boldsymbol{\Sigma}}_i \boldsymbol{a}_j$$

where $\hat{\boldsymbol{\Sigma}}_i$ is the sample covariance matrix of class ω_i, and, defining $\lambda_j = \sum_{i=1}^{C} \lambda_{ij}$, then a discriminatory measure based on the logarithmic entropy function is

$$H_j = -\sum_{i=1}^{C} \frac{\lambda_{ij}}{\lambda_j} \log\left(\frac{\lambda_{ij}}{\lambda_j}\right)$$

The axes giving low entropy values are selected for discrimination.

A further measure that uses the variances is

$$J_j = \prod_{i=1}^{C} \frac{\lambda_{ij}}{\lambda_j}$$

Both of the above measures reach their maximum when the factors λ_{ij}/λ_j are identical, in which case there is no discriminatory information.

KL5: Compression of discriminatory information contained in class means

In the method KL3, the Karhunen–Loève coordinate axes are determined by the eigenvectors of the averaged within-class covariance matrix and the features that are used to represent these data in a reduced-dimension space are determined by ordering the eigenvectors in terms of descending J_j. The quantity J_j is a measure used to represent the discriminatory information contained in the class means. This discriminatory information could be spread over all Karhunen–Loève axes and it is difficult to choose an optimum dimension for the transformed feature space from the values of J_j alone.

The approach considered here (Kittler and Young, 1973) recognises the fact that in a C-class problem, the class means lie in a space of dimension at most $C - 1$, and seeks to find a transformation to a space of at most $C - 1$ giving uncorrelated features.

This is performed in two stages. First of all, a transformation is found that transforms the data to a space in which the averaged within-class covariance matrix is diagonal. This means that the features in the transformed space are uncorrelated, but also any further orthonormal transformation will still produce uncorrelated features for which the *class-centralised vectors* are decorrelated.

If $\boldsymbol{S}_W$ is the average within-class covariance matrix in the original data space, then the transformation that decorrelates the class-centralised vectors is $\boldsymbol{Y} = \boldsymbol{U}^T\boldsymbol{X}$, where $\boldsymbol{U}$ is the matrix of eigenvectors of $\boldsymbol{S}_W$ and the average within-class covariance matrix in the transformed space is

$$\boldsymbol{S}'_W = \boldsymbol{U}^T\boldsymbol{S}_W\boldsymbol{U} = \boldsymbol{\Lambda}$$

where $\boldsymbol{\Lambda} = \text{diag}(\lambda_1, \ldots, \lambda_n)$ is the matrix of variances of the transformed features (eigenvalues of $\boldsymbol{S}_W$). If the rank of $\boldsymbol{S}_W$ is less than p (equal to r, say), then the first stage of dimension reduction is to transform to the r-dimensional space by the transformation $\boldsymbol{U}_r^T\boldsymbol{X}$, $\boldsymbol{U}_r = [\boldsymbol{u}_1, \ldots, \boldsymbol{u}_r]$, so that

$$\boldsymbol{S}'_W = \boldsymbol{U}_r^T\boldsymbol{S}_W\boldsymbol{U}_r = \boldsymbol{\Lambda}_r$$

where $\boldsymbol{\Lambda}_r = \text{diag}(\lambda_1, \ldots, \lambda_r)$. If we wish the within-class covariance matrix to be invariant to further orthogonal transformations, then it should be the identity matrix. This can be achieved by the transformation $\boldsymbol{Y} = \boldsymbol{\Lambda}_r^{-\frac{1}{2}}\boldsymbol{U}_r^T\boldsymbol{X}$ so that

$$\boldsymbol{S}'_W = \boldsymbol{\Lambda}_r^{-\frac{1}{2}}\boldsymbol{U}_r^T\boldsymbol{S}_W\boldsymbol{U}_r\boldsymbol{\Lambda}_r^{-\frac{1}{2}} = \boldsymbol{I}$$

This is the first stage of dimension reduction, and is illustrated in Figure 9.8. It transforms the data so that the average within-class covariance matrix in the new space is the identity matrix. In the new space the between-class covariance matrix, $\boldsymbol{S}'_B$, is given by

$$\boldsymbol{S}'_B = \boldsymbol{\Lambda}_r^{-\frac{1}{2}}\boldsymbol{U}_r^T\boldsymbol{S}_B\boldsymbol{U}_r\boldsymbol{\Lambda}_r^{-\frac{1}{2}}$$

where $\boldsymbol{S}_B$ is the between-class covariance matrix in the data space. The second stage of the transformation is to compress the class mean information; i.e. find the orthogonal transformation that transforms the class mean vectors to a reduced dimension. This transformation, $\boldsymbol{V}$, is determined by the eigenvectors of $\boldsymbol{S}'_B$

$$\boldsymbol{S}'_B\boldsymbol{V} = \boldsymbol{V}\tilde{\boldsymbol{\Lambda}}$$

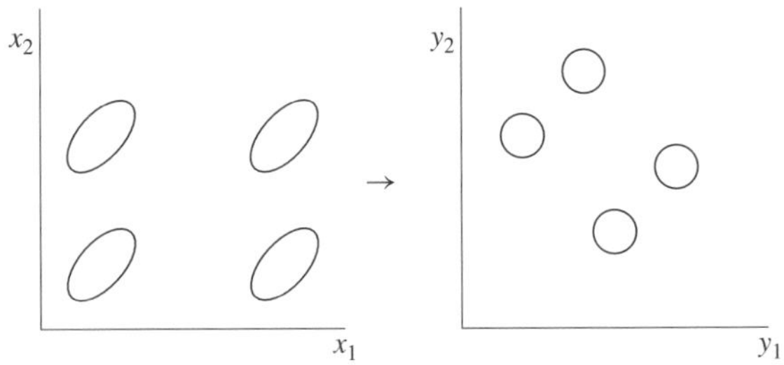

Figure 9.8 Illustration of the first stage of dimension reduction for four groups represented by contours $\boldsymbol{x}^T\hat{\boldsymbol{\Sigma}}_i^{-1}\boldsymbol{x} = \text{constant}$

where $\tilde{\boldsymbol{\Lambda}} = \text{diag}(\tilde{\lambda}_1, \ldots, \tilde{\lambda}_r)$ is the matrix of eigenvalues of $\boldsymbol{S}'_B$. There are at most $C - 1$ non-zero eigenvalues and so the final transformation is $\boldsymbol{Z} = \boldsymbol{V}_\nu^T \boldsymbol{Y}$, where $\boldsymbol{V}_\nu = [\boldsymbol{v}_1, \ldots, \boldsymbol{v}_\nu]$ and ν is the rank of $\boldsymbol{S}'_B$. The optimal feature extractor is therefore

$$\boldsymbol{Z} = \boldsymbol{A}^T \boldsymbol{X}$$

where the $p \times \nu$ linear transformation $\boldsymbol{A}$ is given by

$$\boldsymbol{A}^T = \boldsymbol{V}_\nu^T \boldsymbol{\Lambda}_r^{-\frac{1}{2}} \boldsymbol{U}_r^T$$

In this transformed space

$$\boldsymbol{S}''_W = \boldsymbol{V}_\nu^T \boldsymbol{S}'_W \boldsymbol{V}_\nu = \boldsymbol{V}_\nu^T \boldsymbol{V}_\nu = \boldsymbol{I}$$
$$\boldsymbol{S}''_B = \boldsymbol{V}_\nu^T \boldsymbol{S}'_B \boldsymbol{V}_\nu = \tilde{\boldsymbol{\Lambda}}_\nu$$

where $\tilde{\boldsymbol{\Lambda}}_\nu = \text{diag}(\tilde{\lambda}_1, \ldots, \tilde{\lambda}_\nu)$. Thus, the transformation makes the average within-class covariance matrix equal to the identity and the between-class covariance matrix equal to a diagonal matrix (see Figure 9.9). Usually, all $C - 1$ features are selected, but these can be ordered according to the magnitude of $\tilde{\lambda}_i$ and those with largest eigenvalues selected. The linear transformation can be found by performing two eigenvector decompositions, first of $\boldsymbol{S}_W$ and then of $\boldsymbol{S}'_B$, but an alternative approach based on a QR factorisation can be used (Crownover, 1991).

This two-stage process gives a geometric interpretation of linear discriminant analysis, described in Chapter 4. The feature vectors (columns of the matrix $\boldsymbol{A}$) can be shown to

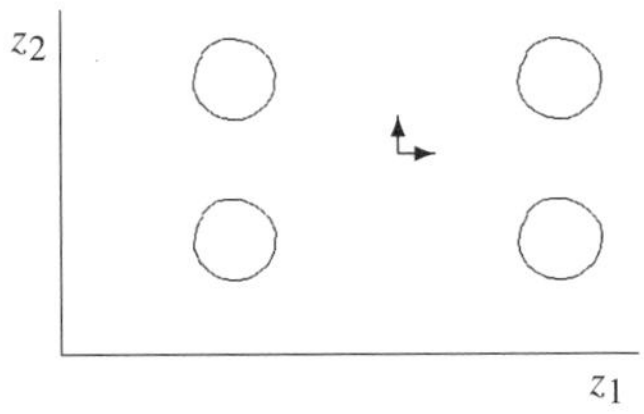

Figure 9.9 Second stage of dimension reduction: orthogonal rotation and projection to diagonalise the between-class covariance matrix

be eigenvectors of the generalised symmetric eigenvector equation (Devijver and Kittler, 1982)

$$\boldsymbol{S}_B \boldsymbol{a} = \lambda \boldsymbol{S}_W \boldsymbol{a}$$

The first stage of the transformation is simply a rotation of the coordinate axes followed by a scaling (assuming that $\boldsymbol{S}_W$ is full rank). The second stage comprises a projection of the data onto the hyperplane defined by the class means in this transformed space (of dimension at most $C - 1$), followed by a rotation of the axes. If we are to use all $C - 1$ coordinates subsequently in a classifier, then this final rotation is irrelevant since any classifier that we construct should be independent of an orthogonal rotation of the axes. Any set of orthogonal axes within the space spanned by the eigenvectors of $\boldsymbol{S}'_B$ with nonzero eigenvalues could be used. We could simply orthogonalise the vectors $\boldsymbol{m}'_1 - \boldsymbol{m}', \ldots, \boldsymbol{m}'_{C-1} - \boldsymbol{m}'$, where $\boldsymbol{m}'_i$ is the mean of class ω_i in the space defined by the first transformation and $\boldsymbol{m}'$ is the overall mean. However, if we wish to obtain a reduced-dimension display of the data, then it is necessary to perform an eigendecomposition of $\boldsymbol{S}'_B$ to obtain a set of coordinate axes that can be ordered using the values of $\tilde{\lambda}_i$, the eigenvalues of $\boldsymbol{S}'_B$.

Example

Figures 9.10 and 9.11 give two-dimensional projections for simulated oil pipeline data. This synthetic data set models non-intrusive measurements on a pipeline transporting a mixture of oil, water and gas. The flow in the pipe takes one out of three possible configurations: horizontally stratified, nested annular or homogeneous mixture flow. The data lie in a 12-dimensional measurement space. Figure 9.11 shows that the Karhunen–Loève transformation, KL5, separates the three classes into (approximately) three spherical clusters. The principal components projection (Figure 9.10) does not separate the classes, but retains some of the structure (for example, class 3 – denoted □ – comprises several subgroups).

Discussion

All of the above methods have certain features in common. All produce a linear transformation to a reduced-dimension space. The transformations are determined using an eigenvector decomposition of a matrix of second-order statistical moments and produce features or components in the new space that are uncorrelated. The features in the new space can be ordered using a measure of discriminability (in the case of labelled data) or approximation error. These methods are summarised in Table 9.1.

We could add to this list the method of common principal components that also determines a coordinate system using matrices of second-order statistical moments. A reduced-dimension representation of the data could be achieved by ordering the principal components using a measure like H_j or J_j in Table 9.1.

The final method (KL5), derived from a geometric argument, produces a transformation that is identical to the linear discriminant transformation (see Chapter 4), obtained by maximising a discriminability criterion. It makes no distributional assumptions, but a nearest class mean type rule will be optimal if normal distributions are assumed for the classes.

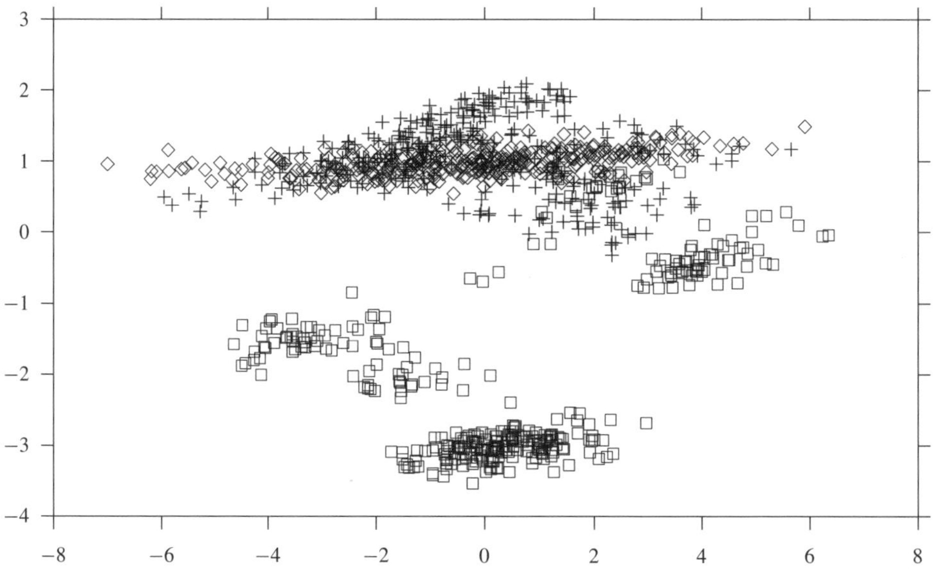

Figure 9.10 Projection onto the first two principal components for the oil pipeline data

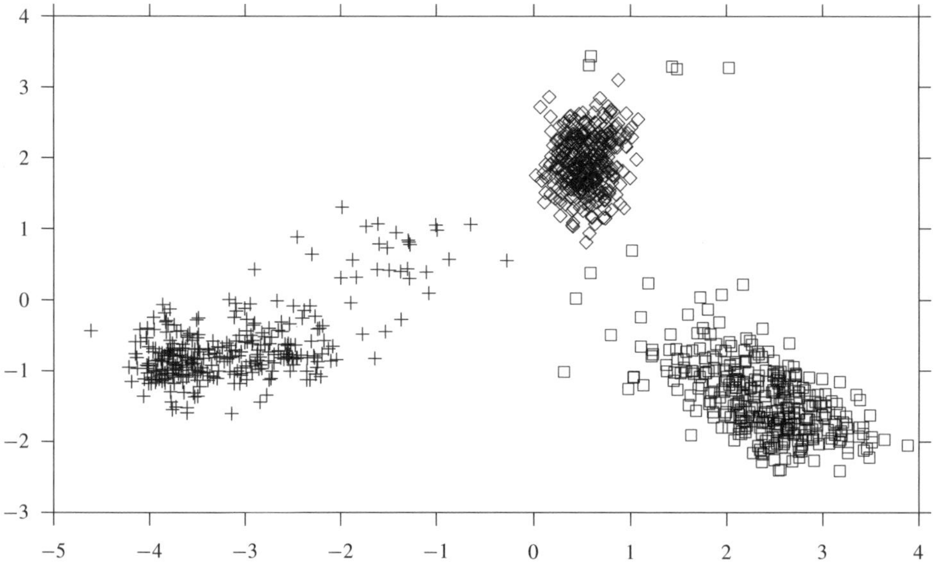

Figure 9.11 Projection onto the two linear discriminant directions for the oil pipeline data

9.3.3 Factor analysis

Factor analysis is a multivariate analysis technique that aims to represent a set of variables in terms of a smaller underlying set of variables called *factors*. Its origin dates back a full the century to work by Charles Spearman, who was concerned with understanding

Table 9.1 Summary of linear transformation methods of feature extraction

Method	Eigenvector decomposition matrix	Ordering function
KL1 (PCA)	$\hat{\boldsymbol{\Sigma}}$	λ_j
KL2	$\boldsymbol{S}_W$	λ_j
KL3	$\boldsymbol{S}_W$	$\boldsymbol{a}_j^T \boldsymbol{S}_B \boldsymbol{a}_j / \lambda_j$
KL4 – (a)	$\boldsymbol{S}_W$	$H_j = -\sum_{i=1}^{C} \frac{\lambda_{ij}}{\lambda_j} \log\left(\frac{\lambda_{ij}}{\lambda_j}\right)$
KL4 – (b)	$\boldsymbol{S}_W$	$J_j = \boldsymbol{a}_j^T \boldsymbol{S}_B \boldsymbol{a}_j / \lambda_j$
KL5	$\boldsymbol{S}_W$, $\boldsymbol{S}_B$	λ_j

intelligence, and the technique was developed for analysing the scores of individuals on a number of aptitude tests. Factor analysis has been developed mainly by psychologists and most applications have been in the areas of psychology and the social sciences though others include medicine, geography and meteorology.

Factor analysis is perhaps the most controversial of the multivariate methods and many of the drawbacks to the approach are given by Chatfield and Collins (1980) who recommend that it should not be used in most practical situations. One criticism has been of the subjectivity involved in interpreting the results of a factor analysis and that one of the reasons for the popularity of the method is that 'it allows the experimenter to impose his preconceived ideas on the data' (Reyment *et al.*, 1984, Chapter 10). It has also been suggested that factor analysis produces a useful result only in cases where a principal component analysis would have yielded the same result. Whether factor analysis is worth the time to understand it and to carry it out (Hills, 1977) is something for you to judge, but it is a technique which has a considerable amount of support among statisticians.

Given these objections (and there are many more – we shall list some of the drawbacks later on), why do we include a section on factor analysis? Can we not be accused of supporting the myth that factor analysis is actually of some use? Certainly we can; there may be other tools that are more reliable than factor analysis and give the same results. Nevertheless, factor analysis may be appropriate for some particular problems. Another reason for including a short section on factor analysis is to highlight the differences from principal components analysis.

Factor analysis, like principal components analysis, is a *variable-directed* technique and has sometimes been confused with principal components analysis. Even though both techniques often yield solutions that are very similar, factor analysis differs from principal components analysis in several important respects. The main one is that, whilst principal components analysis is concerned with determining new variables that account of the maximum variance of the observed variables, in factor analysis the factors are chosen to account for the correlations between variables, rather than the variance. Another difference is that in principal components, the new variables (the principal components) are expressed as a linear function of the observed variables, whilst in factor analysis the observed variables are expressed as a linear combination of the unobserved underlying variables or factors. Further differences and similarities will be made clear in the analysis of the following sections.

The factor analysis model

Suppose that we have p variables $x_1, \ldots, x_p$. The factor analysis model (Harman, 1976) assumes that each variable consists of two parts: a *common* part and a *unique* part. Specifically, we assume that there are m underlying (or latent) variables or *factors* $\xi_1, \ldots, \xi_m$, so that

$$x_i - \mu_i = \sum_{k=1}^{m} \lambda_{ik}\xi_k + \epsilon_i \qquad i = 1, \ldots, p \tag{9.16}$$

where $\boldsymbol{\mu}$ is the mean of the vector $(x_1, \ldots, x_p)^T$. Without loss of generality, we shall take $\boldsymbol{\mu}$ to be zero. The variables ξ_k are sometimes termed the *common factors* since they contribute to all observed variables x_i, and ϵ_i are the *unique* or *specific factors*, describing the residual variation specific to the variable x_i. The weights λ_{ik} are the *factor loadings*. Equation (9.16) is usually written as $\boldsymbol{x} = \boldsymbol{\Lambda}\boldsymbol{\xi} + \boldsymbol{\epsilon}$.

In addition, the basic model makes the following assumptions

1. The specific factors, ϵ_i, are uncorrelated with each other and with the common factors, ξ_k; that is, $\mathrm{E}[\boldsymbol{\epsilon}\boldsymbol{\epsilon}^T] = \boldsymbol{\Psi} = \mathrm{diag}(\psi_1, \ldots, \psi_p)$ and $\mathrm{E}[\boldsymbol{\epsilon}\boldsymbol{\xi}^T] = 0$.

2. The common factors have zero mean and unit variance. We may make the unit variance assumption since the columns of the $p \times m$ matrix $\boldsymbol{\Lambda}$, with (i, k)th element λ_{ik}, may be scaled arbitrarily.

With these assumptions we model the data covariance matrix $\boldsymbol{\Sigma}$ as

$$\boldsymbol{\Sigma} = \boldsymbol{\Lambda}\boldsymbol{\Phi}\boldsymbol{\Lambda}^T + \boldsymbol{\Psi}$$

If, further, we assume that the common factors themselves are independent of one another, then $\boldsymbol{\Phi} = \boldsymbol{I}$ and

$$\boldsymbol{\Sigma} = \boldsymbol{\Lambda}\boldsymbol{\Lambda}^T + \boldsymbol{\Psi} \tag{9.17}$$

Equation (9.17) expresses the covariance matrix of the observed variables as the sum of two terms: a diagonal matrix $\boldsymbol{\Psi}$ and a matrix $\boldsymbol{\Lambda}\boldsymbol{\Lambda}^T$ that accounts for the covariance in the observed variables. In practice, we use the sample covariance matrix or the sample correlation matrix $\boldsymbol{R}$ in place of $\boldsymbol{\Sigma}$. However, a factorisation of the form (9.17) does not necessarily exist, and even if it does it will not be unique. If $\boldsymbol{T}$ is an $m \times m$ orthogonal matrix, then

$$(\boldsymbol{\Lambda}\boldsymbol{T})(\boldsymbol{\Lambda}\boldsymbol{T})^T = \boldsymbol{\Lambda}\boldsymbol{\Lambda}^T$$

Thus, if $\boldsymbol{\Lambda}$ is a solution for the factor loadings, then so is $\boldsymbol{\Lambda}\boldsymbol{T}$, and even though these matrices are different, they can still generate the same covariance structure. Hence, it is always possible to rotate the solution to an alternative 'better' solution. The matrix $\boldsymbol{\Lambda}\boldsymbol{\Lambda}^T$ also contributes to the variance of the observed variables,

$$\mathrm{var}(x_i) = \sum_{k=1}^{m} \lambda_{ik}^2 + \mathrm{var}(\epsilon_i)$$

with the contribution to the variance of the variable x_i due to the factor loadings being termed the *common variance* or *communality* and $\text{var}(\epsilon_i)$ the *unique variance* of x_i.

In addition to the parameters of $\mathbf{\Lambda}$ and $\mathbf{\Phi}$ that must be obtained, the number of factors is not usually known in advance, though the experimenter may wish to find the smallest value of m for which the model fits the data. In practice, this is usually done by increasing m sequentially until a sufficient fit is obtained.

The main theme of this chapter is that of data reduction. In principal components analysis we may project a data vector onto the first few eigenvectors of the data covariance matrix to give the *component scores* (equation (9.13))

$$\boldsymbol{\xi}_r = \boldsymbol{A}_r^T(\boldsymbol{x} - \boldsymbol{\mu})$$

In factor analysis, the situation is different. The basic equation expresses the observed variables $\boldsymbol{x}$ in terms of the underlying variables $\boldsymbol{\xi}$

$$\boldsymbol{x} = \mathbf{\Lambda}\boldsymbol{\xi} + \boldsymbol{\epsilon}$$

For $m < p$, it is not possible to invert this to express $\boldsymbol{\xi}$ in terms of $\boldsymbol{x}$ and hence calculate *factor scores*. There are methods for estimating factor scores for a given individual measurement $\boldsymbol{x}$, and these will be considered later in this section. However, this is yet one more difficulty which makes factor analysis less straightforward to use than principal components analysis.

We summarise the basic factor analysis approach in the following steps:

1. Given a set of observations, calculate the sample covariance matrix and perform a factor analysis for a specified value of m, the number of factors.

2. Carry out a hypothesis test to see if the data fit the model (Dillon and Goldstein, 1984). If they do not, return to step 1.

3. Rotate the factors to give as maximum loading on as few factors as possible for each variable.

4. (This step may be omitted if inappropriate.) Group variables under each factor and interpret the factors.

5. Estimate the factor scores, giving a representation of the data in a reduced dimension.

Techniques for performing each of these steps will be discussed in subsequent sections. We conclude this section with a summary in Table 9.2 of the main differences between principal components analysis and factor analysis.

Factor solutions

In this section, we present two of the available methods of factor extraction. There are other methods available, and a summary of the properties of these different techniques may be found in texts on multivariate analysis (for example, Dillon and Goldstein, 1984). In general, these different methods will give different solutions that depend on various properties of the data, including data sample size, the number of observation variables

Table 9.2 Comparison between factor analysis and principal components analysis

Principal components analysis	Factor analysis
orthogonal set of vectors	not orthogonal
can easily determine component scores	factor scores must be estimated
components are unique	factors are not unique
variable-directed	variable-directed
no underlying statistical model	model for covariance or correlation matrix
explains variance structure	covariance structure
pointless if observed variables are uncorrelated	pointless if observed variables are uncorrelated
scale-dependent	maximum likelihood estimation overcomes scaling problem
nested solutions	not nested solutions

and the magnitude of their communalities. However, for a large number of observations and variables, these methods tend to give similar results, though at the other extreme it is not so clear which is the 'best' method.

The *principal factor method* chooses the first factor to account for the largest possible amount of the total communality. The second factor is chosen to account for as much as possible of the remaining total communality, and so on. This is equivalent to finding the eigenvalues and eigenvectors of the reduced correlation matrix, $\boldsymbol{R}^*$, defined as the correlation matrix $\boldsymbol{R}$ with diagonal elements replaced by the communalities. This is the matrix to factor:

$$\boldsymbol{R}^* = \boldsymbol{R} - \boldsymbol{\Psi} = \boldsymbol{\Lambda\Phi\Lambda}$$

This of course assumes that we know the communalities. There are several methods for estimating the communalities. One is to estimate the communality of the ith variable by the squared multiple correlation of the variable X_i with the remaining $p - 1$ variables.

Once the communalities have been estimated we may perform an eigenanalysis of $\boldsymbol{R}^*$ and determine $\boldsymbol{\Lambda}$. This may then be used (for a predetermined number of factors) to estimate new communalities and a new $\boldsymbol{R}^*$ calculated. The procedure iterates until convergence of communality estimates. However, there is no guarantee that the procedure will converge. Also, it could lead to negative estimates of the specific variances.

The principal factor method, like principal components analysis, makes no assumption as to the underlying distribution of the data. The maximum likelihood method assumes a multivariate normal distribution with covariance matrix $\boldsymbol{\Sigma}$. The sample covariance matrix $\boldsymbol{S}$ is Wishart-distributed and the log-likelihood function we seek to maximise is

$$\log(L) = C - \frac{n}{2}\{\log|\boldsymbol{\Sigma}| + \mathrm{Tr}(\boldsymbol{\Sigma}^{-1}\boldsymbol{S})\}$$

where C is a constant independent of $\mathbf{\Sigma}$. This is equivalent to minimising

$$M = \mathrm{Tr}(\mathbf{\Sigma}^{-1}\boldsymbol{S}) - \log|\mathbf{\Sigma}^{-1}|$$

with respect to the parameters ($\mathbf{\Lambda}$, $\mathbf{\Psi}$ and $\mathbf{\Phi}$) of $\mathbf{\Sigma} = \mathbf{\Lambda\Phi\Lambda} + \mathbf{\Psi}$.

We shall not discuss the details of the method. These may be found in (Lawley and Maxwell, 1971; Jöreskog, 1977). However, we do note two advantages of the maximum likelihood method of estimation over the principal factor method. The first is that the maximum likelihood method enables statistical tests of significance for the number of factors to be carried out. The second advantage is that the estimates of the factor loadings are scale-invariant. Therefore we do not have the problem of having to standardise the variables as we did in principal components analysis. For example, the factor loadings yielded by an analysis of the sample correlation matrix differ from those of the sample covariance matrix of the data by a factor $1/\sqrt{s_{ii}}$.

Rotation of factors

We showed earlier that the model is invariant to an orthogonal transformation of the factors. That is, we may replace the matrix of factor loadings, $\mathbf{\Lambda}$, by $\mathbf{\Lambda}\boldsymbol{T}$, where $\boldsymbol{T}$ is a $m \times m$ orthogonal matrix, without changing the approximation of the model to the covariance matrix. This presents us with a degree of flexibility that allows us to rotate factors and may help in their interpretation.

The main aim of rotation techniques is to rotate the factors so that the variables have high loadings on a small number of factors and very small loadings on the remaining factors. There are two basic approaches to factor rotation: orthogonal rotation in which the transformed factors are still independent (so that the factor axes are perpendicular after rotation), and oblique rotation in which the factors are allowed to become correlated (and the factor axes are not necessarily orthogonal after rotation). An example of the former type is the *varimax* rotation (Kaiser, 1958, 1959). There are various methods for oblique rotation, including *promax* and *oblimin* (Nie *et al.*, 1975).

Estimating the factor scores

In principal components analysis, the principal components are obtained by a linear transformation of the original variables. This linear transformation is determined by the eigenvectors of the covariance matrix or the correlation matrix. Hence it is straightforward to obtain a reduced-dimension representation of an observation. It is also possible to 'invert' the process and obtain an approximation to the original variables given a subset of the principal components. In factor analysis, the original variables are described in terms of the factors (see Figure 9.12) plus an error term. Therefore, to obtain the factor scores this process must be reversed. There are several methods for obtaining the factor scores (Jackson, 1991). One is based on multiple regression analysis (Dillon and Goldstein, 1984). We suppose that the factor scores are a linear combination, denoted by the matrix $\boldsymbol{A}$, of the vector of observations, $\boldsymbol{x} = (x_1, \ldots, x_p)^T$,

$$\boldsymbol{\xi} = \boldsymbol{A}^T\boldsymbol{x}$$

Post-multiplying by $\boldsymbol{x}^T$, taking expectations and noting that $\mathrm{E}[\boldsymbol{x}\boldsymbol{x}^T] = \boldsymbol{R}$, $\mathrm{E}[\boldsymbol{\xi}\boldsymbol{x}^T] = \mathbf{\Lambda}^T$ gives

$$\mathbf{\Lambda}^T = \boldsymbol{A}^T\boldsymbol{R} \text{ or } \boldsymbol{A}^T = \mathbf{\Lambda}^T\boldsymbol{R}^{-1}$$

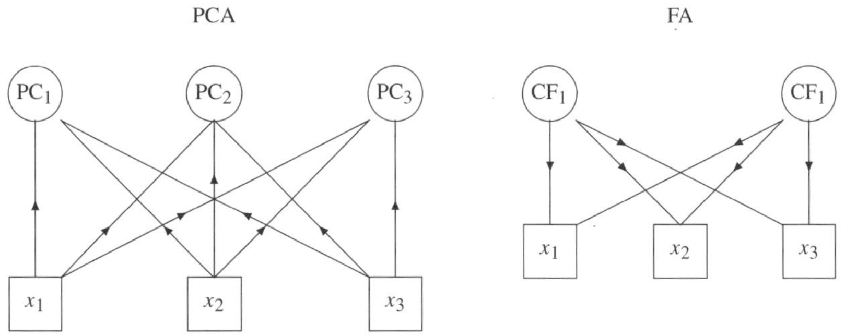

Figure 9.12 Factor analysis and principal components analysis models: PC – principal component variables; CF – common factors

so that an estimate of the factor scores, $\hat{\boldsymbol{\xi}}$ is

$$\hat{\boldsymbol{\xi}} = \boldsymbol{\Lambda R}^{-1}\boldsymbol{x}$$

In practice, the matrix $\boldsymbol{R}$ would be replaced by its sample-based estimate.

How many factors?

So far, we have discussed techniques for estimating the factor loadings and factor scores, but we have not answered the very important question as to how many factors to choose. For the principal factor method, a simple criterion is to choose the number of factors to be equal to the number of eigenvalues of the reduced correlation matrix that are greater than unity.

With the maximum likelihood solution, a more formal procedure may be applied. The null hypothesis is that all the population variance has been extracted by the postulated number of factors. The test statistic is (Everitt and Dunn, 1991)

$$\left(\log\frac{|\boldsymbol{\Sigma}|}{|\boldsymbol{S}|} + \text{Tr}\{\boldsymbol{S\Sigma}^{-1}\} - p\right)\left(n - 1 - \frac{1}{6}(2p+5) - \frac{2}{3}m\right)$$

which is asymptotically distributed as χ^2 with $((p-m)^2 - p - m)/2$ degrees of freedom. If, at a specified probability level, the value of χ^2 is significant, then we conclude that more factors are needed to account for the correlations in the observed variables.

Discussion

Factor analysis is a very widely used multivariate technique that attempts to discover the relationships between a set of observed variables by expressing those variables in terms of a set of underlying, unobservable variables or factors. As we have presented it, it is an exploratory data analysis technique, though another form that we shall not describe is confirmatory factor analysis. This is used when the experimenter wishes to see if the data fit a particular model.

Factor analysis is a much criticised method and some of the objections to its use are:

1. It is complicated. It has been suggested that in many of the situations when factor analysis gives reasonable results, it is only because it is simulating a principal components analysis, which would be simpler to perform.

2. It requires a large number if assumptions. Even the concept of underlying, unobservable variables may be questionable in many applications.

3. There is no unique solution. There are many methods of obtaining the factor loadings. Once these have been obtained, the factors may be rotated, giving yet different solutions and a different interpretation. This is where factor analysis has been heavily criticised as the experimenter can be accused of rotating the factors until the factors that are sought are arrived at.

4. As a method of dimension reduction it has the disadvantage that the factor scores are not easily obtained. Unlike in principal components analysis where the component scores may be obtained as a linear function of the observed variable values, the basic factor analysis equation cannot be inverted.

5. The number of factors is unknown and a test must be carried out to see if the assumed model is 'correct'. Also, because of the arbitrary rotation of the factors, the solution for the factors (as determined by the loadings) for an m-factor model is not a subset of the solution for the factors for an $(m + 1)$-factor model (i.e. the solutions are not nested).

The main point to make is that the model should not be taken too seriously. We would recommend that in most practical situations it is better to use another multivariate analysis technique. For data reduction, leading to subsequent analysis, we recommend that if you require an unsupervised linear technique then you should use principal components analysis. It is much simpler and will do just as well.

Principal components analysis and factor analysis are not the same procedures although there has been some confusion between the two types of analysis, and this is not helped by some statistical software packages (a description of these packages is given by Jackson, 1991). Sometimes the terms 'factor analysis' and 'principal components analysis' are used synonymously.

9.3.4 Example application study

The problem Monitoring changes in land use for planners and government officials (Li and Yeh, 1998).

Summary The application of remote sensing to inventory land resources and to evaluate the impacts of urban developments is addressed. Remote sensing is considered as a fast and efficient means of assessing such developments when detailed 'ground truth' data are unavailable. The aim is to determine the type, amount and location of land use change.

The data The data consist of satellite images, measured in five wavebands, from two images of the same region measured five years apart. A 10-dimensional feature vector is constructed (consisting of pixel measurements in the five wavebands over the two images) and data gathered over the whole region.

The model A standard principal components analysis is performed.

Training procedure Each variable is *standardised* to zero mean and unit variance and a principal components analysis performed. The first few principal components account for 97% of the variance. The data are projected onto the subspace characterised by the principal components, and subregions (identified from a compressed PCA image) labelled manually according to land use change (16 classes of land use, determined by field data).

9.3.5 Further developments

There are various generalisations of the basic linear approaches to feature extraction described in this section. Principal components analysis is still an interesting and active area of research. Common principal components analysis (Flury, 1988), is a generalisation of principal components analysis to the multigroup situation. The common principal components model assumes that the covariance matrices of each group or class, $\mathbf{\Sigma}_i$, can be written

$$\mathbf{\Sigma}_i = \boldsymbol{\beta}\mathbf{\Lambda}_i\boldsymbol{\beta}^T$$

where $\mathbf{\Lambda}_i = \text{diag}(\lambda_{i1}, \ldots, \lambda_{ip})$, the diagonal matrix of eigenvalues of the ith group. Thus, the eigenvectors $\boldsymbol{\beta}$ are common between groups, but the $\mathbf{\Lambda}_i$ are different.

There have been several developments to nonlinear principal components analysis, each taking a particular feature of the linear methods and generalising it. The work described by Gifi (1990) applies primarily to categorical data. Other extensions are those of principal curves (Hastie and Stuetzle, 1989; Tibshirani, 1992), and nonlinear principal components based on radial basis functions (Webb, 1996) and kernel functions (Schölkopf *et al.*, 1999).

Approaches to principal components analysis have been developed for data that may be considered as curves (Ramsay and Dalzell, 1991; Rice and Silverman, 1991; Silverman, 1995). Principal components analysis for categorical data and functions is discussed by Jackson (1991), who also describes robust procedures.

Independent components analysis (Comon, 1994; Hyvärinen and Oja, 2000) aims to find a linear representation of non-Gaussian data so that the components are statistically independent (or as independent as possible). A linear latent variable model is assumed

$$\boldsymbol{x} = \boldsymbol{A}\boldsymbol{s}$$

where $\boldsymbol{x}$ are the observations, $\boldsymbol{A}$ is a mixing matrix and $\boldsymbol{s}$ is the vector of latent variables. Given T realisations of $\boldsymbol{x}$, the problem is to estimate the mixing matrix $\boldsymbol{A}$ and the corresponding realisations of $\boldsymbol{s}$, under the assumption that the components, s_i, are statistically independent. This technique has found widespread application to problems in signal analysis (medical, financial), data compression, image processing and telecommunications.

Further developments of the linear factor model include models that are nonlinear functions of the latent variables, but still linear in the weights or factor loadings (Etezadi-Amoli and McDonald, 1983). A neural network approach, termed *generative topographic mappings*, has been developed by Bishop *et al.* (1998).

9.3.6 Summary

All the procedures described in this section construct *linear* transformations based on matrices of first- and second-order statistics:

1. principal components analysis – an unsupervised method based on a correlation or covariance matrix;
2. Karhunen–Loève transformation – an umbrella term to cover transformations based on within- and between-class covariance matrices;
3. factor analysis – models observed variables as a linear combination of underlying or latent variables.

Algorithms for their implementation are readily available.

9.4 Multidimensional scaling

Multidimensional scaling is a term that is applied to a class of techniques that analyses a matrix of distances or dissimilarities (the proximity matrix) in order to produce a representation of the data points in a reduced-dimension space (termed the *representation space*).

All of the methods of data reduction presented in this chapter so far have analysed the $n \times p$ data matrix $\boldsymbol{X}$ or the sample covariance or correlation matrix. Thus multidimensional scaling differs in the form of the data matrix on which it operates. Of course, given a data matrix, we could construct a dissimilarity matrix (provided we define a suitable measure of dissimilarity between objects) and then proceed with an analysis using multidimensional scaling techniques. However, data often arise already in the form of dissimilarities and so there is no recourse to the other techniques. Also, in the methods previously discussed, the data-reducing transformation derived has, in each case, been a linear transformation. We shall see that some forms of multidimensional scaling permit a nonlinear data-reducing transformation (if indeed we do have access to data samples rather than proximities).

There are many types of multidimensional scaling, but all address the same basic problem: given an $n \times n$ matrix of dissimilarities and a distance measure (usually Euclidean), find a configuration of n points $\boldsymbol{x}_1, \ldots, \boldsymbol{x}_n$ in $\mathbb{R}^e$ so that the distance between a pair of points is close in some sense to the dissimilarity. All methods must find the coordinates of the points and the dimension of the space, e. Two basic types of multidimensional scaling (MDS) are metric and non-metric MDS. Metric MDS assumes that the data are quantitative and metric MDS procedures assume a functional relationship between the interpoint distances and the given dissimilarities. Non-metric MDS assumes that the data are qualitative, having perhaps ordinal significance, and non-metric MDS procedures produce configurations that attempt to maintain the rank order of the dissimilarities.

Metric MDS appears to have been introduced into the pattern recognition literature by Sammon (1969). It has been developed to incorporate class information and has also been used to provide nonlinear transformations for dimension reduction for feature extraction.

We begin our discussion with a description of one form of metric MDS, namely classical scaling.

9.4.1 Classical scaling

Given a set of n points in p-dimensional space, $\boldsymbol{x}_1, \ldots, \boldsymbol{x}_n$, it is straightforward to calculate the Euclidean distance between each pair of points. Classical scaling (or *principal coordinates analysis*) is concerned with the converse problem: given a matrix of distances, which we assume are Euclidean, how can we determine the coordinates of a set of points in a dimension e (also to be determined from the analysis)? This is achieved via a decomposition of the $n \times n$ matrix $\boldsymbol{T}$, the between-individual sums of squares and products matrix

$$\boldsymbol{T} = \boldsymbol{X}\boldsymbol{X}^T \tag{9.18}$$

where $\boldsymbol{X} = [\boldsymbol{x}_1, \ldots, \boldsymbol{x}_n]^T$ is the $n \times p$ matrix of coordinates. The distance between two individuals i and j is

$$d_{ij}^2 = T_{ii} + T_{jj} - 2T_{ij} \tag{9.19}$$

where

$$T_{ij} = \sum_{k=1}^{p} x_{ik}x_{jk}$$

If we impose the constraint that the centroid of the points $\boldsymbol{x}_i, i = 1, \ldots, p$, is at the origin, then (9.19) may be inverted to express the elements of the matrix $\boldsymbol{T}$ in terms of the dissimilarity matrix, giving

$$T_{ij} = -\tfrac{1}{2}[d_{ij}^2 - d_{i.}^2 - d_{.j}^2 + d_{..}^2] \tag{9.20}$$

where

$$d_{i.}^2 = \frac{1}{n}\sum_{j=1}^{n} d_{ij}^2; \quad d_{.j}^2 = \frac{1}{n}\sum_{i=1}^{n} d_{ij}^2; \quad d_{..}^2 = \frac{1}{n^2}\sum_{i=1}^{n}\sum_{j=1}^{n} d_{ij}^2$$

Equation (9.20) allows us to construct $\boldsymbol{T}$ from a given $n \times n$ dissimilarity matrix $\boldsymbol{D}$ (assuming that the dissimilarities are Euclidean distances). All we need to do now is to factorise the matrix $\boldsymbol{T}$ to make it of the form (9.18). Since it is a real symmetric matrix, $\boldsymbol{T}$ can be written in the form (see Appendix C)

$$\boldsymbol{T} = \boldsymbol{U}\boldsymbol{\Lambda}\boldsymbol{U}^T$$

where the columns of $\boldsymbol{U}$ are the eigenvectors of $\boldsymbol{T}$ and $\boldsymbol{\Lambda}$ is a diagonal matrix of eigenvalues, $\lambda_1, \ldots, \lambda_n$. Therefore we take

$$\boldsymbol{X} = \boldsymbol{U}\boldsymbol{\Lambda}^{\frac{1}{2}}$$

as our matrix of coordinates. If the matrix of dissimilarities is indeed a matrix of Euclidean distances between points in $\mathbb{R}^p$, then the eigenvalues may be ordered

$$\lambda_1 \geq \cdots \geq \lambda_n = 0; \ \lambda_{p+1} = \cdots = 0$$

If we are seeking a representation in a reduced dimension then we would use only those eigenvectors associated with the largest eigenvalues. Methods for choosing the number of eigenvalues were discussed in relation to principal components analysis. Briefly, we choose the number r so that

$$\sum_{i=1}^{r-1} \lambda_i < k \sum_{i=1}^{n} \lambda_i < \sum_{i=1}^{r} \lambda_i$$

for some prespecified threshold, k $(0 < k < 1)$; alternatively, we use the 'scree test'.

Then we take

$$\boldsymbol{X} = [\boldsymbol{u}_1, \ldots, \boldsymbol{u}_r]\mathrm{diag}(\lambda_1^{\frac{1}{2}} \ldots \lambda_r^{\frac{1}{2}}) = \boldsymbol{U}_r \boldsymbol{\Lambda}_r^{\frac{1}{2}}$$

as the $n \times r$ matrix of coordinates, where $\boldsymbol{\Lambda}_r$ is the $r \times r$ diagonal matrix with diagonal elements $\lambda_i, i = 1, \ldots, r$.

If the dissimilarities are not Euclidean distances, then $\boldsymbol{T}$ is not necessarily positive semidefinite and there may be negative eigenvalues. Again we may choose the eigenvectors associated with the largest eigenvalues. If the negative eigenvalues are small then this may still lead to a useful representation of the data. In general, the smallest of the set of largest eigenvalues retained should be larger than the magnitude of the most negative eigenvalue. If there is a large number of negative eigenvalues, or some are large in magnitude, then classical scaling may be inappropriate. However, classical scaling appears to be robust to departures from Euclidean distance.

If we were to start with a set of data (rather than a matrix of dissimilarities) and seek a reduced-dimension representation of it using the classical scaling approach (by first forming a dissimilarity matrix and carrying out the procedure above), then the reduced-dimension representation is exactly the same as carrying out a principal components analysis and calculating the component scores (provided we have chosen Euclidean distance as our measure of dissimilarity). Thus, there is no point in carrying out classical scaling *and* a principal components analysis on a data set.

9.4.2 Metric multidimensional scaling

Classical scaling is one particular form of metric multidimensional scaling in which an objective function measuring the discrepancy between the given dissimilarities, δ_{ij}, and the derived distances in $\mathbb{R}^e$, d_{ij}, is optimised. The derived distances depend on the coordinates of the samples that we wish to find. There are many forms that the objective function may take. For example, minimisation of the objective function

$$\sum_{1 \le j < i \le n} (\delta_{ij}^2 - d_{ij}^2)$$

yields a projection onto the first e principal components if δ_{ij} are exactly Euclidean distances. There are other measures of divergence between the sets $\{\delta_{ij}\}$ and $\{d_{ij}\}$ and the major MDS programs are not consistent in the criterion optimised (Dillon and Goldstein, 1984). One particular measure is

$$S = \sum_{ij} a_{ij}(\delta_{ij} - d_{ij})^2 \tag{9.21}$$

for weighting factors a_{ij}. Taking

$$a_{ij} = \left(\sum_{ij} d_{ij}^2 \right)^{-1}$$

gives $\sqrt{S}$ as similar to Kruskal's stress (Kruskal, 1964a, 1964b), defined in the following section. There are other forms for the a_{ij} (Sammon, 1969; Koontz and Fukunaga, 1972; de Leeuw and Heiser, 1977; Niemann and Weiss, 1979). The stress is invariant under rigid transformations of the derived configuration (translations, rotations and reflections) and also under uniform stretching and shrinking.

A more general form of (9.21) is

$$S = \sum_{ij} a_{ij} (\phi(\delta_{ij}) - d_{ij})^2 \tag{9.22}$$

where ϕ is a member of a predefined class of functions; for example, the class of all linear functions, giving

$$S = \sum_{ij} a_{ij} (a + b\delta_{ij} - d_{ij})^2 \tag{9.23}$$

for parameters a and b. In general, there is no analytic solution for the coordinates of the points in the representation space. Minimisation of (9.22) can proceed by an alternating least squares approach (see Gifi, 1990, for further applications of the alternating least squares principle); that is, by alternating minimisation over ϕ and the coordinates. In the linear regression example (9.23), we would minimise with respect to a and b, for a given initial set of coordinates (and hence the derived distances d_{ij}). Then, keeping a and b fixed, minimise with respect to the coordinates of the data points. This process is repeated until convergence.

The expression (9.22) may be normalised by a function $\tau^2(\phi, \boldsymbol{X})$, that is a function of both the coordinates and the function ϕ. Choices for τ^2 are discussed by de Leeuw and Heiser (1977).

In psychology, in particular, the measures of dissimilarity that arise have ordinal significance at best: their numerical values have little meaning and we are interested only in their order. We can say that one stimulus is larger than another, without being able to attach a numerical value to it. In this case, a choice for the function ϕ above is one that belongs to the class of monotone functions. This is the basis of *non-metric* multidimensional scaling or *ordinal scaling*.

9.4.3 Ordinal scaling

Ordinal scaling or *non-metric multidimensional scaling* is a method of finding a configuration of points for which the rank ordering of the interpoint distance is close to the ranking of the values of the given dissimilarities.

In contrast to classical scaling, there is no analytic solution for the configuration of points in ordinal scaling. Further, the procedure is iterative and requires an initial configuration for the points to be specified. Several initial configurations may be tried before an acceptable final solution is achieved.

The desired requirement that the ordering of the distances in the derived configuration is the same as that of the given dissimilarities is of course equivalent to saying that the distances are a monotonic function of the dissimilarities. Figure 9.13 gives a plot of distances d_{ij} (obtained from a classical scaling analysis) against dissimilarities δ_{ij} for the British town data given in Table 9.3. The numbers in the table are the distances in miles between 10 towns in Britain along routes recommended by the RAC. The relationship is clearly not monotonic, though, on the whole, the larger the dissimilarity the larger the

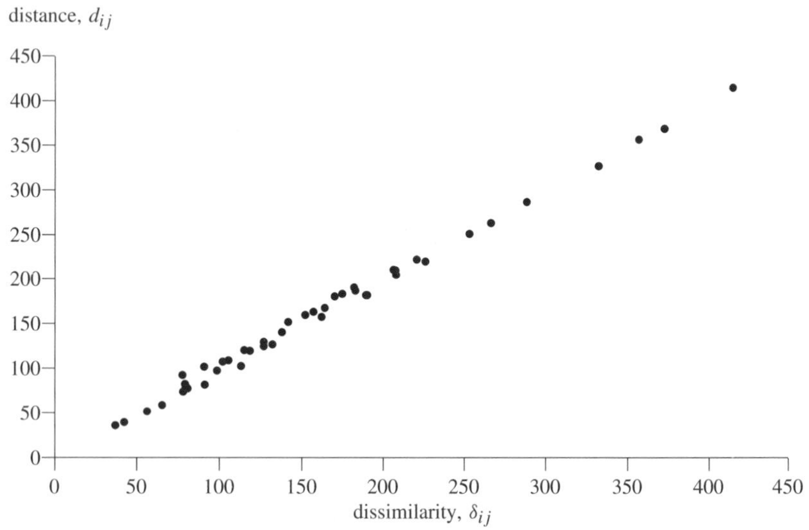

Figure 9.13 Distances v dissimilarities for the British town data

Table 9.3 Dissimilarities between 10 towns in the British Isles (measured as distances in miles along recommended routes)

	Ldn	B'ham	Cmbg	Edin	Hull	Lin	M/c	Nwch	Scar	S'ton
London	0									
Birmingham	111	0								
Cambridge	55	101	0							
Edinburgh	372	290	330	0						
Hull	171	123	124	225	0					
Lincoln	133	85	86	254	39	0				
Manchester	184	81	155	213	96	84	0			
Norwich	112	161	62	360	144	106	185	0		
Scarborough	214	163	167	194	43	81	105	187	0	
Southampton	77	128	130	418	223	185	208	190	266	0

distance. In ordinal scaling, the coordinates of the points in the representation space are adjusted so as to minimise a cost function that is a measure of the degree of deviation from monotonicity of the relationship between d_{ij} and δ_{ij}. It may not be possible to obtain a final solution that is perfectly monotonic but the final ordering on the d_{ij} should be 'as close as possible' to that of the δ_{ij}.

To find a configuration that satisfies the monotonicity requirement, we must first of all specify a definition of monotonicity. Two possible definitions are the *primary monotone condition*

$$\delta_{rs} < \delta_{ij} \Rightarrow \hat{d}_{rs} \leq \hat{d}_{ij}$$

and the *secondary monotone condition*

$$\delta_{rs} \leq \delta_{ij} \Rightarrow \hat{d}_{rs} \leq \hat{d}_{ij}$$

where $\hat{d}_{rs}$ is the point on the fitting line (see Figure 9.14) corresponding to δ_{rs}. The $\hat{d}_{rs}$ are termed the *disparities* or the *pseudo-distances*. The difference between these two conditions is the way in which ties between the δs are treated. In the secondary monotone condition, if $\delta_{rs} = \delta_{ij}$ then $\hat{d}_{rs} = \hat{d}_{ij}$, whereas in the primary condition there is no constraint on $\hat{d}_{rs}$ and $\hat{d}_{ij}$ if $\delta_{rs} = \delta_{ij}$: $\hat{d}_{rs}$ and $\hat{d}_{ij}$ are allowed to differ (which would give rise to vertical lines in Figure 9.14). The secondary condition is usually regarded as too restrictive, often leading to convergence problems.

Given the above definition, we can define a goodness of fit as

$$S_q = \sum_{i<j} (d_{ij} - \hat{d}_{ij})^2$$

and minimising gives the primary (or secondary) least squares monotone regression line. (In fact, it is not a line, only being defined at points δ_{ij}.)

An example of a least squares monotone regression is given in Figure 9.14. The least squares condition ensures that the sum of squares of vertical displacements from the line is a minimum. Practically, this means that for sections of the data where d is actually a monotonic function of δ the line passes through the points. If there is a decrease, the value taken is the mean of a set of samples.

The quantity S_q is a measure of the deviation from monotonicity, but it is not invariant to uniform dilation of the geometric configuration. This can be removed by normalisation

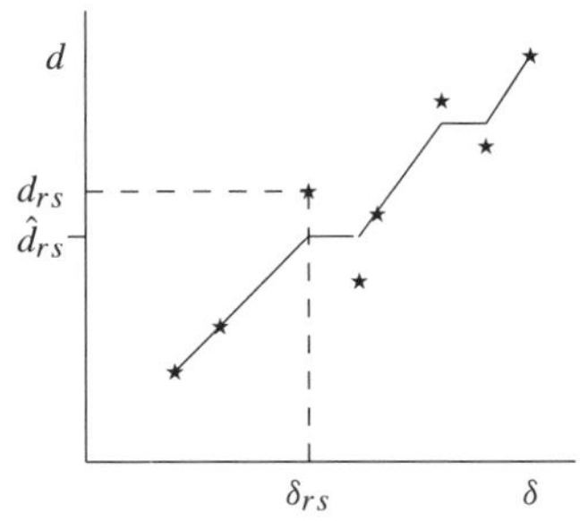

Figure 9.14 Least squares monotone regression line

and the normalised *stress*, given by

$$S = \sqrt{\frac{\sum_{i<j}(d_{ij} - \hat{d}_{ij})^2}{\sum_{i<j} d_{ij}^2}} \tag{9.24}$$

used as the measure of fit. (In some texts the square root factor is omitted in the definition.)

Since S is a differentiable function of the desired coordinates, we use a nonlinear optimisation scheme (see, for example, Press *et al.*, 1992) which requires an initial configuration of data points to be specified. In practice, it has been found that some of the more sophisticated methods do not work so well as steepest descents (Chatfield and Collins, 1980).

The initial configuration of data points could be chosen randomly, or as the result of a principal coordinates analysis. Of course, there is no guarantee of finding the global minimum of S and the algorithm may get trapped in poor local minima. Several initial configurations may be considered before an acceptable value of the minimum stress is achieved.

In the algorithm, a value of the dimension of the representation space, e, is required. This is unknown in general and several values may be tried before a 'low' value of the stress is obtained. The minimum stress depends on n (the dimension of the dissimilarity matrix) and e and it is not possible to apply tests of significance to see if the 'true' dimension has been found (this may not exist). As with principal components analysis, we may plot the stress as a function of dimension and see if there is a change in the slope (elbow in the graph). If we do this, we may find that the stress does not decrease as the dimension increases, because of the problem of finding poor local minima. However, it should always decrease and can be made to do so by initialising the solution for dimension e by that obtained in dimension $e - 1$ (extra coordinates of zero are added).

The summations in the expression for the stress are over all pairwise distances. If the dissimilarity matrix is asymmetric we may include both the pairs (i, j) and (j, i) in the summation. Alternatively, we may carry out ordinal scaling on the symmetric matrix of dissimilarities with

$$\delta_{rs}^* = \delta_{sr}^* = \tfrac{1}{2}(\delta_{rs} + \delta_{sr})$$

Missing values in δ_{ij} can be accommodated by removing the corresponding indices from the summation (9.24) in estimating the stress.

9.4.4 Algorithms

Most MDS programs use standard gradient methods. There is some evidence that sophisticated nonlinear optimisation schemes do not work so well (Chatfield and Collins, 1980). Siedlecki *et al.* (1988) report that steepest descents outperformed conjugate gradients on a metric MDS optimisation problem, but better performance was given by the coordinate descent procedure of Niemann and Weiss (1979).

One approach to minimising the objective function, S, is to use the principle of majorisation (de Leeuw and Heiser, 1977; Heiser, 1991, 1994) as part of the alternating

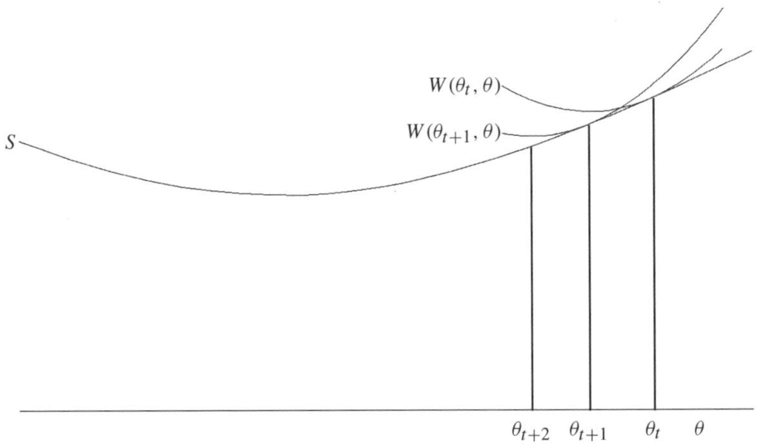

Figure 9.15 Illustration of iterative majorisation principle: minimisation of $S(\theta)$ is achieved through successive minimisations of the majorisation functions, W

least squares process. Given the current values of the coordinates, say θ_t, an upper bound, $W(\theta_t, \theta)$, for the criterion function is defined. It is usually a quadratic form with a single minimum as a function of the coordinates θ. It has the property that $W(\theta_t, \theta)$ is equal to the value of the objective function at θ_t and greater than it everywhere else. Minimising $W(\theta_t, \theta)$ with respect to θ yields a value θ_{t+1} at which the objective function is lower. A new majorising function $W(\theta_{t+1}, \theta)$ is defined and the process repeated (see Figure 9.15). This generates a sequence of estimates $\{\theta_t\}$ for which the objective function decreases and converges to a local minimum.

All algorithms start with an initial configuration and converge to a local minimum. It has been reported that the secondary definition of monotonicity is more likely to get trapped in poor local minima than the primary definition (Gordon, 1999). We recommend that you repeat your experiments for several starting configurations.

9.4.5 Multidimensional scaling for feature extraction

There are several obstacles in applying multidimensional scaling to the pattern recognition problem of feature extraction that we are addressing in this chapter. The first is that usually we are not presented with an $n \times n$ matrix of dissimilarities, but with an $n \times p$ matrix of observations. Although in itself this is not a problem since we can form a dissimilarity matrix using some suitable measure (e.g. Euclidean distance), the number of patterns n can be very large (in some cases thousands). The storage of an $n \times n$ matrix may present a problem. Further, the number of adjustable parameters is $n' = e \times n$, where e is the dimension of the derived coordinates. This may prohibit the use of some nonlinear optimisation methods, particularly those of the quasi-Newton type which either calculate, or iteratively build up, an approximation to the inverse Hessian matrix of size $n' \times n' = e^2 n^2$.

Even if these problems can be overcome, multidimensional scaling does not readily define a transformation that, given a new data sample $\boldsymbol{x} \in \mathbb{R}^p$, produces a result $\boldsymbol{y} \in \mathbb{R}^e$. Further calculation is required.

One approach to this problem is to regard the transformed coordinates $\boldsymbol{y}$ as a nonlinear parametrised function of the data variables

$$\boldsymbol{y} = \boldsymbol{f}(\boldsymbol{x}; \theta)$$

for parameters θ. In this case,

$$\begin{aligned} d_{ij} &= |\boldsymbol{f}(\boldsymbol{x}_i; \theta) - \boldsymbol{f}(\boldsymbol{x}_j; \theta)| \\ \delta_{ij} &= |\boldsymbol{x}_i - \boldsymbol{x}_j| \end{aligned}$$

and we may minimise the criterion function, for example (9.21), *with respect to the parameters*, θ, of $\boldsymbol{f}$ rather than with respect to the coordinates of the data points in the transformed space. This is termed *multidimensional scaling by transformation*. Thus the number of parameters can be substantially reduced. Once the iteration has converged, the function $\boldsymbol{f}$ can be used to calculate the coordinates in $\mathbb{R}^e$ for any new data sample $\boldsymbol{x}$. One approach is to model $\boldsymbol{f}$ as a radial basis function network and determine the network weights using *iterative majorisation*, a scheme that optimises the objective function without gradient calculation (Webb, 1995).

A modification to the distance term, δ, is to augment it with a supervised quantity giving a distance

$$(1 - \alpha)\delta_{ij} + \alpha s_{ij} \tag{9.25}$$

where $0 < \alpha < 1$ and s_{ij} is a separation between objects using labels associated with the data. For example, s_{ij} may represent a class separability term: how separable are the classes to which patterns $\boldsymbol{x}_i$ and $\boldsymbol{x}_j$ belong? A difficulty is the specification of the parameter α.

9.4.6 Example application study

The problem An exploratory data analysis to investigate relationships between research assessment ratings of UK higher education institutions (Lowe and Tipping, 1996).

Summary Several methods of feature extraction, both linear and nonlinear, were applied to high-dimensional data records from the 1992 UK Research Assessment Exercise and projections onto two dimensions produced.

The data Institutions supply information on research activities within different subjects in the form of quantitative indicators of their research activity, such as the number of active researchers, postgraduate students, values of grants and numbers of publications. Together with some qualitative data (for example, publications), this forms part of the data input to committees which provide a research rating, on a scale from 1 to 5. There are over 4000 records for all subjects, but the analysis concentrated on three subjects: physics, chemistry and biological sciences.

Preprocessing included the removal of redundant and repeated variables, accumulating indicators that were given for a number of years and standardisation of variables. The training set consisted of 217 patterns, each with 80 variables.

The model Several models were assessed. These included a principal components analysis, a multidimensional scaling (Sammon mapping) and an MDS by transformation modelled as a radial basis function network.

Training procedure For the multidimensional scaling by transformation, the dissimilarity was augmented with a subjective quantity as in (9.25) where s_{ij} is a separation between objects based on the subjective research rating.

Since the objective function is no longer quadratic, an analytic matrix inversion routine cannot be used for the weights of the radial basis function. A conjugate gradients nonlinear function minimisation routine was used to minimise of the stress criterion.

9.4.7 Further developments

Within the pattern recognition literature, there have been several attempts to use multidimensional scaling techniques for feature extraction both for exploratory data analysis and classification purposes (Sammon, 1969 – see comments by Kruskal, 1971; Koontz and Fukunaga, 1972; Cox and Ferry, 1993).

Approaches that model the nonlinear dimension-reducing transformation as a radial basis function network are described by Webb (1995) and Lowe and Tipping (1996). Mao and Jain (1995) model the transformation as a multilayer perceptron. A comparative study of neural network feature extraction methods has been done by Lerner *et al.* (1999).

9.4.8 Summary

Multidimensional scaling is a name given to a range of techniques that analyse dissimilarity matrices and produce coordinates of points in a 'low' dimension. Three approaches to multidimensional scaling have been presented:

Classical scaling This assumes that the dissimilarity matrix is Euclidean, though it has been shown to be robust if there are small departures from this condition. An eigenvector decomposition of the dissimilarity matrix is performed and the resulting set of coordinates is identical to the principal components analysis scores (to within an orthogonal transformation) if indeed the dissimilarity matrix is a matrix of Euclidean distances. Therefore there is nothing to be gained over a principal components analysis in using this technique as a method of feature extraction, given an $n \times p$ data matrix $\boldsymbol{X}$.

Metric scaling This method regards the coordinates of the points in the derived space as parameters of a stress function that is minimised. This method allows nonlinear reductions in dimensionality. The procedure assumes a functional relationship between the interpoint distances and the given dissimilarities.

Non-metric scaling As with metric scaling, a criterion function (stress) is minimised but the procedure assumes that the data are qualitative, having perhaps ordinal significance at best.

9.5 Application studies

Examples of application studies using feature selection methods include:

- Remote sensing. Bruzzone *et al.* (1995) extend the pairwise Jeffreys–Matusita distance (otherwise known as the Patrick–Fischer distance – see Appendix A) to the multiclass situation and use it as a feature selection criterion in a remote sensing application. The data consist of measurements of six channels in the visible and infrared spectrum on five agricultural classes.
- Hand-printed character recognition. Zongker and Jain (1996) apply the SFFS algorithm to direction features for the 26 lower-case characters.
- Speech. Novovičová *et al.* (1996) apply a feature selection method based on modelling the class densities by finite mixture distributions to a two-class speech recognition problem. The data comprised 15-dimensional feature vectors (containing five segments of three features derived by low-order linear prediction analysis). The approach was compared with a method using SFFS and a Gaussian classifier. For this data set, there was no advantage in mixture modelling.
- Image analysis. Pudil *et al.* (1994c) use an approach to feature selection based on mixture modelling for the classification of granite textures. The features comprise a 26-dimensional vector (eight texture features and 18 colour features). In this case, the data are modelled well by mixtures (in contrast to the speech example above).

Feature extraction application studies include:

- Electroencephalogram (EEG). Jobert *et al.* (1994) use principal components analysis to produce a two-dimensional representation of spectral data (sleep EEG) to view time-dependent variation.
- Positron emission tomography (PET). Pedersen *et al.* (1994) use principal components analysis for data visualisation purposes on dynamic PET images to enhance clinically interesting information.
- Remote sensing. Eklundh and Singh (1993) compare principal components analysis using correlation and covariance matrices in the analysis of satellite remote sensing data. The correlation matrix gives improvement to the signal-to-noise ratio.
- Calibration of near-infrared spectra.
- Structure–activity relationships. Darwish *et al.* (1994) apply principal components analysis (14 variables, nine compounds) in a study to investigate the inhibitory effect of benzine derivatives.
- Target classification. Liu *et al.* (1994) use principal components analysis for feature extraction in a classification study of materials design.
- Face recognition. Principal components analysis has been used in several studies on face recognition to produce 'eigenfaces'. The weights that characterise the expansion of a given facial image in terms of these eigenfaces are features used for face recognition and classification (see the review by Chellappa *et al.*, 1995).
- Speech. Pinkowski (1997) uses principal components analysis for feature extraction on a speaker-dependent data set consisting of spectrograms of 80 sounds representing 20 speaker-dependent words containing English semivowels.

Applications of MDS and Sammon mappings include:

- Medical. Ratcliffe *et al.* (1995) use multidimensional scaling to recover three-dimensional localisation of sonomicrometry transducer elements glued to excised and living ovine hearts. The inter-element distances were measured by the sonomicrometry elements by sequentially activating a single array element followed by eight receiver elements (thus giving inter-transducer distances).
- Bacterial classification. Bonde (1976) uses non-metric multidimensional scaling to produce two and three-dimensional plots of groups of organisms (using a steepest descent optimisation scheme).
- Chemical vapour analysis. For a potential application of an 'artificial nose' (an array of 14 chemical sensors) to atmosphere pollution monitoring, cosmetics, food and defence applications, Lowe (1993) considers a multidimensional scaling approach to feature extraction, where the dissimilarity matrix is determined by class (concentration of the substance).

9.6 Summary and discussion

In this chapter we have considered a variety of techniques for mapping data to a reduced dimension. As you may imagine, there are many more that we have not covered, and we have tried to point to the literature where some of these may be found. A comprehensive account of data transformation techniques requires a volume in itself and in this chapter we have only been able to consider some of the more popular multivariate methods. In common with the following chapter, many of the techniques are used as part of data preprocessing in an exploratory data analysis.

The techniques vary in their complexity – both from mathematical ease of understanding and numerical ease of implementation points of view. Most methods produce linear transformations, but non-metric multidimensional scaling is nonlinear. Some use class information, others are unsupervised, although there are both variants of the Karhunen–Loève transformation. Some techniques, although producing a linear transformation, require the use of a nonlinear optimisation procedure to find the parameters. Others are based on eigenvector decomposition routines, perhaps performed iteratively.

To some extent, the separation of the classifier design into two processes of feature extraction and classification is artificial, but there are many reasons, some of which were enumerated at the beginning of this chapter, why dimension reduction may be advisable. Within the pattern recognition literature, many methods for nonlinear dimension reduction and exploratory data analysis have been proposed.

9.7 Recommendations

If explanation is required of the variables that are used in a classifier, then a feature selection process, as opposed to a feature extraction process, is recommended for dimension reduction. For feature selection, the probabilistic criteria for estimating class separability are complicated, involving estimation of probability density functions and their numerical integration. Even the simple error rate measure is not easy to evaluate for nonparametric density functions. Therefore, we recommend use of the following.

1. The parametric form of the probabilistic distance measures assumes normal distributions. These have the advantage for feature selection that the value of the criterion for a given set of features may be used in the evaluation of the criterion when an additional feature is included. This reduces the amount of computation in some of the feature set search algorithms.
2. The interclass distance measures, J_1 to J_4 (Section 9.2.1).
3. Error rate estimation using a specified classifier.
4. Floating search methods.

Which algorithms should you employ for feature extraction? Whatever your problem, always start with the simplest approach, which for feature extraction is, in our view, a principal components analysis of your data. This will tell you whether your data lie on a linear subspace in the space spanned by the variables and a projection onto the first two principal components, and displaying your data may reveal some interesting and unexpected structure.

It is recommended to apply principal components analysis to standardised data for feature extraction, and to consider it particularly when dimensionality is high. Use a simple heuristic to determine the number of principal components to use, in the first instance. For class-labelled data, use linear discriminant analysis for a reduced-dimension representation.

If you believe there to be nonlinear structure in the data, then techniques based on multidimensional scaling (for example, multidimensional scaling by transformation) are straightforward to implement. Try several starting conditions for the parameters.

9.8 Notes and references

A good description of feature selection techniques for discrimination may be found in the book by Devijver and Kittler (1982). The papers by Kittler (1975b, 1978b) and Siedlecki and Sklansky (1988) also provide reviews of feature selection and extraction methods. Chapter 6 of Hand's (1981a) book on variable selection also discusses several of the methods described in this chapter.

The branch and bound method has been used in many areas of statistics (Hand, 1981b). It was originally proposed for feature subset selection by Narendra and Fukunaga (1977) and receives a comprehensive treatment by Devijver and Kittler (1982) and Fukunaga (1990). Hamamoto *et al.* (1990) evaluate the branch and bound algorithm using a recognition rate measure that does not satisfy the monotonicity condition. Krusińska (1988) describes a semioptimal branch and bound algorithm for feature selection in mixed variable discrimination.

Stepwise procedures have been considered by many authors: Whitney (1971) for the sequential forward selection algorithm; Michael and Lin (1973) for the basis of the l–r algorithm; Stearns (1976) for the l–r algorithm.

Floating search methods were introduced by Pudil *et al.* (1994b; see also Pudil *et al.*, 1994a, for their assessment with non-monotonic criterion functions and Kudo and Sklansky, 2000, for a comparative study). Error-rate-based procedures are described by McLachlan (1992a). Ganeshanandam and Krzanowski (1989) also use error rate as the selection criterion. Within the context of regression, the book by Miller (1990) gives very good accounts of variable selection.

Many of the standard feature extraction techniques may be found in most textbooks on multivariate analysis. Descriptions (with minimal mathematics) may be found in Reyment *et al.* (1984, Chapter 3) and Clifford and Stephenson (1975, Chapter 13).

Thorough treatments of principal components analysis are given in the books by Jolliffe (1986) and Jackson (1991), the latter providing a practical approach and giving many worked examples and illustrations. Common principal components and related methods are described in the book by Flury (1988).

Descriptions of factor analysis appear in many textbooks on multivariate analysis (for example, Dillon and Goldstein, 1984; Kendall, 1975). Jackson (1991) draws out the similarities to and the differences from principal components analysis. There are several specialist books on factor analysis, including those by Harman (1976), and Lawley and Maxwell (1971).

Multidimensional scaling is described in textbooks on multivariate analysis (for example, Chatfield and Collins, 1980; Dillon and Goldstein, 1984) and more detailed treatments are given in the books by Schiffman *et al.* (1981) and Jackson (1991). Cox and Cox (1994) provide an advanced treatment, with details of some of the specialised procedures. An extensive treatment of non-metric MDS can be found in the collection of articles edited by Lingoes *et al.* (1979). There are many computer programs available for performing scaling. The features of some of these are given by Dillon and Goldstein (1984) and Jackson (1991).

Many of the techniques described in this chapter are available in standard statistical software packages. There is some specialised software for multidimensional scaling publicly available.

The website `www.statistical-pattern-recognition.net` contains references and pointers to websites for further information on techniques.

Exercises

Numerical routines for matrix operations, including eigendecomposition, can be found in many numerical packages. Press *et al.* (1992) give descriptions of algorithms.

1. Consider the divergence (see Appendix A),

$$J_D = \int (p(\boldsymbol{x}|\omega_1) - p(\boldsymbol{x}|\omega_2)) \log\left(\frac{p(\boldsymbol{x}|\omega_1)}{p(\boldsymbol{x}|\omega_2)}\right) d\boldsymbol{x}$$

   where $\boldsymbol{x} = (x_1, \ldots, x_p)^T$. Show that under conditions of independence, J_D may be expressed as

$$J_D = \sum_{j=1}^{p} J_j(x_j)$$

2. (Chatfield and Collins, 1980.) Suppose the p-variate random variable $\boldsymbol{x}$ has covariance matrix $\boldsymbol{\Sigma}$ with eigenvalues $\{\lambda_i\}$ and orthonormal eigenvectors $\{\boldsymbol{a}_i\}$. Show that the identity matrix is given by

$$\boldsymbol{I} = \boldsymbol{a}_1\boldsymbol{a}_1^T + \cdots + \boldsymbol{a}_p\boldsymbol{a}_p^T$$

and that

$$\boldsymbol{\Sigma} = \lambda_1\boldsymbol{a}_1\boldsymbol{a}_1^T + \cdots + \lambda_p\boldsymbol{a}_p\boldsymbol{a}_p^T$$

The latter result is called the *spectral decomposition* of $\boldsymbol{\Sigma}$.

3. Given a set of n measurements on p variables, describe the stages in performing a principal components analysis for dimension reduction.

4. Let X_1 and X_2 be two random variables with covariance matrix

$$\boldsymbol{\Sigma} = \begin{bmatrix} 9 & \sqrt{6} \\ \sqrt{6} & 4 \end{bmatrix}$$

Obtain the principal components. What is the percentage of total variance explained by each component?

5. Athletics records for 55 countries comprise measurements made on eight running events. These are each country's record times for (1) 100 m (s); (2) 200 m (s); (3) 400 m (s); 800 m (min); (5) 1500 m (min); (6) 5000 m (min); (7) 10 000 m (min); (8) marathon (min).

Describe how a principal components analysis may be used to obtain a two-dimensional representation of the data.

The results of a principal components analysis are shown in the table (Everitt and Dunn, 1991). Interpret the first two principal components.

	PC1 $\times\ \lambda_1$	PC2 $\times\ \lambda_2$
100 m	0.82	0.50
200 m	0.86	0.41
400 m	0.92	0.21
800 m	0.87	0.15
1500 m	0.94	−0.16
5000 m	0.93	−0.30
10 000 m	0.94	−0.31
Marathon	0.87	−0.42
Eigenvalue	6.41	0.89

What is the percentage of the total variance explained by the first principal component? State any assumptions you make.

6. (Chatfield and Collins, 1980.) Four measurements are made on each of a random sample of 500 animals. The first three variables were different linear dimensions, measured in centimetres, while the fourth variable was the weight of the animal measured in grams. The sample covariance matrix was calculated and its four eigenvalues were found to be 14.1, 4.3, 1.2 and 0.4. The eigenvectors corresponding to the first and second eigenvalues were:

$$\boldsymbol{u}_1^T = [0.39, 0.42, 0.44, 0.69]$$
$$\boldsymbol{u}_2^T = [0.40, 0.39, 0.42, -0.72]$$

   Comment on the use of the sample covariance matrix for the principal components analysis for these data. What is the percentage of variance in the original data accounted for by the first two principal components? Describe the results.

   Suppose the data were stored by recording the eigenvalues and eigenvectors together with the 500 values of the first and second principal components and the mean values for the original variables. Show how to reconstruct the original covariance matrix and an approximation to the original data.

7. Given that

$$\boldsymbol{S} = \begin{bmatrix} \tilde{\boldsymbol{S}} & \boldsymbol{y} \\ \boldsymbol{y}^T & s_{kk} \end{bmatrix}$$

   and assuming that $\tilde{\boldsymbol{S}}^{-1}$ is known, verify that $\boldsymbol{S}^{-1}$ is given by (9.2). Conversely, with $\boldsymbol{S}$ given by the above and assuming that $\boldsymbol{S}^{-1}$ is known,

$$\boldsymbol{S}^{-1} = \begin{bmatrix} \boldsymbol{A} & \boldsymbol{c} \\ \boldsymbol{c}^T & b \end{bmatrix}$$

   show that the inverse of $\tilde{\boldsymbol{S}}$ (the inverse of $\boldsymbol{S}$ after the removal of a feature) can be written as

$$\tilde{\boldsymbol{S}}^{-1} = \boldsymbol{A} - \frac{1}{b}\boldsymbol{c}^T\boldsymbol{c}$$

8. For a given symmetric matrix $\boldsymbol{S}$ of known inverse (of the above form) and symmetric matrix $\boldsymbol{T}$, verify (9.3), where $\tilde{\boldsymbol{T}}$ is the submatrix of $\boldsymbol{T}$ after the removal of a feature. Hence, show that the feature extraction criterion $\mathrm{Tr}(\boldsymbol{S}_W^{-1}\boldsymbol{S}_B)$, where $\boldsymbol{S}_W$ and $\boldsymbol{S}_B$ are the within- and between-class covariance matrices, satisfies the monotonicity property.

9. How could you use floating search methods for radial basis function centre selection? What are the possible advantages and disadvantages of such methods compared with random selection or k-means, for example?

10. Suppose we take classification rate using a nearest class mean classifier as our feature selection criterion. Show by considering the two distributions,

$$p(\boldsymbol{x}|\omega_1) = \begin{cases} 1 & 0 \le x_1 \le 1, 0 \le x_2 \le 1 \\ 0 & \text{otherwise} \end{cases}$$

$$p(\boldsymbol{x}|\omega_2) = \begin{cases} 1 & 1 \le x_1 \le 2, -0.5 \le x_2 \le 0.5 \\ 0 & \text{otherwise} \end{cases}$$

where $\boldsymbol{x} = (x_1, x_2)^T$, that classification rate does not satisfy the monotonicity property.

11. Derive the relationship (9.20) expressing the elements of the sum of squares and products matrix in terms of the elements of the dissimilarity matrix from (9.19) and the definition of T_{ij} and zero-mean data.

12. Given n p-dimensional measurements (in the $n \times p$ data matrix $\boldsymbol{X}$, zero mean, $p < n$) show that a low-dimensional representation in $r < p$ dimensions obtained by constructing the sums of squares and products matrix, $\boldsymbol{T} = \boldsymbol{X}\boldsymbol{X}^T$, and performing a principal coordinates analysis, results in the same projection as principal components (to within an orthogonal transformation).

13. For two classes normally distributed, $N(\boldsymbol{\mu}_1, \boldsymbol{\Sigma})$ and $N(\boldsymbol{\mu}_2, \boldsymbol{\Sigma})$ with common covariance matrix, $\boldsymbol{\Sigma}$, show that the divergence

$$J_D(\omega_1, \omega_2) = \int [p(\boldsymbol{x}|\omega_1) - p(\boldsymbol{x}|\omega_2)] \log \left\{ \frac{p(\boldsymbol{x}|\omega_1)}{p(\boldsymbol{x}|\omega_2)} \right\} d\boldsymbol{x}$$

is given by

$$(\boldsymbol{\mu}_2 - \boldsymbol{\mu}_1)^T \boldsymbol{\Sigma}^{-1} (\boldsymbol{\mu}_2 - \boldsymbol{\mu}_1)$$

the *Mahalanobis distance*

14. Describe how multidimensional scaling solutions that optimise stress can always be constructed that result in a decrease of stress with dimension of the representation space.

10

Clustering

Overview

Clustering methods are used for data exploration and to provide prototypes for use in supervised classifiers. Methods that operate both on dissimilarity matrices and measurements on individuals are described, each implicitly imposing its own structure on the data. Mixtures explicitly model the data structure.

10.1 Introduction

Cluster analysis is the grouping of individuals in a population in order to discover structure in the data. In some sense, we would like the individuals within a group to be close or similar to one another, but dissimilar from individuals in other groups.

Clustering is fundamentally a collection of methods of data exploration. One often uses a method to see if natural groupings are present in the data. If groupings do emerge, these may be named and their properties summarised. For example, if the clusters are compact, then it may be sufficient for some purposes to reduce the information on the original data set to information about a small number of groups, in some cases representing a group of individuals by a single sample. The results of a cluster analysis may produce identifiable structure that can be used to generate hypotheses (to be tested on a separate data set) to account for the observed data.

It is difficult to give a universal definition of the term 'cluster'. All of the methods described in this chapter can produce a *partition* of the data set – a division of the data set into mutually non-overlapping groups. However, different methods will often yield different groupings since each implicitly imposes a structure on the data. Also, the techniques will produce groupings even when there is no 'natural' grouping in the data. The term 'dissection' is used when the data consist of a single homogeneous population that one wishes to partition. Clustering techniques may be used to obtain dissections, but the user must be aware that a structure is being imposed on the data that may not be present. This does not matter in some applications.

Before attempting a classification, it is important to understand the problem you are wishing to address. Different classifications, with consequently different interpretations, can be imposed on a sample and the choice of variables is very important. For example, there are different ways in which books may be grouped on your bookshelf – by subject matter or by size – and different classifications will result from the use of different variables. Each classification may be important in different circumstances, depending on the problem under consideration. Once you understand your problem and data, you must choose your method carefully. An inappropriate match of method to data can give results that are misleading.

Related to clustering is clumping, which allows an object to belong to more than one group. An example often cited is the classification of words according to their meaning: some words have several meanings and can belong to several groups. However, in this chapter we concentrate on clustering methods.

There is a vast literature on clustering. Some of the more useful texts are given at the end of this chapter. There is a wide range of application areas, sometimes with conflicting terminology. This has led to methods being rediscovered in different fields of study. Much of the early work was in the fields of biology and zoology, but clustering methods have also been applied in the fields of psychology, archaeology, linguistics and signal processing. This literature is not without its critics. The paper by Cormack (1971) is worth reading. His remark that 'the real need of the user is to be forced to sit and think' is perhaps even more relevant today. The user often does not want to do so, and is often satisfied with a 'black box' approach to the analysis. However, the first thing to do when wishing to apply a technique is to understand the significance of the data and to understand what a particular technique does.

Five topics are discussed in this chapter:

1. *hierarchical methods*, which derive a clustering from a given dissimilarity matrix;
2. *quick partitions*, methods for obtaining a partition as an initialisation to more elaborate approaches;
3. *mixture models*, which express the probability density function as a sum of component densities;
4. *sum-of-squares methods*, which minimise a sum-of-squares error criterion, including k-means, fuzzy k-means, vector quantisation and stochastic vector quantisation;
5. *cluster validity*, addressing the problem of model selection.

10.2 Hierarchical methods

Hierarchical clustering procedures are the most commonly used method of summarising data structure. A hierarchical tree is a nested set of partitions represented by a tree diagram or *dendrogram* (see Figure 10.1). Sectioning a tree at a particular level produces a partition into g disjoint groups. If two groups are chosen from different partitions (the results of partitioning at different levels) then either the groups are disjoint or one group wholly contains the other. An example of a hierarchical classification is the classification of the animal kingdom. Each species belongs to a series of nested clusters of increasing size with a decreasing number of common characteristics. In producing a tree diagram

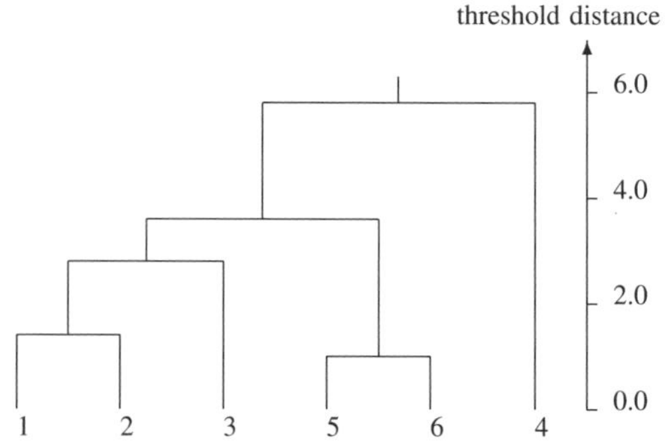

Figure 10.1 Dendrogram

like that in Figure 10.1, it is necessary to order the points so that branches do not cross. This ordering is somewhat arbitrary, but does not alter the structure of the tree, only its appearance. There is a numerical value associated with each position up the tree where branches join. This is a measure of the distance or dissimilarity between two merged clusters. There are many different measures of distances between clusters (some of these are given in Appendix A) and these give rise to different hierarchical structures, as we shall see in later sections of this chapter. Sectioning a tree partitions the data into a number of clusters of comparable homogeneity (as measured by the clustering criterion).

There are several different algorithms for finding a hierarchical tree. An *agglomerative algorithm* begins with n subclusters, each containing a single data point, and at each stage merges the two most similar groups to form a new cluster, thus reducing the number of clusters by one. The algorithm proceeds until all the data fall within a single cluster. A *divisive algorithm* operates by successively splitting groups, beginning with a single group and continuing until there are n groups, each of a single individual. Generally, divisive algorithms are computationally inefficient (except where most of the variables are binary attribute variables).

From the tree diagram a new set of distances between individuals may be defined, with the distance between individual i and individual j being the distance between the two groups that contain them, when these two groups are amalgamated (i.e. the distance level of the lowest link joining them). Thus, the procedure for finding a tree diagram may be viewed as a transformation of the original set of dissimilarities d_{ij} to a new set $\hat{d}_{ij}$, where the $\hat{d}_{ij}$ satisfy the *ultrametric inequality*

$$\hat{d}_{ij} \leq \max(\hat{d}_{ik}, \hat{d}_{jk}) \quad \text{for all objects } i, j, k$$

This means that the distances between three groups can be used to define a triangle that is either equilateral or isosceles (either the three distances are the same or two are equal and the third smaller – see Figure 10.1, for example). A transformation $D : d \rightarrow \hat{d}$ is termed an *ultrametric transformation*. All of the methods in this section produce a clustering from a given dissimilarity matrix.

It is appropriate to introduce here the concept of a *minimum spanning tree*. A minimum spanning tree is not a hierarchical tree, but a tree spanning a set of points such that every

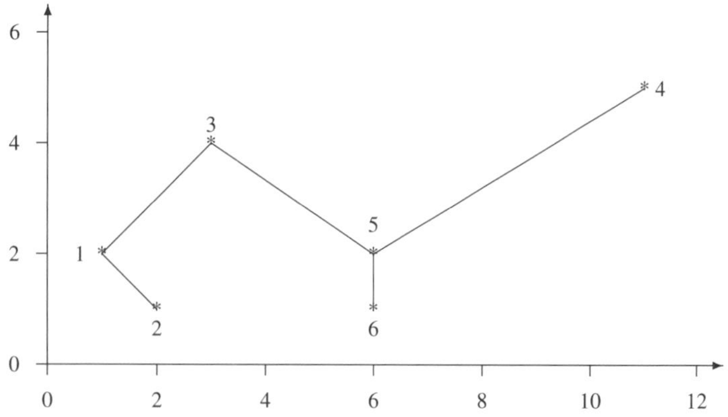

Figure 10.2 Minimum spanning tree

two vertices are connected and there are no closed loops. Associated with each link in the tree is a value or distance, and the minimum spanning tree is the tree such that the sum of all the distances is a minimum. Figure 10.2 shows the minimum spanning tree for the two-dimensional data used to generate the dendrogram of Figure 10.1, where the distance between individuals is taken to be the Euclidean distance. The minimum spanning tree has been used as the basis of a single-link algorithm. Single-link clusters at level h are obtained by deleting from the minimum spanning tree all edges greater than h in length. The minimum spanning tree should also be useful in the identification of clusters, outliers and influential points (points whose removal can alter the derived clustering appreciably).

10.2.1 Single-link method

The single-link method is one of the oldest methods of cluster analysis. It is defined as follows. Two objects a and b belong to the same single-link cluster at level d if there exists a chain of intermediate objects $i_1, \ldots, i_{m-1}$ linking them such that all the distances

$$d_{i_k, i_{k+1}} \leq d \quad \text{for } k = 0, \ldots, m-1$$

where $i_0 = a$ and $i_m = b$. The single-link groups for the data of Figure 10.1 for a threshold of $d = 2.0$, 3.0 and 5.0 are $\{(1, 2), (5, 6), (3), (4)\}$, $\{(1, 2, 3), (5, 6), (4)\}$ and $\{(1, 2, 3, 5, 6), (4)\}$.

We shall illustrate the method by example with an agglomerative algorithm in which, at each stage of the algorithm, the closest two groups are fused to form a new group, where the distance between two groups, A and B, is the distance between their closest members, i.e.

$$d_{AB} = \min_{i \in A, j \in B} d_{ij} \tag{10.1}$$

Consider the dissimilarity matrix for each pair of objects in a set comprising six individuals:

	1	2	3	4	5	6
1	0	4	13	24	12	8
2		0	10	22	11	10
3			0	7	3	9
4				0	6	18
5					0	8.5
6						0

The closest two groups (which contain a single object each at this stage) are those containing the individuals 3 and 5. These are fused to form a new group $\{3, 5\}$ and the distances between this new group and the remaining groups calculated according to (10.1) so that $d_{1,(3,5)} = \min\{d_{13}, d_{15}\} = 12$, $d_{2,(3,5)} = \min\{d_{23}, d_{25}\} = 10$, $d_{4,(3,5)} = 6$, $d_{6,(3,5)} = 8.5$, giving the new dissimilarity matrix

	1	2	(3, 5)	4	6
1	0	4	12	24	8
2		0	10	22	10
(3, 5)			0	6	8.5
4				0	18
6					0

The closest two groups now are those containing objects 1 and 2; therefore these are fused to form a new group (1, 2). We now have four clusters (1, 2), (3, 5), 4 and 6. The distance between the new group and the other three clusters is calculated: $d_{(1,2)(3,5)} = \min\{d_{13}, d_{23}, d_{15}, d_{25}\} = 10$, $d_{(1,2)4} = \min\{d_{14}, d_{24}\} = 22$ $d_{(1,2)6} = \min\{d_{16}, d_{26}\} = 8$. The new dissimilarity matrix is

	(1, 2)	(3, 5)	4	6
(1, 2)	0	10	22	8
(3, 5)		0	6	8.5
4			0	18
6				0

The closest two groups are now those containing 4 and (3, 5). These are fused to form (3, 4, 5) and a new dissimilarity matrix calculated. This is given below with the result of fusing the next two groups. The single-link dendrogram is given in Figure 10.3.

	(1, 2)	(3, 4, 5)	6
(1, 2)	0	10	8
(3, 4, 5)		0	8.5
6			0

	(1, 2, 6)	(3, 4, 5)
(1, 2, 6)	0	8.5
(3, 4, 5)		0

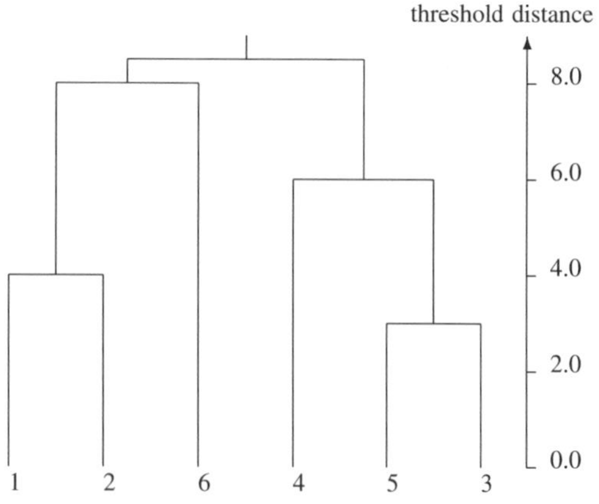

Figure 10.3 Single-link dendrogram

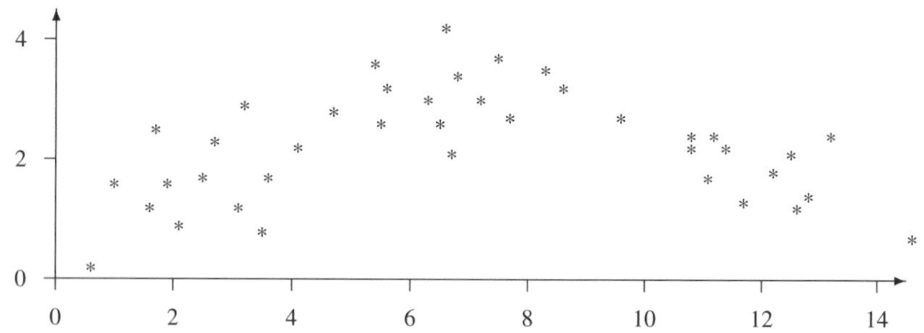

Figure 10.4 Illustration of chaining with the single-link method

The above agglomerative algorithm for a single-link method illustrates the fact that it takes only a single link to join two distinct groups and that the distance between two groups is the distance of their closest neighbours. Hence the alternative name of *nearest-neighbour method.* A consequence of this joining together by a single link is that some groups can become elongated, with some distant points, having little in common, being grouped together because there is a chain of intermediate objects. This drawback of *chaining* is illustrated in Figures 10.4 and 10.5. Figure 10.4 shows a distribution of data samples. Figure 10.5 shows the single-link three-group solution for the data in Figure 10.4. These groups do not correspond to those suggested by the data in Figure 10.4.

There are many algorithms for finding a single-link tree. Some are agglomerative, like the procedure described above, some are divisive; some are based on an ultrametric transformation and others generate the single-link tree via the minimum spanning tree (see Rohlf, 1982 for a review of algorithms). The algorithms vary in their computational efficiency, storage requirements and ease of implementation. Sibson's (1973) algorithm uses the property that only local changes in the reduced dissimilarity result when two clusters are merged, and it has been extended to the complete-link method discussed in the following section. It has computational requirements $\mathcal{O}(n^2)$, for n objects. More time-efficient algorithms are possible if knowledge of the metric properties of the space

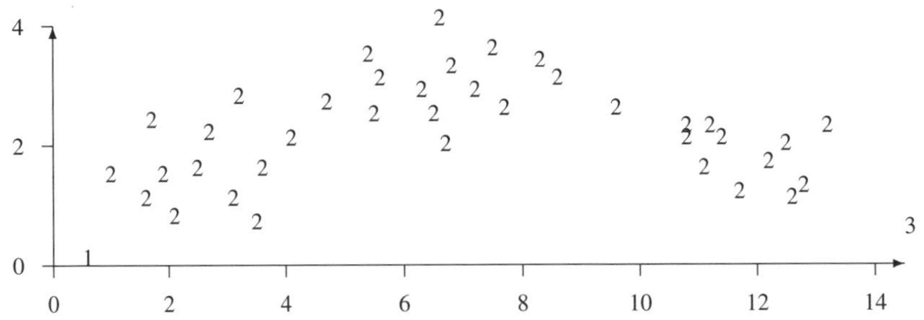

Figure 10.5 Single-link three-group solution for the data of Figure 10.4

in which the data lie is taken into account. In such circumstances, it is not necessary to compute all dissimilarity coefficients. Also, preprocessing the data to facilitate searches for nearest neighbours can reduce computational complexity.

10.2.2 Complete-link method

In the *complete-link* or *furthest-neighbour method* the distance between two groups A and B is the distance between the two furthest points, one taken from each group:

$$d_{AB} = \max_{i \in A, j \in B} d_{ij}$$

In the example used to illustrate the single-link method, the second stage dissimilarity matrix (after merging the closest groups 3 and 5 using the complete-link rule above) becomes

	1	2	(3, 5)	4	6
1	0	4	13	24	8
2		0	11	22	10
(3, 5)			0	7	9
4				0	18
6					0

The final complete-link dendrogram is shown in Figure 10.6. At each stage, the closest groups are merged of course. The difference between this method and the single-link method is the measure of distance between groups. The groups found by sectioning the complete-link dendrogram at level h have the property that $d_{ij} < h$ for all members in the group. The method concentrates on the internal *cohesion* of groups, in contrast to the single-link method, which seeks isolated groups. Sectioning a single-link dendrogram at a level h gives groups with the property that they are separated from each other by at least a 'distance' h.

Defays (1977) provides an algorithm for the complete-link method using the same representation as Sibson. It should be noted that the algorithm is sensitive to the ordering of the data, and consequently has several solutions. Thus it provides only an approximate complete-link clustering.

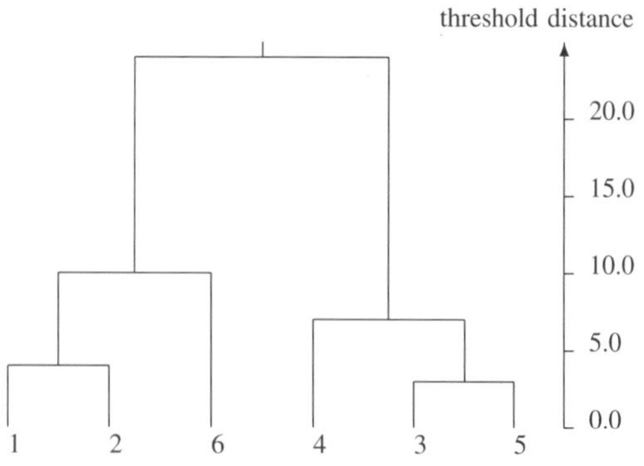

Figure 10.6 Complete-link dendrogram

10.2.3 Sum-of-squares method

The sum-of-squares method is appropriate for the clustering of points in Euclidean space. The aim is to minimise the total within-group sum of squares. *Ward's hierarchical clustering method* (Ward, 1963) uses an agglomerative algorithm to produce a set of hierarchically nested partitions that can be represented by a dendrogram. However, the *optimal* sum-of-squares partitions for different numbers of groups are not necessarily hierarchically nested. Thus the algorithm is suboptimal.

At each stage of the algorithm, the two groups that produce the smallest increase in the total within-group sum of squares are amalgamated. The dissimilarity between two groups is defined to be the increase in the total sum of squares that would result if they were amalgamated. The updating formula for the dissimilarity matrix is

$$d_{i+j,k} = \frac{n_k + n_i}{n_k + n_i + n_j} d_{ik} + \frac{n_k + n_j}{n_k + n_i + n_j} d_{jk} - \frac{n_k}{n_k + n_i + n_j} d_{ij}$$

where $d_{i+j,k}$ is the distance between the amalgamated groups $i + j$ and the group k and n_i is the number of objects in group i. Initially, each group contains a single object and the element of the dissimilarity matrix, d_{ij}, is the squared Euclidean distance between the ith and the jth object.

10.2.4 General agglomerative algorithm

Many agglomerative algorithms for producing hierarchical trees can be expressed as a special case of a single algorithm. The algorithms differ in the way that the dissimilarity matrix is updated. The Lance–Williams recurrence formula expresses the dissimilarity between a cluster k and the cluster formed by joining i and j as

$$d_{i+j,k} = a_i d_{ik} + a_j d_{jk} + b d_{ij} + c|d_{ik} - d_{jk}|$$

Table 10.1 Special cases of the general agglomerative algorithm

	a_i	b	c
Single link	$\frac{1}{2}$	0	$-\frac{1}{2}$
Complete link	$\frac{1}{2}$	0	$\frac{1}{2}$
Centroid	$\frac{n_i}{n_i+n_j}$	$-\frac{n_i n_j}{(n_i+n_j)^2}$	0
Median	$\frac{1}{2}$	$-\frac{1}{4}$	0
Group average link	$\frac{n_i}{n_i+n_j}$	0	0
Ward's method	$\frac{n_i+n_k}{n_i+n_j+n_k}$	$-\frac{n_k}{n_i+n_j+n_k}$	0

where a_i, b and c are parameters that, if chosen appropriately, will give an agglomerative algorithm for implementing some of the more commonly used methods (see Table 10.1).

Centroid distance This defines the distance between two clusters to be the distance between the cluster means or centroids.

Median distance When a small cluster is joined to a larger one, the centroid of the result will be close to the centroid of the larger cluster. For some problems this may be a disadvantage. This measure attempts to overcome this by defining the distance between two clusters to be the distance between the medians of the clusters.

Group average link In the group average method, the distance between two clusters is defined to be the average of the dissimilarities between all pairs of individuals, one from each group:

$$d_{AB} = \frac{1}{n_i n_j} \sum_{i \in A, j \in B} d_{ij}$$

10.2.5 Properties of a hierarchical classification

What desirable properties should a hierarchical clustering method possess? It is difficult to write down a set of properties on which everyone will agree. What might be a set of common-sense properties to one person may be the extreme position of another. Jardine and Sibson (1971) suggest a set of six mathematical conditions that an ultrametric transformation should satisfy; for example, that the results of the method should not depend on the labelling of the data. They show that the single-link method is the only one to satisfy all their conditions and recommend this method of clustering. However, this method has its drawbacks (as do all methods), which has led people to question the plausibility of the set of conditions proposed by Jardine and Sibson. We shall not list the conditions here but refer to Jardine and Sibson (1971) and Williams *et al.* (1971) for further discussion.

10.2.6 Example application study

The problem Weather classification for the study of long-term trends in climate change research (Huth *et al.*, 1993).

The data Daily weather data in winter months (December–February) at Prague-Clementinum were recorded. The data came from two 14-year periods, 1951–1964 and 1965–1978, each consisting of 1263 days. Daily weather was characterised by eight variables – daily mean temperature, temperature amplitude (daily maximum minus minimum), relative humidity, wind speed, zonal and meridional wind components, cloudiness and temperature tendency (the difference between the mean temperature of the day and its predecessor).

These were transformed to five variables through a principal components analysis of the correlation matrix. The five variables accounted for 83% of the total variance in the data.

The model The average-link method was used. At the beginning, each day is an individual cluster. The clusters merge according to the average-link algorithm (Section 10.2.4) until, at the final stage, a single cluster, containing all days, is formed. In the authors' opinion, too many clusters, and one large cluster, are undesirable. Also, it was expected that the number of clusters should correspond roughly with the number of synoptic types of weather for the region (about 30 types).

Training procedure The data were clustered to find groups containing days with the weather as 'uniform as possible'. The average-link procedure terminated when the difference between the properties of merging clusters exceeded a predefined criterion.

Results Removing clusters of sizes less than five days resulted in 28 clusters for 1965–1978 and 29 clusters for 1951–1964 (similar to the number of synoptic types for the region).

10.2.7 Summary

The concept of having a hierarchy of nested clusters was developed primarily in the biological field and may be inappropriate to model the structure in some data. Each hierarchical method imposes its own structure on the data. The single-link method seeks isolated clusters, but is generally not favoured, even though it is the only one satisfying the conditions proposed by Jardine and Sibson. It is subject to the chaining effect, which can result in long straggly groups. This may be useful in some circumstances if the clusters you seek are not homogeneous, but it can mean that distinct groups are not resolved because of intermediate points present between the groups. The group average, complete link and Ward's method tend to concentrate on internal cohesion, producing homogeneous, compact (often spherical) groups.

The centroid and median methods may lead to *inversions* (a reduction in the dissimilarity between an object and a cluster when the cluster increases in size) which make the dendrogram difficult to interpret. Also, ties in the dissimilarities may lead to multiple solutions (*non-uniqueness*) of which the user should be aware (Morgan and Ray, 1995).

Hierarchical agglomerative methods are one of the most common clustering techniques employed. Divisive algorithms are less popular, but efficient algorithms have been proposed, based on recursive partitioning of the cluster with largest diameter (Guénoche *et al.*, 1991).

10.3 Quick partitions

Many of the techniques described subsequently in this chapter are implemented by algorithms that require an initial partition of the data. The normal mixture methods require initial estimates of means and covariance matrices. These could be sample-based estimates derived from an initial partition. The k-means algorithm also requires an initial set of means. Hierarchical vector quantisation (see Section 10.5.3) requires initial estimates of the code vectors, and similarly the topographic mappings require initialisation of the weight vectors. In the context of discrimination, radial basis functions, introduced in Chapter 5, require initial estimates for 'centres'. These could be derived using the quick partition methods of this section or be the result of a more principled clustering approach (which in turn may need to be initialised).

Let us suppose that we have a set of n data samples and we wish to find an initial partition into k groups, or to find k seed vectors. We can always find a seed vector, given a group of objects, by taking the group mean. Also, we can partition a set, given k vectors, using a nearest-neighbour assignment rule. There are many heuristic partition methods. We shall consider a few of them.

1. Random k selection We wish to have k different vectors, so we select one randomly from the whole data set, another from the remaining $n-1$ samples in the data set, and so on. In a supervised classification problem, these vectors should ideally be spread across all classes.

2. Variable division Choose a single variable. This may be selected from one of the measured variables or be a linear combination of variables; for example, the first principal component. Divide it into k equal intervals that span the range of the variable. The data are partitioned according to which bin they fall in and k seed vectors are found from the means of each group.

3. Leader algorithm The leader cluster algorithm (Hartigan, 1975; Späth, 1980) partitions a data set such that for each group there is a leader object and all other objects within the group are within a distance T of the leading example. Figure 10.7 illustrates a partition in two dimensions. The first data point, A, is taken as the centre of the first group. Successive data points are examined. If they fall inside the circle centred at A of radius T then they are assigned to group 1. The first data sample examined to fall outside

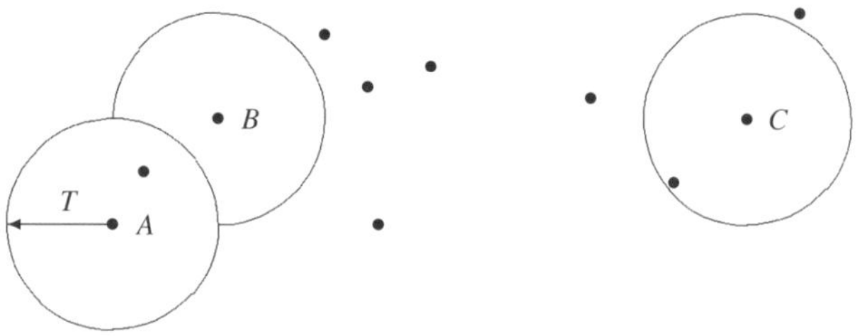

Figure 10.7 Leader clustering

the circle, say at B, is taken as the leader of the second group. Further data points are examined to see if they fall within the first two clusters. The first one to fall outside, say at C, is taken as the centre of the third cluster and so on.

Points to note about the algorithm:

1. All cluster centres are at least a distance T from each other.
2. It is fast, requiring only one pass through the data set.
3. It can be applied to a given dissimilarity matrix.
4. It is dependent on the ordering of the data set. The first point is always a cluster leader. Also, initial clusters tend to be larger than later ones.
5. The distance T is specified, not the number of clusters.

10.4 Mixture models

10.4.1 Model description

In the mixture method of clustering, each different group in the population is assumed to be described by a different probability distribution. These different probability distributions may belong to the same family but differ in the values they take for the parameters of the distribution. Alternatively, mixtures may comprise sums of different component densities (for modelling different effects such as a signal and noise). The population is described by a *finite mixture distribution* of the form

$$p(\boldsymbol{x}) = \sum_{i=1}^{g} \pi_i p(\boldsymbol{x}; \boldsymbol{\theta}_i)$$

where the π_i are the mixing proportions ($\sum_{i=1}^{g} \pi_i = 1$) and $p(\boldsymbol{x}; \boldsymbol{\theta}_i)$ is a p-dimensional probability function depending on a parameter vector $\boldsymbol{\theta}_i$. There are three sets of parameters to estimate: the values of π_i, the components of the vectors $\boldsymbol{\theta}_i$ and the value of g, the number of groups in the population.

Many forms of mixture distributions have been considered and there are many methods for estimating their parameters. An example of a mixture distribution for continuous variables is the mixture of normal distributions

$$p(\boldsymbol{x}) = \sum_{i=1}^{g} \pi_i p(\boldsymbol{x}; \boldsymbol{\Sigma}_i, \boldsymbol{\mu}_i)$$

where $\boldsymbol{\mu}_i$ and $\boldsymbol{\Sigma}_i$ are the means and covariance matrices of a multivariate normal distribution

$$p(\boldsymbol{x}; \boldsymbol{\Sigma}_i, \boldsymbol{\mu}_i) = \frac{1}{(2\pi)^{\frac{p}{2}} |\boldsymbol{\Sigma}_i|^{\frac{1}{2}}} \exp\left\{-\frac{1}{2}(\boldsymbol{x} - \boldsymbol{\mu}_i)^T \boldsymbol{\Sigma}_i^{-1} (\boldsymbol{x} - \boldsymbol{\mu}_i)\right\}$$

and a mixture for binary variables is

$$p(\boldsymbol{x}) = \sum_{i=1}^{g} \pi_i p(\boldsymbol{x}; \boldsymbol{\theta}_i)$$

where

$$p(\boldsymbol{x}; \boldsymbol{\theta}_j) = \prod_{l=1}^{p} \theta_{jl}^{x_l} (1 - \theta_{jl})^{1-x_l}$$

is the multivariate Bernoulli density. The value of θ_{jl} is the probability that variable l in the jth group is unity.

Maximum likelihood procedures for estimating the parameters of normal mixture distributions were given in Chapter 2. Other examples of continuous and discrete mixture distributions, and methods of parameter estimation, can be found in Everitt and Hand (1981) and Titterington *et al.* (1985). Also, in some applications, variables are often of a mixed type – both continuous and discrete.

The usual approach to clustering using finite mixture distributions is first of all to specify the form of the component distributions, $p(\boldsymbol{x}, \boldsymbol{\theta}_i)$. Then the number of clusters, g, is prescribed. The parameters of the model are now estimated and the objects are grouped on the basis of their estimated posterior probabilities of group membership; that is, the object $\boldsymbol{x}$ is assigned to group i if

$$\pi_i p(\boldsymbol{x}; \boldsymbol{\theta}_i) \geq \pi_j p(\boldsymbol{x}; \boldsymbol{\theta}_j) \quad \text{for all } j \neq i;\ j = 1, \ldots, g$$

Clustering using a normal mixture model may be achieved by using the EM algorithm described in Chapter 2, to which we refer for further details.

The main difficulty with the method of mixtures concerns the number of components, g (see Chapter 2). This is the question of model selection we return to many times in this book. Many algorithms require g to be specified before the remaining parameters can be estimated. Several test statistics have been put forward. Many apply to special cases such as assessing the question as to whether or not the data come from a single component distribution or a two-component mixture. However, others have been proposed based on likelihood ratio tests (Everitt and Hand, 1981, Chapter 5; Titterington *et al.*, 1985, Chapter 5).

Another problem with a mixture model approach is that there may be many local minima of the likelihood function and several initial configurations may have to be tried before a satisfactory clustering is produced. In any case, it is worthwhile trying several initialisations, since agreement between the resulting classifications lends more weight to the chosen solution. Celeux and Govaert (1992) describe approaches for developing the basic EM algorithm to overcome the problem of local optima.

There are several forms of the normal mixture model that trade off the number of components against the complexity of each component. For example, we may require that, instead of arbitrary covariance matrices, the covariance matrices are proportional to one another or have common principal components. This reduces the number of parameters per mixture component (Celeux and Govaert, 1995).

10.4.2 Example application study

The problem Flaw detection in textile (denim) fabric before final assembly of the garment (Campbell *et al.*, 1997).

Summary The approach detects a linear pattern in preprocessed images via model-based clustering. It employs an approximate Bayes factor which provides a criterion for assessing the evidence for the presence of a defect.

The data Two-dimensional point pattern data are generated by thresholding and cleaning, using mathematical morphology, images of fabric. Two fabric images are used, each about 500×500 pixels in size.

The model A two-component mixture model was used to model the data gathered from images of fabric: a *Poisson noise* component to model the background and a (possibly highly elliptical) Gaussian cluster to model the anomaly.

Training procedure The model parameters were determined by maximising the likelihood using the EM algorithm (see Chapter 2). Taking $\mathcal{A}$ to be the area of the data region, the *Bayesian information criterion* (BIC) was calculated:

$$\text{BIC} = 2\log(L) + 2n\log(\mathcal{A}) - 6\log(n)$$

where n is the number of samples and L is the value of the likelihood at its maximum. A value of BIC greater than 6 indicates 'strong evidence' for a defect.

Results were presented for some representative examples, and contrasted with a Hough transform.

10.5 Sum-of-squares methods

Sum-of-squares methods find a partition of the data that maximises a predefined clustering criterion based on the within-class and between-class scatter matrices. The methods differ in the choice of clustering criterion optimised and the optimisation procedure adopted. However, the problem all methods seek to solve is given a set of n data samples, to partition the data into g clusters so that the clustering criterion is optimised.

Most methods are suboptimal. Computational requirements prohibit optimal schemes, even for moderate values of n. Therefore we require methods that, although producing a suboptimal partition, give a value of the clustering criterion that is not much greater than the optimal one. First of all, let us consider the various criteria that have been proposed.

10.5.1 Clustering criteria

Let the n data samples be $\boldsymbol{x}_1, \ldots, \boldsymbol{x}_n$. The sample covariance matrix, $\hat{\boldsymbol{\Sigma}}$, is given by

$$\hat{\boldsymbol{\Sigma}} = \frac{1}{n}\sum_{i=1}^{n}(\boldsymbol{x}_i - \boldsymbol{m})(\boldsymbol{x}_i - \boldsymbol{m})^T$$

where $\boldsymbol{m} = \frac{1}{n}\sum_{i=1}^{n}\boldsymbol{x}_i$, the sample mean. Let there be g clusters. The *within-class scatter matrix* or *pooled within-group scatter matrix* is

$$\boldsymbol{S}_W = \frac{1}{n}\sum_{j=1}^{g}\sum_{i=1}^{n} z_{ji}(\boldsymbol{x}_i - \boldsymbol{m}_j)(\boldsymbol{x}_i - \boldsymbol{m}_j)^T,$$

the sum of the sums of squares and cross-products (scatter) matrices over the g groups, where $z_{ji} = 1$ if $\boldsymbol{x}_i \in$ group j, 0 otherwise, $\boldsymbol{m}_j = \frac{1}{n_j}\sum_{i=1}^{n} z_{ji}\boldsymbol{x}_i$ is the mean of cluster j and $n_j = \sum_{i=1}^{n} z_{ji}$, the number in cluster j. The between-class scatter matrix is

$$\boldsymbol{S}_B = \hat{\boldsymbol{\Sigma}} - \boldsymbol{S}_W = \sum_{j=1}^{g}\frac{n_j}{n}(\boldsymbol{m}_j - \boldsymbol{m})(\boldsymbol{m}_j - \boldsymbol{m})^T$$

and describes the scatter of the cluster means about the total mean.

The most popular optimisation criteria are based on univariate functions of the above matrices and are similar to the criteria given in Chapter 9 on feature selection and extraction. The two areas of clustering and feature selection are very much related. In clustering we are seeking clusters that are internally cohesive but isolated from other clusters. We do not know the number of clusters. In feature selection or extraction, we have labelled data from a known number of groups or classes and we seek a transformation that makes the classes distinct. Therefore one that transforms the data into isolated clusters will achieve this.

1. Minimisation of Tr($\boldsymbol{S}_W$)

The trace of $\boldsymbol{S}_W$ is the sum of the diagonal elements

$$\mathrm{Tr}(\boldsymbol{S}_W) = \frac{1}{n}\sum_{j=1}^{g}\sum_{i=1}^{n} z_{ji}|\boldsymbol{x}_i - \boldsymbol{m}_j|^2$$
$$= \frac{1}{n}\sum_{j=1}^{g} S_j$$

where $S_j = \sum_{i=1}^{n} z_{ji}|\boldsymbol{x}_i - \boldsymbol{m}_j|^2$, the within-group sum of squares for group j. Thus, the minimisation of $\mathrm{Tr}(\boldsymbol{S}_W)$ is equivalent to minimising the total within-group sum of

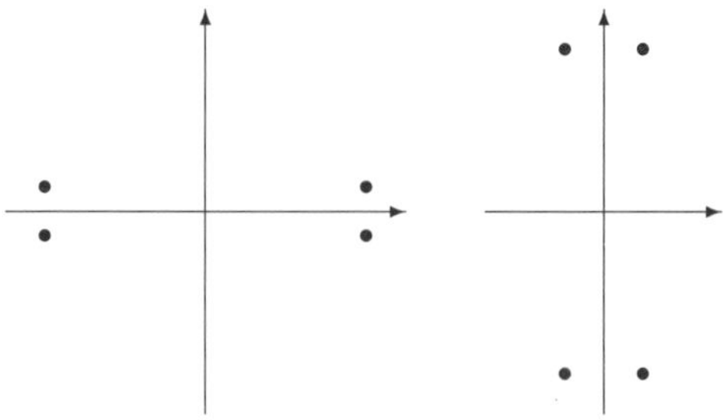

Figure 10.8 Example of effects of scaling in clustering

squares about the g centroids. Clustering methods that minimise this quantity are sometimes referred to as sum-of-squares or *minimum-variance* methods. They tend to produce clusters that are hyperellipsoidal in shape. The criterion is not invariant to the scale of the axes and usually some form of standardisation of the data must be performed prior to application of the method. Alternatively, criteria that are not invariant to linear transformations of the data may be employed.

2. Minimisation of $|S_W|/|\hat{\Sigma}|$

This criterion is invariant to nonsingular linear transformations of the data. For a given data set, it is equivalent to finding the partition of the data that minimises $|S_W|$ (the matrix $\hat{\Sigma}$ is independent of the partition).

3. Maximisation of $Tr(S_W^{-1}S_B)$

This is a generalisation of the sum-of-squares method in that the clusters are no longer hyperspherical, but hyperellipsoidal. It is equivalent to minimising the sum of squares under the Mahalanobis metric. It is also invariant to nonsingular transformations of the data.

4. Minimisation of $Tr(\hat{\Sigma}^{-1}S_W)$

This is identical to minimising the sum of squares for data that have been normalised to make the total scatter matrix equal to the identity.

Note that the two examples in Figure 10.8 would be clustered differently by the sum-of-squares method (criterion 1 above). However, since they only differ from each other by a linear transformation, they must both be local optima of a criterion invariant to linear transformations. Thus, it is not necessarily an advantage to use a method that is invariant to linear transformations of the data since structure may be lost. The final solution will depend very much on the initial assignment of points to clusters.

10.5.2 Clustering algorithms

The problem we are addressing is one in combinatorial optimisation. We seek a non-trivial partition of n objects into g groups for which the chosen criterion is optimised. However, to find the optimum partition requires the examination of every possible partition. The number of non-trivial partitions of n objects into g groups is

$$\frac{1}{g!}\sum_{i=1}^{g}(-1)^{g-i}\binom{g}{i}i^n$$

with the final term in the summation being most significant if $n \gg g$. This increases rapidly with the number of objects. For example, there are $2^{59} - 1 \approx 6 \times 10^{17}$ partitions of 60 objects into two groups. This makes exhaustive enumeration of all possible subsets infeasible. In fact, even the branch and bound procedure described in Chapter 9 is impractical for moderate values of n. Therefore, suboptimal solutions must be derived.

We now describe some of the more popular approaches. Many of the procedures require initial partitions of the data, from which group means may be calculated, or initial estimates of group means (from which an initial partition may be deduced using a nearest class mean rule). These were discussed in Section 10.3.

k-means

The aim of the *k-means* (which also goes by the names of the *c-means* or *iterative relocation* or *basic ISODATA*) algorithm is to partition the data into k clusters so that the within-group sum of squares (criterion 1 of Section 10.5.1) is minimised. The simplest form of the k-means algorithm is based on alternating two procedures. The first is one of assignment of objects to groups. An object is usually assigned to the group to whose mean it is closest in the Euclidean sense. The second procedure is the calculation of new group means based on the assignments. The process terminates when no movement of an object to another group will reduce the within-group sum of squares. Let us illustrate with a very simple example. Consider the two-dimensional data shown in Figure 10.9. Let us set $k = 2$ and choose two vectors from the data set as initial cluster mean vectors. Those selected are points 5 and 6. We now cycle through the data set and allocate individuals to groups A and B represented by the initial vectors 5 and 6 respectively. Individuals 1, 2, 3, 4 and 5 are allocated to A and individual 6 to B. New means are calculated and the within-group sum of squares is evaluated, giving 6.4. The results of this iteration are summarised in Table 10.2. The process is now repeated, using the new mean vectors as the reference vectors. This time, individuals 1, 2, 3, and 4 are allocated to group A and

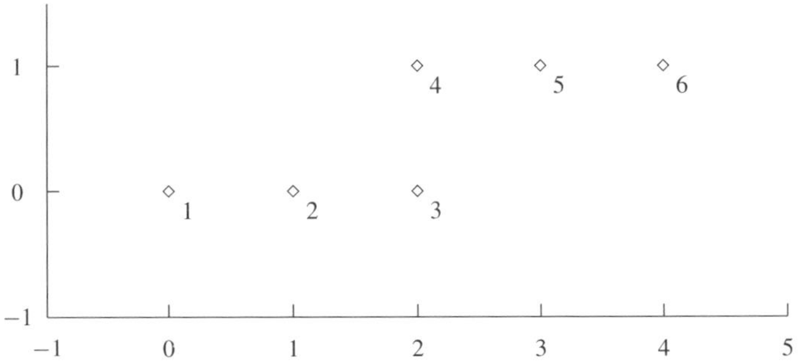

Figure 10.9 Data to illustrate the k-means procedure

Table 10.2 Summary of k-means iterations

Step	Group A		Group B		Tr($\boldsymbol{W}$)
	Membership	Mean	Membership	Mean	
1.	1, 2, 3, 4, 5	(1.6, 0.4)	6	(4.0, 1.0)	6.4
2.	1, 2, 3, 4	(1.25, 0.25)	5, 6	(3.5, 1.0)	4.0
3.	1, 2, 3, 4	(1.25, 0.25)	5, 6	(3.5, 1.0)	4.0

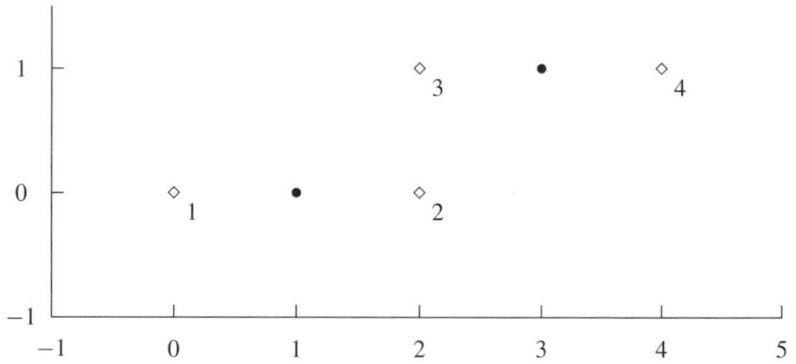

Figure 10.10 Data to illustrate the k-means local optimum

5 and 6 to group B. The within-group sum of squares has now decreased to 4.0. A third iteration produces no change in the within-group sum of squares.

The iterative procedure of allocating objects to groups on a nearest group mean basis, followed by recalculation of group means, gives the version of the k-means called HMEANS by Späth (1980). It is also termed Forgy's method or the basic ISODATA method.

There are two main problems with HMEANS. It may lead to empty groups and it may lead to a partition for which the sum-squared error could be reduced by moving an individual from one group to another. Thus the partition of the data by HMEANS is not necessarily one for which the within-group sum of squares is a minimum (see Selim and Ismail, 1984a, for a treatment of the convergence of this algorithm). For example, in Figure 10.10 four data points and two groups are illustrated. The means are at positions (1.0, 0.0) and (3.0, 1.0), with a sum-squared error of 4.0. Repeated iterations of the algorithm HMEANS will not alter that allocation. However, if we allocate object 2 to the group containing objects 3 and 4, the means are now at (0.0, 0.0) and (8/3, 2/3), and the sum-squared error is reduced to 10/3. This suggests an iterative procedure that cycles through the data points and allocates each to a group for which the within-group sum of squares is reduced the most. Allocation takes place on a sample-by-sample basis, rather than after a pass through the entire data set. An individual $\boldsymbol{x}_i$ (in group l) is assigned to group r if

$$\frac{n_l}{n_l - 1} d_{il}^2 > \frac{n_r}{n_r + 1} d_{ir}^2$$

where d_{il} is the distance to the lth centroid and n_l is the number in group l. The greatest decrease in the sum-squared error is achieved by choosing the group for which $n_r d_{ir}^2/(n_r + 1)$ is a minimum. This is the basis of the k-means algorithm.

There are many variants of the k-means algorithm to improve efficiency of the algorithm in terms of computing time and of achieving smaller error. Some algorithms allow new clusters to be created and existing ones deleted during the iterations. Others may move an object to another cluster on the basis of the best improvement in the objective function. Alternatively, the first encountered improvement during the pass through the data set could be used.

Nonlinear optimisation

The within-groups sum-of-squares criterion may be written in the form

$$\mathrm{Tr}(\boldsymbol{S}_W) = \frac{1}{n}\sum_{i=1}^{n}\sum_{k=1}^{g} z_{ki} \sum_{j=1}^{p}(x_{ij} - m_{kj})^2 \qquad (10.2)$$

where x_{ij} is the jth coordinate of the ith point ($i = 1, \ldots, n$; $j = 1, \ldots, p$), m_{kj} is the jth coordinate of the mean of the kth group and $z_{ki} = 1$ if the ith point belongs to the kth group and 0 otherwise. The mean quantities m_{kj} may be written as

$$m_{kj} = \frac{\sum_{i=1}^{n} z_{ki} x_{ij}}{\sum_{i=1}^{n} z_{ki}} \qquad (10.3)$$

for z_{ki} as defined above. To obtain an optimal partition, we must find the values of z_{ki} (either 0 or 1) for which (10.2) is a minimum.

The approach of Gordon and Henderson (1977) is to regard the $g \times n$ matrix $\boldsymbol{Z}$ with (i, j)th element z_{ij} as consisting of real-valued quantities (as opposed to binary quantities) with the property

$$\sum_{k=1}^{g} z_{ki} = 1 \quad \text{and} \quad z_{ki} \geq 0 \quad (i = 1, \ldots, n;\ k = 1, \ldots, g) \qquad (10.4)$$

Minimisation of (10.2) with respect to z_{ki} ($i = 1, \ldots, n$; $k = 1, \ldots, g$), subject to the constraints above, yields a final solution for $\boldsymbol{Z}$ with elements that are all 0 or 1. Therefore we can obtain a partition by minimising (10.2) subject to the constraints (10.4) and assigning objects to groups on the basis of the values z_{ik}. Thus, m_{kj} is not equal to a group mean until the iteration has converged.

The problem can be transformed to one of unconstrained optimisation by writing z_{ji} as

$$z_{ji} = \frac{\exp(\nu_{ji})}{\sum_{k=1}^{g} \exp(\nu_{ki})} \qquad (j = 1, \ldots, g;\ i = 1, \ldots, n)$$

that is, we regard $\mathrm{Tr}\{\boldsymbol{S}_W\}$ as a nonlinear function of parameters $\nu_{ki}, i = 1, \ldots, n$; $k = 1, \ldots, g$, and seek a minimum of $\mathrm{Tr}\{\boldsymbol{S}_W(\nu)\}$. Other forms of transformation to unconstrained optimisation are possible. However, for the particular form given above, the gradient of $\mathrm{Tr}\{\boldsymbol{S}_W(\nu)\}$ with respect to ν_{ab} has the simple form

$$\frac{\partial \mathrm{Tr}\{\boldsymbol{S}_W(\nu)\}}{\partial \nu_{ab}} = \frac{1}{n} \sum_{k=1}^{g} z_{kb}(\delta_{ka} - z_{ab})|\boldsymbol{x}_b - \boldsymbol{m}_k|^2 \tag{10.5}$$

where $\delta_{ka} = 0, k \neq a$, and 1 otherwise. There are many nonlinear optimisation schemes that can be used. The parameters ν_{ij} must be given initial values. Gordon and Henderson (1977) suggest choosing an initial set of random values z_{ji} uniformly distributed in the range $[1, 1 + a]$ and scaled so that their sum is unity. A value of about 2 is suggested for the parameter a. Then ν_{ji} is given by $\nu_{ji} = \log(z_{ji})$.

Fuzzy k-means

The partitioning methods described so far in this chapter have the property that each object belongs to one group only, though the mixture model can be regarded as providing degrees of cluster membership. Indeed, the early work on fuzzy clustering was closely related to multivariate mixture models. The basic idea of the fuzzy clustering method is that patterns are allowed to belong to all clusters with different degrees of membership. The first generalisation of the k-means algorithm was presented by Dunn (1974). The *fuzzy k-means* (or fuzzy c-means) algorithm attempts to find a solution for parameters y_{ji} $(i = 1, \ldots, n;\ j = 1, \ldots, g)$ for which

$$J_r = \sum_{i=1}^{n} \sum_{j=1}^{g} y_{ji}^r |\boldsymbol{x}_i - \boldsymbol{m}_j|^2 \tag{10.6}$$

is minimised subject to the constraints

$$\sum_{j=1}^{g} y_{ji} = 1 \quad 1 \leq i \leq n$$

$$y_{ji} \geq 0 \quad i = 1, \ldots, n;\ j = 1, \ldots, g$$

The parameter y_{ji} represents the *degree of association* or *membership function* of the ith pattern or object with the jth group. In (10.6), r is a scalar termed the *weighting exponent* which controls the 'fuzziness' of the resulting clusters ($r \geq 1$) and $\boldsymbol{m}_j$ is the 'centroid' of the jth group

$$\boldsymbol{m}_j = \frac{\sum_{i=1}^{n} y_{ji}^r \boldsymbol{x}_i}{\sum_{i=1}^{n} y_{ji}^r} \tag{10.7}$$

A value of $r = 1$ gives the same problem as the nonlinear optimisation scheme presented earlier. In that case, we know that a minimum of (10.6) gives values for the y_{ji} that are either 0 or 1.

The basic algorithm is iterative and can be stated as follows (Bezdek, 1981).

1. Select r $(1 < r < \infty)$; initialise the membership function values $y_{ji}, i = 1, \ldots, n;$ $j = 1, \ldots, g$.
2. Compute the cluster centres $\boldsymbol{m}_j$, $j = 1, \ldots, g$, according to (10.7).

3. Compute the distances d_{ij}, $i = 1, \ldots, n$; $j = 1, \ldots, g$, where $d_{ij} = |\boldsymbol{x}_i - \boldsymbol{m}_j|$.

4. Compute the membership function: if $d_{il} = 0$ for some l, $y_{li} = 1$, and $y_{ji} = 0$, for all $j \neq l$; otherwise

$$y_{ji} = \frac{1}{\sum_{k=1}^{g} \left(\frac{d_{ij}}{d_{ik}}\right)^{\frac{2}{r-1}}}$$

5. If not converged, go to step 2.

As $r \to 1$, this algorithm tends to the basic k-means algorithm. Improvements to this basic algorithm, resulting in faster convergence, are described by Kamel and Selim (1994).

Several stopping rules have been proposed (Ismail, 1988). One is to terminate the algorithm when the relative change in the centroid values becomes small; that is, terminate when

$$D_z \triangleq \left\{ \sum_{j=1}^{g} |\boldsymbol{m}_j(k) - \boldsymbol{m}_j(k-1)|^2 \right\}^{\frac{1}{2}} < \epsilon$$

where $\boldsymbol{m}_j(k)$ is the value of the jth centroid on the kth iteration and ϵ is a user-specified threshold. Alternative stopping rules are based on changes in the membership function values, y_{ji}, or the cost function, J_r. Another condition based on the local optimality of the cost function is given by Selim and Ismail (1986). It is proposed to stop when

$$\max_{1 \le i \le n} \alpha_i < \epsilon$$

where

$$\alpha_i = \max_{1 \le j \le g} y_{ji}^{r-1} |\boldsymbol{x}_i - \boldsymbol{m}_j|^2 - \min_{1 \le j \le g} y_{ji}^{r-1} |\boldsymbol{x}_i - \boldsymbol{m}_j|^2$$

since at a local minimum, $\alpha_i = 0, i = 1, \ldots, n$.

Complete search

Complete search of the space of partitions of n objects into g groups is impractical for all but very small data sets. The branch and bound method (described in a feature subset selection context in Chapter 9) is one approach for finding the partition that results in the minimum value of the clustering criterion, without exhaustive enumeration. Nevertheless, it may still be impractical. Koontz *et al.* (1975) have developed an approach that extends the range of problems to which branch and bound can be applied. The criterion they seek to minimise is $\mathrm{Tr}\{\boldsymbol{S}_W\}$. Their approach is to divide the data set into 2^m independent sets. The branch and bound method is applied to each set separately and then sets are combined in pairs (to give 2^{m-1} sets) and the branch and bound method applied to each of these combined sets, using the results obtained from the branch and bound application to the constituent parts. This is continued until the branch and bound procedure is applied to the entire set. This hierarchical approach results in a considerable saving in computer time.

Other approaches based on global optimisation algorithms such as simulated annealing have also been proposed. Simulated annealing is a stochastic relaxation technique in which a randomly selected perturbation to the current configuration is accepted or rejected probabilistically. Selim and Al-Sultan (1991) apply the method to the minimisation of $\text{Tr}\{\boldsymbol{S}_W\}$. Generally the method is slow, but it can lead to effective solutions.

10.5.3 Vector quantisation

Vector quantisation (VQ) is not a *method* of producing clusters or partitions of a data set but rather an *application* of many of the clustering algorithms already presented. Indeed, many clustering techniques have been rediscovered in the vector quantisation literature. On the other hand, there are some important algorithms in the VQ literature that are not found in the standard texts on clustering. This section is included in the section on optimisation methods since in VQ a distortion measure (often, but by no means exclusively, based on the Euclidean distance) is optimised during training. A comprehensive and very readable account of the fundamentals of VQ is given by Gersho and Gray (1992).

VQ is the encoding of a p-dimensional vector $\boldsymbol{x}$ as one from a *codebook* of g vectors, $z_1, \ldots, z_g$, termed the *code vectors* or the *codewords*. The purpose of VQ is primarily to perform data compression. A vector quantiser consists of two components: an *encoder* and a *decoder* (see Figure 10.11).

The encoder maps an input vector, $\boldsymbol{x}$, to a scalar variable, y, taking discrete values $1, \ldots, g$. After transmission of the *index*, y, the inverse operation of reproducing an approximation $\boldsymbol{x}'$ to the original vector takes place. This is termed decoding and is a mapping from the index set $\mathcal{I} = \{1, \ldots, g\}$ to the codebook $\mathcal{C} = \{z_1, \ldots, z_g\}$. Codebook design is the problem of determining the codebook entries given a set of training samples. From a clustering point of view, we may regard the problem of codebook design as one of clustering the data and then choosing a representative vector for each cluster. These vectors could be cluster means, for example, and they form the entries in the codebook. They are indexed by integer values. Then the code vector for a given input vector $\boldsymbol{x}$ is the representative vector, say z, of the cluster to which $\boldsymbol{x}$ belongs. Membership of a cluster may be determined on a nearest-to-cluster-mean basis. The distortion or error in approximation is then $d(\boldsymbol{x}, z)$, the distance between $\boldsymbol{x}$ and z (see Figure 10.12).

The problem in VQ is to find a set of codebook vectors that characterise a data set. This is achieved by choosing the set of vectors for which a distortion measure between an input vector, $\boldsymbol{x}$, and its quantised vector, $\boldsymbol{x}'$, is minimised. Many distortion measures have been proposed, the most common being based on the squared error measure giving average distortion, D_2,

$$\begin{aligned} D_2 &= \int p(\boldsymbol{x})\, d(\boldsymbol{x}, \boldsymbol{x}')\, d\boldsymbol{x} \\ &= \int p(\boldsymbol{x}) \|\boldsymbol{x}'(y(\boldsymbol{x})) - \boldsymbol{x}\|^2\, d\boldsymbol{x} \end{aligned} \tag{10.8}$$

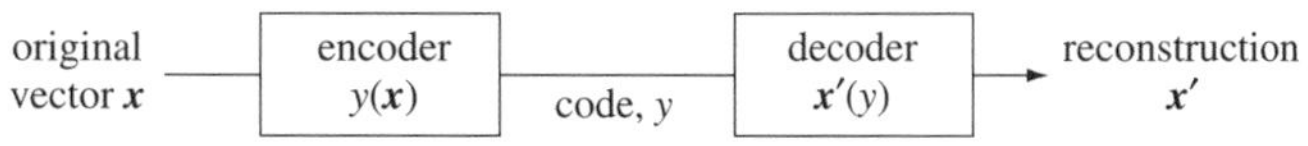

Figure 10.11 The encoding–decoding operation in vector quantisation

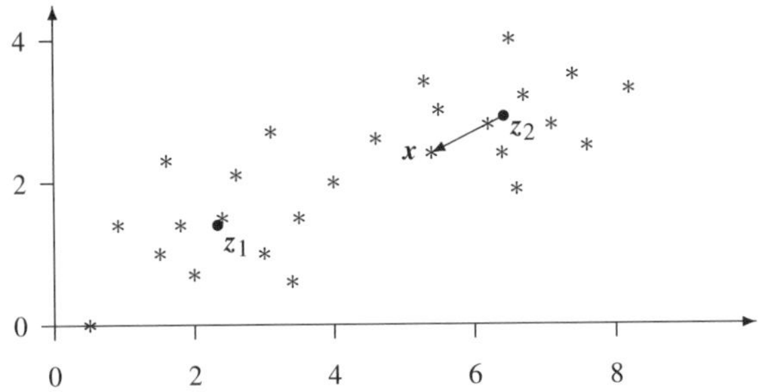

Figure 10.12 VQ distortion for two code vectors. The reconstruction of $\boldsymbol{x}$ after encoding and decoding is z_2, the nearest code vector. The distortion is $d(\boldsymbol{x}, z_2)$

Table 10.3 Some distortion measures used in vector quantisation

Type of norm	$d(\boldsymbol{x}, \boldsymbol{x}')$
L_2, Euclidean	$\lvert\boldsymbol{x}' - \boldsymbol{x}\rvert$
L_ν	$\left\{\sum_{i=1}^{p} \lvert\boldsymbol{x}' - \boldsymbol{x}\rvert^{\nu}\right\}^{\frac{1}{\nu}}$
Minkowski	$\max_{1 \leq i \leq p} \lvert x_i' - x_i\rvert$
Quadratic (for positive definite symmetric $\boldsymbol{B}$)	$(\boldsymbol{x}' - \boldsymbol{x})^T \boldsymbol{B}(\boldsymbol{x}' - \boldsymbol{x})$

where $p(\boldsymbol{x})$ is the probability density function over samples $\boldsymbol{x}$ used to train the vector quantiser and $\|.\|$ denotes the norm of a vector. Other distortion measures are given in Table 10.3.

For a finite number of training samples, $\boldsymbol{x}_1, \ldots, \boldsymbol{x}_n$, we may write the distortion as

$$D = \sum_{j=1}^{g} \sum_{\boldsymbol{x}_i \in S_j} d(\boldsymbol{x}_i, z_j)$$

where S_j is the set of training vectors for which $y(\boldsymbol{x}) = j$, i.e. those that map onto the jth code vector, z_j. For a given set of code vectors z_j, the partition that minimises the average distortion is constructed by mapping each $\boldsymbol{x}_i$ to the z_j for which $d(\boldsymbol{x}_i, z_j)$ is a minimum over all z_j – i.e. choosing the minimum distortion or nearest-neighbour code vector. Alternatively, for a given partition the code vector of a set S_j, z_j, is defined to be the vector for which

$$\sum_{\boldsymbol{x}_i \in S_j} d(\boldsymbol{u}, \boldsymbol{x}_i)$$

is a minimum with respect to $\boldsymbol{u}$. This vector is called the centroid (for the squared error measure it is the mean of the vectors $\boldsymbol{x}_i$).

This suggests an iterative algorithm for a vector quantiser:

1. Initialise the code vectors.
2. Given a set of code vectors, determine the minimum distortion partition.
3. Find the optimal set of code vectors for a given partition.
4. If the algorithm has not converged, then go to step 2.

This is clearly a variant of the k-means algorithms given earlier. It is identical to the basic k-means algorithm provided that the distortion measure used is the squared error distortion since all the training vectors are considered at each iteration rather than making an adjustment of code vectors by considering each in turn. This is known as the *generalised Lloyd algorithm* in the VQ literature (Gersho and Gray, 1992) or the LBG algorithm in the data compression literature. One of the main differences between the LBG algorithm and some of the k-means implementations is the method of initialisation of the centroid vectors. The LBG algorithm (Linde *et al.*, 1980) given below starts with a one-level quantiser (a single cluster) and, after obtaining a solution for the code vector z, 'splits' the vector z into two close vectors that are used as seed vectors for a two-level quantiser. This is run until convergence and a solution is obtained for the two-level quantiser. Then these two codewords are split to give four seed vectors for a four-level quantiser. The process is repeated so that finally quantisers of $1, 2, 4, \ldots, N$ levels are obtained (see Figure 10.13).

1. Initialise a code vector z_1 to be the group mean; initialise ϵ.
2. Given a set of m code vectors, 'split' each vector z_i to form $2m$ vectors, $z_i + \epsilon$ and $z_i - \epsilon$. Set $m = 2m$; relabel the code vectors as x'_i, $i = 1, \ldots, m$.
3. Given the set of code vectors, determine the minimum distortion partition.
4. Find the optimal set of code vectors for a given partition.

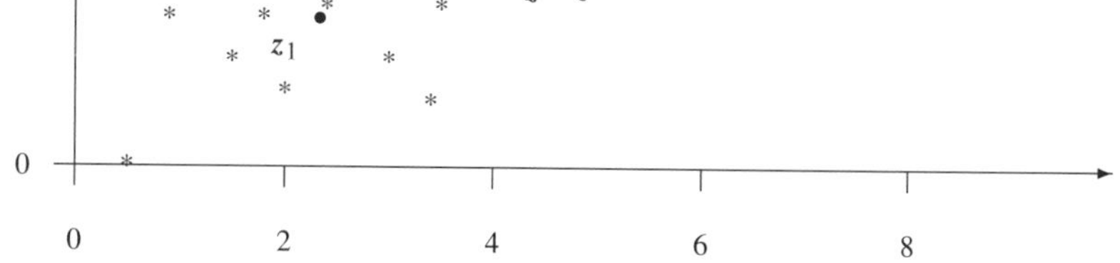

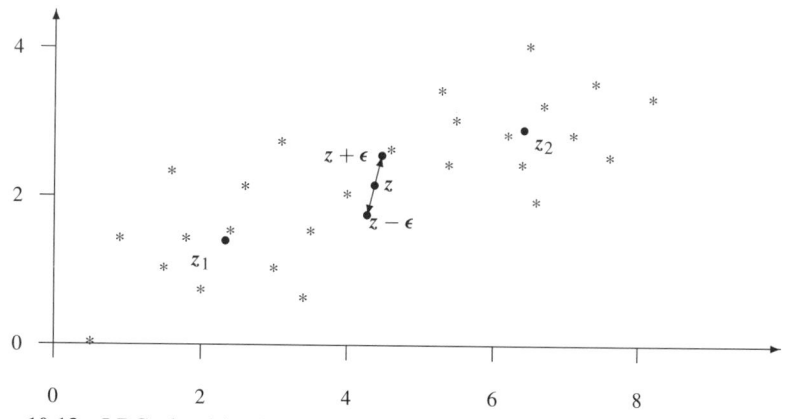

Figure 10.13 LBG algorithm illustration: z denotes the group centroid; z_1 and z_2 denote the code vectors for a two-level quantiser

5. Repeat steps 3 and 4 until convergence.

6. If $m \neq N$, the desired number of levels, go to step 2.

Although it appears that all we have achieved with the introduction of VQ in this chapter is yet another version of the k-means algorithm, the VQ framework allows us to introduce two important concepts: that of tree-structured codebook search that reduces the search complexity in VQ; and that of topographic mappings in which a topology is imposed on the code vectors.

Tree-structured vector quantisation

Tree-structured vector quantisation is a way of structuring the codebook in order to reduce the amount of computation required in the encoding operation. It is a special case of the classification trees or decision trees discussed in a discrimination context in Chapter 7. Here we shall consider *fixed-rate* coding, in which there are the same number of bits used to represent each code vector. *Variable-rate coding*, which allows pruning of the tree, will not be addressed. Pruning methods for classification trees are described in Chapter 7, and in the VQ context by Gersho and Gray (1992).

We shall begin our description of tree-structured VQ with a simple *binary tree* example. The first stage in the design procedure is to run the k-means algorithm on the entire data set to partition the set into two parts. This leads to two code vectors (the means of each cluster) (see Figure 10.14). Each group is considered in turn and the k-means algorithm applied to each group, partitioning each group into two parts again. This second stage then produces four code vectors and four associated clusters. The mth stage produces 2^m code vectors. The *total* number of code vectors produced in an m-stage design algorithm is $\sum_{i=1}^{m} 2^i = 2^{m+1} - 2$. This process produces a hierarchical clustering in which two clusters are disjoint or one wholly contains the other.

Encoding of a given vector, $\boldsymbol{x}$, proceeds by starting at the root of the tree (labelled A_0 in Figure 10.15) and comparing $\boldsymbol{x}$ with each of the two level 1 code vectors, identifying the nearest. We then proceed along the branch to A_1 and compare the vector $\boldsymbol{x}$ with the two code vectors at this level which were generated from members of the training in this group. Thus there are m comparisons in an m-stage encoder. This compares with 2^m

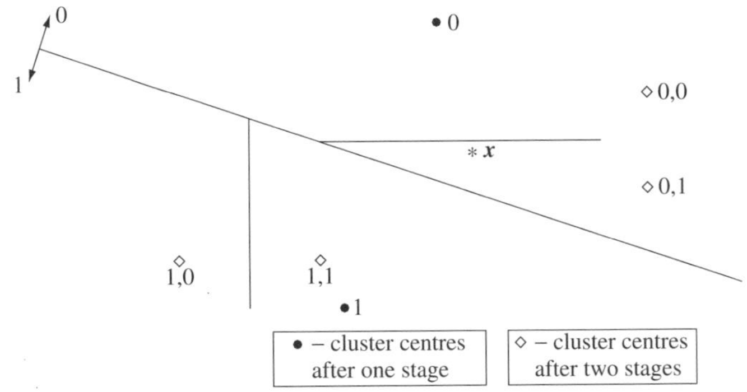

Figure 10.14 Tree-structured vector quantisation

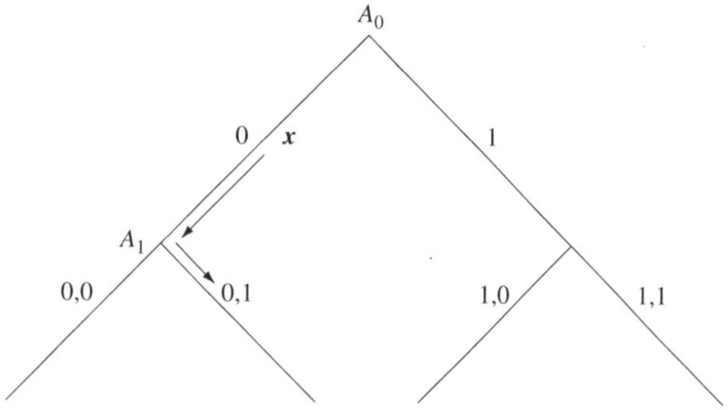

Figure 10.15 Tree-structured vector quantisation tree

code vectors at the final level. Tree-structured VQ may not be optimal in the sense that the nearest neighbour of the final level code vectors is not necessarily found (the final partition in Figure 10.14 does not consist of nearest-neighbour regions). However, the code has the property that it is a progressively closer approximation as it is generated and the method can lead to a considerable saving in encoding time.

Self-organising feature maps

Self-organising feature maps are a special kind of vector quantisation in which there is an ordering or topology imposed on the code vectors. The aim of self-organisation is to represent high-dimensional data as a low-dimensional array of numbers (usually a one- or two-dimensional array) that captures the structure in the original data. Distinct clusters of data points in the data space will map to distinct clusters of code vectors in the array, although the converse is not necessarily true: separated clusters in the array do not necessarily imply separated clusters of data points in the original data. In some ways, self-organising feature maps may be regarded as a method of exploratory data analysis in keeping with those described in Chapter 9. The basic algorithm has the k-means algorithm as a special case.

Figures 10.16 and 10.17 illustrate the results of the algorithm applied to data in two dimensions.

- In Figure 10.16, 50 data samples are distributed in three groups in two dimensions and we have used a self-organisation process to obtain a set of nine ordered cluster centres in one dimension. By a set of ordered cluster centres we mean that centre z_i is close in some sense to z_{i-1} and z_{i+1}. In the k-means algorithm, the order that the centres are stored in the computer is quite arbitrary and depends on the initialisation of the procedure.

- In Figure 10.17, the data (not shown) comprise 500 samples drawn from a uniform distribution over a square ($[-1 \leq x, y \leq 1]$) and do not lie on (or close to) a one-dimensional manifold in the two-dimensional space. Again, we have imposed a one-dimensional topology on the cluster centres, which are joined by straight lines. In this case, we have obtained a space-filling curve.

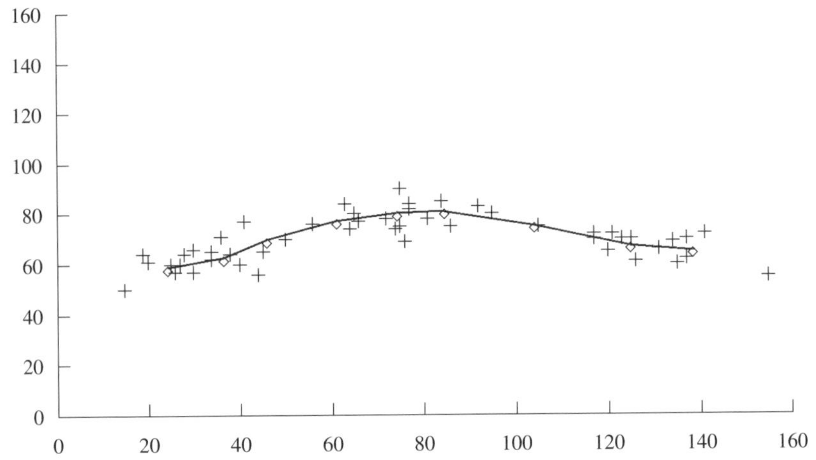

Figure 10.16 Topographic mapping. Adjacent cluster centres (◇) in the stored array of code vectors are joined

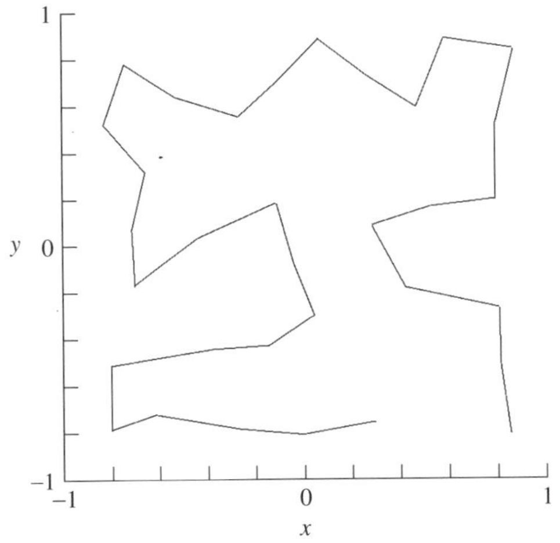

Figure 10.17 Topographic mapping for data uniformly distributed over a square. Thirty-three centres are determined and again adjacent cluster centres in the stored array are joined

In each of the above illustrations, a transformation to a reduced dimension is achieved using a topographic mapping in which there is an ordering on the cluster centres. Each point in the data space is mapped to the ordered index of its nearest cluster centre. The mapping is nonlinear, and for purposes of illustration we considered mappings to one dimension only. If the data do lie on a reduced-dimension manifold within a high-dimensional space, then it is possible for topographic mappings to capture the structure in the data and present it in a form that may aid interpretation. In a supervised classification problem, it is possible to label each cluster centre with a class label according to the majority of the objects for which that cluster centre is the nearest. Of course we can do this even if there were no ordering on the cluster centres, but the ordering does allow the relationships between classes (according to decision boundaries) to be viewed easily.

The algorithm for determining the cluster centres may take many forms. The basic approach is to cycle through the data set and adjust the cluster centres in the neighbourhood (suitably defined) of each data point. The algorithm is often presented as a function of time, where time refers to the number of presentations of a data sample. One algorithm is as follows.

1. Decide on the topology of the cluster centres (code vectors). Initialise α, the neighbourhood and the cluster centres $z_1, \ldots, z_N$.

2. Repeat until convergence:

   (a) Select a data sample $\boldsymbol{x}$ (one of the training samples) and find the closest centre: let $d_{j^*} = \min j(d_j)$, where

   $$d_j = |\boldsymbol{x} - z_j| \quad j = 1, \ldots, N$$

   (b) Update the code vectors in the neighbourhood, $\mathcal{N}_{j^*}$ of code vector z_{j^*}

   $$z(t+1) = z(t) + \alpha(t)(\boldsymbol{x}(t) - z(t)) \quad \text{for all centres } z \in \mathcal{N}_{j^*}$$

   where α is a learning rate that decreases with iteration number, t $(0 \leq \alpha \leq 1)$.

   (c) Decrease the neighbourhood and the learning parameter, α.

In order to apply the algorithm an initial set of weight vectors, the learning rate $\alpha(t)$ and the change with t of the neighbourhoods must be chosen.

Definition of topology The choice of topology of the cluster centres requires some prior knowledge of the data structure. For example, if you suspect circular topology in your data, then the topology of your cluster centres should reflect this. Alternatively, if you wish to map your data onto a two-dimensional surface, then a regular lattice structure for the code vectors may be sufficient.

Learning rate The learning rate, α, is a slowly decreasing function of t. It is suggested by Kohonen (1989) that it could be a linear function of t, stopping when α reaches 0, but there are no hard and fast rules for choosing $\alpha(t)$. It could be linear, inversely proportional to t or exponential. Haykin (1994) describes two phases: the *ordering* phase, of about 1000 iterations, when α starts close to unity and decreases, but remains above 0.1; and the *convergence* phase, when α decreases further and is maintained at a small value – 0.01 or less – for typically thousands of iterations.

Initialisation of code vectors Code vectors z_i are initialised to $\boldsymbol{m} + \boldsymbol{\epsilon}_i$, where $\boldsymbol{m}$ is the sample mean and $\boldsymbol{\epsilon}_i$ is a vector of small random values.

Decreasing the neighbourhood The topological neighbourhood $\mathcal{N}_j$ of a code vector z_j is itself a function of t and decreases as the number of iterations proceeds. Initially the neighbourhood may cover most of the code vectors ($z_{j-r}, \ldots, z_{j-1}, z_{j+1}, \ldots, z_{j+r}$ for large r), but towards the end of the iterations it covers the nearest (topological) neighbours z_{j-1} and z_{j+1} only. Finally it shrinks to zero. The problem is how to initialise the neighbourhood and how to reduce it as a function of t. During the *ordering* phase, the neighbourhood is decreased to cover only a few neighbours.

An alternative approach proposed by Luttrell (1989) is to fix the neighbourhood size and to start off with a few code vectors. The algorithm is run until convergence, and then the number of vectors is increased by adding vectors intermediate to those already calculated. The process is repeated and continued until a mapping of the desired size has been grown. Although the neighbourhood size is fixed, it starts off by covering a large area (since there are few centres) and the physical extent is reduced as the mapping grows. Specifically, given a data sample $\boldsymbol{x}$, if the nearest neighbour is z^*, then all code vectors z in the neighbourhood of z^* are updated according to

$$z \rightarrow z + \pi(z, z^*)(\boldsymbol{x} - z) \tag{10.9}$$

where π (> 0) is a function that depends on the position of z in the neighbourhood of z^*. For example, with a one-dimensional topology, we may take

$$\pi(z, z^*) = \begin{cases} 0.1 & \text{for } z = z^* \\ 0.01 & \text{for } z \text{ a topographic neighbour of } z^* \end{cases}$$

The Luttrell algorithm for a one-dimensional topographic mapping is as follows.

1. Initialise two code vectors, z_1 and z_2; set $m = 2$. Define the neighbourhood function, π; set the number of updates per code vector, u.
2. Repeat until the distortion is small enough or the maximum number of code vectors is reached:
   (a) For $j = 1$ to $m \times u$ do
   - Sample from the data set $\boldsymbol{x}_1, \ldots, \boldsymbol{x}_n$, say $\boldsymbol{x}$.
   - Determine the nearest-neighbour code vector, say z^*.
   - Update the code vectors according to: $z \rightarrow z + \pi(z, z^*)(\boldsymbol{x} - z)$.

   (b) Define $2m - 1$ new code vectors: for $j = m - 1$ down to 1 do
   - $z_{2j+1} = z_{j+1}$
   - $z_{2j} = \dfrac{z_j + z_{j+1}}{2}$

   (c) Set $m = 2m - 1$.

Topographic mappings have received widespread use as a means of exploratory data analysis (for example, Kraaijveld *et al.*, 1992; Roberts and Tarassenko, 1992) They have also been misused and applied when the ordering of the resulting cluster centres is irrelevant in any subsequent data analysis and a simple k-means approach could have been adopted. An assessment of the method and its relationship to other methods of multivariate analysis is provided by Murtagh and Hernández-Pajares (1995). Luttrell (1989) has derived an approximation to the basic learning algorithm from a VQ approach assuming a minimum distortion (Euclidean) and a robustness to noise on the codes. This puts the approach on a firmer mathematical footing. Also, the requirement for *ordered* cluster centres is demonstrated for a hierarchical vector quantiser.

Learning vector quantisation

Vector quantisation or clustering (in the sense of partitioning a data set, not seeking meaningful groupings of objects) is often performed as a preprocessor for supervised classification. There are several ways in which vector quantisers or self-organising maps have been used with labelled training data. In the radar target classification example of Luttrell (1995), each class is modelled separately using a self-organising map. This was chosen to feed in prior knowledge that the underlying manifold was a circle. Test data are classified by comparing each pattern with the prototype patterns in each of the self-organising maps (codebook entries) and classifying on a nearest-neighbour rule basis.

An alternative approach that uses vector quantisers in a supervised way is to model the whole of the training data with a single vector quantiser (rather than each class separately). Each training pattern is assigned to the nearest code vector, which is then labelled with the class of the majority of the patterns assigned to it. A test pattern is then classified using a nearest-neighbour rule using the labelled codebook entries.

Learning vector quantisation is a supervised generalisation of vector quantisation that takes account of class labels *in the training process*. The basic algorithm is given below.

1. Initialise cluster centres (or code vectors), $z_1, \dots, z_N$, and labels of cluster centres, $\omega_1, \dots, \omega_N$.

2. Select a sample $\boldsymbol{x}$ from the training data set with associated class $\omega_{\boldsymbol{x}}$ and find the closest centre: let $d_{j^*} = \min j(d_j)$, where

$$d_j = |\boldsymbol{x} - z_j| \quad j = 1, \dots, N$$

with corresponding centre z_{j^*} and class ω_{j^*}.

3. If $\omega_{\boldsymbol{x}} = \omega_{j^*}$ then update the nearest vector, z_{j^*}, according to

$$z_{j^*}(t+1) = z_{j^*}(t) + \alpha(t)(\boldsymbol{x}(t) - z_{j^*}(t))$$

where $0 < \alpha_t < 1$ and decreases with t, starting at about 0.1.

4. If $\omega_{\boldsymbol{x}} \neq \omega_{j^*}$ then update the nearest vector, z_{j^*}, according to

$$z_{j^*}(t+1) = z_{j^*}(t) - \alpha(t)(\boldsymbol{x}(t) - z_{j^*}(t))$$

5. Go to 1 and repeat until several passes have been made through the data set.

Correct classification of a pattern in the data set leads to a refinement of the codeword in the direction of the pattern. Incorrect classification leads to a movement of the codeword away from the training pattern.

Stochastic vector quantisation

In the approach to vector quantisation described in the previous section, a codebook is used to encode each input vector $\boldsymbol{x}$ as a code index y which is then decoded to produce an approximation, $\boldsymbol{x}'(y)$, to the original input vector. Determining the codebook vectors that characterise a data set is achieved by optimising an objective function, the most common being based on the squared error measure (equation (10.8))

$$D_2 = \int p(\boldsymbol{x})\|\boldsymbol{x}'(y(\boldsymbol{x})) - \boldsymbol{x}\|^2 \, d\boldsymbol{x}$$

Optimisation is achieved through an iterative process using a k-means algorithm or variants (e.g. the LBG algorithm). Encoding is the deterministic process of finding the nearest entry in the codebook.

Stochastic vector quantisation (SVQ, Luttrell, 1997, 1999a) is a generalisation of the standard approach in which an input vector $\boldsymbol{x}$ is encoded as a *vector* of code indices $\boldsymbol{y}$ (rather than as a *single* code index) that are stochastically sampled from a probability distribution $p(\boldsymbol{y}|\boldsymbol{x})$ that depends on the input vector $\boldsymbol{x}$. The decoding operation that produces a reconstruction, $\boldsymbol{x}'$, is also probabilistic, with $\boldsymbol{x}'$ being a sample drawn from $p(\boldsymbol{x}|\boldsymbol{y})$ given by

$$p(\boldsymbol{x}|\boldsymbol{y}) = \frac{p(\boldsymbol{y}|\boldsymbol{x})p(\boldsymbol{x})}{\int p(\boldsymbol{y}|\boldsymbol{z})p(\boldsymbol{z})\, d\boldsymbol{z}} \tag{10.10}$$

One of the key factors motivating the development of the SVQ approach is that of scalability to high dimensions. A problem with standard VQ is that the codebook grows exponentially in size as the dimensionality of the input vector is increased, assuming that the contribution to the reconstruction error from each dimension is held constant. This means that such vector quantisers are not appropriate for encoding extremely high-dimensional input vectors, such as images. An advantage of using the stochastic approach is that it automates the process of splitting high-dimensional input vectors into low-dimensional blocks before encoding them, because minimising the mean Euclidean reconstruction error can encourage different stochastically sampled code indices to become associated with different input subspaces.

SVQ provides a unifying framework for many of the methods of the previous subsections, with standard VQ (k-means), fuzzy k-means and topographic mappings emerging as special cases.

We denote the number of groups (i.e. the number of codebook entries) by g and let r be the number of samples in the code index vector, $\boldsymbol{y}$ (i.e. $\boldsymbol{y} = (y_1, \ldots, y_r)$). The objective function is taken to be the mean Euclidean reconstruction error measure defined by

$$D = \int d\boldsymbol{x}\, p(\boldsymbol{x}) \sum_{y_1=1}^{g} \sum_{y_2=1}^{g} \cdots \sum_{y_r=1}^{g} p(\boldsymbol{y}|\boldsymbol{x}) \int d\boldsymbol{x}'\, p(\boldsymbol{x}'|\boldsymbol{y})|\boldsymbol{x} - \boldsymbol{x}'(\boldsymbol{y})|^2 \tag{10.11}$$

Integrating over $\boldsymbol{x}'$ yields

$$D = 2\int d\boldsymbol{x}\, p(\boldsymbol{x}) \sum_{y_1=1}^{g} \sum_{y_2=1}^{g} \cdots \sum_{y_r=1}^{g} p(\boldsymbol{y}|\boldsymbol{x})|\boldsymbol{x} - \boldsymbol{x}'(\boldsymbol{y})|^2 \tag{10.12}$$

where the reconstruction vector is defined as

$$\begin{aligned} \boldsymbol{x}' &= \int d\boldsymbol{x}\, p(\boldsymbol{x}|\boldsymbol{y})\boldsymbol{x} \\ &= \frac{\displaystyle\int d\boldsymbol{x}\, p(\boldsymbol{y}|\boldsymbol{x})p(\boldsymbol{x})\boldsymbol{x}}{\displaystyle\int dz\, p(\boldsymbol{y}|z)p(z)} \end{aligned} \tag{10.13}$$

using (10.10). Equations (10.12) and (10.13) are simply the integral analogues of (10.2) and (10.3) where the role of z_{ki} is taken by $p(\boldsymbol{y}_k|\boldsymbol{x}_i)$, $\boldsymbol{y}_k$ being the kth state of $\boldsymbol{y}$ (there are g^r states). Unconstrained minimisation of (10.12) with respect to $p(\boldsymbol{y}|\boldsymbol{x})$ gives values for $p(\boldsymbol{y}_k|\boldsymbol{x}_i)$ that are either 0 or 1.

A key step in the SVQ approach is to constrain the minimisation of (10.12) in such a way as to encourage the formation of code schemes in which each component of the code vector codes a different subspace of the input vector $\boldsymbol{x}$ – an essential requirement for scalability to high dimensions. Luttrell (1999a) imposes two constraints on $p(\boldsymbol{y}|\boldsymbol{x})$ and $\boldsymbol{x}'(\boldsymbol{y})$, namely

$$\begin{aligned} p(\boldsymbol{y}|\boldsymbol{x}) &= \prod_{i=1}^{r} p(y_i|\boldsymbol{x}) \\ \boldsymbol{x}'(\boldsymbol{y}) &= \frac{1}{r}\sum_{i=1}^{r} \boldsymbol{x}'(y_i) \end{aligned} \tag{10.14}$$

The first constraint states that each component y_i is an independent sample drawn from the codebook using $p(y_i|\boldsymbol{x})$; the second states that $\boldsymbol{x}'(\boldsymbol{y})$ is assumed to be a superposition of r contributions $\boldsymbol{x}'(y_i)$ $(i = 1, \dots, r)$. These constraints encourage the formation of coding schemes in which independent subspaces are separately coded (Luttrell, 1999a) and allow (10.12) to be bounded above by the sum of two terms $D_1 + D_2$ where

$$\begin{aligned} D_1 &= \frac{2}{r}\int d\boldsymbol{x} \sum_{y=1}^{g} p(y|\boldsymbol{x})|\boldsymbol{x} - \boldsymbol{x}'(y)|^2 \\ D_2 &= \frac{2(r-1)}{r}\int d\boldsymbol{x}\, p(\boldsymbol{x}) \left|\boldsymbol{x} - \sum_{y=1}^{g} p(y|\boldsymbol{x})\boldsymbol{x}'(y)\right|^2 \end{aligned} \tag{10.15}$$

The term D_1 is a stochastic version of the standard VQ term (see (10.8)) and dominates for small values of r. D_2 is a nonlinear principal components type term, which describes the integrated squared error between $\boldsymbol{x}$ and a fitting surface modelled as a linear combination (specified by $\boldsymbol{x}'(y)$) of nonlinear basis functions ($p(y|\boldsymbol{x})$). Compare this with the lines and planes of closest fit definition of principal components analysis in Chapter 9.

The expression $D_1 + D_2$ must now be minimised with respect to $\boldsymbol{x}'(y)$ and $p(y|\boldsymbol{x})$. The parameters r (the number of samples drawn from the codebook using $p(y|\boldsymbol{x})$) and g (the size of the codebook) are model order parameters whose values determine the nature of the optimum solution. All that remains now is to specify a suitable form for $p(y|\boldsymbol{x})$. It can be shown (Luttrell, 1999b) that the optimal form of $p(y|\boldsymbol{x})$ is piecewise linear in $\boldsymbol{x}$ for regions of the data space that contain a non-vanishing probability density. A convenient approximation is

$$p(y|\boldsymbol{x}) = \frac{Q(y|\boldsymbol{x})}{\sum_{y'=1}^{g} Q(y'|\boldsymbol{x})}$$

where $Q(y|\boldsymbol{x})$ is taken to be of the form

$$Q(y|\boldsymbol{x}) = \frac{1}{1 + \exp(-\boldsymbol{w}^T(y)\boldsymbol{x} - b(y))}$$

Thus, $D_1 + D_2$ is minimised with respect to $\{\boldsymbol{w}(y), b(y);\ y = 1, \ldots, g\}$, the parameters of $p(y|\boldsymbol{x})$ and $\{\boldsymbol{x}'(y);\ y = 1, \ldots, g\}$ using some suitable nonlinear optimisation scheme.

Illustration Four-dimensional data, $\boldsymbol{x}$, are generated to lie on a torus: $\boldsymbol{x} = (\boldsymbol{x}_1, \boldsymbol{x}_2)$, where $\boldsymbol{x}_1 = (\cos(\theta_1), \sin(\theta_1))$ and $\boldsymbol{x}_2 = (\cos(\theta_2), \sin(\theta_2))$, for θ_1 and θ_2 uniformly distributed over $[1, 2\pi]$. An SVQ is trained using $g = 8$ codebook entries (8 'clusters') and for two values of r, the number of samples in the code index vector. Figure 10.18 shows a density plot of $p(y|\boldsymbol{x})$ for each value of the code index, y, and for $r = 5$ in the (θ_1, θ_2) space (of course, in practice, the underlying variables describing the data are unavailable to us). This type of coding, termed *joint encoding*, has produced approximately circular receptive fields in the space on which the data lie.

Figure 10.19 shows a density plot of $p(y|\boldsymbol{x})$ for $r = 50$ in the (θ_1, θ_2) space. This type of coding, termed *factorial encoding*, has produced receptive fields that respond to independent directions. Factorial coding is the key to scalability in high dimensions.

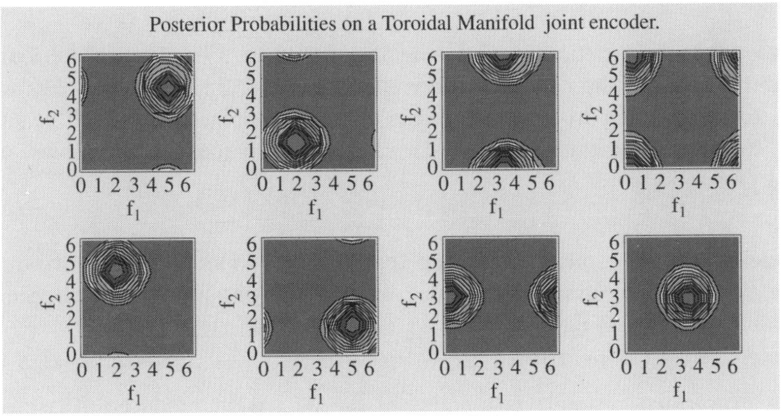

Figure 10.18 Joint encoding

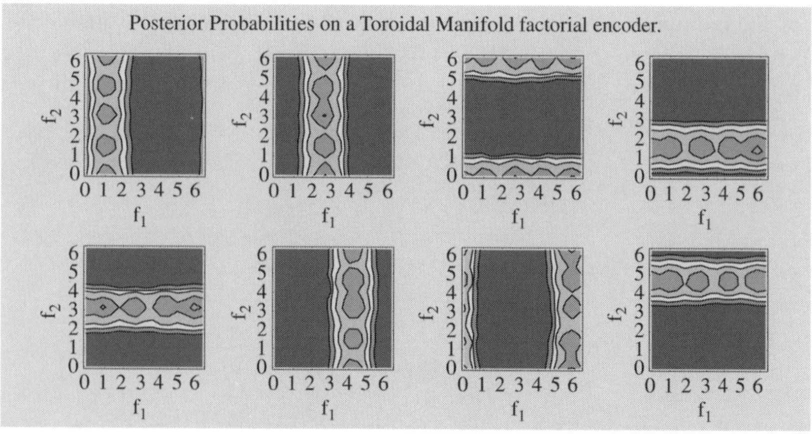

Figure 10.19 Factorial encoding

10.5.4 Example application study

The problem A study of the influence of early diagenesis (the conversion, by compaction or chemical reaction, of sediment into rock) on the natural remanent magnetisation in sediments from the Calabrian ridge in the central Mediterranean (Dekkers *et al.*, 1994).

Summary A fuzzy *k*-means algorithm (Section 10.5.2) was applied to measured data and the clusters plotted in two dimensions using a nonlinear mapping (multidimensional scaling–Chapter 9). The approach appeared to be useful, linking rock parameters to the geochemical environment.

The data Palaeomagnetic samples were taken from a 37 m long piston core from the Calabrian ridge in the central Mediterranean. In total, 337 samples, taken at 10 cm intervals (corresponding to an average resolution of approximately 3000 years), were analysed and measurements made on six variables (two magnetic variables and four chemical variables).

The model A fuzzy *k*-means clustering approach was adopted. Also, a two-dimensional projection of the six-dimensional data was derived using a nonlinear mapping based on multidimensional scaling.

Training procedure Simple histograms of the data were produced. These revealed approximately log-normal distributions. Therefore, before applying either the clustering or mapping procedures, the data were logarithmically transformed. Models were developed with an increasing number of clusters, and attempts were made to interpret these in a chemical and magnetic context. Clearly, in problems like these, domain knowledge is important, but care must be taken that the results are not biased by the investigator's prejudices.

Results General trends appeared to be best expressed with an eight-cluster model. Models with six or less clusters did not have sufficiently homogeneous clusters. These clusters could be divided into two main categories, one expressing mainly lithological features and the other expressing mainly diagenesis.

10.5.5 Further developments

Procedures for reducing the computational load of the k-means algorithm are discussed by Venkateswarlu and Raju (1992). Further developments of k-means procedures to other metric spaces (with l_1 and l_∞ norms) are described by Bobrowski and Bezdek (1991). Juan and Vidal (1994) propose a fast k-means algorithm (based on the approximating and eliminating search algorithm, AESA) for the case when data arise in the form of a dissimilarity. That is, the data cannot be represented in a suitable vector space (without performing a multidimensional scaling procedure), though the dissimilarity between points is available. Termed the *k-centroids* procedure, it determines the 'most centred sample' as a centroid of a cluster.

There have been many developments of the basic fuzzy clustering approach and many algorithms have been proposed. Sequential approaches are described by de Mántaras and Aguilar-Martín (1985). In 'semi fuzzy' or 'soft' clustering (Ismail, 1988; Selim and Ismail, 1984b) patterns are considered to belong to some, though not necessarily all, clusters. In thresholded fuzzy clustering (Kamel and Selim, 1991) membership values below a threshold are set to zero, with the remaining being normalised, and an approach that performs a fuzzy classification with prior assumptions on the number of clusters is reported by Gath and Geva (1989). Developments of the fuzzy clustering approach to data arising in the form of dissimilarity matrices are described by Hathaway and Bezdek (1994).

Developments of the stochastic vector quantiser approach include developments to hierarchical schemes (*folded Markov chains*) for encoding data (Luttrell, 2002).

10.5.6 Summary

The techniques described in this section minimise a squared error objective function. The k-means procedure is a special case of all of the techniques. It is widely used in pattern recognition, forming the basis for many supervised classification techniques. It produces a 'crisp' coding of the data in that a pattern belongs to one cluster only.

Fuzzy k-means is a development that allows a pattern to belong to more than one cluster. This is controlled by a membership function. The clusters resulting from a fuzzy k-means approach are softly overlapping, in general, with the degree of overlap controlled by a user-specified parameter. The learning procedure determines the partition.

The vector quantisation approaches are application-driven. The aim is to produce a crisp coding, and algorithms such as tree-structured vector quantisation are motivated by the need for fast coding. Self-organising feature maps produce an ordered coding.

Stochastic vector quantisation is a procedure that learns both the codebook entries and the membership function. Different codings may result, depending on a user-specified parameter. Of particular interest is the factorial encoding that scales linearly with the intrinsic dimensionality of the data.

10.6 Cluster validity

10.6.1 Introduction

Cluster validity is an issue fraught with difficulties and rarely straightforward. A clustering algorithm will partition a data set of objects even if there are no natural clusters within the data. Different clustering methods may produce different classifications. How do we know whether the structure is a property of the data set and not imposed by the particular method that we have chosen? In some applications of clustering techniques we may not be concerned with groupings in the data set. For example, in vector quantisation we may be concerned with the average distortion in reconstructing the original data or in the performance of any subsequent analysis technique. This may be measured by the error rate in a discrimination problem or diagnostic performance in image reconstruction (see the examples in the following section). In these situations, clustering is simply a means of obtaining a partition, not of discovering structure in the data.

Yet, if we are concerned with discovering groupings within a data set, how do we validate the clustering? A simple approach is to view the clustering in a low-dimensional representation of the data. Linear and nonlinear projection methods have been discussed in Chapter 9. Alternatively, we may perform several analyses using different clustering methods and compare the resulting classifications to see whether the derived structure is an artefact of a particular method. More formal procedures may also be applied and we discuss some approaches in this section.

There are several related issues in cluster validity. The first concerns the goodness of fit of the derived classification to the given data: how well does the clustering reflect the true data structure? What appropriate *measures of distortion* or *internal criterion measures* should be used? A second issue concerns the determination of the 'correct' number of groups within the data. This is related to the first problem since it often requires the calculation of distortion measures. Finally, we address the issue of identifying genuine clusters in a classification based on work by Gordon (1996a).

There are three main classes of null model for the complete absence of structure in a data set (Gordon 1994b, 1996a).

1. *Poisson model* (Bock, 1985). Objects are represented as points uniformly distributed in some region A of the p-dimensional data space. The main problem with this model is the specification of A. Standard definitions include the unit hypercube and the hypersphere.

2. *Unimodal model* (Bock, 1985). The variables have a unimodal distribution. The difficulty here is the specification of this density function. Standard definitions include the spherical multivariate normal.

3. *Random dissimilarity matrix* (Ling, 1973). This is based on data arising in the form of dissimilarities. The elements of the (lower triangle of the) dissimilarity matrix are ranked in random order, all orderings being regarded as equally likely. One of the problems with this is that if objects i and j are close (d_{ij} is small), you would expect d_{ik} and d_{jk} to have similar ranks for each object k.

The Poisson model and the unimodal model were used as part of the study described in Section 10.6.4.

10.6.2 Distortion measures

In assessing particular hierarchical schemes we must consider how well the structure in the original data can be described by a dendrogram. However, since the structure in the data is not known (this is precisely what we are trying to determine) and since each clustering is simply a method of exploratory data analysis that imposes its own structure on the data, this is a difficult question to address. One approach is to examine various measures of distortion. Many measures have been proposed (see, for example, Cormack, 1971, for a summary). They are based on differences between the dissimilarity matrix $\boldsymbol{d}$ and the matrix of ultrametric dissimilarity coefficients $\boldsymbol{d}^*$, where d^*_{ij} is the distance between the groups containing i and j when the groups are amalgamated. Jardine and Sibson (1971) propose several goodness-of-fit criteria. One scale-free measure of classifiability is defined by

$$\Delta_1 = \frac{\sum_{i<j} |d_{ij} - d^*_{ij}|}{\sum_{i<j} d_{ij}}$$

Small values of Δ_1 are indicative that the data are amenable to the classification method that produced $\boldsymbol{d}^*$.

There are many other measures of distortion, both for hierarchical and nonhierarchical schemes. Milligan (1981) performed an extensive Monte Carlo study of 30 internal criterion measures applied to the results of hierarchical clusterings, although the results may also apply to non-hierarchical methods.

10.6.3 Choosing the number of clusters

The problem of deciding how many clusters are present in the data is one common to all clustering methods. There are numerous techniques for cluster validity reported in the literature. Many methods are very subjective and can be unreliable, with some criteria indicating clusters present when analysing unstructured data. Many methods have been proposed for determining the number of groups. When applied to hierarchical schemes, these are sometimes referred to as *stopping rules*. Many intuitive schemes have been proposed for hierarchical methods; for example, we may examine the plot of fusion level against the number of groups, g (see Figure 10.20), and look for a flattening of the curve showing that little improvement in the description of the data structure is to be gained above a particular value of g. Defining α_j, $j = 0, \ldots, n-1$, to be the fusion level

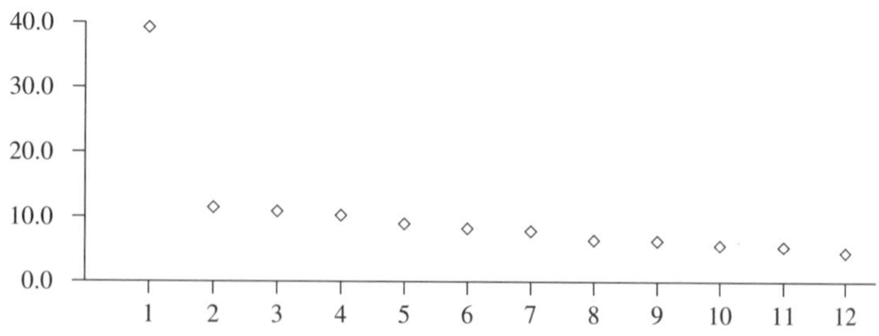

Figure 10.20 Fusion level against the number of clusters

corresponding to the stage with $n - j$ clusters, Mojena (1977) proposes a stopping rule that selects the number of groups as g such that α_{n-g} is the lowest value of α for which

$$\alpha_{n-g} > \overline{\alpha} + ks_{\alpha}$$

where $\overline{\alpha}$ and s_{α} are the mean and the unbiased standard deviation of the fusion levels α; k is a constant, suggested by Mojena to be in the range 2.75–3.5.

Milligan and Cooper (1985) examine 30 procedures applied to classifications of data sets containing 2, 3, 4, or 5 distinct non-overlapping clusters by four hierarchical schemes. They find that Mojena's rule performs poorly with only two groups present in the data and the best performance is for 3, 4 or 5 groups with a value of k of 1.25. One of the better criteria that Milligan and Cooper (1985) assess is that of Calinski and Harabasz (1974). The number of groups is taken to be the value of g that corresponds to the maximum of C, given by

$$C = \frac{\text{Tr}(\boldsymbol{S}_B)}{\text{Tr}(\boldsymbol{S}_W)} \left(\frac{n-g}{g-1} \right)$$

This is evaluated further by Atlas and Overall (1994) who compare it with a split-sample replication rule of Overall and Magee (1992). This gave improved performance over the Calinski and Harabasz criterion.

Dubes (1987) also reports the results of a Monte Carlo study on the effectiveness of two internal criterion measures in determining the number of clusters. Jain and Moreau (1987) propose a method to estimate the number of clusters in a data set using the bootstrap technique. A clustering criterion based on the within- and between-group scatter matrices (developing a criterion of Davies and Bouldin, 1979) is proposed and the k-means and three hierarchical algorithms are assessed. The basic method of determining the number of clusters using a bootstrap approach can be used with any cluster method.

Several authors have considered the problem of testing for the number of components of a normal mixture (see Chapter 2). This is not a trivial problem and depends on many factors, including shape of clusters, separation, relative sizes, sample size and dimension of data. Wolfe (1971) proposes a modified likelihood ratio test in which the null hypothesis $g = g_0$ is tested against the alternative hypothesis $g = g_1$. The quantity

$$-\frac{2}{n}\left(n - 1 - p - \frac{g_1}{2}\right)\log(\lambda)$$

where λ is the likelihood ratio, is tested as chi-square with degrees of freedom being twice the difference in the number of parameters in the two hypotheses (Everitt *et al.*,

2001), excluding mixing proportions. For components of a normal mixture with arbitrary covariance matrices, the number of parameters, n_p, is

$$n_p = 2(g_1 - g_0)\frac{p(p+3)}{2}$$

and with common covariance matrices (the case studied by Wolfe, 1971), $n_p = 2(g_1 - g_0)p$. McLachlan and Basford (1988) recommend that Wolfe's modified likelihood ratio test be used as a guide to structure rather than rigidly interpreted. Ismail (1988) reports the results of cluster validity studies within the context of soft clustering and lists nine validity functionals that provide useful tools in determining cluster structure (see also Pal and Bezdek 1995).

It is not reasonable to expect a single statistic to be suitable for all problems in cluster validity. Many different factors are involved and since clustering is essentially a method of exploratory data analysis, we should not put too much emphasis on the results of a single classification, but perform several clusterings using different algorithms and measures of fit.

10.6.4 Identifying genuine clusters

We wish to identify whether a cluster $\mathcal{C}$ (of size c) defined by

$$\mathcal{C} = \{i : d_{ij} < d_{ik} \text{ for all } j \in \mathcal{C}, k \notin \mathcal{C}\}$$

is a valid cluster. We describe here the procedure of Gordon (1994a) who develops an approach to cluster validation based on a U statistic:

$$U_{ijkl} = \begin{cases} 0 & \text{if } d_{ij} < d_{kl} \\ \frac{1}{2} & \text{if } d_{ij} = d_{kl} \\ 1 & \text{if } d_{ij} > d_{kl} \end{cases}$$

and

$$U = \sum_{(i,j)\in W} \sum_{(k,l)\in B} U_{ijkl}$$

for subsets W and B of ordered pairs $(i, j) \in W$ and $(k, l) \in B$. W is taken to be those pairs where objects i and j both belong to the cluster $\mathcal{C}$ and B comprises pairs where one element belongs to $\mathcal{C}$ and the other does not.

The basic algorithm is defined as follows:

1. Evaluate U for the cluster $\mathcal{C}$; denote it by U^*.

2. Generate a random $n \times p$ pattern matrix and cluster it using the same algorithm used to produce $\mathcal{C}$.

3. Calculate $U(k)$ for each cluster of size $k = 2, \ldots, n-1$ (arising through the partitioning of a dendrogram, for example). If there is more than one of a given size, select one of that size randomly.

4. Repeat steps 2 and 3 until there are $m - 1$ values of $U(k)$ for each value of k.

5. If U^* is less than the jth smallest value of $U(k)$, the null hypothesis of randomness is rejected at the $100(j/m)\%$ level of significance.

Note that the values of $U(k)$ are independent of the data set.

Gordon takes $m = 100$ and evaluates the above approach under both the Poisson and unimodal (spherical multivariate normal) models. The clusterings using Ward's method are assessed on four data sets. Results for the approach are encouraging. Further refinements could include the use of other test statistics and developments of the null models.

10.7 Application studies

Applications of hierarchical methods of cluster analysis include the following:

- Flight monitoring. Eddy *et al.* (1996) consider single-link clustering of large data sets (more than 40 000 observations) of high-dimensional data relating to aircraft flights over the United States.

- Clinical data. D'Andrea *et al.* (1994) apply the nearest centroid method to data relating to adult children of alcoholics.

- In a comparative study of seven methods of hierarchical cluster analysis on 20 data sets, Morgan and Ray (1995) examine the extent of inversions in dendrograms and non-uniqueness. They conclude that inversions are expected to be encountered and recommend against the use of the median and centroid methods. Also, non-uniqueness is a real possibility for many data sets.

The k-means clustering approach is widely used as a preprocessor for supervised classification to reduce the number of prototypes:

- Coal petrography. In a study to classify the different constituents (macerals) of coal (Mukherjee *et al.*, 1994), the k-means algorithm was applied to training images (training vectors consist of RGB level values) to determine four clusters of known types (vitrinite, inertinite, exinite and background). These clusters are labelled and test images are classified using the labelled training vectors.

- Crop classification. Conway *et al.* (1991) use a k-means algorithm to segment synthetic aperture radar images as part of a study into crop classification. k-means is used to identify sets of image regions that share similar attributes prior to labelling. Data were gathered from a field of five known crop types and could be clearly separated into two clusters – one containing the broad-leaved crops and the other the narrow-leaved crops.

k-means is also used for image and speech coding applications (see below).

Examples of fuzzy k-means applications are as follows:

- Medical diagnosis. Li *et al.* (1993) use a fuzzy k-means algorithm for image segmentation in a study on automatic classification and tissue labelling of two-dimensional magnetic resonance images of the human brain.
- Acoustic quality control. Meier *et al.* (1994) describe the application of fuzzy k-means to cluster six-dimensional feature vectors as part of a quality control system for ceramic tiles. The signals are derived by hitting the tiles and digitising and filtering the recorded signal. The resulting classes are interpreted as good or bad tiles.
- Water quality. Mukherjee *et al.* (1995) compared fuzzy k-means with two alternative approaches to image segmentation in a study to identify and count bacterial colonies from images.

See also the survey on fuzzy clustering by Yang (1993) for further references to applications of fuzzy k-means.

One example of a Bayesian approach to mixture modelling is worthy of note (other applications of mixture models are given in Chapter 2). Dellaportas (1998) considers the application of mixture modelling to the classification of neolithic ground stone tools. A Bayesian methodology is adopted and developed in three main ways to apply to data (147 measurements on four variables) consisting of variables of mixed type – continuous and categorical. Missing values and measurement errors (errors in variables) in the continuous variables are also treated. Gibbs sampling is used to generate samples from the posterior densities of interest and classification to one of two classes is based on the mean of the indicator variable, z_i. After a 'burn-in' of 4000 iterations, 4000 further samples were used as the basis for posterior inference.

There are several examples of self-organising feature map applications:

- Engineering applications. Kohonen *et al.* (1996) review the self-organising map algorithm and describe several engineering applications, including fault detection, process analysis and monitoring, computer vision, speech recognition, robotic control and in the area of telecommunications.
- Human protein analysis. Ferrán *et al.* (1994) use a self-organising map to cluster protein sequences into families. Using 1758 human protein sequences, they cluster using two-dimensional maps of various sizes and label the nodes in the grid using proteins belonging to known sequences.
- Radar target classification. Stewart *et al.* (1994) develop a self-organising map and a learning vector quantisation approach to radar target classification using turntable data of four target types. The data consist of 33-dimensional feature vectors (33 range gates) and 36 000 patterns per target were used. Performance as a function of the number of cluster centres is reported, with the performance of learning vector quantisation better than that of a simplistic nearest-neighbour algorithm.
- Fingerprint classification. Halici and Ongun (1996), in a study on automatic fingerprint classification, use a self-organising map, and one modified by preprocessing the feature vectors by combining them with 'certainty' vectors that encode uncertainties in the fingerprint images. Results show an improvement on previous studies using a multilayer perceptron on a database of size 2000.

Sum-of-squares methods have been applied to language disorders. Powell *et al.* (1979) use a normal mixture approach and a sum-of-squares method in a study of 86 aphasic cases referred to a speech therapy unit. Four groups are found, which are labelled as severe, high–moderate, low–moderate and mild aphasia.

Vector quantisation has been widely applied as a preprocessor in many studies:

- Speech recognition. Zhang *et al.* (1994) assess three different vector quantisers (including the LBG algorithm and an algorithm based on normal mixture modelling) as preprocessors for a hidden Markov model based recogniser in a small speech recognition problem. They found that the normal mixture model gave the best performance of the subsequent classifier. See also Bergh *et al.* (1985).
- Medical diagnosis. Cosman *et al.* (1993) assess the quality of tree-structured vector quantisation images by the diagnostic performance of radiologists in a study on lung tumour and lymphadenopathy identification. Initial results suggest that a 12 bits per pixel (bpp) computerised tomography chest scan image can be compressed to between 1 bpp and 2 bpp with no significant change in diagnostic accuracy: subjective quality seems to degrade sooner than diagnostic accuracy falls off.
- Speaker recognition. Recent advances in speaker recognition are reviewed by Furui (1997). Vector quantisation methods are used to compress the training data and produce codebooks of representative feature vectors characterising speaker-specific features. A codebook is generated for each speaker by clustering training feature vectors. At the speaker recognition stage, an input utterance is quantised using the codebook of each speaker and recognition performed by assigning the utterance to the speaker whose codebook produces minimum distortion. A tutorial on vector quantisation for speech coding is given by Makhoul *et al.* (1985).

10.8 Summary and discussion

In this chapter we have covered a wide range of techniques for partitioning a data set. This has included approaches based on cluster analysis methods and vector quantisation methods. Although both approaches have much in common – they both produce a dissection of a given data set – there are differences. In cluster analysis, we tend to look for 'natural' groupings in the data that may be labelled in terms of the subject matter of the data. In contrast, the vector quantisation methods are developed to optimise some appropriate criterion from communication theory. One area of common ground we have discussed in this chapter is that of optimisation methods with specific implementations in terms of the k-means algorithm in cluster analysis and the LBG algorithm in vector quantisation.

As far as cluster analysis or classification is concerned, there is no single best technique. Different clustering methods can yield different results and some methods will fail to detect obvious clusters. The reason for this is that each method implicitly forces a structure on the given data. For example, the sum-of-squares methods will tend to produce hyperspherical clusters. Also, the fact that there is a wide range of available methods partly stems from the lack of a single definition of the word 'cluster'. There is no universal agreement as to what constitutes a cluster and so a single definition is insufficient.

A further difficulty with cluster analysis is in deciding the number of clusters present. This is a trade-off between parsimony and some measure of increase in within-cluster homogeneity. This problem is partly due to the difficulty in deciding what a cluster actually is and partly because clustering algorithms tend to produce clusters even when applied to random data.

Both of the above difficulties may be overcome to some degree by considering several possible classifications or comparing classifications on each half of a data set (for example, McIntyre and Blashfield, 1980; Breckenridge, 1989). The interpretation of these is more important than a rigid inference of the number of groups. But which methods should we employ? There are advantages and disadvantages of all the approaches we have described. The optimisation methods tend to require a large amount of computer time (and consequently may be infeasible for large data sets, though this is becoming less critical these days). Of the hierarchical methods, the single link is preferred by many users. It is the only one to satisfy the Jardine–Sibson conditions, yet with noisy data it can join separate clusters (chaining effect). It is also invariant under monotone transformations of the dissimilarity measure. Ward's method is also popular. The centroid and median methods should be avoided since inversions may make the resulting classification difficult to interpret.

There are several aspects of cluster analysis that we have mentioned only briefly in this chapter and we must refer the reader to the literature on cluster analysis for further details. An important problem is the choice of technique for mixed mode data. Everitt and Merette (1990) (see also Everitt, 1988) propose a finite mixture model approach for clustering mixed mode data, but computational considerations may mean that it is not practically viable when the data sets contain a large number of categorical variables.

Clumping methods (or methods of overlapping classification in which an object can belong to more than one group) have not been considered explicitly, though the membership function in the fuzzy clustering approach may be regarded as allowing an object partial membership of several groups in some sense. An algorithm for generating overlapping clusters is given by Cole and Wishart (1970).

The techniques described in this chapter all apply to the clustering of objects. However, there may be some situations where clustering of variables, or simultaneous clustering of objects and variables, is required. In clustering of variables, we seek subsets of variables that are so highly correlated that each can be replaced by any one of the subset, or perhaps a (linear or nonlinear) combination of the members. Many of the techniques described in this chapter can be applied to the clustering of variables, and therefore we require a measure of similarity or dissimilarity between variables. Of course, techniques for feature extraction (for example, principal components analysis) perform this process.

Another point to reiterate about cluster analysis is that it is essentially an exploratory method of multivariate data analysis providing a description of the measurements. Once a solution or interpretation has been obtained then the investigator must re-examine and assess the data set. This may allow further hypotheses (perhaps concerning the variables used in the study, the measures of dissimilarity and the choice of technique) to be generated. These may be tested on a new sample of individuals.

Both cluster analysis and vector quantisation are means of reducing a large amount of data to a form in which it is either easier to describe or represent in a machine. Clustering of data may be performed prior to supervised classification. For example, the number of stored prototypes in a k-nearest-neighbour classifier may be reduced by a clustering procedure. The new prototypes are the cluster means, and the class of the new prototype is decided on a majority basis of the members of the cluster. A development of this approach that adjusts the decision surface by modifying the prototypes is *learning vector quantisation* (Kohonen, 1989).

Self-organising maps may be viewed as a form of *constrained classification*: clustering in which there is some form of constraint on the solution. In this particular case, the constraint is an ordering on the cluster centres. Other forms of constraint may be that objects within a cluster are required to comprise a spatially contiguous set of objects (for example, in some texture segmentation applications). This is an example of *contiguity-constrained clustering* (Gordon, 1999; Murtagh, 1992; see also Murtagh, 1995, for an application to the outputs of the self-organising map). Other forms of constraint may be on the topology of the dendrogram or the size or composition of the classes. We refer to Gordon (1996b) for a review.

Gordon (1994b, 1996a) provides reviews of approaches to cluster validation; see also Bock (1989) and Jain and Dubes (1988).

10.9 Recommendations

It has been said that 'automatic classification is rapidly replacing factor analysis and principal component analysis as a "heffalump" trap for the innocent scientist' (Jardine, 1971). Certainly, there is a large number of techniques to choose from and the availability of computer packages means that analyses can be readily performed. Nevertheless, there are some general guidelines to follow when carrying out a classification.

1. Detect and remove outliers. Many clustering techniques are sensitive to the presence of outliers. Therefore, some of the techniques discussed in Chapter 11 should be used to detect and possibly remove these outliers.

2. Plot the data in two dimensions if possible in order to understand structure in the data. It might be helpful to use the first two principal components.

3. Carry out any preprocessing of the data. This may include a reduction in the number of variables or standardisation of the variables to zero mean and unit variance.

4. If the data are not in the form of a dissimilarity matrix then a dissimilarity measure must be chosen (for some techniques) and a dissimilarity matrix formed.

5. Choose an appropriate technique. Optimisation techniques are more computationally expensive. Of the hierarchical methods, some studies favour the use of the average-link method, but the single-link method gives solutions that are invariant to a monotone transformation measure. It is the only one to satisfy all the conditions laid down by Jardine and Sibson (1971) and is their preferred method. We advise against the use of the centroid and median methods since inversions are likely to arise.

6. Evaluate the method. Assess the results of the clustering method you have employed. How do the clusters differ? We recommend that you split the data set into two parts and

compare the results of a classification on each subset. Similar results would suggest that useful structure has been found. Also, use several methods and compare results. With many of the methods some parameters must be specified and it is worthwhile carrying out a classification over a range of parameter values to assess stability.

7. In using vector quantisation as a preprocessor for a supervised classification problem, model each class separately rather than the whole data set, and label the resulting codewords.

8. If you require some representative prototypes, we recommend using the k-means algorithm.

Finally, we reiterate that cluster analysis is usually the first stage in an analysis and unquestioning acceptance of the results of a classification is extremely unwise.

10.10 Notes and references

There is a vast literature on cluster analysis. A very good starting point is the book by Everitt *et al.* (2001). Now in its fourth edition, this book is a mainly non-mathematical account of the most common clustering techniques, together with their advantages and disadvantages. The practical advice is supported by several empirical investigations. Another good introduction to methods and assessments of classification is the book by Gordon (1999). McLachlan and Basford (1988) discuss the mixture model approach to clustering in some detail.

Of the review papers, that by Cormack (1971) is worth reading and provides a good summary of the methods and problems of cluster analysis. The article by Diday and Simon (1976) gives a more mathematical treatment of the methods, together with descriptive algorithms for their implementation.

Several books have an orientation towards biological and ecological matters. Jardine and Sibson (1971) give a mathematical treatment. The book by Sneath and Sokal (1973) is a comprehensive account of cluster analysis and the biological problems to which it can be applied. The book by Cliffrd and Stephenson (1975) is a non-mathematical general introduction to the ideas and principles of numerical classification and data analysis, though it does not cover many of the approaches described in this chapter, concentrating on hierarchical classificatory procedures. The book by Jain and Dubes (1988) has a pattern recognition emphasis. McLachlan (1992b) reviews cluster analysis in medical research.

A more specialist book is that of Zupan (1982). This monograph is concerned with the problem of implementing hierarchical techniques on large data sets.

The literature on fuzzy techniques in cluster analysis is reviewed by Bezdek and Pal (1992). This book contains a collection of some of the important papers on fuzzy models for pattern recognition, including cluster analysis and supervised classifier design, together with fairly extensive bibliographies. A survey of fuzzy clustering and its applications is provided by Yang (1993). An interesting probabilistic perspective of fuzzy methods is provided by Laviolette *et al.* (1995).

Tutorials and surveys of self-organising maps are given by Kohonen (1990, 1997), Kohonen *et al.* (1996) and Ritter *et al.* (1992).

There are various books on techniques for implementing methods, which give algorithms in the form of Fortran code or pseudo-code. The books by Anderberg (1973), Hartigan (1975), Späth (1980) and Jambu and Lebeaux (1983) all provide a description of a clustering algorithm, Fortran source code and a supporting mathematical treatment, sometimes with case studies. The book by Murtagh (1985) covers more recent developments and is also concerned with implementation on parallel machines.

There are many software packages publicly available for cluster analysis. The website `www.statistical-pattern-recognition.net` contains references and pointers to websites for further information on techniques.

Exercises

Data set 1: Generate $n = 500$ samples $(x_i, y_i), i = 1, \ldots, n$, according to

$$x_i = \frac{i}{n}\pi + n_x$$

$$y_i = \sin\left(\frac{i}{n}\pi\right) + n_y$$

where n_x and n_y are normally distributed with mean 0.0 and variance 0.01.

Data set 2: Generate n samples from a multivariate normal (p variables) of diagonal covariance matrix $\sigma^2\boldsymbol{I}$, $\sigma^2 = 1$, and zero mean. Take $n = 40$, $p = 2$.

Data set 3: This consist of data comprising two classes: class ω_1 is distributed as $0.5N((0,0), \boldsymbol{I}) + 0.5N((2,2), \boldsymbol{I})$ and class $\omega_2 \sim N((2,0), \boldsymbol{I})$ (generate 500 samples for training and test sets, $p(\omega_1) = p(\omega_2) = 0.5$).

1. Is the square of the Euclidean distance a metric? Does it matter for any clustering algorithm?

2. Observations on six variables are made for seven groups of canines and given in Table 10.4 (Krzanowski and Marriott, 1994; Manly, 1986). Construct a dissimilarity matrix using Euclidean distance after standardising each variable to unit variance. Carry out a single-link cluster analysis.

3. Compare the single-link method of clustering with k-means, discussing computational requirements, storage, and applicability of the methods.

4. A mixture of two normals divided by a normal density having the same mean and variance as the mixed density is always bimodal. Prove this for the univariate case.

5. Implement a k-means algorithm and test it on two-dimensional normally distributed data (data set 2 with $n = 500$). Also, use the algorithm within a tree-structured vector quantiser and compare the two methods.

Table 10.4 Data on mean mandible measurements (from Manly, 1986)

Group	x_1	x_2	x_3	x_4	x_5	x_6
Modern Thai dog	9.7	21.0	19.4	7.7	32.0	36.5
Golden jackal	8.1	16.7	18.3	7.0	30.3	32.9
Chinese wolf	13.5	27.3	26.8	10.6	41.9	48.1
Indian wolf	11.5	24.3	24.5	9.3	40.0	44.6
Cuon	10.7	23.5	21.4	8.5	28.8	37.6
Dingo	9.6	22.6	21.1	8.3	34.4	43.1
Prehistoric dog	10.3	22.1	19.1	8.1	32.3	35.0

6. Using data set 1, code the data using the Luttrell algorithm (plot positions of centres for various numbers of code vectors). Compute the average distortion as a function of the number of code vectors. How would you modify the algorithm for data having circular topology?

7. Using data set 1, construct a tree-structured vector quantiser, partitioning the clusters with the largest sum-squared error at each stage. Compute the average distortion.

8. Using data set 2, cluster the data using Ward's method and Euclidean distance. Using Gordon's approach for identifying genuine clusters (unimodal null model), how many clusters are valid at the 5% level of significance?

9. Implement a learning vector quantisation algorithm on data set 3. Plot performance as a function of the number of cluster centres. What would be the advantages and disadvantages of using the resulting cluster centres as centres in a radial basis function network?

10. Show that the single-link dendrogram is invariant to a nonlinear monotone transformation of the dissimilarities.

11. For a distance between two clusters A and B of objects given by $d_{AB} = |\boldsymbol{m}_A - \boldsymbol{m}_B|^2$, where $\boldsymbol{m}_A$ is the mean of the objects in cluster A, show that the formula expressing the distance between a cluster k and a cluster formed by joining i and j is

$$d_{i+j,k} = \frac{n_i}{n_i + n_j} d_{ik} + \frac{n_j}{n_i + n_j} d_{jk} - \frac{n_i n_j}{(n_i + n_j)^2} d_{ij}$$

where there are n_i objects in group i. This is the update rule for the centroid method.

11

Additional topics

Overview

Two main issues in classifier design are addressed. The first concerns model selection–choosing the appropriate type and complexity of classifier. The second concerns problems with data–mixed variables, outliers, missing values and unreliable labelling.

11.1 Model selection

In many areas of pattern recognition we are faced with the problem of model selection; that is, how complex should we allow our model to be, measured perhaps in terms of the number of free parameters to estimate? The optimum complexity of the model depends on the quantity and the quality of the training data. If we choose a model that is too complex, then we may be able to model the training data very well (and also any noise in the training data), but it is likely to have poor generalisation performance on unseen data, drawn from the same distribution as the training set was drawn from (thus the model *over-fits* the data). If the model is not complex enough, then it may fail to model structure in the data adequately. Model selection is inherently a part of the process of determining optimum model parameters. In this case, the complexity of the model is a parameter to determine. As a consequence, many model selection procedures are based on optimising a criterion that penalises a goodness-of-fit measure by a model-complexity measure.

The problem of model selection arises with many of the techniques described in this book. Some examples are:

1. How many components in a mixture model should be chosen to model the data adequately?

2. How is the optimum tree structure found in a decision-tree approach to discrimination? This is an example in determining an appropriate number of basis functions in an expansion, where the basis functions in this case are hyperrectangles, with sides

parallel to the coordinate axes. Other examples include the number of projections in projection pursuit and the number of hinges in the hinging hyperplane model.

3. How many hidden units do we take in a multilayer perceptron, or centres in a radial basis function network?

4. How many clusters describe the data set?

In this section, we give some general procedures that have been widely used for model selection. A comprehensive review of model selection methods, particularly in the context of time series analysis though with much wider applicability, is given by Glendinning (1993). Anders and Korn (1999) examine model selection procedures in the context of neural networks, comparing five strategies in a simulation study.

11.1.1 Separate training and test sets

In the separate training and test set approach, both training and test sets are used for model selection. A test set used in this way is often termed the *validation set*. The training set is used to optimise a goodness-of-fit criterion and the performance recorded on the validation set. As the complexity of the model is increased, it is expected that performance on the training set will improve (as measured in terms of the goodness-of-fit criterion), while the performance on the validation set will begin to deteriorate beyond a certain model complexity.

This is one approach (though not the preferred approach) used for the classification and regression tree training models (Breiman *et al.*, 1984). A large tree is grown to over-fit the data, and pruned back until the performance on a separate validation set fails to improve. A similar approach may be taken for neural network models.

The separate training and validation set procedure may also be used as part of the optimisation process for a model of a given complexity, particularly when the optimisation process is carried out iteratively, as in a nonlinear optimisation scheme. The values of the parameters of the model are chosen, not as the ones that minimise the given criterion on the training set, but those for which the validation set performance is minimum. Thus, as training proceeds, the performance on the validation set is monitored (by evaluating the goodness-of-fit criterion using the validation data) and training is terminated when the validation set performance begins to deteriorate.

Note that in this case the validation set is not an independent test set that may be used for error rate estimation. It is part of the training data. A third data set is required for an independent estimate of the error rate.

11.1.2 Cross-validation

Cross-validation as a method of error rate estimation was described in Section 8.2. It is a simple idea. The data set of size n samples is partitioned into two parts. The model parameters are estimated using one set (by minimising some optimisation criterion) and the goodness-of-fit criterion evaluated on the second set. The usual version of cross-validation is the simple leave-one-out method in which the second set consists of a single sample. The cross-validation estimate of the goodness-of-fit criterion, CV, is then the average over all possible training sets of size $n - 1$.

As a means of determining an appropriate model, the cross-validation error, CV, is calculated for each member of the family of candidate models, $\{M_k, k = 1, \ldots, K\}$, and the model $M_{\hat{k}}$ chosen, where

$$\hat{k} = \text{argmin CV}(k)$$

Cross-validation tends to over-fit when selecting a correct model; that is, it chooses an over-complex model for the data set. There is some evidence that multifold cross-validation, when $d > 1$ samples are deleted from the training set, does better than simple leave-one-out cross-validation for model selection purposes (Zhang, 1993). The use of cross-validation to select a classification method is discussed further by Schaffer (1993).

11.1.3 The Bayesian viewpoint

In the Bayesian approach, prior knowledge about models, M_k, and parameters, $\boldsymbol{\theta}_k$, is incorporated into the model selection process. Given a data set $\boldsymbol{X}$, the distribution of the models may be written using Bayes' theorem as

$$\begin{aligned} p(M_k|\boldsymbol{X}) &\propto p(\boldsymbol{X}|M_k)p(M_k) \\ &= p(M_k)\int p(\boldsymbol{X}|M_k, \boldsymbol{\theta}_k)p(\boldsymbol{\theta}_k|M_k)d\boldsymbol{\theta}_k \end{aligned}$$

and we are therefore required to specify a prior distribution $p(M_k, \boldsymbol{\theta}_k)$. If a single model is required, we may choose $M_{\hat{k}}$, where

$$\hat{k} = \text{argmax } p(M_k|\boldsymbol{X})$$

However, in a Bayesian approach, all models are considered. Over-complex models are penalised since they predict the data poorly. Yet, there are difficulties in specifying priors and several methods have been suggested that use data-dependent 'priors' (Glendinning, 1993). Nevertheless, there are good applications of the Bayesian approach in pattern recognition.

11.1.4 Akaike's information criterion

Akaike, in a series of papers including Akaike (1973, 1974, 1977, 1981, 1985), used ideas from information theory to suggest a model selection criterion. A good introduction to the general principles is given by Bozdogan (1987; see also Sclove, 1987).

Suppose that we have a family of candidate models $\{M_k, k = 1, \ldots, K\}$, with the kth model depending on a parameter vector $\boldsymbol{\theta}_k = (\theta_{k,1}, \ldots, \theta_{k,\epsilon(k)})^T$, where $\epsilon(k)$ is the number of free parameters of model k. Then the information criterion proposed by Akaike (AIC) is given by

$$\text{AIC}(k) = -2\log[L(\hat{\boldsymbol{\theta}}_k)] + 2\epsilon(k) \tag{11.1}$$

where $\hat{\boldsymbol{\theta}}_k$ is the maximum likelihood estimate of $\boldsymbol{\theta}_k$, $L[.]$ is the likelihood function, and the model M_k is chosen as that model, $M_{\hat{k}}$, where

$$\hat{k} = \text{argmin AIC}(k)$$

Equation (11.1) represents the unbiased estimate of minus twice the expected log-likelihood,

$$-2\text{E}[\log(p(\boldsymbol{X}_n|\boldsymbol{\theta}_k))]$$

where $\boldsymbol{X}_n$ is the set of observations $\{\boldsymbol{x}_1, \ldots, \boldsymbol{x}_n\}$, characterised by $p(\boldsymbol{x}|\boldsymbol{\theta})$.

There are a number of difficulties in applying (11.1) in practice. The main problem is that the correction for the bias of the log-likelihood, $\epsilon(k)$, is only valid asymptotically. Various other corrections have been proposed that have the same asymptotic performance, but different finite-sample performance.

11.2 Learning with unreliable classification

All of the supervised classification procedures that we have considered in this book have assumed that we have a training set of independent pairs of observations and corresponding labels. Further, we have assumed that the 'teacher' who provides the labels is perfect; that is, the teacher never makes mistakes and classifies each object with certainty. Thus, the problem that we have addressed is as follows. Given a random variable pair (X, Y), denote the observation $\boldsymbol{x}$ and the class label y. Given a set of measurements $\mathcal{S} = \{(\boldsymbol{x}_1, y_1), \ldots, (\boldsymbol{x}_n, y_n)\}$, design a classifier to estimate y given $\boldsymbol{x}$ and $\mathcal{S}$.

There are situations in which we do not know the class labels with certainty. This can arise in a number of different ways. Firstly, we may have errors in the labels. Thus the training set comprises $\mathcal{S}' = \{(\boldsymbol{x}_1, z_1), \ldots, (\boldsymbol{x}_n, z_n)\}$, where z_i are the erroneous labels. The problem now is to estimate y given x and $\mathcal{S}'$, the misclassified design set. Misclassification may or may not depend on the feature vector $\boldsymbol{x}$ (termed non-random and random misclassification respectively). Much of the early work in this area assumed patterns that were randomly mislabelled (Lachenbruch, 1966; Chitteneni, 1980, 1981; Michalek and Tripathi, 1980), but it is logical that if the teacher is allowed to be imperfect then it will be more likely that those patterns that are difficult to classify will be mislabelled. Grayson (1987) considers non-random misclassification.

Lugosi (1992) investigates the error rate of two nonparametric classifiers (nearest-neighbour and an L_1 estimator of the posterior probabilities). Three types of dependence of z on $\boldsymbol{x}$ and y are considered:

1. Discrete memoryless channel. The teacher is communicating with a student over a noisy channel so that the true values y_i are transmitted, but the training labels z_i are received as the output of the channel. The channel is determined by the transition probabilities

$$a_{ji} = P(Z = i|Y = j)$$

2. Misprints in the training sequence. Here, the labels z_i can take some arbitrary value with probability p.

3. The 'consequently lying' teacher. The training set value z is the result of a decision $z = h(\boldsymbol{x}), h : \mathbb{R}^p \rightarrow \{0, 1\}$ (a two-class problem is considered).

This latter model may apply in the medical domain, for example, when a physician makes a diagnosis that may be incorrect. However, in such circumstances it is those patterns that are difficult to recognise that are more likely to be mislabelled. It would be better for the 'teacher' (the physician in this case) to be allowed to be an imperfect recogniser, indicating the reliability of a decision rather than choosing one class with certainty. This is another way in which we do not know the class labels with certainty and is the problem addressed by Aitchison and Begg (1976). In many areas it is difficult to classify with certainty, and removing uncertain data may give misleading results. Is it far better to have a classifier that can allow some expression of doubt to be given for the class.

11.3 Missing data

Many classification techniques assume that we have a set of observations with measurements made on each of p variables. However, missing values are common. For example, questionnaires may be returned incomplete; in an archaeological study, it may not be possible to make a complete set of measurements on an artefact because of missing parts; in a medical problem, a complete set of measurements may not be made on a patient, perhaps owing to forgetfulness by the physician or being prevented by the medical condition of the patient.

How missing data are handled depends on a number of factors: how much is missing; why data are missing; whether the missing values can be recovered; whether values are missing in both the design and test set. There are several approaches to this problem.

1. We may omit all incomplete vectors from our analysis. This may be acceptable in some circumstances, but not if there are many observations with missing values. For example, in the head injury study of Titterington *et al.* (1981) referred to in Chapter 3, 206 out of the 500 training patterns and 199 out of the 500 test patterns have at least one observation missing. Neglecting an observation because perhaps one out of 100 variables has not been measured means that we are throwing away potentially useful information for classifier design. Also, in an incomplete observation, it may be that the variables that have been measured are the important ones for classification anyway.

2. We may use all the available information. The way that we would do this depends on the analysis that we are performing. In estimating means and covariances, we would use only those observations for which measurements have been made on the relevant variables. Thus, the estimates would be made on different numbers of samples. This can give poor results and may lead to covariance matrices that are not positive definite. Other approaches must be used to estimate principal components when data are missing (Jackson, 1991). In clustering, we would use a similarity measure that takes missing values into account. In density estimation using an independence assumption, the marginal density estimates will be based on different numbers of samples.

3. We may substitute for the missing values and proceed with our analysis as if we had a complete data set.

There are many approaches to missing value estimation, varying in sophistication and computational complexity. The simplest and perhaps the crudest method is to substitute mean values of the corresponding components. This has been used in many studies. In a supervised classification problem, class means may be substituted for the missing values in the training set and the sample mean for the missing values in the test set since in this case we do not know the class. In a clustering problem, we would estimate the missing values of the group of objects to which the object belongs.

A thorough treatment of missing data in statistical analysis is given by Little and Rubin (1987). A review in the context of regression is given by Little (1992) and a discussion in the classification context by Liu *et al.* (1997).

11.4 Outlier detection and robust procedures

We now consider the problem of detecting outliers in multivariate data. This is one of the aims of *robust statistics*. Outliers are observations that are not consistent with the rest of the data. They may be genuine observations (termed *discordant observations* by Beckman and Cook, 1983) that are surprising to the investigator. Perhaps they may be the most valuable, indicating certain structure in the data that shows deviations from normality. Alternatively, outliers may be *contaminants*, errors caused by copying and transferring the data. In this situation, it may be possible to examine the original data source and correct for any transcription errors.

In both of the above cases, it is important to detect the outliers and to treat them appropriately. Many of the techniques we have discussed in this book are sensitive to outlying values. If the observations are atypical on a single variable, it may be possible to apply univariate methods to the variable. Outliers in multivariate observations can be difficult to detect, particularly when there are several outliers present. A classical procedure is to compute the Mahalanobis distance for each sample $\boldsymbol{x}_i$ $(i = 1, \ldots, n)$

$$D_i = \left\{(\boldsymbol{x}_i - \boldsymbol{m})^T \hat{\boldsymbol{\Sigma}}^{-1} (\boldsymbol{x}_i - \boldsymbol{m})\right\}^{\frac{1}{2}}$$

where $\boldsymbol{m}$ is the sample mean and $\hat{\boldsymbol{\Sigma}}$ the sample covariance matrix. Outliers may be identified as those samples yielding large values of the Mahalanobis distance. This approach suffers from two problems in practice:

1. *Masking*. Multiple outliers in a cluster will distort $\boldsymbol{m}$ and $\hat{\boldsymbol{\Sigma}}$, attracting $\boldsymbol{m}$ and inflating $\hat{\boldsymbol{\Sigma}}$ in their direction, thus giving lower values for the Mahalanobis distance.

2. *Swamping*. This refers to the effect that a cluster of outliers may have on some observations that are consistent with the majority. The cluster could cause the covariance matrix to be distorted so that high values of D are found for observations that are not outliers.

One way of overcoming these problems is to use robust estimates for the mean and covariance matrices. Different estimators have different *breakdown points*, the fraction of outliers that they can tolerate. Rousseeuw (1985) has proposed a *minimum volume ellipsoid* (MVE) estimator that has a high breakdown point of approximately 50%. It can be computationally expensive, but approximate algorithms have been proposed.

Outlier detection and robust procedures have represented an important area of research, investigated extensively in the statistical literature. Chapter 1 of Hampel *et al.* (1986) gives a good introduction to and background on robust procedures. Robust estimates of mean and covariance matrices are reviewed by Rocke and Woodruff (1997); further procedures for the detection of outliers in the presence of appreciable masking are given by Atkinson and Mulira (1993). Krusińska (1988) reviews robust methods within the context of discrimination.

11.5 Mixed continuous and discrete variables

In many areas of pattern recognition involving multivariate data, the variables may be of mixed type, comprising perhaps continuous, ordered categorical, unordered categorical and binary variables. If the discrete variable is ordered, and it is important to retain this information, then the simplest approach is to treat the variable as continuous and to use a technique developed for multivariate continuous data. Alternatively, a categorical variable with k states can be coded as $k - 1$ dummy binary variables. All take the value zero except the jth if the observed categorical variable is in the jth state, $j = 1, \dots, k-1$. All are zero if the variable is in the kth state. This allows some of the techniques developed for mixtures of binary and continuous variables to be used (with some modifications, since not all combinations of binary variables are observable).

The above approaches attempt to distort the data to fit the model. Alternatively, we may apply existing methods to mixed variable data with little or no modification. These include:

1. nearest-neighbour methods (Chapter 3) with a suitable choice of metric;

2. independence models where each univariate density estimate is chosen to be appropriate for the particular variable (Chapter 3);

3. kernel methods using product kernels, where the choice of kernel depends on the variable type (Chapter 3);

4. dependence tree models and Bayesian networks, where the conditional densities are modelled appropriately (for example, using product kernels) (Chapter 3);

5. recursive partitioning methods such as CART and MARS (Chapter 7).

The *location model*, introduced by Olkin and Tate (1961), was developed specifically with mixed variables in mind and applied to discriminant analysis by Chang and Afifi (1974). Consider the problem of classifying a vector $\boldsymbol{v}$ which may be partitioned into two parts, $\boldsymbol{v} = (\boldsymbol{z}^T, \boldsymbol{y}^T)^T$, where $\boldsymbol{z}$ is a vector of r binary variables and $\boldsymbol{y}$ is a vector of p

continuous variables. The random vector z gives rise to 2^r different cells. We may order the cells such that, given a measurement z, the cell number of z, cell(z), is given by

$$\text{cell}(z) = 1 + \sum_{i=1}^{r} z_i 2^{i-1}$$

The location model assumes a multivariate normal distribution for $\boldsymbol{y}$, whose mean depends on the state of z and the class from which $\boldsymbol{v}$ is drawn. It also assumes that the covariance matrix is the same for both classes and for all states. Thus, given a measurement $(z, \boldsymbol{y})$ such that $m = \text{cell}(z)$, the probability density for class ω_i $(i = 1, 2)$ is

$$p(\boldsymbol{y}|z) = \frac{1}{(2\pi)^{p/2}|\boldsymbol{\Sigma}|} \exp\left\{-\frac{1}{2}(\boldsymbol{y} - \boldsymbol{m}_i^m)^T \boldsymbol{\Sigma}^{-1} (\boldsymbol{y} - \boldsymbol{m}_i^m)\right\}$$

Thus, the means of the distribution depend on z and the class. Then, if the probability of observing z in cell m for class ω_i is p_{im}, then $\boldsymbol{v}$ may be assigned to ω_1 if

$$(\boldsymbol{m}_1^m - \boldsymbol{m}_2^m)^T \boldsymbol{\Sigma}^{-1} \left(\boldsymbol{y} - \tfrac{1}{2}(\boldsymbol{m}_1^m + \boldsymbol{m}_2^m)\right) \geq \log(p_{2m}/p_{1m})$$

The maximum likelihood estimates for the parameters p_{im}, $\boldsymbol{m}_i^m$ and $\boldsymbol{\Sigma}$ are

$$\hat{p}_{im} = \frac{n_{im}}{n}$$

$$\hat{\boldsymbol{m}}_i^m = \sum_{j=1}^{n} v_{imj} \frac{1}{n_{im}} \boldsymbol{y}_j$$

$$\boldsymbol{\Sigma} = \frac{1}{n} \sum_{i=1}^{2} \sum_{m=1}^{k} \sum_{j=1}^{n} v_{imj} (\boldsymbol{y}_j - \boldsymbol{m}_i^m)(\boldsymbol{y}_j - \boldsymbol{m}_i^m)^T$$

where $v_{imj} = 1$ if $\boldsymbol{y}_j$ is in cell m of class ω_i, 0 otherwise; and n_{im} is the number of observations in cell m of class ω_i equal to $\sum_j v_{imj}$.

If the sample size n is very large relative to the number of cells, then these naïve estimates may be sufficient. However, in practice there will be too many parameters to estimate. Some of the cells may not be populated, giving poor estimates for $\hat{p}_{im}$. There have been several developments of the basic approach. For a review, see Krzanowski (1993).

11.6 Structural risk minimisation and the Vapnik–Chervonenkis dimension

11.6.1 Bounds on the expected risk

There are general bounds in statistical learning theory (Vapnik, 1998) that govern the relationship between the capacity of a learning system and its performance and thus can provide some guidance in the design of such systems.

We assume that we have a training set of independently and identically distributed samples $\{(\boldsymbol{x}_i, y_i), i = 1, \ldots, n\}$ drawn from a distribution $p(\boldsymbol{x}, y)$. We wish to learn a mapping $\boldsymbol{x} \rightarrow y$ and we have a set of possible functions (classifiers) indexed by parameters $\boldsymbol{\alpha}$, namely $f(\boldsymbol{x}; \boldsymbol{\alpha})$. A particular choice for $\boldsymbol{\alpha}$ results in a particular classifier or *trained machine*. We would like to choose $\boldsymbol{\alpha}$ to minimise the classification error. If y_i takes the value $+1$ for patterns in class ω_1 and -1 for patterns in class ω_2, then the expected value of the test error (the true error–see Chapter 8) is

$$\mathcal{R}(\boldsymbol{\alpha}) = \frac{1}{2} \int |y - f(\boldsymbol{x}; \boldsymbol{\alpha})| \, p(\boldsymbol{x}, y) \, d\boldsymbol{x} \, dy$$

This is sometimes termed the *expected risk* (note that differs from the definition of risk in Chapter 1). This is not known in general, but we may estimate it based on a training set, to give the *empirical risk*,

$$\mathcal{R}_K(\boldsymbol{\alpha}) = \frac{1}{2n} \sum_{i=1}^{n} |y_i - f(\boldsymbol{x}_i; \boldsymbol{\alpha})|$$

For any value of η, $0 \leq \eta \leq 1$, the following bound of statistical learning theory holds (Vapnik, 1998)

$$\mathcal{R}(\boldsymbol{\alpha}) \leq \mathcal{R}_K + \sqrt{\frac{h(\log(2n/h) + 1) - \log(\eta/4)}{n}} \qquad (11.2)$$

with probability $1 - \eta$, where h is a non-negative integer called the *Vapnik–Chervonenkis (VC) dimension*. The first term on the right-hand side of the inequality above depends on the particular function f chosen by the training procedure. The second term, the *VC confidence*, depends on the class of functions.

11.6.2 The Vapnik–Chervonenkis dimension

The VC dimension is a property of the set of functions $f(\boldsymbol{x}; \boldsymbol{\alpha})$. If a given set of m points can be labelled in all possible 2^m ways using the functions $f(\boldsymbol{x}; \boldsymbol{\alpha})$, then the set of points is said to be *shattered* by the set of functions; that is, for any labelling of the set of points, there is a function $f(\boldsymbol{x}; \boldsymbol{\alpha})$ that correctly classifies the patterns.

The VC dimension of a set of functions is defined as the maximum number of training points that can be shattered. Note that if the VC dimension is m, then there is at least one set of m points that can be shattered, but not necessarily every set of m points can be shattered. For example, if $f(\boldsymbol{x}; \boldsymbol{\alpha})$ is the set of all lines in the plane, then every set of two points can be shattered, and most sets of three (see Figure 11.1), but no sets of four points can be shattered by a linear model. Thus the VC dimension is 3. More generally, the VC dimension of a set of hyperplanes in r-dimensional Euclidean space is $r + 1$.

Inequality (11.2) shows that the risk may be controlled through a balance of optimising a fit to the data and the capacity of functions used in learning. In practice, we would consider sets of models, f, with each set of a fixed VC dimension. For each set, minimise the empirical risk and choose the model over all sets for which the sum of the empirical risk and VC confidence is a minimum. However, the inequality (11.2) is only a guide. There may be models with equal empirical risk but with different VC dimensions. The

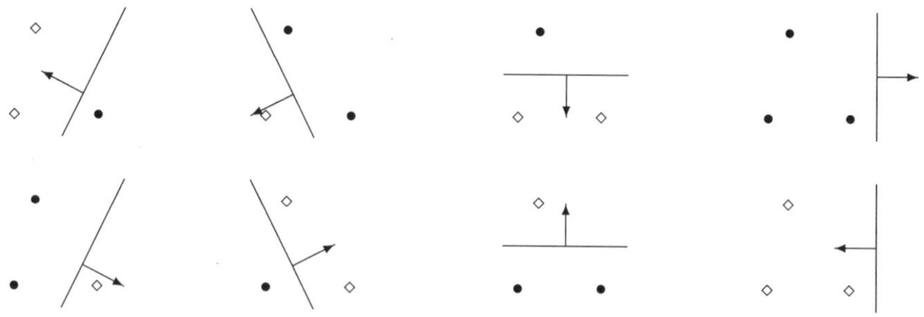

Figure 11.1 Shattering three points in two dimensions

one with higher VC dimension does not necessarily have poorer performance. A k-nearest-neighbour classifier has zero empirical risk (any labelling of a set of points will be correctly classified) and infinite VC dimension.

A

Measures of dissimilarity

A.1 Measures of dissimilarity

Patterns or objects analysed using the techniques described in this book are usually represented by a vector of measurements. Many of the techniques require some measure of dissimilarity or distance between two pattern vectors, although sometimes data can arise directly in the form of a dissimilarity matrix.

A particular class of dissimilarity functions called *dissimilarity coefficients* are required to satisfy the following conditions. If d_{rs} is the dissimilarity of object s from object r, then

$$\begin{aligned} d_{rs} &\geq 0 && \text{for every } r, s \\ d_{rr} &= 0 && \text{for every } r \\ d_{rs} &= d_{sr} && \text{for every } r, s \end{aligned}$$

The symmetry condition is not always satisfied by some dissimilarity functions. If the dissimilarity between two places in a city centre is the distance travelled by road between them, then because of one-way systems the distance may be longer in one direction than the other. Measures of dissimilarity can be transformed to similarity measures using various transformations, for example, $s_{ij} = 1/(1 + d_{ij})$ or $s_{ij} = c - d_{ij}$ for some constant c, where s_{ij} is the similarity between object i and object j.

If, in addition to the three conditions above, the dissimilarity measure satisfies the triangle inequality

$$d_{rt} + d_{ts} \geq d_{rs} \quad \text{for every } r, s, t \tag{A.1}$$

then the dissimilarity measure is a *metric* and the term *distance* is usually used.

A.1.1 Numeric variables

Many dissimilarity measures have been proposed for numeric variables. Table A.1 gives some of the more common measures. The choice of a particular metric depends on the application. Computational considerations aside, for feature selection and extraction purposes you would choose the metric that gives the best performance (perhaps in terms of classification error on a validation set).

Table A.1 Dissimilarity measures for numeric variables (between $\boldsymbol{x}$ and $\boldsymbol{y}$)

Dissimilarity measure	Mathematical form
Euclidean distance	$d_e = \left\{ \sum_{i=1}^{p} (x_i - y_i)^2 \right\}^{\frac{1}{2}}$
City-block distance	$d_{cb} = \sum_{i=1}^{p} \lvert x_i - y_i \rvert$
Chebyshev distance	$d_{ch} = \max_i \lvert x_i - y_i \rvert$
Minkowski distance of order m	$d_M = \left\{ \sum_{i=1}^{p} (x_i - y_i)^m \right\}^{\frac{1}{m}}$
Quadratic distance	$d_q = \sum_{i=1}^{p} \sum_{j=1}^{p} (x_i - y_i) Q_{ij} (x_j - y_j)$, Q positive definite
Canberra distance	$d_{ca} = \sum_{i=1}^{p} \frac{\lvert x_i - y_i \rvert}{x_i + y_i}$
Nonlinear distance	$d_n = \begin{cases} H & d_e > D \\ 0 & d_e \leq D \end{cases}$
Angular separation	$\frac{\sum_{i=1}^{p} x_i y_i}{\left[\sum_{i=1}^{p} x_i^2 \sum_{i=1}^{p} y_i^2 \right]^{1/2}}$

Euclidean distance

$$d_e = \sqrt{\sum_{i=1}^{p} (x_i - y_i)^2}$$

The contours of equal Euclidean distance from a point are hyperspheres (circles in two dimensions). It has the (perhaps undesirable) property of giving greater emphasis to larger differences on a single variable.

Although we may wish to use a dissimilarity measure that is a metric, some of the methods do not require the metric condition (A.1) above. Therefore in some cases a monotonic function of the Euclidean metric, which will still be a dissimilarity coefficient but not necessarily a metric, will suffice. For example, squared Euclidean distance is a dissimilarity coefficient but not a metric.

City-block distance

$$d_{cb} = \sum_{i=1}^{p} \lvert x_i - y_i \rvert$$

Also known as the *Manhattan* or *box-car* or *absolute value* distance, this metric uses a distance calculation which would be suitable for finding the distances between points in a city consisting of a grid of intersecting thoroughfares (hence the names used). The contours of equal distance from a point for the city-block metric are diamonds in two dimensions. The city-block metric is a little cheaper to compute than the Euclidean distance so it may be used if the speed of a particular application is important.

Chebyshev distance

$$d_{ch} = \max_i |x_i - y_i|$$

The Chebyshev or *maximum value* distance is often used in cases where the execution speed is so critical that the time involved in calculating the Euclidean distance is unacceptable. The Chebyshev distance, like the city-block distance, examines the absolute magnitude of the elementwise differences in the pair of vectors. The contour lines of equal Chebyshev distance from a point are squares in two dimensions. Figure A.1 plots the contours of equal distance in $\mathbb{R}^2$ for the Euclidean, city-block and Chebyshev metrics.

If the user needs an approximation to Euclidean distance but with a cheaper computational load then the first line of approach is to use either the Chebyshev or the city-block metrics. A better approximation can be gained by using a combination of these two distances:

$$d = \max(\tfrac{2}{3}d_{cb}, d_{ch})$$

In two dimensions the contours of equal distance form octagons.

Minkowski distance The Minkowski distance is a more general form of the Euclidean and city-block distances. The Minkowski distance of order m is

$$d_M = \left(\sum_{i=1}^{p} |x_i - y_i|^m\right)^{1/m}$$

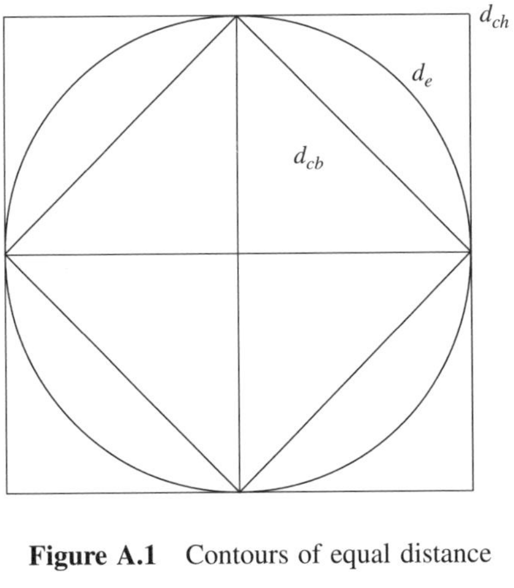

Figure A.1 Contours of equal distance

The Minkowski distance of the first order is the same as the city-block metric and the Minkowski distance of the second order is the Euclidean distance. The contours of equal distance for such metrics form squared-off circles which gradually obtain more abrupt vertices as m increases. The choice of an appropriate value for m depends on the amount of emphasis you would like to give to the larger differences: larger values of m give progressively more emphasis to the larger differences $|x_i - y_i|$, and as m tends to infinity the metric tends to the Chebyshev distance (and square contours).

Quadratic distance

$$d_q^2 = (\boldsymbol{x} - \boldsymbol{y})^T \boldsymbol{Q}(\boldsymbol{x} - \boldsymbol{y})$$

A choice for $\boldsymbol{Q}$ is the inverse of the within-group covariance matrix. This is sometimes referred to as the Mahalanobis distance, because of the similarity to the distance measure between two distributions (see below).

Canberra metric

$$d_{ca} = \sum_{i=1}^{p} \frac{|x_i - y_i|}{x_i + y_i}$$

The Canberra metric is a sum of a series of fractions and is suitable for variables taking non-negative values. If both x_i and y_i are zero the ratio of the difference to the sum is taken to be zero. If only one value is zero, the term is unity, independent of the other value. Thus, 0 and 1 are equally dissimilar to a pair of elements 0 and 10^6. Sometimes values of 0 are replaced by small positive numbers (smaller than the recorded values of that variable).

Nonlinear distance

$$d_N = \begin{cases} 0 & \text{if } d_e(\boldsymbol{x}, \boldsymbol{y}) < D \\ H & \text{if } d_e(\boldsymbol{x}, \boldsymbol{y}) \geq D \end{cases}$$

where D is a threshold and H is a constant. Kittler (1975a) shows that an appropriate choice for H and D for feature selection is that they should satisfy

$$H = \frac{\Gamma(p/2)}{D^p 2\sqrt{\pi^p}}$$

and that D satisfies the unbiasedness and consistency conditions of the Parzen estimator, namely $D^p n \to \infty$ and $D \to 0$ as $n \to \infty$, where n is the number of samples in the data set.

Angular separation

$$\frac{\sum_{i=1}^{p} x_i y_i}{\left[\sum_{i=1}^{p} x_i^2 \sum_{i=1}^{p} y_i^2\right]}$$

The angular separation is a similarity rather than a dissimilarity measure that measures the angle between the unit vectors in the direction of the two pattern vectors of interest. This is appropriate when data are collected for which only the relative magnitudes are important.

The choice of a particular proximity measure depends on the application and may depend on several factors, including distribution of data and computational considerations. It is not possible to make recommendations, and studies in this area have been largely empirical, but the method you choose should be the one that you believe will capture the essential differences between objects.

A.1.2 Nominal and ordinal variables

Nominal and ordinal variables are usually represented as a set of binary variables. For example, a nominal variable with s states is represented as s binary variables. If it is in the mth state, then each of the s binary variables has value 0 except the mth, which has the value unity. The dissimilarity between two objects can be obtained by summing the contributions from the individual variables.

For ordinal variables, the contribution to the dissimilarity between two objects from a single variable does not simply depend on whether or not the values are identical. If the contribution for one variable in state m and one in state l ($m < l$) is δ_{ml}, then we require

$$\begin{aligned} \delta_{ml} &\geq \delta_{ms} \quad \text{for } s < l \\ \delta_{ml} &\geq \delta_{sl} \quad \text{for } s > m \end{aligned}$$

that is, δ_{ml} is monotonic down each row and across each column of the half-matrix of distances between states ($\delta_{14} > \delta_{13} > \delta_{12}$ etc.; $\delta_{14} > \delta_{24} > \delta_{34}$). The values chosen for δ_{ml} depend very much on the problem. For example, we may have a variable describing fruits of a plant that can take the values short, long or very long. We would want the dissimilarity between a plant with very long fruit and one with short fruit to be greater than that between one with long fruit and one with short fruit (all other attributes having equal values). A numeric coding of 1, 2, 3 would achieve this, but so would 1, 10, 100.

A.1.3 Binary variables

Various dissimilarity measures have been proposed for binary variables. For vectors of binary variables $\boldsymbol{x}$ and $\boldsymbol{y}$ these may be expressed in terms of quantities a, b, c, and d where

a is equal to the number of occurrences of $x_i = 1$ and $y_i = 1$
b is equal to the number of occurrences of $x_i = 0$ and $y_i = 1$
c is equal to the number of occurrences of $x_i = 1$ and $y_i = 0$
d is equal to the number of occurrences of $x_i = 0$ and $y_i = 0$

This is summarised in Table A.2. Note that $a+b+c+d = p$, the total number of variables (attributes). It is customary to define a similarity measure rather than a dissimilarity measure. Table A.3 summarises some of the more commonly-used similarity measures for binary data.

Table A.2 Co-occurrence table for binary variables

		x_i	
		1	0
y_i	1	a	b
	0	c	d

Table A.3 Similarity measures for binary data

Similarity measure	Mathematical form
Simple matching coefficient	$d_{sm} = \frac{a+d}{a+b+c+d}$
Russell and Rao	$d_{rr} = \frac{a}{a+b+c+d}$
Jaccard	$d_j = \frac{a}{a+b+c}$
Czekanowski	$d_{Cz} = \frac{2a}{2a+b+c}$

Simple matching coefficient The simple matching coefficient is the proportion of variables for which two variables have the same value. The dissatisfaction with this measure has been with the term d representing *conjoint absences*. The fact that two sites in an ecological survey both lack something should not make them more similar. The dissimilarity measure defined by $d_{xy} = 1 - s_{xy} = (b+c)/p$ is proportional to the square of the Euclidean distance, $b+c$, which is the Hamming distance in communication theory.

Russell and Rao This does not involve the term d in the numerator and is appropriate in certain circumstances. The quantity $1 - s_{xy}$ is not a dissimilarity coefficient since the dissimilarity between an object and itself is not necessarily zero.

Jaccard This does not involve the quantity d at all and is used extensively by ecologists. The term $d_{xy} = 1 - s_{xy}$ is a metric dissimilarity coefficient.

Czekanowski This is similar to the Jaccard measure except that coincidences carry double weight.

Many other coefficients have been proposed that handle the conjoint absences in various ways (Clifford and Stephenson, 1975; Diday and Simon, 1976).

A.1.4 Summary

We have listed some of the measures of proximity which can be found in the pattern processing and classification literature. A general similarity coefficient between two objects $\boldsymbol{x}$ and $\boldsymbol{y}$ encompassing variables of mixed type has been proposed by Gower

(1971). Of course, there is no such thing as a best measure. Some will be more appropriate for certain tasks than others. Therefore, we cannot make recommendations. However, the user should consider the following points when making a choice: (1) simplicity and ease of understanding; (2) ease of implementation; (3) speed requirements; (4) knowledge of data.

A.2 Distances between distributions

All of the distance measures described so far have been defined between two patterns or objects. We now turn to measures of distances between groups of objects or distributions. These measures are used to determine the discriminatory power of a feature set, discussed in Chapter 9. Many measures have been proposed in the pattern recognition literature, and we introduce two basic types here. The first uses prototype vectors for each class together with the distance metrics of the previous section. The second uses knowledge of the class-conditional probability density functions. Many methods of this type are of academic interest only. Their practical application is rather limited since they involve numerical integration and estimation of the probability density functions from samples. They do simplify if the density functions belong to a family of parametric functions such as the exponential family, which includes the normal distribution. The use of both of these approaches for feature selection is described in Chapter 9.

A.2.1 Methods based on prototype vectors

There are many measures of inter-group dissimilarities based on prototype vectors. In the context of clustering, these give rise to different hierarchical schemes, which are discussed in Chapter 10. Here we introduce the average separation, defined to be the average distance between all pairs of points, with one point in each pair coming from each distribution. That is, for n_1 points in ω_1 ($\boldsymbol{x}_i, i = 1, \ldots, n_1$) and n_2 points in ω_2 ($\boldsymbol{y}_i, i = 1, \ldots, n_2$),

$$J_{as}(\omega_1, \omega_2) = \frac{1}{n_1 n_2} \sum_{i=1}^{n_1} \sum_{j=1}^{n_2} d(\boldsymbol{x}_i, \boldsymbol{y}_j)$$

where d is a distance between $\boldsymbol{x}_i$ and $\boldsymbol{y}_j$.

A.2.2 Methods based on probabilistic distance

These measures use the complete information about the structure of the classes provided by the conditional densities. The distance measure, J, satisfies the following conditions:

1. $J = 0$ if the probability density functions are identical, $p(x|\omega_1) = p(x|\omega_2)$;

2. $J \geq 0$;

3. J attains its maximum when the classes are disjoint, i.e. when $p(x|\omega_1) = 0$ and $p(x|\omega_2) \neq 0$.

Many measures satisfying these conditions have been proposed (Chen, 1976; Devijver and Kittler, 1982). As an introduction, consider two overlapping distributions with conditional densities $p(\boldsymbol{x}|\omega_1)$ and $p(\boldsymbol{x}|\omega_2)$. The classification error, e (see Chapter 8), is given by

$$e = \frac{1}{2}\left\{1 - \int |p(\omega_1|\boldsymbol{x}) - p(\omega_2|\boldsymbol{x})|p(\boldsymbol{x})\,d\boldsymbol{x}\right\}$$

The integral in the equation above,

$$J_K = \int |p(\omega_1|\boldsymbol{x}) - p(\omega_2|\boldsymbol{x})|p(\boldsymbol{x})\,d\boldsymbol{x}$$

is called the *Kolmogorov variational distance* and has the important property that it is directly related to the classification error. Other measures cannot be expressed in terms of the classification error, but can be used to provide bounds on the error. Three of these are given in Table A.4. A more complete list can be found in the books by Chen (1976) and Devijver and Kittler (1982).

One of the main disadvantages of the probabilistic dependence criteria is that they require an estimate of a probability density function and its numerical integration. This restricts their usefulness in many practical situations. However, under certain assumptions regarding the form of the distributions, the expressions can be evaluated analytically.

First of all, we shall consider a specific parametric form for the distributions, namely normally distributed with means $\boldsymbol{\mu}_1$ and $\boldsymbol{\mu}_2$ and covariance matrices $\boldsymbol{\Sigma}_1$ and $\boldsymbol{\Sigma}_2$. Under these assumptions, the distance measures can be written down as follows.

Table A.4 Probabilistic distance measures

Dissimilarity measure	Mathematical form
Average separation	$\frac{1}{n_a n_b}\sum_{i=1}^{n_a}\sum_{j=1}^{n_b} d(\boldsymbol{x}_i, \boldsymbol{y}_j), \boldsymbol{x}_i \in \omega_A, \boldsymbol{y}_j \in \omega_B,$ d any distance metric
Chernoff	$J_c = -\log\int p^s(\boldsymbol{x}|\omega_1)p^{1-s}(\boldsymbol{x}|\omega_2)\,d\boldsymbol{x}$
Bhattacharyya	$J_B = -\log\int (p(\boldsymbol{x}|\omega_1)p(\boldsymbol{x}|\omega_2))^{\frac{1}{2}}\,d\boldsymbol{x}$
Divergence	$J_D = \int [p(\boldsymbol{x}|\omega_1) - p(\boldsymbol{x}|\omega_2)]\log\left(\frac{p(\boldsymbol{x}|\omega_1)}{p(\boldsymbol{x}|\omega_2)}\right)d\boldsymbol{x}$
Patrick–Fischer	$J_P = \left\{\int [p(\boldsymbol{x}|\omega_1)p_1 - p(\boldsymbol{x}|\omega_2)p_2]^2\,d\boldsymbol{x}\right\}^{\frac{1}{2}}$

Chernoff

$$J_c = \frac{1}{2}s(1-s)(\boldsymbol{\mu}_2 - \boldsymbol{\mu}_1)^T[\boldsymbol{\Sigma}_s]^{-1}(\boldsymbol{\mu}_2 - \boldsymbol{\mu}_1) + \frac{1}{2}\log\left(\frac{|\boldsymbol{\Sigma}_s|}{|\boldsymbol{\Sigma}_1|^{1-s}|\boldsymbol{\Sigma}_2|^s}\right)$$

where $\boldsymbol{\Sigma}_s = (1-s)\boldsymbol{\Sigma}_1 + s\boldsymbol{\Sigma}_2$ and $s \in [0, 1]$. For $s = 0.5$, we have the Bhattacharyya distance.

Bhattacharyya

$$J_B = \frac{1}{4}(\boldsymbol{\mu}_2 - \boldsymbol{\mu}_1)^T\left[\boldsymbol{\Sigma}_1 + \boldsymbol{\Sigma}_2\right]^{-1}(\boldsymbol{\mu}_2 - \boldsymbol{\mu}_1) + \frac{1}{2}\log\left(\frac{|\boldsymbol{\Sigma}_1 + \boldsymbol{\Sigma}_2|}{2(|\boldsymbol{\Sigma}_1||\boldsymbol{\Sigma}_2|)^{\frac{1}{2}}}\right)$$

Divergence

$$J_D = \tfrac{1}{2}(\boldsymbol{\mu}_2 - \boldsymbol{\mu}_1)^T(\boldsymbol{\Sigma}_1^{-1} + \boldsymbol{\Sigma}_2^{-1})(\boldsymbol{\mu}_2 - \boldsymbol{\mu}_1) + \mathrm{Tr}\{\boldsymbol{\Sigma}_1^{-1}\boldsymbol{\Sigma}_2 + \boldsymbol{\Sigma}_1^{-1}\boldsymbol{\Sigma}_2 - 2\boldsymbol{I}\}$$

Patrick–Fischer

$$J_P = (2\pi)^{-p/2}\left[|2\boldsymbol{\Sigma}_1|^{-\frac{1}{2}} + |2\boldsymbol{\Sigma}_2|^{-\frac{1}{2}} - 2|\boldsymbol{\Sigma}_1 + \boldsymbol{\Sigma}_2|^{-\frac{1}{2}}\exp\left\{-\frac{1}{2}(\boldsymbol{\mu}_2 - \boldsymbol{\mu}_1)^T(\boldsymbol{\Sigma}_1 + \boldsymbol{\Sigma}_2)^{-1}(\boldsymbol{\mu}_2 - \boldsymbol{\mu}_1)\right\}\right]$$

Finally, if the covariance matrices are equal, $\boldsymbol{\Sigma}_1 = \boldsymbol{\Sigma}_2 = \boldsymbol{\Sigma}$, the Bhattacharyya and divergence distances simplify to

$$J_M = J_D = 8J_B = (\boldsymbol{\mu}_2 - \boldsymbol{\mu}_1)^T\boldsymbol{\Sigma}^{-1}(\boldsymbol{\mu}_2 - \boldsymbol{\mu}_1)$$

which is the Mahalanobis distance.

Of course, the means and covariance matrices are not known in practice and must be estimated from the available training data.

The above parametric forms are useful both in feature selection and extraction. In feature selection, the set of features at the kth stage of an algorithm is constructed from the set of features at the $(k-1)$th stage by the addition or subtraction of a small number of features. The value of the feature selection criterion at stage $k+1$ may be computed from that at stage k rather than evaluating the above expressions directly. This saves on computation. Recursive calculation of separability measures is discussed in Chapter 9.

Probabilistic distance measures can also be extended to the multigroup case by evaluating all pairwise distances between classes,

$$J = \sum_{i=1}^{C}\sum_{j=1}^{C} p_i p_j J_{ij}$$

where J_{ij} is the chosen distance measure evaluated for class ω_i and class ω_j.

A.2.3 Probabilistic dependence

The probabilistic distance measures are based on discrimination between a pair of classes, using the class-conditional densities to describe each class. Probabilistic dependence measures are multiclass feature selection criteria that measure the distance between the class-conditional density and the mixture probability density function (see Figure A.2). If $p(\boldsymbol{x}|\omega_i)$ and $p(\boldsymbol{x})$ are identical then we gain no information about class by observing $\boldsymbol{x}$, and the 'distance' between the two distributions is zero. Thus, $\boldsymbol{x}$ and ω_i are independent. If the distance between $p(\boldsymbol{x}|\omega_i)$ and $p(\boldsymbol{x})$ is large, then the observation $\boldsymbol{x}$ is dependent on ω_i. The greater the distance, the greater the dependence of $\boldsymbol{x}$ on the class ω_i. Table A.5

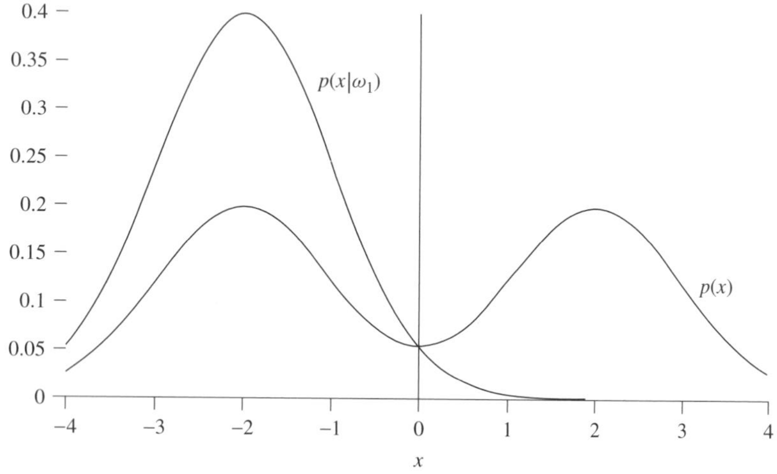

Figure A.2 Probabilistic dependence

Table A.5 Probabilistic dependence measures

Dissimilarity measure	Mathematical form
Chernoff	$J_c = \sum_{i=1}^{C} p_i \left\{ -\log \int p^s(\boldsymbol{x}\|\omega_i) p^{1-s}(\boldsymbol{x})\, d\boldsymbol{x} \right\}$
Bhattacharyya	$J_B = \sum_{i=1}^{C} p_i \left\{ -\log \int (p(\boldsymbol{x}\|\omega_i) p(\boldsymbol{x}))^{\frac{1}{2}}\, d\boldsymbol{x} \right\}$
Joshi	$J_D = \sum_{i=1}^{C} p_i \int [p(\boldsymbol{x}\|\omega_i) - p(\boldsymbol{x})] \log \left(\frac{p(\boldsymbol{x}\|\omega_i)}{p(\boldsymbol{x})} \right) d\boldsymbol{x}$
Patrick–Fischer	$J_P = \sum_{i=1}^{C} p_i \left\{ \int [p(\boldsymbol{x}\|\omega_i) - p(\boldsymbol{x})]^2\, d\boldsymbol{x} \right\}^{\frac{1}{2}}$

gives the probabilistic dependence measures corresponding to the probabilistic distance measures in Table A.4. In practice, application of probabilistic dependence measures is limited because, even for normally distributed classes, the expressions given in Table A.5 cannot be evaluated analytically since the mixture distribution $p(\boldsymbol{x})$ is not normal.

A.3 Discussion

This appendix has reviewed some of the distance and dissimilarity measures used in Chapter 9 on feature selection and extraction and Chapter 10 on clustering. Of course the list is not exhaustive and those that are presented may not be the best for your problem. We cannot make rigid recommendations as to which ones you should use since the choice is highly problem-specific. However, it may be advantageous from a computational point of view to use one that simplifies for normal distributions even if your data are not normally distributed.

The book by Gordon (1999) provides a good introduction to classification methods. The chapter on dissimilarity measures also highlights difficulties encountered in practice with real data sets. There are many other books and papers on clustering which list other measures of dissimilarity: for example, Diday and Simon (1976), Cormack (1971) and Clifford and Stephenson (1975), which is written primarily for biologists but the issues treated occur in many areas of scientific endeavour. The papers by Kittler (1975b, 1986) provide very good introductions to feature selection and list some of the more commonly used distance measures. Others may be found in Chen (1976). A good account of probabilistic distance and dependence measures can be found in the book by Devijver and Kittler (1982).

B

Parameter estimation

B.1 Parameter estimation

B.1.1 Properties of estimators

Perhaps before we begin to discuss some of the desirable properties of estimators, we ought to define what an estimator is. For example, in a measurement experiment we may assume that the observations are normally distributed but with unknown mean, μ, and variance, σ^2. The problem then is to estimate the values of these two parameters from the set of observations. Therefore, an *estimator* $\hat{\theta}$ of θ is defined as any function of the sample values which is calculated to be close in some sense to the true value of the unknown parameter θ. This is a problem in *point estimation*, by which is meant deriving a single-valued function of a set of observations to represent the unknown parameter (or a function of the unknown parameters), without explicitly stating the precision of the estimate. The estimation of the confidence interval (the limits within which we expect the parameter to lie) is an exercise in *interval estimation*. For a detailed treatment of estimation we refer to Stuart and Ord (1991).

Unbiased estimate The estimator $\hat{\theta}$ of the parameter θ is *unbiased* if the expectation over the sampling distribution is equal to θ, i.e.

$$\mathrm{E}[\hat{\theta}] \triangleq \int \hat{\theta} p(\boldsymbol{x}_1, \ldots, \boldsymbol{x}_n)\, d\boldsymbol{x}_1 \ldots d\boldsymbol{x}_n = \theta$$

where $\hat{\theta}$ is a function of the sample vectors $\boldsymbol{x}_1, \ldots, \boldsymbol{x}_n$ drawn from the distribution $p(\boldsymbol{x}_1, \ldots, \boldsymbol{x}_n)$. We might always want estimators to be approximately unbiased, but there is no reason why we should insist on exact unbiasedness.

Consistent estimate An estimator $\hat{\theta}$ of a parameter θ is *consistent* if it *converges in probability* (or *converges stochastically*) to θ as the number of observations, $n \to \infty$. That is, for all $\delta, \epsilon > 0$

$$p(\|\hat{\theta} - \theta\| < \epsilon) > 1 - \delta \qquad \text{for } n > n_0$$

or

$$\lim_{n \to \infty} p(\|\hat{\theta} - \theta\| > \epsilon) = 0$$

Efficient estimate The efficiency, η, of one estimator $\hat{\theta}_2$ relative to another $\hat{\theta}_1$ is defined as the ratio of the variance of the estimators

$$\eta = \frac{\mathrm{E}[\|\hat{\theta}_1 - \theta\|^2]}{\mathrm{E}[\|\hat{\theta}_2 - \theta\|^2]}$$

$\hat{\theta}_1$ is an *efficient* estimator if it has the smallest variance (in large samples) compared to all other estimators, i.e. $\eta \leq 1$ for all $\hat{\theta}_2$.

Sufficient estimate A statistic $\hat{\theta}_1 = \hat{\theta}_1(\boldsymbol{x}_1, \ldots, \boldsymbol{x}_n)$ is termed a *sufficient statistic* if, for any other statistic $\hat{\theta}_2$,

$$p(\theta|\hat{\theta}_1, \hat{\theta}_2) = p(\theta|\hat{\theta}_1) \tag{B.1}$$

that is, all the relevant information for the estimation of θ is contained in $\hat{\theta}_1$ and the additional knowledge of $\hat{\theta}_2$ makes no contribution. An equivalent condition for a distribution to possess a sufficient statistic is the factorability of the likelihood function (Stuart and Ord, 1991; Young and Calvert, 1974):

$$p(\boldsymbol{x}_1, \ldots, \boldsymbol{x}_n|\theta) = g(\hat{\theta}|\theta)h(\boldsymbol{x}_1, \ldots, \boldsymbol{x}_n) \tag{B.2}$$

where h is a function of $\boldsymbol{x}_1, \ldots, \boldsymbol{x}_n$ and is essentially $p(\boldsymbol{x}_1, \ldots, \boldsymbol{x}_n|\hat{\theta})$ and does not depend on θ, and g is a function of the statistic $\hat{\theta}$ and θ. Equation (B.2) is also the condition for *reproducing densities* (Spragins, 1976; Young and Calvert, 1974) or *conjugate priors* (Lindgren, 1976): a probability density of θ, $p(\theta)$, is a reproducing density with respect to the conditional density $p(\boldsymbol{x}_1, \ldots, \boldsymbol{x}_n|\theta)$ if the posterior density $p(\theta|\boldsymbol{x}_1, \ldots, \boldsymbol{x}_n)$ and the prior density $p(\theta)$ are of the same functional form. The family is called *closed under sampling* or *conjugate* with respect to $p(\boldsymbol{x}_1, \ldots, \boldsymbol{x}_n|\theta)$. Conditional densities that admit sufficient statistics of fixed dimension for any sample size (and hence reproducing densities) include the normal, binomial and Poisson density functions (Spragins, 1976).

Example 1 The sample mean $\overline{x}_n = \frac{1}{n}\sum_{i=1}^n x_i$ is an unbiased estimator of the population mean, μ, since

$$\mathrm{E}[\overline{x}_n] = \mathrm{E}\left[\frac{1}{n}\sum_{i=1}^n x_i\right] = \frac{1}{n}\sum_{i=1}^n \mathrm{E}[x_i] = \mu$$

but the sample variance

$$\begin{aligned}\mathrm{E}\left[\frac{1}{n}\sum_{i=1}^n (x_i - \overline{x}_n)^2\right] &= \mathrm{E}\left[\frac{1}{n}\sum_{i=1}^n \left(x_i - \frac{1}{n}\sum_{j=1}^n x_j\right)^2\right] \\ &= \frac{1}{n}\mathrm{E}\left[\frac{n-1}{n}\sum_{j=1}^n x_j^2 - \frac{1}{n}\sum_j \sum_{k,k\neq j} x_j x_k\right] \\ &= \frac{n-1}{n}\sigma^2\end{aligned}$$

is not an unbiased estimator of the variance σ^2. Therefore, the unbiased estimator

$$s = \frac{1}{n-1}\sum_{i=1}^{n}(x_i - \overline{x}_n)^2$$

is usually preferred. □

Example 2 The sample mean is a consistent estimator of the mean of a normal population. The sample mean is normally distributed as

$$p(\overline{x}_n) = \left(\frac{n}{2\pi}\right)^{\frac{1}{2}} \exp\left(-\frac{1}{2}n(\overline{x}_n - \theta)^2\right)$$

with mean θ (the mean of the population with unit variance) and variance $1/n$. That is, $(\overline{x}_n - \theta)n^{\frac{1}{2}}$ is normally distributed with zero mean and unit variance. Thus, the probability that $|(\overline{x}_n - \theta)n^{\frac{1}{2}}| \leq \epsilon n^{\frac{1}{2}}$ (i.e. $|(\overline{x}_n - \theta)n| \leq \epsilon n$) is the value of the normal integral between the limits $\pm\epsilon n^{\frac{1}{2}}$. By choosing n sufficiently large, this can always be larger than $1 - \eta$ for any given η. □

B.1.2 Maximum likelihood

The likelihood function is the joint density of a set of samples $\boldsymbol{x}_1, \ldots, \boldsymbol{x}_n$ from a distribution $p(\boldsymbol{x}_i|\theta)$, i.e.

$$L(\theta) = p(\boldsymbol{x}_1, \ldots, \boldsymbol{x}_n|\theta)$$

regarded as a function of the parameters, θ, rather than the data samples. The method of maximum likelihood is a general method of point estimation in which the estimate of the parameter θ is taken to be that value for which $L(\theta)$ is a maximum. That is, we are choosing the value of θ which is 'most likely to give rise to the observed data'. Thus, we seek a solution to the equation

$$\frac{\partial L}{\partial \theta} = 0$$

or, equivalently,

$$\frac{\partial \log(L)}{\partial \theta} = 0$$

since any monotonic function of the likelihood, L, will also be a minimum at the same value of θ as the function L. Under very general conditions (Stuart and Ord, 1991, Chapter 18), the maximum likelihood estimator is consistent, asymptotically normal and asymptotically efficient. The estimator is not, in general, unbiased though it will be asymptotically unbiased if it is consistent and if the asymptotic distribution has finite mean. However, for an unbiased estimator, $\hat{\theta}$, of a parameter θ, the lower bound on the variance is given by the Cramér–Rao bound

$$\mathrm{E}[(\hat{\theta} - \theta)^2] = \frac{1}{\sigma_n^2}$$

where

$$\sigma_n^2 = \mathrm{E}\left[\left(\frac{\partial}{\partial\theta}\log[p(\boldsymbol{x}_1,\ldots,\boldsymbol{x}_n|\theta)]\right)^2\right]$$

is called the *Fisher information* in the sample. It follows from the definition of the efficient estimator that any unbiased estimator that satisfies this bound is efficient.

B.1.3 Problems with maximum likelihood

The main difficulty with maximum likelihood parameter estimation is obtaining a solution of the equations

$$\frac{\partial L}{\partial\boldsymbol{\theta}} = 0$$

for a vector of parameters, $\boldsymbol{\theta}$. Unlike in the normal case, these are not always tractable and iterative techniques must be employed. These may be a nonlinear optimisation scheme using gradient information such as conjugate gradient methods or quasi-Newton methods or, for certain parametric forms of the likelihood function, expectation–maximisation (EM) methods may be employed. The problem with the former approach is that maximisation of the likelihood function is not simply an exercise in unconstrained optimisation. That is, the quantities being estimated are often required to satisfy some (inequality) constraint. For example, in estimating the covariance matrix of a normal distribution, the elements of the matrix are constrained so that they satisfy the requirements of a covariance matrix (positive definiteness, symmetry). The latter methods have been shown to converge for particular likelihood functions.

Maximum likelihood is the most extensively used statistical estimation technique. It may be regarded as an approximation to the Bayesian approach, described below, in which the prior probability, $p(\theta)$, is assumed uniform, and is arguably more appealing since it has no subjective or uncertain element represented by $p(\theta)$. However, the Bayesian approach is in more agreement with the foundations of probability theory. A detailed treatment of maximum likelihood estimation may be found in Stuart and Ord (1991).

B.1.4 Bayesian estimates

The maximum likelihood method is a method of *point estimation* of an unknown parameter θ. It may be considered to be an approximation to the Bayesian method described in this section and is used when one has no prior knowledge concerning the distribution of θ.

Given a set of observations $\boldsymbol{x}_1,\ldots,\boldsymbol{x}_n$, the probability of obtaining $\{\boldsymbol{x}_i\}$ under the assumption that the probability density is $p(\boldsymbol{x}|\theta)$ is

$$p(\boldsymbol{x}_1,\ldots,\boldsymbol{x}_n|\theta) = \prod_{i=1}^{n} p(\boldsymbol{x}_i|\theta)$$

if the $\boldsymbol{x}_i$ are independent. By Bayes' theorem, we may write the distribution of the parameter θ as

$$p(\theta|\boldsymbol{x}_1,\ldots,\boldsymbol{x}_n) = \frac{p(\boldsymbol{x}_1,\ldots,\boldsymbol{x}_n|\theta)p(\theta)}{\int p(\boldsymbol{x}_1,\ldots,\boldsymbol{x}_n|\theta)p(\theta)\,d\theta}$$

where $p(\theta)$ is the *prior* probability density of θ. The quantity above is the probability density of the parameter, given the data samples and is termed the posterior density. Given this quantity, how can we choose a single estimate for the parameter (assuming that we wish to)? There are obviously many ways in which the distribution could be used to generate a single value. For example, we could take the median or the mean as estimates for the parameter. Alternatively, we could select that value of θ for which the distribution is a maximum:

$$\hat{\theta} = \underset{\theta}{\arg\max}\ p(\theta|\boldsymbol{x}_1, \ldots, \boldsymbol{x}_n) \tag{B.3}$$

the value of θ which occurs with greatest probability. This is termed the *maximum a posteriori* (MAP) estimate or the *Bayesian estimate*.

One of the problems with this estimate is that it assumes knowledge of $p(\theta)$, the prior probability of θ. Several reasons may be invoked for neglecting this term. For example, $p(\theta)$ may be assumed to be uniform (though why should $p(\theta)$ be uniform on a scale of θ rather than, say, a scale of θ^2?). Nevertheless, if it can be neglected, then the value of θ which maximises $p(\theta|\boldsymbol{x}_1, \ldots, \boldsymbol{x}_n)$ is equal to the value of θ which maximises

$$p(\boldsymbol{x}_1, \ldots, \boldsymbol{x}_n|\theta)$$

This is the *maximum likelihood* estimate described earlier.

More generally, the *Bayes estimate* is that value of the parameter that minimises the *Bayes risk* or *average risk*, R, defined by

$$\begin{aligned} R &= \mathrm{E}[C(\theta, \hat{\theta})] \\ &= \int C(\theta, \hat{\theta}) p(\boldsymbol{x}_1, \ldots, \boldsymbol{x}_n, \theta)\, d\boldsymbol{x}_1 \ldots d\boldsymbol{x}_n\, d\theta \end{aligned} \tag{B.4}$$

where $C(\theta, \hat{\theta})$ is a *loss function* which depends on the true value θ of a parameter and its estimate, $\hat{\theta}$.

Two particular forms for $C(\theta, \hat{\theta})$ are the quadratic and the uniform loss functions. The Bayes estimate for the quadratic loss function (minimum mean square estimate)

$$C(\theta, \hat{\theta}) = \|\theta - \hat{\theta}\|^2$$

is the conditional expectation (expected error of the *a posteriori* density) (Young and Calvert, 1974)

$$\hat{\theta} = \mathrm{E}_{\theta|\boldsymbol{x}_1,\ldots,\boldsymbol{x}_n}[\theta] = \int \theta p(\theta|\boldsymbol{x}_1, \ldots, \boldsymbol{x}_n)\, d\theta \tag{B.5}$$

This is also true for cost functions that are symmetric functions of θ and convex, and if the posterior density is symmetric about its mean. Choosing a uniform loss function

$$C(\theta, \hat{\theta}) = \begin{cases} 0 & |\theta - \hat{\theta}| \le \delta \\ 1 & |\theta - \hat{\theta}| > \delta \end{cases}$$

leads to the maximum *a posteriori* estimate derived above as $\delta \to 0$.

Minimising the Bayes risk (B.4) yields a single estimate for θ and if we are only interested in obtaining a single estimate for θ, then the Bayes estimate is one we might consider.

C

Linear algebra

C.1 Basic properties and definitions

Throughout this book, we assume that the reader is familiar with standard operations on vectors and matrices. This appendix is included as a reference for terminology and notation used elsewhere in the book. All of the properties of matrices that we give are stated without proof. Proofs may be found in any good book on elementary linear algebra (see Section C.2). We also provide information on sources for software for some of the standard matrix operations.

Given an $n \times m$ matrix, $\boldsymbol{A}$, we denote the element of the ith row and jth column by a_{ij}. The *transpose* of $\boldsymbol{A}$ is denoted $\boldsymbol{A}^T$, and we note that

$$(\boldsymbol{AB})^T = \boldsymbol{B}^T \boldsymbol{A}^T$$

The square matrix $\boldsymbol{A}$ is *symmetric* if $a_{ij} = a_{ji}$ for all i, j. Symmetric matrices occur frequently in this book.

The *trace* of a square matrix $\boldsymbol{A}$, denoted $\mathrm{Tr}\{\boldsymbol{A}\}$, is the sum of its diagonal elements,

$$\mathrm{Tr}\{\boldsymbol{A}\} = \sum_{i=1}^{n} a_{ii}$$

and satisfies $\mathrm{Tr}\{\boldsymbol{AB}\} = \mathrm{Tr}\{\boldsymbol{BA}\}$ provided that $\boldsymbol{AB}$ is a square matrix, though neither $\boldsymbol{A}$ nor $\boldsymbol{B}$ need be square.

The *determinant* of a matrix $\boldsymbol{A}$, written $|\boldsymbol{A}|$, is the sum

$$|\boldsymbol{A}| = \sum_{j=1}^{n} a_{ij} A_{ij} \quad \text{for } i = 1, \ldots, n$$

where the *cofactor*, A_{ij}, is the determinant of the matrix formed by deleting the ith row and the jth column of $\boldsymbol{A}$, multiplied by $(-1)^{i+j}$. The matrix of cofactors, $\boldsymbol{C}$ ($c_{ij} = A_{ij}$), is called the *adjoint* of $\boldsymbol{A}$. If $\boldsymbol{A}$ and $\boldsymbol{B}$ are square matrices of the same order, then $|\boldsymbol{AB}| = |\boldsymbol{A}||\boldsymbol{B}|$

The *inverse* of a matrix $\boldsymbol{A}$ is that unique matrix $\boldsymbol{A}^{-1}$ with elements such that

$$\boldsymbol{A}^{-1}\boldsymbol{A} = \boldsymbol{A}\boldsymbol{A}^{-1} = \boldsymbol{I}$$

where $\boldsymbol{I}$ is the identity matrix. If the inverse exists, the matrix is said to be *nonsingular*. If the inverse does not exist, the matrix is *singular* and $|\boldsymbol{A}| = 0$. We shall frequently use the properties $(\boldsymbol{A}^T)^{-1} = (\boldsymbol{A}^{-1})^T$, $(\boldsymbol{AB})^{-1} = \boldsymbol{B}^{-1}\boldsymbol{A}^{-1}$ and if $\boldsymbol{A}$ is symmetric, then so is $\boldsymbol{A}^{-1}$.

A set of k vectors (of equal dimension) are *linearly dependent* if there exists a set of scalars $c_1, \ldots, c_k$, not all zero, such that

$$c_1\boldsymbol{x}_1 + \cdots + c_k\boldsymbol{x}_k = 0$$

If it is impossible to find such a set $c_1, \ldots, c_k$ then the vectors $\boldsymbol{x}_1, \ldots, \boldsymbol{x}_k$ are said to be linearly independent. The *rank* of a matrix is the maximum number of linearly independent rows (or equivalently, the maximum number of linearly independent columns). An $n \times n$ matrix is of *full rank* if the rank is equal to n. In this case, the determinant is non-zero, i.e. the inverse exists. For a rectangular matrix $\boldsymbol{A}$ of order $m \times n$, $\text{rank}(\boldsymbol{A}) \leq \min(m, n)$ and

$$\text{rank}(\boldsymbol{A}) = \text{rank}(\boldsymbol{A}^T) = \text{rank}(\boldsymbol{A}^T\boldsymbol{A}) = \text{rank}(\boldsymbol{A}\boldsymbol{A}^T)$$

and the rank is unchanged by pre- or post-multiplication of $\boldsymbol{A}$ by a nonsingular matrix.

A square matrix is *orthogonal* if

$$\boldsymbol{A}\boldsymbol{A}^T = \boldsymbol{A}^T\boldsymbol{A} = \boldsymbol{I},$$

that is, the rows and the columns of $\boldsymbol{A}$ are *orthonormal* ($\boldsymbol{x}^T\boldsymbol{y} = 0$ and $\boldsymbol{x}^T\boldsymbol{x} = 1$, $\boldsymbol{y}^T\boldsymbol{y} = 1$ for two different columns $\boldsymbol{x}$ and $\boldsymbol{y}$). An orthogonal matrix represents a linear transformation that preserves distances and angles, consisting of a rotation and/or reflection. It is clear from the above definition that an orthogonal matrix is nonsingular and the inverse of an orthogonal matrix is its transpose: $\boldsymbol{A}^{-1} = \boldsymbol{A}^T$. Also the determinant of an orthogonal matrix is ± 1 (-1 indicates a reflection, $+1$ is a pure rotation).

A square matrix $\boldsymbol{A}$ is *positive definite* if the quadratic form $\boldsymbol{x}^T\boldsymbol{A}\boldsymbol{x} > 0$ for all $\boldsymbol{x} \neq \boldsymbol{0}$. The matrix is *positive semidefinite* if $\boldsymbol{x}^T\boldsymbol{A}\boldsymbol{x} \geq 0$ for all $\boldsymbol{x} \neq \boldsymbol{0}$. Positive definite matrices are of full rank.

The *eigenvalues* (or *characteristic roots*) of a $p \times p$ matrix $\boldsymbol{A}$ are solutions of the *characteristic equation*

$$|\boldsymbol{A} - \lambda\boldsymbol{I}| = 0$$

which is a pth-order polynomial in λ. Thus, there are p solutions, which we denote $\lambda_1, \ldots, \lambda_p$. They are not necessarily distinct and may be real or complex. Associated with each eigenvalue λ_i is an *eigenvector* $\boldsymbol{u}_i$ with the property

$$\boldsymbol{A}\boldsymbol{u}_i = \lambda_i\boldsymbol{u}_i$$

These are not unique, since any scalar multiple of $\boldsymbol{u}_i$ also satisfies the above equation. Therefore, the eigenvectors are usually normalised so that $\boldsymbol{u}_i^T\boldsymbol{u}_i = 1$.

In this book we use the following properties of eigenvalues and eigenvectors:

1. The product of the eigenvalues is equal to the determinant, i.e. $\prod_{i=1}^{p} \lambda_i = |\boldsymbol{A}|$. Thus, it follows that if the eigenvalues are all non-zero, then the inverse of $\boldsymbol{A}$ exists.

2. The sum of the eigenvalues is equal to the trace of the matrix, $\sum_{i=1}^{p} \lambda_i = \text{Tr}\{\boldsymbol{A}\}$.

3. If $\boldsymbol{A}$ is a real symmetric matrix, the eigenvalues and eigenvectors are all real.

4. If $\boldsymbol{A}$ is positive definite, the eigenvalues are all greater than zero.

5. If $\boldsymbol{A}$ is positive semidefinite of rank m, then there will be m non-zero eigenvalues and $p - m$ eigenvalues with the value zero.

6. Every real symmetric matrix has a set of orthonormal characteristic vectors. Thus, the matrix $\boldsymbol{U}$, whose columns are the eigenvectors of a real symmetric matrix ($\boldsymbol{U} = [\boldsymbol{u}_1, \dots, \boldsymbol{u}_p]$), is orthogonal, $\boldsymbol{U}^T\boldsymbol{U} = \boldsymbol{U}\boldsymbol{U}^T = \boldsymbol{I}$ and $\boldsymbol{U}^T\boldsymbol{A}\boldsymbol{U} = \boldsymbol{\Lambda}$ where $\boldsymbol{\Lambda} = \text{diag}(\lambda_1, \dots, \lambda_p)$, the diagonal matrix with diagonal elements the eigenvalues λ_i. Alternatively, we may write

$$\boldsymbol{A} = \boldsymbol{U}\boldsymbol{\Lambda}\boldsymbol{U}^T = \sum_{i=1}^{p} \lambda_i \boldsymbol{u}_i \boldsymbol{u}_i^T$$

If $\boldsymbol{A}$ is positive definite then $\boldsymbol{A}^{-1} = \boldsymbol{U}\boldsymbol{\Lambda}^{-1}\boldsymbol{U}^T$, where $\boldsymbol{\Lambda}^{-1} = \text{diag}(1/\lambda_1, \dots, 1/\lambda_p)$.

The general symmetric eigenvector equation,

$$\boldsymbol{A}\boldsymbol{u} = \lambda \boldsymbol{B}\boldsymbol{u}$$

where $\boldsymbol{A}$ and $\boldsymbol{B}$ are real symmetric matrices, arises in linear discriminant analysis (described in Chapter 4) and other areas of pattern recognition. If $\boldsymbol{B}$ is positive definite, then the equation above has p eigenvectors, $(\boldsymbol{u}_1, \dots, \boldsymbol{u}_p)$, that are orthonormal with respect to $\boldsymbol{B}$, that is

$$\boldsymbol{u}_i^T \boldsymbol{B} \boldsymbol{u}_j = \begin{cases} 0 & i \neq j \\ 1 & i = j \end{cases}$$

and consequently

$$\boldsymbol{u}_i^T \boldsymbol{A} \boldsymbol{u}_j = \begin{cases} 0 & i \neq j \\ \lambda_j & i = j \end{cases}$$

These may be written

$$\boldsymbol{U}^T\boldsymbol{B}\boldsymbol{U} = \boldsymbol{I} \quad \boldsymbol{U}^T\boldsymbol{A}\boldsymbol{U} = \boldsymbol{\Lambda}$$

where $\boldsymbol{U} = [\boldsymbol{u}_1, \dots, \boldsymbol{u}_p]$ and $\boldsymbol{\Lambda} = \text{diag}(\lambda_1, \dots, \lambda_p)$.

Many problems in this book involve the minimisation of a squared error measure. The general linear least squares problem may be solved using the *singular value decomposition* of a matrix. An $m \times n$ matrix $\boldsymbol{A}$ may be written in the form

$$\boldsymbol{A} = \boldsymbol{U}\boldsymbol{\Sigma}\boldsymbol{V}^T = \sum_{i=1}^{r} \sigma_i \boldsymbol{u}_i \boldsymbol{v}_i^T$$

where r is the rank of $\boldsymbol{A}$; $\boldsymbol{U}$ is an $m \times r$ matrix with columns $\boldsymbol{u}_1, \dots, \boldsymbol{u}_r$, the *left singular vectors* and $\boldsymbol{U}^T\boldsymbol{U} = \boldsymbol{I}_r$, the $r \times r$ identity matrix; $\boldsymbol{V}$ is an $n \times r$ matrix with columns $\boldsymbol{v}_1, \dots, \boldsymbol{v}_r$, the *right singular vectors* and $\boldsymbol{V}^T\boldsymbol{V} = \boldsymbol{I}_r$ also; $\boldsymbol{\Sigma} = \text{diag}(\sigma_1, \dots, \sigma_r)$, the diagonal matrix of *singular values* $\sigma_i, i = 1, \dots, r$.

The singular values of $\boldsymbol{A}$ are the square roots of the non-zero eigenvalues of $\boldsymbol{A}\boldsymbol{A}^T$ or $\boldsymbol{A}^T\boldsymbol{A}$. The *pseudo-inverse* or *generalised inverse* is the $n \times m$ matrix $\boldsymbol{A}^\dagger$

$$\boldsymbol{A}^\dagger = \boldsymbol{V}\boldsymbol{\Sigma}^{-1}\boldsymbol{U}^T = \sum_{i=1}^{r} \frac{1}{\sigma_i}\boldsymbol{v}_i\boldsymbol{u}_i^T \tag{C.1}$$

and the solution for $\boldsymbol{x}$ that minimises the squared error

$$\|\boldsymbol{A}\boldsymbol{x} - \boldsymbol{b}\|^2$$

is given by

$$\boldsymbol{x} = \boldsymbol{A}^\dagger\boldsymbol{b}$$

If the rank of $\boldsymbol{A}$ is less than n, then there is not a unique solution for $\boldsymbol{x}$ and singular value decomposition delivers the solution with minimum norm.

The pseudo-inverse has the following properties:

$$\boldsymbol{A}\boldsymbol{A}^\dagger\boldsymbol{A} = \boldsymbol{A} \qquad (\boldsymbol{A}\boldsymbol{A}^\dagger)^T = \boldsymbol{A}\boldsymbol{A}^\dagger$$
$$\boldsymbol{A}^\dagger\boldsymbol{A}\boldsymbol{A}^\dagger = \boldsymbol{A}^\dagger \qquad (\boldsymbol{A}^\dagger\boldsymbol{A})^T = \boldsymbol{A}^\dagger\boldsymbol{A}$$

Finally, in this section, we introduce some results about derivatives. We shall denote the partial derivative operator by

$$\frac{\partial}{\partial \boldsymbol{x}} = \left(\frac{\partial}{\partial x_1}, \dots, \frac{\partial}{\partial x_p}\right)^T$$

Thus, the derivative of the scalar function f of the vector $\boldsymbol{x}$ is the vector

$$\frac{\partial f}{\partial \boldsymbol{x}} = \left(\frac{\partial f}{\partial x_1}, \dots, \frac{\partial f}{\partial x_p}\right)^T$$

Similarly, the derivative of a scalar function of a matrix is denoted by the matrix $\partial f/\partial \boldsymbol{A}$, where

$$\left[\frac{\partial f}{\partial \boldsymbol{A}}\right]_{ij} = \frac{\partial f}{\partial a_{ij}}$$

In particular, we have

$$\frac{\partial |\boldsymbol{A}|}{\partial \boldsymbol{A}} = (\text{adjoint of } \boldsymbol{A})^T = |\boldsymbol{A}|(\boldsymbol{A}^{-1})^T \text{ if } \boldsymbol{A}^{-1} \text{ exists}$$

and for a symmetric matrix

$$\frac{\partial}{\partial \boldsymbol{x}}\boldsymbol{x}^T\boldsymbol{A}\boldsymbol{x} = 2\boldsymbol{A}\boldsymbol{x}$$

Also, an important derivative involving traces of matrices is

$$\frac{\partial}{\partial \boldsymbol{A}}(\mathrm{Tr}\{\boldsymbol{A}^T\boldsymbol{M}\boldsymbol{A}\}) = \boldsymbol{M}\boldsymbol{A} + \boldsymbol{M}^T\boldsymbol{A}$$

C.2 Notes and references

In this appendix we have introduced the necessary matrix terminology that will be useful throughout most of this book. We have necessarily been brief on detail and proofs of some of the assertions, together with additional explanation, may be found in most books on linear algebra; for example, the book by Stewart (1973) provides a very good introduction to matrix computations. Also, Press *et al.* (1992) give clear descriptions of the numerical procedures.

The book by Thisted (1988) gives a good introduction to elements of statistical computing, including numerical linear algebra and nonlinear optimisation schemes for function minimisation that are required by some pattern processing algorithms.

D

Data

D.1 Introduction

Many of the methods described in this book for discrimination and classification often have been, and no doubt will continue to be, employed in isolation from other essential stages of a statistical investigation. This is partly due to that fact that data are often collected without the involvement of a statistician. Yet it is the early stages of a statistical investigation that are the most important and may prove critical in producing a satisfactory conclusion. These preliminary stages, if carried out carefully, may indeed be sufficient, in many cases, to answer the questions being addressed. The stages of an investigation may be characterised as follows.

1. Formulation of the problem.

2. Data collection. This relates to questions concerning the type of data, the amount, and the method and cost of collection.

3. Initial examination of the data. Assess the quality of the data and get a feel for its structure.

4. Data analysis. Apply discrimination and classification methods as appropriate.

5. Assessment of the results.

6. Interpretation.

This is necessarily an iterative process. In particular, the interpretation may pose questions for a further study or lead to the conclusion that the original formulation of the problem needs re-examination.

D.2 Formulating the problem

The success of any pattern recognition investigation depends to a large extent on how well the investigator understands the problem. A clear understanding of the aims of the study is essential. It will enable a set of objectives to be specified in a way that will allow the problem to be formulated in precise statistical terms.

Although certainly the most important part of a study, problem formulation is probably the most difficult. Care must be taken not to be over-ambitious, and the investigator must try to look ahead, beyond the current study, possibly to further investigations, and understand how the results of the present study will be used and what will be the possible consequences of various outcomes.

Thought should be given to the data collection process. In a discrimination problem, priors and costs will need to be specified or estimated. Often the most important classes are underrepresented.

A great deal can be learned from past approaches to the particular problem under investigation, and related ones. Data from previous studies may be available. Some preliminary experiments with these data may give an indication as to the choice of important variables for measurement. Previous studies may also highlight good and bad strategies to follow.

Another factor to take into account at the planning stage is cost. More data means a greater cost, both in terms of collection and analysis. Taking too many observations, with measurements on every possible variable, is not a good strategy. Cost is also strongly related to performance and a strategy that meets the desired performance level with acceptable cost is required.

Finally, thought must be given to the interpretation and presentation of the results to the user. For example, in a discrimination problem, it may be inappropriate to use the classifier giving the best results if it is unable to 'explain' its decisions. A suboptimal classifier that is easier to understand may be preferable.

D.3 Data collection

In a pattern recognition problem in which a classifier is being designed to automate some particular task, it is important that the data collected are representative of the expected operating conditions. If this is not so, then you must say how and why the data differ.

There are several aspects of data collection that must be addressed. These include the collection of calibration or 'ground truth' data; the variables measured and their accuracy; the sampling strategy, including the total sample size and the proportions within each class (if sampling from each class separately); the costs involved; and the principle of randomisation.

When collecting data it is important to record details of the procedure and the equipment used. This may include specifying a type of sensor, calibration values, and a description of the digitising and recording equipment. Conditions prevailing at the time of the experiment may be important for classifier design, particularly when generalisation to other conditions is required.

The choice of which variables to measure is crucial to successful pattern recognition and is based on knowledge of the problem at hand and previous experience. Use knowledge of previous experiments whenever this is available. It has been found that increasing the number of variables measured does not necessarily increase classification performance on a data set of finite size. This trade-off between sample size and dimensionality, discussed more fully below, is something to consider when taking measurements. Although variables may be pruned using a variable selection technique, such as those described in Chapter 9, if measurements are expensive, then it would be better not to make unnecessary ones in the first place. However, there is sometimes a need for redundant variables as an aid to error checking.

Once the measurement variables have been prescribed, the designer must decide on the sampling strategy and select an appropriate sample size. In a discrimination problem, in which a measurement vector $\boldsymbol{x}$ has an associated class label, perhaps coded as a vector z, there are two main sampling strategies under which the training data $\{(\boldsymbol{x}_i, z_i), i = 1, \ldots, p\}$ may be realized. The first of these is *separate* or class-conditional sampling. Here, the feature vectors are sampled from each class separately and therefore this strategy does not give us estimates of the prior probabilities of the classes. Of course, if these are known, then we may sample from the classes in these proportions. The second sampling design is *joint* or *mixture* sampling. In this scheme, the feature vector and the class are recorded on the samples drawn from a mixture of groups. The proportions in each group emerge from the data.

The sample size depends on a number of factors: the number of features, the desired performance, the complexity of the classification rule (in terms of the number of parameters to estimate) and the asymptotic probability of misclassification. It is very difficult to obtain theoretical results about the effects of finite sample size on classifier performance. However, if the number of features is large and the classifier is complex then a large number of measurements should be made. Yet, it is difficult to know, before the data collection, how complex the resulting classifier needs to be. In addition, a large number of data samples is necessary if the separation between classes is small or high confidence in the error rate estimate is desired. Further, if little knowledge about the problem is available, necessitating the use of nonparametric methods, then generally larger data sets than those used for parametric approaches are required. This must be offset against factors limiting the sample size such as measurement cost. Again, much can be learned from previous work (for example, Fukunaga and Hayes, 1989c; Jain and Chandrasekaran, 1982). Several papers suggest that the ratio of sample size to number of features is a very important factor in the design of a pattern recognition system (Jain and Chandrasekaran, 1982; Kalayeh and Landgrebe, 1983) giving the general guidance of having 5–10 times more samples per class than feature measurements.

Once gathered, the data are often partitioned into *training* (or *design*) and *test sets*. The training set should be a random sample from the total data set. There are two main reasons for partitioning the data. In one instance, the classifier is trained using the training set and the test set is used to provide an independent estimate of its performance. This makes inefficient use of the data for training a classifier (see Chapter 11). The second way in which training and test sets are used is in classifier design. It applies to classifiers of perhaps differing complexity or classifiers of the same complexity (in terms of the architecture and the number of parameters) but with different initial conditions in a nonlinear optimisation procedure. These are trained using the training set and the performance on the test set monitored. The classifier giving the best performance on the test set is adopted. Using the test set in this manner means that it is being used as part of the training process and cannot be used to provide an independent error estimate. Other methods of error rate estimation must be employed.

The test set used in the above manner is more properly termed a *validation set*, which is used to tune classifier parameters. A third independent data set may required for performance evaluation purposes.

Whatever way we design our classifier, in many practical applications we shall desire good generalisation; that is, good performance on unseen data representative of the true operational conditions in which the classifier is to be deployed. Of course, the aim should be to collect data representative of those conditions (this is the test set). Nevertheless, if your budget allows, it is worthwhile collecting a set of data, perhaps at a different location, perhaps at a different time of year, or by a different group of researchers (if your classifier is meant to be invariant to these factors) and validating your conclusions on this data set.

A final point to mention with respect to data collection is *randomisation*. This reduces the effects of bias. The order of the data should be randomised, *not* collecting all the examples from class 1, then all the examples from class 2 and so on. This is particularly important if the measurement equipment or the background or environmental conditions change with time. Complete randomisation is not possible and some form of restricted randomisation could be employed (Chatfield, 1988).

The following points summarise the data-collection strategy.

1. Choose the variables. If prior knowledge is available then this must be used.

2. Decide on the sample size and proportions within each class (for separate sampling).

3. Record the details of the procedure and measuring equipment.

4. Measure an independent test set.

5. Randomise the data collection.

D.4 Initial examination of data

Once the data have been collected it is tempting to rush into an analysis perhaps using complicated multivariate analysis techniques. With the wide availability of computer packages and software, it is relatively easy to perform such analyses without first having a careful or even cursory look at the data. The initial examination of the data is one of the most important parts of the data analysis cycle (and we emphasise again the iterative aspect to an investigation). Termed an initial data analysis (IDA) by Chatfield (1985, 1988) or exploratory data analysis by Tukey (1977), it constitutes the first phase of the analysis and comprises three parts:

1. checking the quality of the data;

2. calculating summary statistics;

3. producing plots of the data in order to get a feel for their structure.

Checking the data

There are several factors that degrade data quality, the main ones being due to errors, outliers and missing observations. Errors may occur in several ways. They may be due to malfunctions in recording equipment, for example transcription errors, or they may even be deliberate if a respondent in a survey gives false replies. Some errors may be difficult to detect, particularly if the value in error is consistent with other observations. Alternatively, if the error gives rise to an outlier (an observation that appears to be inconsistent with the remainder of the data) then a simple range test on each variable may pick it up.

Missing values can arise in a number of different ways and it is important to know how and why they occur. Extreme care must be taken in the coding of missing values, not treating them as special numerical values if possible. Procedures for dealing with missing observations are also discussed more fully in Chapter 11.

Summary statistics

Summary statistics should be calculated for the whole data set and for each class individually. The most widely used measures of location and spread are the sample mean and the standard deviation. The sample mean should be calculated on each variable and can be displayed, along with the standard deviation and range, in a table comparing these values with the class-conditional ones. This might provide important clues to the variables important for discriminating particular classes.

Plotting the data

Graphical views of the data are extremely useful in providing an insight into the nature of multivariate data. They can prove to be a help as a means of detecting clusters in the data, indicating important (or unimportant) variables, suggesting appropriate transformations of the variables and for detecting outliers.

Histograms of the individual variables and scatterplots of pairwise combinations of variables are easy to produce. Scatterplots produce projections of the data onto various planes and may show up outliers. Different classes can be plotted with different symbols. Scatterplots do not always reflect the true multidimensional structure and if the number of variables, p, is large then it might be difficult to draw conclusions from the correspondingly large number, $p(p-1)$, of scatterplots.

There are many other plots, some of which project the data linearly or nonlinearly onto two dimensions. These techniques, such as multidimensional scaling, Sammon plots, and principal components analysis, are very useful for exploratory data analysis. They involve more than simply 'looking' at the data and are explored more fully in Chapter 9.

Summary

The techniques briefly described in this section form an essential part of data analysis in the early stages before submitting the data to a computer package. Initial data analysis provides aids for data description, data understanding and for model formation, perhaps giving vital clues as to which subsequent methods of analysis should be undertaken. In many cases it will save a lot of wasted effort.

D.5 Data sets

The book by Andrews and Herzberg (1985) provides a range of data sets from the widely used Fisher–Anderson iris data to the more unusual data set of number of deaths by falling off horseback in the Prussian army. Some of the data sets in the book can be used for discrimination problems. The *Handbook of Data Sets* of Hand *et al.* (1993) consists of around 500 small data sets that are very useful for illustration purposes.

There are several electronic sources. The UCI repository of machine learning databases and domain theories (Murphy and Aha, 1995) contains over 70 data sets, most of them documented and for which published results of various analyses are available.

In addition to assessing methods on real data, which is an essential part of technique development, simulated data can prove extremely valuable. We can obtain as much (subject to the properties of the random number generation process) or as little as we wish for nominal cost. This enables asymptotic error rates to be estimated. These can be used to assess performance of error rate estimation procedures. Also, if we know the model that has generated the data, then we have a means of assessing our model order selection procedures. For example, if we design a classifier that we believe is optimal (in the sense of producing the Bayes minimum error rate) for the particular case of normally distributed classes, then we ought to be able to test our numerical procedure using such data.

D.6 Notes and references

The book by Chatfield (1988) provides an excellent account of the general principles involved in statistical investigation. Emphasis is placed on initial data analysis (see also Chatfield, 1985; Tukey, 1977), but all stages from problem formulation to interpretation of results are addressed. On experimental design, the classic texts by Cochran and Cox (1957) and Cox (1958) are still very relevant today. Further guidelines are given by Hahn (1984), based on experiences of six case studies. The book by Everitt and Hay (1992) provides a very good introduction to the design and analysis of experiments through a case study in psychology. It also presents many recent developments in statistics and is much more widely applicable than the particular application domain described. A further discussion in mapping a scientific question to a statistical one is given by Hand (1994).

Techniques for visualising multivariate data are described in the books by Chatfield and Collins (1980), Dillon and Goldstein (1984) and Everitt and Dunn (1991). Wilkinson (1992) gives a review of graphical displays.

E

Probability theory

E.1 Definitions and terminology

The rudiments of probability theory used in the development of decision theory are now presented. Firstly, we introduce the idea of an experiment or observation. The set of all possible outcomes is called the *sample space*, Ω, of the model, with each possible outcome being a *sample point*. An event, A, is a set of experimental outcomes, and corresponds to a subset of points in Ω.

A *probability measure* is a function, $P(A)$, with a set, A, as argument. It can be regarded as the expected proportion of times that A actually occurs and has the following properties:

1. $0 \leq P(A) \leq 1$.
2. $P(\Omega) = 1$.
3. If A and B are mutually exclusive events (disjoint sets) then

$$P(A \cup B) = P(A) + P(B)$$

More generally, when A and B are not necessarily exclusive,

$$P(A \cup B) = P(A) + P(B) - P(A \cap B)$$

A random variable is a function that associates a number with each possible outcome $\omega \in \Omega$. We denote random variables by upper-case letters here, but in the main body of work we generally use the same symbol for a random variable and a measurement, the meaning being clear from context. Although the argument of P is a set, it is usual to use a description of the event as the argument, regarding this as equivalent to the corresponding set. Thus, we write $P(X > Y)$ for the probability that X is greater than Y, rather than $P(\{\omega : X(\omega) > Y(\omega)\})$.

The *cumulative distribution function*, sometimes simply called the distribution function, of a random variable is the probability that the random variable is less than or equal to some specified value x; that is,

$$P_X(x) = \text{probability that } X \leq x \tag{E.1}$$

Usually, when there is no ambiguity, we drop the subscript X. The cumulative distribution function is a monotonic function of its argument with the property that $P(-\infty) = 0$; $P(\infty) = 1$. The derivative of the distribution function,

$$p(x) = \frac{dP}{dx}$$

is the *probability density function* of the random variable X. For sample spaces such as the entire real line, the probability density function, $p(x)$, has the following properties:

$$\begin{aligned}\int_{-\infty}^{\infty} p(x)\,dx &= 1\\ \int_a^b p(x)\,dx &= \text{probability that } X \text{ lies between } a \text{ and } b\\ &= P(a \le X \le b)\\ p(x) &\ge 0\end{aligned}$$

Much of the discussion in this book will relate to vector quantities, since the inputs to many pattern classification systems may be expressed in vector form. Random vectors are defined similarly to random variables, associating each point in the sample space, Ω, to a point in $\mathbb{R}^p$:

$$\boldsymbol{X} : \Omega \longrightarrow \mathbb{R}^p$$

The *joint distribution* of $\boldsymbol{X}$ is the p-dimensional generalisation of (E.1) above:

$$P_X(\boldsymbol{x}) = P_X(x_1, \ldots, x_p) = \text{probability that } X_1 \le x_1, \ldots, X_p \le x_p$$

and the *joint density function* is similarly given by

$$p(\boldsymbol{x}) = \frac{\partial^p P(\boldsymbol{x})}{\partial x_1 \ldots \partial x_p}$$

Given the joint density of a set of random variables $X_1, \ldots, X_p$, then a smaller set $X_1, \ldots, X_m (m < p)$ also possesses a probability density function determined by

$$p(x_1, \ldots, x_m) = \int_{-\infty}^{\infty} \cdots \int_{-\infty}^{\infty} p(x_1, \ldots, x_p)\,dx_{m+1} \ldots dx_p$$

This is sometimes known as the *marginal density* of $X_1, \ldots X_m$, although the expression is more usually applied to the single-variable densities, $p(x_1), p(x_2), \ldots, p(x_p)$ given by

$$p(x_i) = \int_{-\infty}^{\infty} \cdots \int_{-\infty}^{\infty} p(x_1, \ldots, x_p)\,dx_1 \ldots dx_{i-1}\,dx_{i+1} \ldots dx_p$$

The *expected vector*, or mean vector, of a random variable $\boldsymbol{x}$, is defined by

$$\boldsymbol{m} = \mathrm{E}[\boldsymbol{X}] = \int \boldsymbol{x}\,p(\boldsymbol{x})\,d\boldsymbol{x}$$

where $\boldsymbol{dx}$ denotes $dx_1 \ldots dx_p$ and the integral is over the entire space (and unless otherwise stated $\int = \int_{-\infty}^{\infty}$) and E[.] denotes the expectation operator. The ith component of the mean can be calculated by

$$m_i = \mathrm{E}[X_i] = \int \ldots \int x_i p(x_1, \ldots, x_p)\, dx_1 \ldots dx_p = \int x_i p(\boldsymbol{x})\, \boldsymbol{dx}$$
$$= \int_{-\infty}^{\infty} x_i p(x_i)\, dx_i$$

where $p(x_i)$ is the marginal density of the single variable X_i given above.

The *covariance* of two random variables provides a measure of the extent to which the deviations of the random variables from their respective mean values tend to vary together. The covariance of random variables x_i and x_j, denoted by C_{ij}, is given by

$$C_{ij} = \mathrm{E}[(X_j - \mathrm{E}[X_j])(X_i - \mathrm{E}[X_i])]$$

which may be expressed as

$$C_{ij} = \mathrm{E}[X_i X_j] - \mathrm{E}[X_i]\mathrm{E}[X_j]$$

where $\mathrm{E}[X_i X_j]$ is the *autocorrelation*. The matrix with (i, j)th component C_{ij} is the covariance matrix

$$\boldsymbol{C} = \mathrm{E}[(\boldsymbol{X} - \boldsymbol{m})(\boldsymbol{X} - \boldsymbol{m})^T]$$

Two random variables X_i and X_j are *uncorrelated* if the covariance between the two variables is zero, $C_{ij} = 0$, which implies

$$\mathrm{E}[X_i X_j] = \mathrm{E}[X_i]\mathrm{E}[X_j]$$

Two vectors, $\boldsymbol{X}$ and $\boldsymbol{Y}$, are uncorrelated if

$$\mathrm{E}[\boldsymbol{X}^T \boldsymbol{Y}] = \mathrm{E}[\boldsymbol{X}]^T \mathrm{E}[\boldsymbol{Y}] \qquad \text{(E.2)}$$

In the special case where the means of the vectors are zero, so that the relation above becomes $\mathrm{E}[\boldsymbol{X}^T \boldsymbol{Y}] = 0$, then the random variables are said to be *mutually orthogonal*.

Two events, A and B, are *statistically independent* if

$$P(A \cap B) = P(A)P(B)$$

and two random variables X_i and X_j are independent if

$$p(x_i, x_j) = p(x_i)p(x_j)$$

If the random variables $X_1, X_2, \ldots, X_p$ are independent then the joint density function may be written as a product of the individual densities:

$$p(x_1, \ldots, x_p) = p(x_1) \ldots p(x_p)$$

If two variables are independent then the expectation of $X_1 X_2$ is given by

$$\mathrm{E}[X_1 X_2] = \int \int x_1 x_2 p(x_1, x_2)\, dx_1\, dx_2$$

and, using the independence property,

$$\mathrm{E}[X_1X_2] = \int x_1 p(x_1)\,dx_1 \int x_2 p(x_2)\,dx_2$$
$$= \mathrm{E}[X_1]\mathrm{E}[X_2]$$

This shows that X_1 and X_2 are uncorrelated. However, this does not imply that two variables that are uncorrelated are statistically independent.

Often we shall want to consider a functional transformation from a given set of random variables $\{X_1, X_2, \ldots, X_p\}$ represented by the vector $\boldsymbol{X}$ to a set $\{Y_1, Y_2, \ldots, Y_p\}$ represented by the vector $\boldsymbol{Y}$. How do probability density functions change under such a transformation? Let the transformation be given by $\boldsymbol{Y} = \boldsymbol{g}(\boldsymbol{X})$, where $\boldsymbol{g} = (g_1, g_2, \ldots, g_p)^T$. Then the density functions of $\boldsymbol{X}$ and $\boldsymbol{Y}$ are related by

$$p_Y(\boldsymbol{y}) = \frac{p_X(\boldsymbol{x})}{|\boldsymbol{J}|}$$

where $|\boldsymbol{J}|$ is the absolute value of the Jacobian determinant

$$\boldsymbol{J}(x_1, \ldots, x_p) = \begin{vmatrix} \dfrac{\partial g_1}{\partial x_1} & \cdots & \dfrac{\partial g_1}{\partial x_p} \\ \vdots & \ddots & \vdots \\ \dfrac{\partial g_p}{\partial x_1} & \cdots & \dfrac{\partial g_p}{\partial x_p} \end{vmatrix}$$

A simple transformation is the linear one

$$\boldsymbol{Y} = \boldsymbol{A}\boldsymbol{X} + \boldsymbol{B}$$

Then if $\boldsymbol{X}$ has the probability density $p_X(\boldsymbol{x})$, the probability density of $\boldsymbol{Y}$ is

$$p_Y(\boldsymbol{y}) = \frac{p_X(\boldsymbol{A}^{-1}(\boldsymbol{y} - \boldsymbol{B}))}{|\boldsymbol{A}|} \tag{E.3}$$

where $|\boldsymbol{A}|$ is the absolute value of the determinant of the matrix $\boldsymbol{A}$.

Given a random system and any two events A and B that can occur together, we can form a new system by taking only those trials in which B occurs. The probability of A in this new system is called the *conditional probability* of A given B and is denoted by $P(A|B)$; if $P(B) > 0$ it is given by

$$P(A|B) = \frac{P(A \cap B)}{P(B)}$$

or

$$P(A \cap B) = P(A|B)P(B) \tag{E.4}$$

This is the *total probability theorem*.

Now, since $P(A \cap B) = P(B \cap A)$, we have from (E.4)

$$P(A|B)P(B) = P(B|A)P(A)$$

or

$$P(A|B) = \frac{P(B|A)P(A)}{P(B)}$$

This is *Bayes' theorem.*

Now, if $A_1, A_2, \ldots, A_N$ are events which partition the sample space (that is, they are mutually exclusive and their union is Ω) then

$$\begin{aligned} P(B) &= \sum_{i=1}^{N} P(B \cap A_i) \\ &= \sum_{i=1}^{N} P(B|A_i)P(A_i) \end{aligned} \tag{E.5}$$

and we obtain a more practical form of Bayes' theorem

$$P(A_j|B) = \frac{P(B|A_j)P(A_j)}{\sum_{i=1}^{N} P(B|A_i)P(A_i)} \tag{E.6}$$

In pattern classification problems, B is often an observation event and the A_j are pattern classes. The term *a priori* probability is often used for the quantity $P(A_i)$ and the objective is to find $P(A_i|B)$, which is termed the *a posteriori* probability of A_i.

The *conditional distribution*, $P_X(x|A)$, of a random variable X given the event A is defined as the conditional probability of the event $\{X \leq x\}$

$$P(x|A) = \frac{P(\{X \leq x\}, A)}{P(A)}$$

and $P(\infty|A) = 1$; $P(-\infty|A) = 0$. The *conditional density* $p(x|A)$ is the derivative of $P(x|A)$

$$p(x|A) = \frac{dP}{dx} = \lim_{\Delta x \to 0} \frac{P(x \leq X \leq x + \Delta x|A)}{\Delta x}$$

The extension of the result (E.5) to the continuous case gives

$$p(x) = \sum_{i=1}^{N} p(x|A_i)p(A_i)$$

where $p(x)$ is the *mixture* density and we have taken $B = \{X \leq x\}$, and the continuous version of Bayes' theorem may be written

$$p(x|A) = \frac{p(A|x)p(x)}{p(A)} = \frac{p(A|X = x)p(x)}{\int_{-\infty}^{\infty} p(A|X = x)p(x)\,dx}$$

The conditional density of x given that the random vector Y has some specified value, y, is obtained by letting $A = \{y \leq Y \leq y + \Delta y\}$ and taking the limit

$$\lim_{\Delta y \to 0} p(x|\{y \leq Y \leq y + \Delta y\}) = \lim_{\Delta y \to 0} \frac{p(x, \{y \leq Y \leq y + \Delta y\})}{p(\{y \leq Y \leq y + \Delta y\})}$$

giving

$$p(x|y) = \frac{p(x, y)}{p(y)} \tag{E.7}$$

where $p(x, y)$ is the joint density of X and Y and $p(y)$ is the marginal density

$$p(y) = \int p(x, y)\, dx \tag{E.8}$$

Equations (E.7) and (E.8) lead to the density form of Bayes' theorem

$$\begin{aligned} p(x|y) &= \frac{p(y|x)p(x)}{p(y)} \\ &= \frac{p(y|x)p(x)}{\int p(y|x)p(x)\, dx} \end{aligned}$$

A generalisation of (E.7) is the conditional density of the random variables $X_{k+1}, \ldots, X_p$ given $X_1, \ldots, X_k$:

$$p(x_{k+1}, \ldots, x_p | x_1, \ldots, x_k) = \frac{p(x_1, \ldots, x_p)}{p(x_1, \ldots, x_k)} \tag{E.9}$$

This leads to the *chain rule*

$$p(x_1, \ldots, x_p) = p(x_p|x_1, \ldots, x_{p-1})p(x_{p-1}|x_1, \ldots, x_{p-2}) \ldots p(x_2|x_1)p(x_1)$$

The results of (E.8) and (E.9) allow unwanted variables in a conditional density to be removed. If they occur to the left of the vertical line, then integrate with respect to them. If they occur to the right, then multiply by the conditional density of the variables given the remaining variables on the right and integrate. For example,

$$p(a|l, m, n) = \int p(a, b, c|l, m, n)\, db\, dc$$

$$p(a, b, c|m) = \int p(a, b, c|l, m, n)p(l, n|m)\, dl\, dn$$

E.2 Normal distribution

We shall now illustrate some of the definitions and results of this section using the *Gaussian* or *normal* distribution (we use the two terms interchangeably in this book). It is a distribution to which we often refer in our discussion of pattern recognition algorithms.

The *standard normal density* of a random variable X has zero mean and unit variance, and has the form

$$p(x) = \frac{1}{\sqrt{2\pi}} \exp\left\{-\frac{x^2}{2}\right\} \qquad -\infty < x < \infty$$

The distribution function is given by

$$P(X) = \int_{-\infty}^{X} p(x)\,dx = \frac{1}{\sqrt{2\pi}} \int_{-\infty}^{X} \exp\left\{-\frac{1}{2}x^2\right\} dx = \frac{1}{2} + \frac{1}{2}\mathrm{erf}\left\{\frac{X}{\sqrt{2}}\right\}$$

where erf{x} is the *error function*, $\frac{2}{\sqrt{\pi}} \int_0^x \exp(-x^2)\,dx$.

For the function $Y = \mu + \sigma X$ of the random variable X, the density function of Y is

$$p(y) = \frac{1}{\sqrt{2\sigma^2\pi}} \exp\left[-\frac{1}{2}\left(\frac{y-\mu}{\sigma}\right)^2\right],$$

which has mean μ and variance σ^2, and we write $Y \sim N(\mu, \sigma^2)$.

If $X_1, X_2, \ldots, X_p$ are independently and identically distributed, each following the standard normal distribution, then the joint density is given by

$$p(x_1, x_2, \ldots, x_p) = \prod_{i=1}^{p} p(x_i) = \frac{1}{(2\pi)^{p/2}} \exp\left\{-\frac{1}{2}\sum_{i=1}^{p} x_i^2\right\}$$

The transformation $\boldsymbol{Y} = \boldsymbol{AX} + \boldsymbol{\mu}$ leads to the density function for $\boldsymbol{Y}$ (using (E.3))

$$p(\boldsymbol{y}) = \frac{1}{(2\pi)^{p/2}|\boldsymbol{A}|} \exp\left\{-\frac{1}{2}(\boldsymbol{y}-\boldsymbol{\mu})^T(\boldsymbol{A}^{-1})^T\boldsymbol{A}^{-1}(\boldsymbol{y}-\boldsymbol{\mu})\right\} \tag{E.10}$$

and since the covariance matrix, $\boldsymbol{\Sigma}$, of $\boldsymbol{Y}$ is

$$\boldsymbol{\Sigma} = \mathrm{E}[(\boldsymbol{Y}-\boldsymbol{\mu})(\boldsymbol{Y}-\boldsymbol{\mu})^T] = \boldsymbol{AA}^T$$

Equation (E.10) is usually written

$$p(\boldsymbol{y}) = N(\boldsymbol{y}|\boldsymbol{\mu}, \boldsymbol{\Sigma}) = \frac{1}{(2\pi)^{p/2}|\boldsymbol{\Sigma}|^{\frac{1}{2}}} \exp\left\{-\frac{1}{2}(\boldsymbol{y}-\boldsymbol{\mu})^T\boldsymbol{\Sigma}^{-1}(\boldsymbol{y}-\boldsymbol{\mu})\right\}$$

This is the *multivariate normal distribution*.

Recall from the previous section that if two variables are independent then they are uncorrelated, but that the converse is not necessarily true. However, a special property of the normal distribution is that if two variables are joint normally distributed and uncorrelated, then they are independent.

The marginal densities and conditional densities of a joint normal distribution are all normal.

E.3 Probability distributions

We introduce some of the more commonly used distributions with pointers to some places where they are used in the book (further probability distributions are listed by Bernardo and Smith, 1994). If x has a specific probability density function, $\mathrm{R}(x|\alpha)$, where α is the set of parameters of the specific functional form R, then for shorthand notation we may write $x \sim \mathrm{R}(\alpha)$; similarly, we use $x|\beta \sim \mathrm{R}(f(\beta))$, for some function f, to mean $p(x|\beta) = \mathrm{R}(x|f(\beta))$.

$N(\boldsymbol{x}|\boldsymbol{\mu}, \boldsymbol{\Sigma})$, pp. 30, 34, 52, 68 **Normal 1**

$$p(\boldsymbol{x}) = \frac{1}{|\boldsymbol{\Sigma}|^{1/2}(2\pi)^{p/2}} \exp\left\{-\frac{1}{2}(\boldsymbol{x}-\boldsymbol{\mu})^T \boldsymbol{\Sigma}^{-1}(\boldsymbol{x}-\boldsymbol{\mu})\right\}$$

$\boldsymbol{\Sigma}$, symmetric positive definite; $\mathrm{E}[\boldsymbol{x}] = \boldsymbol{\mu}$; $\mathrm{V}[\boldsymbol{x}] = \boldsymbol{\Sigma}$.

Sometimes it is convenient to express the normal with the inverse of the covariance matrix as a parameter.

$\mathrm{N}_p(\boldsymbol{x}|\boldsymbol{\mu}, \boldsymbol{\lambda})$, pp. 53, 53, 61, 67 **Normal 2**

$$p(\boldsymbol{x}) = \frac{|\boldsymbol{\lambda}|^{1/2}}{(2\pi)^{p/2}} \exp\left\{-\frac{1}{2}(\boldsymbol{x}-\boldsymbol{\mu})^T \boldsymbol{\lambda}(\boldsymbol{x}-\boldsymbol{\mu})\right\}$$

$\boldsymbol{\lambda}$, symmetric positive definite; $\mathrm{E}[\boldsymbol{x}] = \boldsymbol{\mu}$; $\mathrm{V}[\boldsymbol{x}] = \boldsymbol{\lambda}^{-1}$.

$\mathrm{Wi}_p(\boldsymbol{x}|\alpha, \boldsymbol{\beta})$, pp. 53, 53 **Wishart**

$$p(\boldsymbol{x}) = \left[\pi^{p(p-1)/4} \prod_{i=1}^{p} \left(\frac{1}{2}(2\alpha + 1 - i)\right)\right]^{-1} |\boldsymbol{\beta}|^{\alpha} |\boldsymbol{x}|^{\alpha-(p+1)/2} \exp(-\mathrm{Tr}(\boldsymbol{\beta}\boldsymbol{x}))$$

$\boldsymbol{x}$, symmetric positive definite; $\boldsymbol{\beta}$, symmetric nonsingular; $2\alpha > p - 1$; $\mathrm{E}[\boldsymbol{x}] = \alpha\boldsymbol{\beta}^{-1}$; $\mathrm{E}[\boldsymbol{x}^{-1}] = (\alpha - (p+1)/2)^{-1}\boldsymbol{\beta}$.

$\mathrm{St}_p(\boldsymbol{x}|\boldsymbol{\mu}, \boldsymbol{\lambda}, \alpha)$, p. 53 **Multivariate Student**

$$p(\boldsymbol{x}) = \frac{\Gamma(\frac{1}{2}(\alpha + p))}{\Gamma(\frac{\alpha}{2})(\alpha\pi)^{p/2}} |\boldsymbol{\lambda}|^{\frac{1}{2}} \left[1 + \frac{1}{\alpha}(\boldsymbol{x}-\boldsymbol{\mu})^T \boldsymbol{\lambda}(\boldsymbol{x}-\boldsymbol{\mu})\right]^{-(\alpha+p)/2}$$

$\alpha > 0$, $\boldsymbol{\lambda}$ symmetric positive definite; $\mathrm{E}[\boldsymbol{x}] = \boldsymbol{\mu}$; $\mathrm{V}[\boldsymbol{x}] = \boldsymbol{\lambda}^{-1}(\alpha - 2)^{-1}\alpha$.

$\mathrm{Di}_C(\boldsymbol{x}|\boldsymbol{a})$, pp. 54, 70 **Dirichlet**

$$p(\boldsymbol{x}) = \left(\frac{\Gamma\left(\sum_{i=1}^{C} a_i\right)}{\prod_{i=1}^{C} a_i}\right) \prod_{i=1}^{C} x_i^{a_i - 1}$$

$0 < x_i < 1$, $a_i > 0$, $\boldsymbol{a} = (a_1, \ldots, a_C)$, $\boldsymbol{x} = (x_1, \ldots, x_C)$, $\sum_{i=1}^{C} x_i = 1$; $\mathrm{E}[x_i] = a_i / \sum_i a_i$.

Gamma

$\mathrm{Ga}(x|\alpha, \beta)$, pp. 61, 61

$$p(x) = \frac{\beta^\alpha}{\Gamma(\alpha)} x^{\alpha-1} \exp(-\beta x)$$

$\alpha > 0, \beta > 0$; $\mathrm{E}[x] = \alpha\beta^{-1}$; $\mathrm{V}[x] = \alpha\beta^{-2}$.

Inverted gamma

$\mathrm{Ig}(x|\alpha, \beta)$, pp. 67, 68

$$p(x) = \frac{\beta^\alpha}{\Gamma(\alpha)} x^{-(\alpha+1)} \exp(-\beta/x)$$

$\alpha > 0, \beta > 0$; $\mathrm{E}[x] = \beta/(\alpha - 1)$.

If $1/y \sim \mathrm{Ga}(\alpha, \beta)$, then $y \sim \mathrm{Ig}(\alpha, \beta)$.

Beta

$\mathrm{Be}(x|\alpha, \beta)$, p. 253

$$p(x) = \frac{\Gamma(\alpha+\beta)}{\Gamma(\alpha)\Gamma(\beta)} x^{\alpha-1}(1-x)^{\beta-1}$$

$\alpha > 0, \beta > 0, 0 < x < 1$; $\mathrm{E}[x] = \alpha/(\alpha+\beta)$.

Binomial

$\mathrm{Bi}(x|\theta, n)$, p. 253

$$p(x) = \binom{n}{x} \theta^x (1-\theta)^{n-x}$$

$0 < \theta < 1, n = 1, 2, \ldots, x = 0, 1, \ldots, n$.

Multinomial

$\mathrm{Mu}_k(\boldsymbol{x}|\boldsymbol{\theta}, n)$, pp. 54, 292

$$p(x) = \frac{n!}{\prod_{i=1}^k x_i!} \prod_{i=1}^k \theta_i^{x_i}$$

$0 < \theta_i < 1, \sum_{i=1}^k \theta_i = 1, \sum_{i=1}^k x_i = n, x_i = 0, 1, \ldots, n$.

References

Abramson, I.S. (1982) On bandwidth variation in kernel estimates – a square root law. *Annals of Statistics*, 10:1217–1223.

Abu-Mostafa, Y.S., Atiya, A.F., Magdon-Ismail, M. and White, H., eds (2001) Special issue on ‘Neural Networks in Financial Engineering’. *IEEE Transactions on Neural Networks*, 12(4).

Adams, N.M. and Hand, D.J. (1999) Comparing classifiers when the misallocation costs are uncertain. *Pattern Recognition*, 32:1139–1147.

Adams, N.M. and Hand, D.J. (2000) Improving the practice of classifier performance assessment. *Neural Computation*, 12:305–311.

Aeberhard, S., Coomans, D. and de Vel, O. (1993) Improvements to the classification performance of RDA. *Chemometrics*, 7:99–115.

Aeberhard, S., Coomans, D. and de Vel, O. (1994) Comparative analysis of statistical pattern recognition methods in high dimensional settings. *Pattern Recognition*, 27(8):1065–1077.

Aitchison, J. and Begg, C.B. (1976) Statistical diagnosis when basic cases are not classified with certainty. *Biometrika*, 63(1):1–12.

Aitchison, J., Habbema, J.D.F. and Kay, J.W. (1977) A critical comparison of two methods of statistical discrimination. *Applied Statistics*, 26:15–25.

Akaike, H. (1973) Information theory and an extension of the maximum likelihood principle. In B.N. Petrov and B.F. Csaki, eds, *Second International Symposium on Information Theory*, pp. 267–281, Akadémiai Kiadó, Budapest.

Akaike, H. (1974) A new look at the statistical model identification. *IEEE Transactions on Automatic Control*, 19:716–723.

Akaike, H. (1977) On entropy maximisation principle. In P.R. Krishnaiah, ed., *Proceedings of the Symposium on Applications of Statistics*, pp. 27–47. North Holland, Amsterdam.

Akaike, H. (1981) Likelihood of a model and information criteria. *Journal of Econometrics*, 16:3–14.

Akaike, H. (1985) Prediction and entropy. In A.C. Atkinson and S.E. Fienberg, eds, *A Celebration of Statistics*, pp. 1–24. Springer-Verlag, New York.

Al-Alaoui, M. (1977) A new weighted generalized inverse algorithm for pattern recognition. *IEEE Transactions on Computers*, 26(10):1009–1017.

Aladjem, M. (1991) Parametric and nonparametric linear mappings of multidimensional data. *Pattern Recognition*, 24(6):543–553.

Aladjem, M. and Dinstein, I. (1992) Linear mappings of local data structures. *Pattern Recognition Letters*, 13:153–159.

Albert, A. and Lesaffre, E. (1986) Multiple group logistic discrimination. *Computers and Mathematics with Applications*, 12A(2):209–224.

Anderberg, M.R. (1973) *Cluster Analysis for Applications*. Academic Press, New York.

Anders, U. and Korn, O. (1999) Model selection in neural networks. *Neural Networks*, 12:309–323.

Anderson, J.A. (1974) Diagnosis by logistic discriminant function: further practical problems and results. *Applied Statistics*, 23:397–404.

Anderson, J.A. (1982) Logistic discrimination. In P.R. Krishnaiah and L.N. Kanal, eds, *Handbook of Statistics*, vol. 2, pp. 169–191. North Holland, Amsterdam.

Anderson, J.J. (1985) Normal mixtures and the number of clusters problem. *Computational Statistics Quarterly*, 2:3–14.

Andrews, D.F. and Herzberg, A.M. (1985) *Data: A Collection of Problems from Many Fields for the Student and Research Worker*. Springer-Verlag, New York.

Andrews, H.C. (1972) *Introduction to Mathematical Techniques in Pattern Recognition*. Wiley Interscience, New York.

Andrieu, C. and Doucet, A. (1999) Joint Bayesian model selection and estimation of noisy sinusoids via reversible jump MCMC. *IEEE Transactions on Signal Processing*, 47(10):2667–2676.

Apté, C., Sasisekharan, R., Seshadri, S. and Weiss, S.M. (1994) Case studies of high-dimensional classification. *Journal of Applied Intelligence*, 4:269–281.

Arimura, K. and Hagita, N. (1994) Image screening based on projection pursuit for image recognition. In *Proceedings of the 12th International Conference on Pattern Recognition*, vol. 2, pp. 414–417, IEEE, Los Alamitos, CA.

Ashikaga, T. and Chang, P.C. (1981) Robustness of Fisher's linear discriminant function under two-component mixed normal models. *Journal of the American Statistical Association*, 76:676–680.

Atkinson, A.C. and Mulira, H.-M. (1993) The stalactite plot for the detection of multivariate outliers. *Statistics and Computing*, 3:27–35.

Atlas, L., Connor, J., Park, D., El-Sharkawi, M., Marks, R., Lippman, A., Cole, R. and Muthusamy, Y. (1989) A performance comparison of trained multilayer perceptrons and trained classification trees. In *Proceedings of the 1989 IEEE Conference on Systems, Man and Cybernetics*, pp. 915–920, IEEE, New York.

Atlas, R.S. and Overall, J.E. (1994) Comparative evaluation of two superior stopping rules for hierarchical cluster analysis. *Psychometrika*, 59(4):581–591.

Babich, G.A. and Camps, O.I. (1996) Weighted Parzen windows for pattern classification. *IEEE Transactions on Pattern Analysis and Machine Intelligence*, 18(5):567–570.

Bailey, T.L. and Elkan, C. (1995) Unsupervised learning of multiple motifs in biopolymers using expectation maximization. *Machine Learning Journal*, 21:51–83.

Balakrishnan, N. and Subrahmaniam, K. (1985) Robustness to nonnormality of the linear discriminant function: mixtures of normal distributions. *Communications in Statistics – Theory and Methods*, 14(2):465–478.

Barron, A.R. and Barron, R.L. (1988) Statistical learning networks: a unifying view. In E.J. Wegman, D.T. Gantz and J.J. Miller, eds, *Statistics and Computer Science: 1988 Symposium at the Interface*, pp. 192–203. American Statistical Association, Faisfax, VA.

Bauer, E. and Kohavi, R. (1999) An empirical comparison of voting classification algorithms: bagging, boosting and variants. *Machine Learning*, 36:105–139.

Baxter, M.J. (1995) Standardisation and transformation in principal component analysis with applications to archaeometry. *Applied Statistics*, 4(4):513–527.

Beckman, R.J. and Cook, R.D. (1983) Outlier . . . . s. *Technometrics*, 25(2):119–163.

Bedworth, M.D. (1988) Improving on standard classifiers by implementing them as a multilayer perceptron. RSRE Memorandum, DERA, St Andrews Road, Malvern, Worcs, WR14 3PS, UK.

Bedworth, M.D., Bottou, L., Bridle, J.S., Fallside, F., Flynn, L., Fogelman, F., Ponting, K.M. and Prager, R.W. (1989) Comparison of neural and conventional classifiers on a speech recognition problem. In *IEE International Conference on Artificial Neural Networks*, pp. 86–89, IEE, London.

Bengio, Y., Buhmann, J.M., Embrechts, M. and Zurada, J.M., eds (2000) Special issue on 'Neural Networks for Data Mining and Knowledge Discovery'. *IEEE Transactions on Neural Networks*, 11(3).

Bensmail, H. and Celeux, G. (1996) Regularized Gaussian discriminant analysis through eigenvalue decomposition. *Journal of the American Statistical Association*, 91:1743–1748.

Bergh, A.F., Soong, F.K. and Rabiner, L.R. (1985) Incorporation of temporal structure into a vector-quantization-based preprocessor for speaker-independent, isolated-word recognition. *AT&T Technical Journal*, 64(5):1047–1063.

Bernardo, J.M. and Smith, A.F.M. (1994) *Bayesian Theory*. Wiley, Chichester.

Bezdek, J.C. (1981) *Pattern Recognition with Fuzzy Objective Function Algorithms*. Plenum Press, New York.

Bezdek, J.C. and Pal, S.K., eds (1992) *Fuzzy Models for Pattern Recognition. Methods that Search for Structure in Data*. IEEE Press, New York.

Bishop, C.M. (1993) Curvature-driven smoothing: a learning algorithm for feedforward networks. *IEEE Transactions on Neural Networks*, 4(5):882–884.

Bishop, C.M. (1995) *Neural Networks for Pattern Recognition*. Oxford University Press, Oxford.

Bishop, C.M., Svensén, C.M. and Williams, C.K.I. (1998) GTM: The generative topographic mapping. *Neural Computation*, 10:215–234.

Blacknell, D. and White, R.G. (1994) Optimum classification of non-Gaussian processes using neural networks. *IEE Proceedings on Vision, Image and Signal Processing*, 141(1):56–66.

Blue, J.L., Candela, G.T., Grother, P.J., Chellappa, R. and Wilson, C.L. (1994) Evaluation of pattern classifiers for fingerprint and OCR applications. *Pattern Recognition*, 27(4):485–501.

Bobrowski, L. and Bezdek, J.C. (1991) *c*-means clustering with the l_1 and l_∞ norms. *IEEE Transactions on Systems, Man, and Cybernetics*, 21(3):545–554.

Bock, H.H. (1985) On some significance tests in cluster analysis. *Journal of Classification*, 2: 77–108.

Bock, H.H. (1989) Probabilistic aspects in cluster analysis. In O. Opitz, ed., *Conceptual and Numerical Analysis of Data*, pp. 12–44. Springer-Verlag, Berlin.

Bonde, G.J. (1976) Kruskal's non-metric multidimensional scaling – applied in the classification of bacteria. In J. Gordesch and P. Naeve, eds, *Proceedings in Computational Statistics*, pp. 443–449. Physica-Verlag, Vienna.

Booth, J.G. and Hall, P. (1994) Monte Carlo approximation and the iterated bootstrap. *Biometrika*, 81(2):331–340.

Bowman, A.W. (1985) A comparative study of some kernel-based nonparametric density estimators. *Journal of Statistical Computation and Simulation*, 21:313–327.

Bozdogan, H. (1987) Model selection and Akaike's information criterion (AIC): the general theory and its analytical extensions. *Psychometrika*, 52(3):345–370.

Bozdogan, H. (1993) Choosing the number of component clusters in the mixture-model using a new informational complexity criterion of the inverse-Fisher information matrix. In O. Opitz, B. Lausen and R. Klar eds, *Information and Classification*, pp. 40–54. Springer-Verlag, Heidelberg.

Bradley, A.P. (1997) The use of the area under the ROC curve in the evaluation of machine learning algorithms. *Pattern Recognition*, 30(7):1145–1159.

Breckenridge, J.N. (1989) Replicating cluster analysis: method, consistency, and validity. *Multivariate Behavioral Research*, 24(32):147–161.

Breiman, L. (1996) Bagging predictors. *Machine Learning*, 26(2):123–140.

Breiman, L. (1998) Arcing classifiers. *Annals of Statistics*, 26(3):801–849.

Breiman, L. and Friedman, J.H. (1988) Discussion on article by Loh and Vanichsetakul: 'Tree-structured classification via generalized discriminant analysis'. *Journal of the American Statistical Association*, 83:715–727.

Breiman, L. and Ihaka, R. (1984) Nonlinear discriminant analysis via scaling and ACE. Technical Report 40, Department of Statistics, University of California, Berkeley.

Breiman, L., Meisel, W. and Purcell, E. (1977) Variable kernel estimates of multivariate densities. *Technometrics*, 19(2):135–144.

Breiman, L., Friedman, J.H., Olshen, R.A. and Stone, C.J. (1984) *Classification and Regression Trees*. Wadsworth International Group, Belmont, CA.

Brent, R.P. (1991) Fast training algorithms for multilayer neural nets. *IEEE Transactions on Neural Networks*, 2(3):346–354.

Brill, F.Z., Brown, D.E. and Martin, W.N. (1992) Fast genetic selection of features for neural network classifiers. *IEEE Transactions on Neural Networks*, 3(2):324–328.

Broomhead, D.S. and Lowe, D. (1988) Multi-variable functional interpolation and adaptive networks. *Complex Systems*, 2(3):269–303.

Brown, D.E., Corruble, V. and Pittard, C.L. (1993) A comparison of decision tree classifiers with backpropagation neural networks for multimodal classification problems. *Pattern Recognition*, 26(6):953–961.

Brown, M., Lewis, H.G. and Gunn, S.R. (2000) Linear spectral mixture models and support vector machines for remote sensing. *IEEE Transactions on Geoscience and Remote Sensing*, 38(5):2346–2360.

Brown, P.J. (1993) *Measurement, Regression, and Calibration*. Clarendon Press, Oxford.

Bruzzone, L., Roli, F. and Serpico, S.B. (1995) An extension of the Jeffreys–Matusita distance to multiclass cases for feature selection. *IEEE Transactions on Geoscience and Remote Sensing*, 33(6):1318–1321.

Bull, S.B. and Donner, A. (1987) The efficiency of multinomial logistic regression compared with multiple group discriminant analysis. *Journal of the American Statistical Association*, 82:1118–1121.

Buntine, W.L. (1992) Learning classification trees. *Statistics and Computing*, 2:63–73.

Buntine, W.L. (1996) A guide to the literature on learning probabilistic networks from data. *IEEE Transactions on Knowledge and Data Engineering*, 8(2):195–210.

Buntine, W.L. and Weigend, A.S. (1991) Bayesian back-propagation. *Complex Systems*, 5:603–643.

Burbidge, R., Trotter, M., Buxton, B. and Holden, S. (2001) Drug design by machine learning: support vector machines for pharmaceutical data analysis. *Computers and Chemistry*, 26:5–14.

Burges, C.J.C. (1998) A tutorial on support vector machines for pattern recognition. *Data Mining and Knowledge Discovery*, 2:121–167.

Burrell, P.R. and Folarin, B.O. (1997) The impact of neural networks in finance. *Neural Computing and Applications*, 6:193–200.

Buturović, L.J. (1993) Improving k-nearest neighbor density and error estimates. *Pattern Recognition*, 26(4):611–616.

Calinski, R.B. and Harabasz, J. (1974) A dendrite method for cluster analysis. *Communications in Statistics*, 3:1–27.

Campbell, J.G., Fraley, C., Murtagh, F. and Raftery, A.E. (1997) Linear flaw detection in woven textiles using model-based clustering. *Pattern Recognition Letters*, 18:1539–1548.

Cao, L. and Tay, F.E.H. (2001) Financial forecasting using support vector machines. *Neural Computing and Applications*, 10:184–192.

Cao, R., Cuevas, A. and Manteiga, W.G. (1994) A comparative study of several smoothing methods in density estimation. *Computational Statistics and Data Analysis*, 17:153–176.

Carter, C. and Catlett, J. (1987) Assessing credit card applications using machine learning. *IEEE Expert*, 2:71–79.

Casella, G. and George, E.I. (1992) Explaining the Gibbs sampler. *American Statistician*, 46(3):167–174.

Celeux, G. and Govaert, G. (1992) A classification EM algorithm for clustering and two stochastic versions. *Computational Statistics and Data Analysis*, 14:315–332.

Celeux, G. and Govaert, G. (1995) Gaussian parsimonious clustering models. *Pattern Recognition*, 28(5):781–793.

Celeux, G. and Mkhadri, A. (1992) Discrete regularized discriminant analysis. *Statistics and Computing*, 2(3):143–151.

Celeux, G. and Soromenho, G. (1996) An entropy criterion for assessing the number of clusters in a mixture model. *Journal of Classification*, 13:195–212.

Čencov, N.N. (1962) Evaluation of an unknown distribution density from observations. *Soviet Mathematics*, 3:1559–1562.

Chair, Z. and Varshney, P.R. (1986) Optimal data fusion in multiple sensor detection systems. *IEEE Transactions on Aerospace and Electronic Systems*, 22:98–101.

Chan, C.-W.J., Huang, C. and DeFries, R. (2001) Enhanced algorithm performance for land cover classification from remotely sensed data using bagging and boosting. *IEEE Transactions on Geoscience and Remote Sensing*, 39(3):693–695.

Chang, C.-Y. (1973) Dynamic programming as applied to feature subset selection in a pattern recognition system. *IEEE Transactions on Systems, Man, and Cybernetics*, 3(2):166–171.

Chang, E.I. and Lippmann, R.P. (1991) Using genetic algorithms to improve pattern classification performance. In R.P. Lippmann, J.E. Moody and D.S. Touretzky, eds, *Advances in Neural Information Processing Systems*, vol. 3, pp. 797–803. Morgan Kaufmann, San Mateo, CA.

Chang, E.I. and Lippmann, R.P. (1993) A boundary hunting radial basis function classifier which allocates centers constructively. In S.J. Hanson, J.D. Cowan and C.L. Giles, eds, *Advances in Neural Information Processing Systems*, vol. 5, pp. 139–146. Morgan Kaufmann, San Mateo, CA.

Chang, P.C. and Afifi, A.A. (1974) Classification based on dichotomous and continuous variables. *Journal of the American Statistical Association*, 69:336–339.

Chatfield, C. (1985) The initial examination of data (with discussion). *Journal of the Royal Statistical Society Series A*, 148:214–253.

Chatfield, C. (1988) *Problem Solving. A Statistician's Guide*. Chapman & Hall, London.

Chatfield, C. and Collins, A.J. (1980) *Introduction to Multivariate Analysis*. Chapman & Hall, London.

Chatzis, V., Borş, A.G. and Pitas, I. (1999) Multimodal decision-level fusion for person authentication. *IEEE Transactions on Systems, Man, and Cybernetics – Part A: Systems and Humans*, 29(6):674–680.

Chellappa, R., Fukushima, K., Katsaggelos, A.K., Kung, S.-Y., LeCun, Y., Nasrabadi, N.M. and Poggio, T., eds (1998) Special issue on 'Applications of Artificial Neural Networks to Image Processing'. *IEEE Transactions on Image Processing*, 7(8).

Chellappa, R., Wilson, C.L. and Sirohey, S. (1995) Human and machine recognition of faces: a survey. *Proceedings of the IEEE*, 83(5):705–740.

Chen, C.H. (1973) *Statistical Pattern Recognition*. Hayden, Washington, DC.

Chen, C.H. (1976) On information and distance measures, error bounds, and feature selection. *Information Sciences*, 10:159–173.

Chen, J.S. and Walton, E.K. (1986) Comparison of two target classification schemes. *IEEE Transactions on Aerospace and Electronic Systems*, 22(1):15–22.

Chen, K., Xu, L. and Chi, H. (1999) Improved learning algorithms for mixture of experts in multiclass classification. *Neural Networks*, 12:1229–1252.

Chen, L.-F., Liao, H.-Y.M., Ko, M.-T., Lin, J.-C. and Yu, G.-J. (2000a) A new LDA-based face recognition system which can solve the small sample size problem. *Pattern Recognition*, 33:1713–1726.

Chen, S., Cowan, C.F.N. and Grant, P.M. (1991) Orthogonal least squares learning algorithm for radial basis function networks. *IEEE Transactions on Neural Networks*, 2(2):302–309.

Chen, S., Grant, P.M. and Cowan, C.F.N. (1992) Orthogonal least-squares algorithm for training multioutput radial basis function networks. *IEE Proceedings, Part F*, 139(6):378–384.

Chen, S., Chng, E.S. and Alkadhimi, K. (1996) Regularised orthogonal least squares algorithm for constructing radial basis function networks. *International Journal of Control*, 64(5):829–837.

Chen, S., Gunn, S. and Harris, C.J. (2000) Decision feedback equaliser design using support vector machines. *IEE Proceedings on Vision, Image and Signal Processing*, 147(3):213–219.

Chen, T. and Chen, H. (1995) Approximation capability to functions of several variables, nonlinear functionals, and operators by radial basis function neural networks. *IEEE Transactions on Neural Networks*, 6(4):904–910.

Cheng, B. and Titterington, D.M. (1994) Neural networks: a review from a statistical perspective (with discussion). *Statistical Science*, 9(1):2–54.

Cheng, Y.-Q., Zhuang, Y.-M. and Yang, J.-Y. (1992) Optimal Fisher discriminant analysis using rank decomposition. *Pattern Recognition*, 25(1):101–111.

Chernick, M.R., Murthy, V.K. and Nealy, C.D. (1985) Application of bootstrap and other resampling techniques: evaluation of classifier performance. *Pattern Recognition Letters*, 3:167–178.

Chiang, S.-S., Chang, C.-I. and Ginsberg, I.W. (2001) Unsupervised target detection in hyperspectral images using projection pursuit. *IEEE Transactions on Geoscience and Remote Sensing*, 39(7):1380–1391.

Chien, Y.T. and Fu, K.S. (1967) On the generalized Karhunen–Loève expansion. *IEEE Transactions on Information Theory*, 13:518–520.

Chingánda, E.F. and Subrahmaniam, K. (1979) Robustness of the linear discriminant function to nonnormality: Johnson's system. *Journal of Statistical Planning and Inference*, 3:69–77.

Chitteneni, C.B. (1980) Learning with imperfectly labeled patterns. *Pattern Recognition*, 12: 281–291.

Chitteneni, C.B. (1981) Estimation of probabilities of label imperfections and correction of mislabels. *Pattern Recognition*, 13:257–268.

Chou, P.A. (1991) Optimal partitioning for classification and regression trees. *IEEE Transactions on Pattern Analysis and Machine Intelligence*, 13(4):340–354.

Chow, C.K. (1970) On optimum recognition error and reject tradeoff. *IEEE Transactions on Information Theory*, 16(1):41–46.

Chow, C.K. and Liu, C.N. (1968) Approximating discrete probability distributions with dependence trees. *IEEE Transactions on Information Theory*, 14(3):462–467.

Chow, M.-Y., ed. (1993) Special issue on 'Applications of Intelligent Systems to Industrial Electronics'. *IEEE Transactions on Industrial Electronics*, 40(2).

Clifford, H.T. and Stephenson, W. (1975) *An Introduction to Numerical Classification*. Academic Press, New York.

Cochran, W.G. and Cox, G.M. (1957) *Experimental Designs*. Wiley, New York.

Cole, A.J. and Wishart, D. (1970) An improved algorithm for the Jardine–Sibson method of generating overlapping clusters. *Computer Journal*, 13(2):156–163.

Comon, P. (1994) Independent component analysis, a new concept? *Signal Processing*, 36:287–314.

Constantinides, A.G., Haykin, S., Hu, Y.H., Hwang, J.-N., Katagiri, S., Kung, S.-Y. and Poggio, T.A., eds (1997) Special issue on 'Neural Networks for Signal Processing'. *IEEE Transactions on Signal Processing*, 45(11).

Conway, J.A., Brown, L.M.J., Veck, N.J. and Cordey, R.A. (1991) A model-based system for crop classification from radar imagery. *GEC Journal of Research*, 9(1):46–54.

Cooper, G.F. and Herskovits, E. (1992) A Bayesian method for the induction of probabilistic networks from data. *Machine Learning*, 9:309–347.

Copsey, K.D. and Webb, A.R. (2001) Bayesian approach to mixture models for discrimination. In F.J. Ferri, J.M. Iñesta, A. Amin and P. Pudil, eds, *Advances in Pattern Recognition*, pp. 491–500. Springer, Berlin.

Cormack, R.M. (1971) A review of classification (with discussion). *Journal of the Royal Statistical Society Series A*, 134:321–367.

Cortes, C. and Vapnik, V. (1995) Support-vector networks. *Machine Learning*, 20:273–297.

Cosman, P.C., Tseng, C., Gray, R.M., Olshen, R.A., Moses, L.E., Davidson, H.C., Bergin, C.J. and Riskin, E.A. (1993) Tree-structured vector quantization of CT chest scans: image quality and diagnostic accuracy. *IEEE Transactions on Medical Imaging*, 12(4):727–739.

Courant, R. and Hilbert, D. (1959) *Methods of Mathematical Physics.* Wiley, New York.

Cover, T.M. and Hart, P.E. (1967) Nearest neighbour pattern classification. *IEEE Transactions on Information Theory*, 13:21–27.

Cowell, R.G., Dawid, A.P., Hutchinson, T. and Spiegelhalter, D.J. (1991) A Bayesian expert system for the analysis of an adverse drug reaction. *Artificial Intelligence in Medicine*, 3:257–270.

Cox, D.R. (1958) *The Planning of Experiments.* Wiley, New York.

Cox, T.F. and Cox, M.A.A. (1994) *Multidimensional Scaling*. Chapman & Hall, London.

Cox, T.F. and Ferry, G. (1993) Discriminant analysis using non-metric multidimensional scaling. *Pattern Recognition*, 26(1):145–153.

Cox, T.F. and Pearce, K.F. (1997) A robust logistic discrimination model. *Statistics and Computing*, 7:155–161.

Craven, P. and Wahba, G. (1979) Smoothing noisy data with spline functions. *Numerische Mathematik*, 31:317–403.

Crawford, S.L. (1989) Extensions to the CART algorithm. *International Journal of Man–Machine Studies*, 31:197–217.

Cristianini, N. and Shawe-Taylor, J. (2000) *An Introduction to Support Vector Machines.* Cambridge University Press, Cambridge.

Crownover, R.M. (1991) A least squares approach to linear discriminant analysis. *SIAM*, 12(3): 595–606.

Curram, S.P. and Mingers, J. (1994) Neural networks, decision tree induction and discriminant analysis: an empirical comparison. *Journal of the Operational Research Society*, 45(4): 440–450.

D'Andrea, L.M., Fisher, G.L. and Harrison, T.C. (1994) Cluster analysis of adult children of alcoholics. *International Journal of Addictions*, 29(5):565–582.

Darwish, Y., Cserháti, T. and Forgács, E. (1994) Use of principal component analysis and cluster analysis in quantitative structure-activity relationships: a comparative study. *Chemometrics and Intelligent Laboratory Systems*, 24:169–176.

Dasarathy, B.V. (1991) NN concepts and techniques. an introductory survey. In B.V. Dasarathy, ed., *Nearest Neighbour Norms: NN Pattern Classification Techniques*, pp. 1–30. IEEE Computer Society Press, Los Alamitos, CA.

Dasarathy, B.V. (1994a) Minimal consistent set (MCS) identification for nearest neighbour decision systems design. *IEEE Transactions on Systems, Man, and Cybernetics*, 24(3):511–517.

Dasarathy, B.V. (1994b) *Decision Fusion.* IEEE Computer Society Press, Los Alamitos, CA.

Davies, D.L. and Bouldin, D. (1979) A cluster separation measure. *IEEE Transactions on Pattern Analysis and Machine Intelligence*, 1:224–227.

Davison, A.C. and Hall, P. (1992) On the bias and variability of bootstrap and cross-validation estimates of error rate in discrimination problems. *Biometrika*, 79:279–284.

Davison, A.C., Hinkley, D.V. and Schechtman, E. (1986) Efficient bootstrap simulation. *Biometrika*, 73(3):555–556.

Dawant, B.M. and Garbay, C., eds (1999) Special topic section on 'Biomedical Data Fusion'. *IEEE Transactions on Biomedical Engineering*, 46(10).

Day, N.E. and Kerridge, D.F. (1967) A general maximum likelihood discriminant. *Biometrics*, 23:313–323.

De Gooijer, J.G., Ray, B.K. and Horst, K. (1998) Forecasting exchange rates using TSMARS. *Journal of International Money and Finance*, 17(3):513–534.

De Jager, O.C., Swanepoel, J.W.H. and Raubenheimer, B.C. (1986) Kernel density estimators applied to gamma ray light curves. *Astronomy and Astrophysics*, 170:187–196.

de Leeuw, J. (1977) Applications of convex analysis to multidimensional scaling. In J.R. Barra, F. Brodeau, G. Romier and B. van Cutsem, eds, *Recent Developments in Statistics*, pp. 133–145. North Holland, Amsterdam.

de Leeuw, J. and Heiser, W.J. (1977) Convergence of correction matrix algorithms for multidimensional scaling. In J.C. Lingoes, ed., *Geometric Representations of Relational Data*, pp. 735–752. Mathesis Press, Ann Arbor, MI.

de Leeuw, J. and Heiser, W.J. (1980) Multidimensional scaling with restrictions on the configuration. In P.R. Krishnaiah, ed., *Multivariate Analysis*, vol. V, pp. 501–522. North Holland, Amsterdam.

de Mántaras, R.L. and Aguilar-Martín, J. (1985) Self-learning pattern classification using a sequential clustering technique. *Pattern Recognition*, 18(3/4):271–277.

De Veaux, R.D., Gordon, A.L., Comiso, J.C. and Bacherer, N.E. (1993) Modeling of topographic effects on Antarctic sea ice using multivariate adaptive regression splines. *Journal of Geophysical Research*, 98(C11):20307–20319.

de Vel, O. Anderson, A., Corney, M. and Mohay, G. (2001) Mining e-mail content for author identification forensics. *SIGMOD Record*, 30(4):55–64.

Defays, D. (1977) An efficient algorithm for a complete link method. *Computer Journal*, 20(4):364–366.

Dekkers, M.J., Langereis, C.G., Vriend, S.P., van Santvoort, P.J.M. and de Lange, G.J. (1994) Fuzzy *c*-means cluster analysis of early diagenetic effects on natural remanent magnetisation acquisition in a 1.1 Myr piston core from the Central Mediterranean. *Physics of the Earth and Planetary Interiors*, 85:155–171.

Dellaportas, P. (1998) Bayesian classification of neolithic tools. *Applied Statistics*, 47(2):279–297.

Dempster, A.P., Laird, N.M. and Rubin, D.B. (1977) Maximum likelihood from incomplete data via the EM algorithm. *Journal of the Royal Statistical Society Series B*, 39:1–38.

Denison, D.G.T., Mallick, B.K. and Smith, A.F.M. (1998a) A Bayesian CART algorithm. *Biometrika*, 85(2):363–377.

Denison, D.G.T., Mallick, B.K. and Smith, A.F.M. (1998b) Bayesian MARS. *Statistics and Computing*, 8:337–346.

Devijver, P.A. (1973) Relationships between statistical risks and the least-mean-square error criterion in pattern recognition. In *Proceedings of the First International Joint Conference on Pattern Recognition*, pp. 139–148.

Devijver, P.A. and Kittler, J. (1982) *Pattern Recognition, A Statistical Approach.* Prentice Hall, London.

Devroye, L. (1986) *Non-uniform Random Variate Generation.* Springer-Verlag, New York.

Devroye, L. and Györfi, L. (1985) *Nonparametric Density Estimation. The L_1 View*. Wiley, New York.

Diday, E. and Simon, J.C. (1976) Clustering analysis. In K.S. Fu, ed., *Digital Pattern Recognition*, pp. 47–94. Springer-Verlag, Berlin.

Dietterich, T.G. (1998) Approximate statistical tests for comparing supervised classification learning algorithms. *Neural Computation*, 10:1895–1923.

Diggle, P.J. and Hall, P. (1986) The selection of terms in an orthogonal series density estimator. *Journal of the American Statistical Association*, 81:230–233.

Dillon, T., Arabshahi, P. and Marks, R.J., eds (1997) Special issue on 'Everyday Applications of Neural Networks'. *IEEE Transactions on Neural Networks*, 8(4).

Dillon, W.R. and Goldstein, M. (1984) *Multivariate Analysis Methods and Applications.* Wiley, New York.

Djouadi, A. and Bouktache, E. (1997) A fast algorithm for the nearest-neighbor classifier. *IEEE Transactions on Pattern Analysis and Machine Intelligence*, 19(3):277–282.

Domingos, P. and Pazzani, M. (1997) On the optimality of the simple Bayesian classifier under zero–one loss. *Machine Learning*, 29:103–130.

Dony, R.D. and Haykin, S. (1995) Neural network approaches to image compression. *Proceedings of the IEEE*, 83(2):288–303.

Doucet, A., De Freitas, N. and Gordon, N., eds (2001) *Sequential Monte Carlo Methods in Practice.* Springer-Verlag, New York.

Dracopoulos, D.C. and Rosin, P.L., eds (1998) Special issue on 'Machine Vision Using Neural Networks'. *Neural Computing and Applications*, 7(3).

Drake, K.C., Kim, Y., Kim, T.Y. and Johnson, O.D. (1994) Comparison of polynomial network and model-based target recognition. In N. Nandhakumar, ed., *Sensor Fusion and Aerospace Applications II*, vol. 2233, pp. 2–11. SPIE.

Drucker, H. and Le Cun, Y. (1992) Improving generalization performance using double back-propagation. *IEEE Transactions on Neural Networks*, 3(6):991–997.

Dubes, R.C. (1987) How many clusters are best? – an experiment. *Pattern Recognition*, 20(6): 645–663.

Dubuisson, B. and Lavison, P. (1980) Surveillance of a nuclear reactor by use of a pattern recognition methodology. *IEEE Transactions on Systems, Man and Cybernetics*, 10(10):603–609.

Duchene, J. and Leclercq, S. (1988) An optimal transformation for discriminant analysis and principal component analysis. *IEEE Transactions on Pattern Analysis and Machine Intelligence*, 10(6):978–983.

Duchon, J. (1976) Interpolation des fonctions de deux variables suivant le principe de la flexion des plaques minces. *R.A.I.R.O. Analyse Numérique*, 10(12):5–12.

Duda, R.O., Hart, P.E. and Stork, D.G. (2001) *Pattern Classification.* 2nd edn., Wiley, New York.

Dudani, S.A. (1976) The distance-weighted k-nearest-neighbour rule. *IEEE Transactions on Systems, Man, and Cybernetics*, 6(4):325–327.

Duffy, D., Yuhas, B., Jain, A. and Buja, A. (1994) Empirical comparisons of neural networks and statistical methods for classification and regression. In B. Yuhas and N. Ansari, eds, *Neural Networks in Telecommunications*, pp. 325–349. Kluwer Academic Publishers, Norwell, MA.

Duin, R.P.W. (1976) On the choice of smoothing parameters for Parzen estimators of probability density functions. *IEEE Transactions on Computers*, 25:1175–1179.

Duin, R.P.W. (1996) A note on comparing classifiers. *Pattern Recognition Letters*, 17:529–536.

Dunn, J.C. (1974) A fuzzy relative of the ISODATA process and its use in detecting compact well-separated clusters. *Journal of Cybernetics*, 3(3):32–57.

Eddy, W.F., Mockus, A. and Oue, S. (1996) Approximate single linkage cluster analysis of large data sets in high-dimensional spaces. *Computational Statistics and Data Analysis*, 23:29–43.

Efron, B. (1979) Bootstrap methods: Another look at the jackknife. *Annals of Statistics*, 7:1–26.

Efron, B. (1982) *The Jackknife, the Bootstrap, and Other Resampling Plans.* Society for Industrial and Applied Mathematics, Philadelphia.

Efron, B. (1983) Estimating the error rate of a prediction rule: Improvement on cross-validation. *Journal of the American Statistical Association*, 78:316–331.

Efron, B. (1990) More efficient bootstrap computations. *Journal of the American Statistical Association*, 85:79–89.

Efron, B. and Tibshirani, R.J. (1986) Bootstrap methods for standard errors, confidence intervals, and other measures of statistical accuracy (with discussion). *Statistical Science*, 1:54–77.

Eklundh, L. and Singh, A. (1993) A comparative analysis of standardised and unstandardised principal components analysis in remote sensing. *International Journal of Remote Sensing*, 14(7):1359–1370.

Enas, G.G. and Choi, S.C. (1986) Choice of the smoothing parameter and efficiency of k-nearest neighbor classification. *Computers and Mathematics with Applications*, 12A(2):235–244.

Esposito, F., Malerba, D. and Semeraro, G. (1997) A comparative analysis of methods for pruning decision trees. *IEEE Transactions on Pattern Analysis and Machine Intelligence*, 19(5): 476–491.

Etezadi-Amoli, J. and McDonald, R.P. (1983) A second generation nonlinear factor analysis. *Psychometrika*, 48:315–342.

Everitt, B.S. (1981) A Monte Carlo investigation of the likelihood ratio test for the number of components in a mixture of normal distributions. *Multivariate Behavioral Research*, 16:171–180.

Everitt, B.S. (1988) A finite mixture model for the clustering of mixed-mode data. *Statistics and Probability Letters*, 6:305–309.

Everitt, B.S. and Dunn, G. (1991) *Applied Multivariate Data Analysis.* Edward Arnold, London.

Everitt, B.S. and Hand, D.J. (1981) *Finite Mixture Distributions.* Chapman & Hall, London.

Everitt, B.S. and Hay, D.F. (1992) *Talking about Statistics. A Psychologists Guide to Design and Analysis.* Edward Arnold, London.

Everitt, B.S. and Merette, C. (1990) The clustering of mixed-mode data: a comparison of possible approaches. *Journal of Applied Statistics*, 17:283–297.

Everitt, B.S., Landau, S. and Leese, M. (2001) *Cluster Analysis.* 4th edn., Arnold, London.

Falconer, J.A., Naughton, B.J., Dunlop, D.D., Roth, E.J., Strasser, D.C., and Sinacore, J.M. (1994) Predicting stroke in patient rehabilitation outcome using a classification tree approach. *Archives of Physical Medicine and Rehabilitation*, 75:619–625.

Faragó, A. and Lugosi, G. (1993) Strong universal consistency of neural network classifiers. *IEEE Transactions on Information Theory*, 39(4):1146–1151.

Feng, C. and Michie, D. (1994) Machine learning of rules and trees. In D. Michie, D.J. Spiegelhalter and C.C. Taylor, eds, *Machine Learning, Neural and Statistical Classification.* Ellis Horwood, Hemel Hempstead.

Fernández de Cañete, J. and Bulsari, A.B., eds (2000) Special issue on 'Neural Networks in Process Engineering'. *Neural Computing and Applications*, 9(3).

Ferrán, E.A., Pflugfelder, B. and Ferrara, P. (1994) Self-organized neural maps of human protein sequences. *Protein Science*, 3:507–521.

Ferré, L. (1995) Selection of components in principal components analysis: a comparison of methods. *Computational Statistics and Data Analysis*, 19:669–682.

Ferri, F.J. and Vidal, E. (1992a) Small sample size effects in the use of editing techniques. In *Proceedings of the 11th IAPR International Conference on Pattern Recognition*, pp. 607–610, The Hague, IEEE Computer Society Press, Los Alamitos, CA.

Ferri, F.J. and Vidal, E. (1992b) Colour image segmentation and labeling through multiedit-condensing. *Pattern Recognition Letters*, 13:561–568.

Ferri, F.J., Albert, J.V. and Vidal, E. (1999) Considerations about sample-size sensitivity of a family of nearest-neighbor rules. *IEEE Transactions on Systems, Man, and Cybernetics*, 29(5).

Fitzmaurice, G.M., Krzanowski, W.J. and Hand, D.J. (1991) A Monte Carlo study of the 632 bootstrap estimator of error rate. *Journal of Classification*, 8:239–250.

Fletcher, R. (1988) *Practical Methods of Optimization.* Wiley, New York.

Flury, B. (1987) A hierarchy of relationships between covariance matrices. In A.K. Gupta, ed., *Advances in Multivariate Analysis.* D. Reidel, Dordrecht.

Flury, B. (1988) *Common Principal Components and Related Multivariate Models.* Wiley, New York.

Foley, D.H. and Sammon, J.W. (1975) An optimal set of discriminant vectors. *IEEE Transactions on Computers*, 24(3):281–289.

French, S. and Smith, J.Q. (1997) *The Practice of Bayesian Analysis.* Arnold, London.

Freund, Y. and Schapire, R. (1996) Experiments with a new boosting algorithm. In *Machine Learning: Proceedings of the 13th International Conference*, pp. 148–156, Morgan Kaufmann, San Francisco.

Freund, Y. and Schapire, R. (1999) A short introduction to boosting. *Journal of the Japanese Society for Artificial Intelligence*, 14(5):771–780.

Friedl, M.A., Brodley, C.E. and Strahler, A.H. (1999) Maximising land cover classification accuracies produced by decision trees at continental to global scales. *IEEE Transactions on Geoscience and Remote Sensing*, 37(2):969–977.

Friedman, J.H. (1987) Exploratory projection pursuit. *Journal of the American Statistical Association*, 82:249–266.

Friedman, J.H. (1989) Regularized discriminant analysis. *Journal of the American Statistical Association*, 84:165–175.

Friedman, J.H. (1991) Multivariate adaptive regression splines. *Annals of Statistics*, 19(1):1–141.

Friedman, J.H. (1993) Estimating functions of mixed ordinal and categorical variables using adaptive splines. In S. Morgenthaler, E.M.D. Ronchetti and W.A. Stahel, eds, *New Directions in Statistical Data Analysis and Robustness*, pp. 73–113. Birkhäuser-Verlag, Basel.

Friedman, J.H. (1994) Flexible metric nearest neighbor classification. Report, Department of Statistics, Stanford University.

Friedman, J.H. and Stuetzle, W. (1981) Projection pursuit regression. *Journal of the American Statistical Association*, 76:817–823.

Friedman, J.H. and Tukey, J.W. (1974) A projection pursuit algorithm for exploratory data analysis. *IEEE Transactions on Computers*, 23(9):881–889.

Friedman, J.H., Bentley, J.L. and Finkel, R.A. (1977) An algorithm for finding best matches in logarithmic expected time. *ACM Transactions on Mathematical Software*, 3(3):209–226.

Friedman, J.H., Stuetzle, W. and Schroeder, A. (1984) Projection pursuit density estimation. *Journal of the American Statistical Association*, 79:599–608.

Friedman, J.H., Hastie, T.J. and Tibshirani, R.J. (1998) Additive logistic regression: a statistical view of boosting. Technical report available from the authors' website: `http://www-stat.stanford.edu/\~jhf/`, Department of Statistics, Stanford University.

Friedman, N., Geiger, D. and Goldszmidt, M. (1997) Bayesian network classifiers. *Machine Learning*, 29:131–163.

Fu, K.S. (1968) *Sequential Methods in Pattern Recognition and Machine Learning*. Academic Press, New York.

Fukunaga, K. (1990) *Introduction to Statistical Pattern Recognition*, 2nd edn., Academic Press, London.

Fukunaga, K. and Flick, T.E. (1984) An optimal global nearest neighbour metric. *IEEE Transactions on Pattern Analysis and Machine Intelligence*, 6:314–318.

Fukunaga, K. and Hayes, R.R. (1989a) The reduced Parzen classifier. *IEEE Transactions on Pattern Analysis and Machine Intelligence*, 11(4):423–425.

Fukunaga, K. and Hayes, R.R. (1989b) Estimation of classifier performance. *IEEE Transactions on Pattern Analysis and Machine Intelligence*, 11(10):1087–1101.

Fukunaga, K. and Hayes, R.R. (1989c) Effects of sample size in classifier design. *IEEE Transactions on Pattern Analysis and Machine Intelligence*, 11(8):873–885.

Fukunaga, K. and Hummels, D.M. (1987a) Bias of nearest neighbor error estimates. *IEEE Transactions on Pattern Analysis and Machine Intelligence*, 9(1):103–112.

Fukunaga, K. and Hummels, D.M. (1987b) Bayes error estimation using Parzen and k-NN procedures. *IEEE Transactions on Pattern Analysis and Machine Intelligence*, 9(5):634–643.

Fukunaga, K. and Hummels, D.M. (1989) Leave-one-out procedures for nonparametric error estimates. *IEEE Transactions on Pattern Analysis and Machine Intelligence*, 11(4):421–423.

Fukunaga, K. and Kessell, D.L. (1971) Estimation of classification error. *IEEE Transactions on Computers*, 20:1521–1527.

Fukunaga, K. and Narendra, P.M. (1975) A branch and bound algorithm for computing k-nearest neighbors. *IEEE Transactions on Computers*, 24(7).

Furey, T.S., Cristianini, N., Duffy, N., Bednarski, D.W., Schummer, M. and Haussler, D. (2000) Support vector machine classification and validation of cancer tissue samples using microarray expression data. *Bioinformatics*, 16(10):906–914.

Furui, S. (1997) Recent advances in speaker recognition. *Pattern Recognition Letters*, 18:859–872.

Ganeshanandam, S. and Krzanowski, W.J. (1989) On selecting variables and assessing their performance in linear discriminant analysis. *Australian Journal of Statistics*, 31(3):433–447.

Gath, I. and Geva, A.B. (1989) Unsupervised optimal fuzzy clustering. *IEEE Transactions on Pattern Analysis and Machine Intelligence*, 11(7):773–781.

Geisser, S. (1964) Posterior odds for multivariate normal classifications. *Journal of the Royal Statistical Society Series B*, 26:69–76.

Gelfand, A.E. (2000) Gibbs sampling. *Journal of the American Statistical Association*, 95:1300–1304.

Gelfand, A.E. and Smith, A.F.M. (1990) Sampling-based approaches to calculating marginal densities. *Journal of the American Statistical Association*, 85:398–409.

Gelfand, S.B. and Delp, E.J. (1991) On tree structured classifiers. In I.K. Sethi and A.K. Jain, eds, *Artificial Neural Networks and Statistical Pattern Recognition*, pp. 51–70. North Holland, Amsterdam.

Gelman, A. (1996) Inference and monitoring convergence. In W.R. Gilks, S. Richardson and D.J. Spiegelhalter, eds, *Markov Chain Monte Carlo in Practice*, pp. 131–143. Chapman & Hall, London.

Gersho, A. and Gray, R.M. (1992) *Vector Quantization and Signal Compression*. Kluwer Academic, Dordrecht.

Giacinto, F., Roli, G. and Bruzzone, L. (2000) Combination of neural and statistical algorithms for supervised classification of remote-sensing images. *Pattern Recognition Letters*, 21:385–397.

Gifi, A. (1990) *Nonlinear Multivariate Analysis*. Wiley, Chichester.

Gilks, W.R., Richardson, S. and Spiegelhalter, D.J., eds (1996) *Markov Chain Monte Carlo in Practice*. Chapman & Hall, London.

Gini, F. (1997) Optimal multiple level decision fusion with distributed sensors. *IEEE Transaction on Aerospace and Electronic Systems*, 33(3):1037–1041.

Glendinning, R.H. (1993) Model selection for time series. Part I: a review of criterion based methods. DRA Memorandum 4730, QinetiQ, St Andrews Road, Malvern, Worcs, WR14 3PS, UK.

Golub, G.H., Heath, M. and Wahba, G. (1979) Generalised cross-validation as a method of choosing a good ridge parameter. *Technometrics*, 21:215–223.

Goodman, R.M. and Smyth, P. (1990) Decision tree design using information theory. *Knowledge Acquisition*, 2:1–19.

Gordon, A.D. (1994a) Identifying genuine clusters in a classification. *Computational Statistics and Data Analysis*, 18:561–581.

Gordon, A.D. (1994b) Clustering algorithms and cluster validity. In P. Dirschedl and R. Ostermann, eds, *Computational Statistics*, pp. 497–512. Physica-Verlag, Heidelberg.

Gordon, A.D. (1996a) Null models in cluster validation. In W. Gaul and D. Pfeifer, eds, *From Data to Knowledge: Theoretical and Practical Aspects of Classification, Data Analysis and Knowledge Organization*, pp. 32–44. Springer-Verlag, Berlin.

Gordon, A.D. (1996b) A survey of constrained classification. *Computational Statistics and Data Analysis*, 21:17–29.

Gordon, A.D. (1999) *Classification*, 2nd edn., Chapman & Hall/CRC, Boca Raton, FL.

Gordon, A.D. and Henderson, J.T. (1977) An algorithm for Euclidean sum of squares classification. *Biometrics*, 33:355–362.

Gower, J.C. (1971) A general coefficient of similarity and some of its properties. *Biometrics*, 27:857–874.

Grayson, D.A. (1987) Statistical diagnosis and the influence of diagnostic error. *Biometrics*, 43:975–984.

Green, P.J. and Silverman, B.W. (1994) *Nonlinear Regression and Generalized Linear Models. A Roughness Penalty Approach*. Chapman & Hall, London.

Greene, T. and Rayens, W. (1989) Partially pooled covariance estimation in discriminant analysis. *Communications in Statistics*, 18(10):3679–3702.

Gu, C. and Qiu, C. (1993) Smoothing spline density estimation: theory. *Annals of Statistics*, 21(1):217–234.

Guénoche, A., Hansen, P. and Jaumard, B. (1991) Efficient algorithms for divisive hierarchical clustering with the diameter criterion. *Journal of Classification*, 8:5–30.

Guyon, I., Makhoul, J., Schwartz, R. and Vapnik, V. (1998) What size test set gives good error rate estimates? *IEEE Transactions on Pattern Analysis and Machine Intelligence*, 20(1):52–64.

Guyon, I. and Stork, D.G. (1999) Linear discriminant and support vector classifiers. In A. Smola, P. Bartlett, B. Schölkopf and C. Schuurmans, eds, *Large Margin Classifiers*, pp. 147–169. MIT Press, Cambridge, MA.

Guyon, I., Weston, J., Barnhill, S. and Vapnik, V. (2002) Gene selection for cancer classification using support vector machines. *Machine Learning*, 46:389–422.

Hahn, G.J. (1984) Experimental design in the complex world. *Technometrics*, 26(1):19–31.

Hajian-Tilaki, K.O., Hanley, J.A., Joseph, L. and Collett, J.-P. (1997a) A comparison of parametric and nonparametric approaches to ROC analysis of quantitative diagnostic tests. *Medical Decision Making*, 17(1):94–102.

Hajian-Tilaki, K.O., Hanley, J.A., Joseph, L. and Collett, J.-P. (1997b) Extension of receiver operating characteristic analysis to data concerning multiple signal detection. *Academic Radiology*, 4(3):222–229.

Halici, U. and Ongun, G. (1996) Fingerprint classification through self-organising feature maps modified to treat uncertainties. *Proceedings of the IEEE*, 84(10):1497–5112.

Hall, P. (1992) *The Bootstrap and Edgeworth Expansion*. Springer-Verlag, New York.

Hall, P., Hu, T.-C. and Marron, J.S. (1995) Improved variable window kernel estimates of probability densities. *Annals of Statistics*, 23(1):1–10.

Hamamoto, Y., Uchimura, S., Matsuura, Y., Kanaoka, T. and Tomita, S. (1990) Evaluation of the branch and bound algorithm for feature selection. *Pattern Recognition Letters*, 11:453–456.

Hamamoto, Y., Matsuura, Y., Kanaoka, T. and Tomita, S. (1991) A note on the orthonormal discriminant vector method for feature extraction. *Pattern Recognition*, 24(7):681–684.

Hamamoto, Y., Kanaoka, T. and Tomita, S. (1993) On a theoretical comparison between the orthonormal discriminant vector method and discriminant analysis. *Pattern Recognition*, 26(12):1863–1867.

Hamamoto, Y., Fujimoto, Y. and Tomita, S. (1996) On the estimation of a covariance matrix in designing Parzen classifiers. *Pattern Recognition*, 29(10):1751–1759.

Hamamoto, Y., Uchimura, S. and Tomita, S. (1997) A bootstrap technique for nearest neighbor classifier design. *IEEE Transactions on Pattern Analysis and Machine Intelligence*, 19(1):73–79.

Hampel, F.R., Ronchetti, E.M., Rousseuw, P.J. and Stahel, W.A. (1986) *Robust Statistics. The Approach Based on Influence Functions*. Wiley, New York.

Hand, D.J. (1981a) *Discrimination and Classification*. Wiley, New York.

Hand, D.J. (1981b) Branch and bound in statistical data analysis. *The Statistician*, 30:1–13.

Hand, D.J. (1982) *Kernel Discriminant Analysis*. Research Studies Press, Letchworth.

Hand, D.J. (1986) Recent advances in error rate estimation. *Pattern Recognition Letters*, 4:335–346.

Hand, D.J. (1992) Statistical methods in diagnosis. *Statistical Methods in Medical Research*, 1(1):49–67.

Hand, D.J. (1994) Assessing classification rules. *Journal of Applied Statistics*, 21(3):3–16.

Hand, D.J. (1997) *Construction and Assessment of Classification Rules*. Wiley, Chichester.

Hand, D.J. and Batchelor, B.G. (1978) Experiments on the edited condensed nearest neighbour rule. *Information Sciences*, 14:171–180.

Hand, D.J. and Till, R.J. (2001) A simple generalisation of the area under the ROC curve for multiple class classification problems. *Machine Learning*, 45:171–186.

Hand, D.J. and Yu, K. (2001) Idiot's Bayes – not so stupid after all? *International Statistical Review*, 69(3):385–398.

Hand, D.J., Daly, F., McConway, K., Lunn, D. and Ostrowski, E. (1993) *Handbook of Data Sets*. Chapman & Hall, London.

Hand, D.J., Adams, N.M. and Kelly, M.G. (2001) Multiple classifier systems based on interpretable linear classifiers. In J. Kittler and F. Roli, eds, *Multiple Classifier Systems*, pp. 136–147. Springer-Verlag, Berlin.

Hansen, P.L. and Salamon, P. (1990) Neural network ensembles. *IEEE Transactions on Pattern Analysis and Machine Intelligence*, 12:993–1001.

Harkins, L.S., Sirel, J.M., McKay, P.J., Wylie, R.C., Titterington, D.M., and Rowan, R.M. (1994) Discriminant analysis of macrocytic red cells. *Clinical and Laboratory Haematology*, 16:225–234.

Harman, H.H. (1976) *Modern Factor Analysis*, 3rd edn., University of Chicago Press, Chicago.

Harris, C.J., Bailey, A. and Dodd, T.J. (1997) Multi-sensor data fusion in defence and aerospace. *Aeronautical Journal*, 102:229–244.

Hart, J.D. (1985) On the choice of a truncation point in Fourier series density estimation. *Journal of Statistical Computation and Simulation*, 21:95–116.

Hart, P.E. (1968) The condensed nearest neighbor rule. *IEEE Transactions on Information Theory*, 14:515–516.

Hartigan, J.A. (1975) *Clustering Algorithms*. Wiley, New York.

Hasselblad, V. (1966) Estimation of parameters for a mixture of normal distributions. *Technometrics*, 8:431–444.

Hastie, T.J. and Stuetzle, W. (1989) Principal curves. *Journal of the American Statistical Association*, 84:502–516.

Hastie, T.J. and Tibshirani, R.J. (1990) *Generalized Additive Models*. Chapman & Hall, London.

Hastie, T.J. and Tibshirani, R.J. (1996) Discriminant analysis by Gaussian mixtures. *Journal of the Royal Statistical Society Series B*, 58(1):155–176.

Hastie, T.J., Tibshirani, R.J. and Buja, A. (1994) Flexible discriminant analysis by optimal scoring. *Journal of the American Statistical Association*, 89:1255–1270.

Hastie, T.J., Buja, A. and Tibshirani, R.J. (1995) Penalized discriminant analysis. *Annals of Statistics*, 23(1):73–102.

Hastie, T.J., Tibshirani, R.J. and Friedman, J.H. (2001) *The Elements of Statistical Learning: Data Mining, Inference, and Prediction*. Springer, New York.

Hathaway, R.J. and Bezdek, J.C. (1994) NERF c-means: non-Euclidean relational fuzzy clustering. *Pattern Recognition*, 27(3):429–437.

Haykin, S. (1994) *Neural Networks. A Comprehensive Foundation*. Macmillan College Publishing, New York.

Haykin, S., Stehwien, W., Deng, C., Weber, P. and Mann, R. (1991) Classification of radar clutter in an air traffic control environment. *Proceedings of the IEEE*, 79(6):742–772.

Heckerman, D. (1999) A tutorial on learning with Bayesian networks. In M.I. Jordan, ed., *Learning in Graphical Models*, pp. 301–354. MIT Press, Cambridge, MA.

Heckerman, D., Geiger, D. and Chickering, D.M. (1995) Learning Bayesian networks: the combination of knowledge and statistical data. *Machine Learning*, 20:197–243.

Heiser, W.J. (1991) A generalized majorization method for least squares multidimensional scaling of pseudodistances that may be negative. *Psychometrika*, 56(1):7–27.

Heiser, W.J. (1994) Convergent computation by iterative majorization: theory and applications in multidimensional data analysis. In W.J. Krzanowski, ed., *Recent Advances in Descriptive Multivariate Analysis*, pp. 157–189. Clarendon Press, Oxford.

Henley, W.E. and Hand, D.J. (1996) A k-nearest-neighbour classifier for assessing consumer credit risk. *The Statistician*, 45(1):77–95.

Highleyman, W.H. (1962) The design and analysis of pattern recognition experiments. *Bell System Technical Journal*, 41:723–744.

Hills, M. (1977) Book review. *Applied Statistics*, 26:339–340.

Hinkley, D.V. (1988) Bootstrap methods. *Journal of the Royal Statistical Society Series B*, 50(3): 321–337.

Hjort, N.L. and Glad, I.K. (1995) Nonparametric density estimation with a parametric start. *Annals of Statistics*, 23(3):882–904.

Hjort, N.L. and Jones, M.C. (1996) Locally parametric nonparametric density estimation. *Annals of Statistics*, 24(4):1619–1647.

Ho, Y.-C. and Agrawala, A.K. (1968) On pattern classification algorithms. Introduction and survey. *Proceedings of the IEEE*, 56(12):2101–2114.

Holmes, C.C. and Mallick, B.K. (1998) Bayesian radial basis functions of variable dimension. *Neural Computation*, 10:1217–1233.

Holmström, L. and Koistinen, P. (1992) Using additive noise in back-propagation training. *IEEE Transactions on Neural Networks*, 3(1):24–38.

Holmström, L., Koistinen, P., Laaksonen, J. and Oja, E. (1997) Neural and statistical classifiers – taxonomy and two case studies. *IEEE Transactions on Neural Networks*, 8(1):5–17.

Holmström, L. and Sain, S.R. (1997) Multivariate discrimination methods for top quark analysis. *Technometrics*, 39(1):91–99.

Hong, Z.-Q. and Yang, J.-Y. (1991) Optimal discriminant plane for a small number of samples and design method of classifier on the plane. *Pattern Recognition*, 24(4):317–324.

Hornik, K. (1993) Some new results on neural network approximation. *Neural Networks*, 6:1069–1072.

Hotelling, H. (1933) Analysis of a complex of statistical variables into principal components. *Journal of Educational Psychology*, 24:417–444.

Hsu, C.W. and Lin, C.J. (2002) A comparison on methods for multi-class support vector machines. *IEEE Transactions on Neural Networks*. To appear.

Hua, S. and Sun, Z. (2001) A novel method of protein secondary structure prediction with high segment overlap measure: support vector machine approach. *Journal of Molecular Biology*, 308:397–407.

Huang, Y.S. and Suen, C.Y. (1995) A method of combining multiple experts for the recognition of unconstrained handwritten numerals. *IEEE Transactions on Pattern Analysis and Machine Intelligence*, 17(1):90–94.

Huber, P.J. (1985) Projection pursuit (with discussion). *Annals of Statistics*, 13(2):435–452.

Hush, D.R., Horne, W. and Salas, J.M. (1992) Error surfaces for multilayer perceptrons. *IEEE Transactions on Systems, Man, and Cybernetics*, 22(5):1151–1161.

Huth, R., Nemešová, I. and Klimperová, N. (1993) Weather categorization based on the average linkage clustering technique: an application to European mid-latitudes. *International Journal of Climatology*, 13:817–835.

Hwang, J.-N., Lay, S.-R. and Lippman, A. (1994a) Nonparametric multivariate density estimation: a comparative study. *IEEE Transactions on Signal Processing*, 42(10):2795–2810.

Hwang, J.-N., Lay, S.-R., Maechler, M., Martin, D. and Schimert, J. (1994b) Regression modeling in back-propagation and projection pursuit learning. *IEEE Transactions on Neural Networks*, 5(3):342–353.

Hyvärinen, A. and Oja, E. (2000) Independent component analysis: algorithms and applications. *Neural Networks*, 13:411–430.

Ifarraguerri, A. and Chang, C.-I. (2000) Unsupervised hyperspectral image analysis with projection pursuit. *IEEE Transactions on Geoscience and Remote Sensing*, 38(6):2529–2538.

Ingrassia, S. (1992) A comparison between the simulated annealing and the EM algorithms in normal mixture decompositions. *Statistics and Computing*, 2:203–211.

Ismail, M.A. (1988) Soft clustering: algorithms and validity of solutions. In M.M. Gupta and T. Yamakawa, eds, *Fuzzy Computing*, pp. 445–472. Elsevier Science (North Holland), Amsterdam.

Izenman, A.J. (1991) Recent developments in nonparametric density estimation. *Journal of the American Statistical Association*, 86:205–223.

Izenman, A.J. and Sommer, C.J. (1988) Philatelic mixtures and multimodal densities. *Journal of the American Statistical Association*, 83:941–953.

Jackson, D.A. (1993) Stopping rules in principal components analysis: a comparison of heuristical and statistical approaches. *Ecology*, 74(8):2204–2214.

Jackson, J.E. (1991) *A User's Guide to Principal Components*. Wiley, New York.

Jacobs, R.A., Jordan, M.I., Nowlan, S.J. and Hinton, G.E. (1991) Adaptive mixtures of local experts. *Neural Computation*, 3:79–87.

Jain, A.K. and Chandrasekaran, B. (1982) Dimensionality and sample size considerations in pattern recognition practice. In P.R. Krishnaiah and L.N. Kanal, eds, *Handbook of Statistics*, pp. 835–855. North Holland, Amsterdam.

Jain, A.K. and Dubes, R.C. (1988) *Algorithms for Clustering Data*. Prentice Hall, London.

Jain, A.K. and Moreau, J.V. (1987) Bootstrap technique in cluster analysis. *Pattern Recognition*, 20(5):547–568.

Jain, A.K., Duin, R.P.W. and Mao, J. (2000) Statistical pattern recognition: A review. *IEEE Transactions on Pattern Analysis and Machine Intelligence*, 22(1):4–37.

Jain, S. and Jain, R.K. (1994) Discriminant analysis and its application to medical research. *Biomedical Journal*, 36(2):147–151.

Jambu, M. and Lebeaux, M.-O. (1983) *Cluster Analysis and Data Analysis*. North Holland, Amsterdam.

James, G.M. and Hastie, T.J. (2001) Functional linear discriminant analysis for irregularly sampled curves. *Journal of the Royal Statistical Society Series B*, 63(3):533–550.

Jamshidian, M. and Jennrich, R.I. (1993) Conjugate gradient acceleration of the EM algorithm. *Journal of the American Statistical Association*, 88:221–228.

Jamshidian, M. and Jennrich, R.I. (1997) Acceleration of the EM algorithm by using quasi-Newton methods. *Journal of the Royal Statistical Society Series B*, 59(3):569–587.

Jansen, R.C. and Den Nijs, A.P.M. (1993) A statistical mixture model for estimating the proportion of unreduced pollen grains in perennial ryegrass (*Lolium perenne* l.) via the size of pollen grains. *Euphytica*, 70:205–215.

Jardine, N. (1971) In discussion on Cormack's paper: A review of classification. *Journal of the Royal Statistical Society Series A*, 134:321–367.

Jardine, N. and Sibson, R. (1971) *Mathematical Taxonomy*. Wiley, London.

Jensen, F.V. (1996) *An Introduction to Bayesian Networks*. UCL Press, London.

Jeon, B. and Landgrebe, D.A. (1994) Fast Parzen density estimation using clustering-based branch and bound. *IEEE Transactions on Pattern Analysis and Machine Intelligence*, 16(9):950–954.

Jiang, Q. and Zhang, W. (1993) An improved method for finding nearest neighbours. *Pattern Recognition Letters*, 14:531–535.

Jobert, M., Escola, H., Poiseau, E. and Gaillard, P. (1994) Automatic analysis of sleep using two parameters based on principal component analysis of electroencephalography spectral data. *Biological Cybernetics*, 71:197–207.

Jolliffe, I.T. (1986) *Principal Components Analysis*. Springer-Verlag, New York.

Jones, M.C. and Lotwick, H.W. (1984) A remark on algorithm AS176. Kernel density estimation using the fast Fourier transform. *Applied Statistics*, 33:120–122.

Jones, M.C. and Sibson, R. (1987) What is projection pursuit? (with discussion). *Journal of the Royal Statistical Society Series A*, 150:1–36.

Jones, M.C. and Signorini, D.F. (1997) A comparison of higher order bias kernel density estimators. *Journal of the American Statistical Association*, 92:1063–1073.

Jones, M.C. and Sheather, S.J. (1991) Using non-stochastic terms to advantage in kernel-based estimation of integrated squared density derivatives. *Statistics and Probability Letters*, 11: 511–514.

Jones, M.C., McKay, I.J. and Hu, T.-C. (1994) Variable location and scale kernel density estimation. *Annals of the Institute of Statistical Mathematics*, 46(3):521–535.

Jones, M.C., Marron, J.S. and Sheather, S.J. (1996) A brief survey of bandwidth selection for density estimation. *Journal of the American Statistical Association*, 91:401–407.

Jordan, M.I. and Jacobs, R.A. (1994) Hierarchical mixtures of experts and the EM algorithm. *Neural Computation*, 6:181–214.

Jöreskog, K.G. (1977) Factor analysis by least-squares and maximum-likelihood methods. In K. Enslein, A. Ralston and H.S. Wilf, eds, *Statistical Methods for Digital Computers*, pp. 125–153. Wiley Interscience, New York.

Juan, A. and Vidal, E. (1994) Fast *k*-means-like clustering in metric spaces. *Pattern Recognition Letters*, 15:19–25.

Juang, B.-H. and Rabiner, L.R. (1985) Mixture autoregressive hidden Markov models for speech signals. *IEEE Transactions on Acoustics, Speech and Signal Processing*, 33(6):1404–1413.

Kaiser, H.F. (1958) The varimax criterion for analytic rotation in factor analysis. *Psychometrika*, 23:187–200.

Kaiser, H.F. (1959) Computer program for varimax rotation in factor analysis. *Educational and Psychological Measurement*, 19:413–420.

Kalayeh, H.M. and Landgrebe, D.A. (1983) Predicting the required number of training samples. *IEEE Transactions on Pattern Analysis and Machine Intelligence*, 5(6):664–667.

Kam, M., Zau, Q. and Gray, W.S. (1992) Optimal data fusion of correlated local decisions in multiple sensor detection systems. *IEEE Transactions on Aerospace and Electronic Systems*, 28(3):916–920.

Kam, M., Rorres, C., Chang, W. and Zhu, X. (1999) Performance and geometric interpretation for decision fusion with memory. *IEEE Transactions on Systems, Man, and Cybernetics. Part A: Systems and Humans*, 29(1):52–62.

Kamel, M.S. and Selim, S.Z. (1991) A thresholded fuzzy *c*-means algorithm for semi-fuzzy clustering. *Pattern Recognition*, 24(9):825–833.

Kamel, M.S. and Selim, S.Z. (1994) New algorithms for solving the fuzzy clustering problem. *Pattern Recognition*, 27(3):421–428.

Karayiannis, N.B. and Venetsanopolous, A.N. (1993) Efficient learning algorithms for neural networks (ELEANNE). *IEEE Transactions on Systems, Man, and Cybernetics*, 23(5):1372–1383.

Karayiannis, N.B. and Wi, G.M. (1997) Growing radial basis neural networks: merging supervised and unsupervised learning with network growth techniques. *IEEE Transactions on Neural Networks*, 8(6):1492–1506.

Kashyap, R.L. (1970) Algorithms for pattern classification. In J.M. Mendel and K.S. Fu, eds, *Adaptive, Learning and Pattern Recognition Systems. Theory and Applications*, pp. 81–113. Academic Press, New York.

Kay, J.W. (1997) Comments on paper by Esposito *et al.* *IEEE Transactions on Pattern Analysis and Machine Intelligence*, 19(5):492–493.

Kendall, M. (1975) *Multivariate Analysis*. Griffin, London.

Kirkwood, C.A., Andrews, B.J. and Mowforth, P. (1989) Automatic detection of gait events: a case study using inductive learning techniques. *Journal of Biomedical Engineering*, 11:511–516.

Kittler, J. (1975a) A nonlinear distance metric criterion for feature selection in the measurement space. *Information Sciences*, 9:359–363.

Kittler, J. (1975b) Mathematical methods of feature selection in pattern recognition. *International Journal of Man–Machine Studies*, 7:609–637.

Kittler, J. (1978a) Une généralisation de quelques algorithmes sous-optimaux de recherche d'ensembles d'attributs. In *Proc. Congrès AFCET/IRIA Reconnaissance des Formes et Traitement des Images*, pp. 678–686.

Kittler, J. (1978b) Feature set search algorithms. In C.H. Chen, ed., *Pattern Recognition and Signal Processing*, pp. 41–60. Sijthoff and Noordhoff, Alphen aan den Rijn, Netherlands.

Kittler, J. (1986) Feature selection and extraction. In T.Y. Young and K.S. Fu, eds, *Handbook of Pattern Recognition and Image Processing*, pp. 59-83. Academic Press, London.

Kittler, J. and Alkoot, F.M. (2001) Relationship of sum and vote fusion strategies. In J. Kittler and F. Roli, eds, *Multiple Classifier SystemsLecture Notes in Computer Science 2096*, pp. 339–348. Springer-Verlag, Berlin.

Kittler, J. and Young, P.C. (1973) A new approach to feature selection based on the Karhunen–Loève expansion. *Pattern Recognition*, 5:335–352.

Kittler, J., Matas, J., Jonsson, K. and Ramos Sánchez, M.U. (1997) Combining evidence in personal identity verification systems. *Pattern Recognition Letters*, 18:845–852.

Kittler, J., Hatef, M., Duin, R.P.W. and Matas, J. (1998) On combining classifiers. *IEEE Transactions on Pattern Analysis and Machine Intelligence*, 20(3):226–239.

Kohonen, T. (1989) *Self-organization and Associative Memory*, 3rd edn., Springer-Verlag, Berlin.

Kohonen, T. (1990) The self-organizing map. *Proceedings of the IEEE*, 78:1464–1480.

Kohonen, T. (1997) *Self-organizing Maps*, 2nd edn., Springer-Verlag, Berlin.

Kohonen, T., Oja, E., Simula, O., Visa, A. and Kangas, J. (1996) Engineering applications of the self-organising map. *Proceedings of the IEEE*, 84(10):1358–1384.

Konishi, S. and Honda, M. (1990) Comparison of procedures for estimation of error rates in discriminant analysis under nonnormal populations. *Journal of Statistical Computation and Simulation*, 36:105–115.

Koontz, W.L.G. and Fukunaga, K. (1972) A nonlinear feature extraction algorithm using distance transformation. *IEEE Transactions on Computers*, 21(1):56–63.

Koontz, W.L.G., Narendra, P.M. and Fukunaga, K. (1975) A branch and bound clustering algorithm. *IEEE Transactions on Computers*, 24(9):908–915.

Kraaijveld, M.A. (1996) A Parzen classifier with an improved robustness against deviations between training and test data. *Pattern Recognition Letters*, 17:679–689.

Kraaijveld, M.A., Mao, J. and Jain, A.K. (1992) A non-linear projection method based on Kohonen's topology preserving maps. In *Proceedings of the 11th IAPR International Conference on Pattern Recognition*, The Hague. IEEE Computer Society Press, Los Alamitos, CA.

Kreithen, D.E., Halversen, S.D. and Owirka, G.J. (1993) Discriminating targets from clutter. *Lincoln Laboratory Journal*, 6(1):25–51.

Kronmal, R.A. and Tarter, M. (1962) The estimation of probability densities and cumulatives by Fourier series methods. *Journal of the American Statistical Association*, 63:925–952.

Krusińska, E. (1988) Robust methods in discriminant analysis. *Rivista di Statistica Applicada*, 21(3):239–253.

Kruskal, J.B. (1964a) Multidimensional scaling by optimizing goodness-of-fit to a nonmetric hypothesis. *Psychometrika*, 29:1–28.

Kruskal, J.B. (1964b) Nonmetric multidimensional scaling: a numerical method. *Psychometrika*, 29(2):115–129.

Kruskal, J.B. (1971) Comments on 'A nonlinear mapping for data structure analysis'. *IEEE Transactions on Computers*, 20:1614.

Kruskal, J.B. (1972) Linear transformation of multivariate data to reveal clustering. In R.N. Shepard, A.K. Romney and S.B. Nerlove, eds, *Multidimensional Scaling: Theory and Applications in the Behavioural Sciences.*, vol. 1, pp. 179–191. Seminar Press, London.

Krzanowski, W.J. (1993) The location model for mixtures of categorical and continuous variables. *Journal of Classification*, 10(1):25–49.

Krzanowski, W.J. and Marriott, F.H.C. (1994) *Multivariate Analysis. Part 1: Distributions, Ordination and Inference.* Edward Arnold, London.

Krzanowski, W.J. and Marriott, F.H.C. (1996) *Multivariate Analysis. Part 2: Classification, Covariance Structures and Repeated Measurements.* Edward Arnold, London.

Krzanowski, W.J., Jonathan, P., McCarthy, W.V. and Thomas, M.R. (1995) Discriminant analysis with singular covariance matrices: methods and applications to spectroscopic data. *Applied Statistics*, 44(1):101–115.

Krzyżak, A. (1983) Classification procedures using multivariate variable kernel density estimate. *Pattern Recognition Letters*, 1:293–298.

Kudo, M. and Sklansky, J. (2000) Comparison of algorithms that select features for pattern classifiers. *Pattern Recognition*, 33:25–41.

Kwoh, C.-K. and Gillies, D.F. (1996) Using hidden nodes in Bayesian networks. *Artificial Intelligence*, 88:1–38.

Kwok, T.-Y. and Yeung, D.-Y. (1996) Use of bias term in projection pursuit learning improves approximation and convergence properties. *IEEE Transactions on Neural Networks*, 7(5):1168–1183.

Lachenbruch, P.A. (1966) Discriminant analysis when the initial samples are misclassified. *Technometrics*, 8:657–662.

Lachenbruch, P.A. and Mickey, M.R. (1968) Estimation of error rates in discriminant analysis. *Technometrics*, 10:1–11.

Lachenbruch, P.A., Sneeringer, C. and Revo, L.T. (1973) Robustness of the linear and quadratic discriminant function to certain types of non-normality. *Communications in Statistics*, 1(1):39–56.

Lam, L. and Suen, C.Y. (1995) Optimal combinations of pattern classifiers. *Pattern Recognition Letters*, 16:945–954.

Lampinen, J. and Vehtari, A. (2001) Bayesian approach for neural networks – review and case studies. *Neural Networks*, 14:257–274.

Lange, K. (1995) A gradient algorithm locally equivalent to the EM algorithm. *Journal of the Royal Statistical Society Series B*, 57(2):425–437.

Lauritzen, S.L. and Spiegelhalter, D.J. (1988) Local computations with probabilities on graphical structures and their application to expert systems (with discussion). *Journal of the Royal Statistical Society Series B*, 50:157–224.

Lavine, M. and West, M. (1992) A Bayesian method for classification and discrimination. *Canadian Journal of Statistics*, 20(4):451–461.

Laviolette, M., Seaman, J.W., Barrett, J.D. and Woodall, W.H. (1995) A probabilistic and statistical view of fuzzy methods (with discussion). *Technometrics*, 37(3):249–292.

Lawley, D.N. and Maxwell, A.E. (1971) *Factor Analysis as a Statistical Method*, 2nd edn., Butterworths, London.

Lee, J.S., Grunes, M.R. and Kwok, R. (1994) Classification of multi-look polarimetric SAR imagery based on complex Wishart distribution. *International Journal of Remote Sensing*, 15(11):2299–2311.

Lerner, B., Guterman, H., Aladjem, M. and Dinstein, I. (1999) A comparative study of neural network based feature extraction paradigms. *Pattern Recognition Letters*, 20:7–14.

Leshno, M., Lin, V.Y., Pinkus, A. and Schocken, S. (1993) Multilayer feedforward networks with a nonpolynomial activation function can approximate any function. *Neural Networks*, 6:861–867.

Li, C., Goldgof, D.B. and Hall, L.O. (1993) Knowledge-based classification and tissue labeling of MR images of human brain. *IEEE Transactions on Medical Imaging*, 12(4):740–750.

Li, T. and Sethi, I.K. (1993) Optimal multiple level decision fusion with distributed sensors. *IEEE Transaction on Aerospace and Electronic Systems*, 29(4):1252–1259.

Li, X. and Yeh, A.G.O. (1998) Principal component analysis of stacked multi-temporal images for the monitoring of rapid urban expansion in the Pearl River Delta. *International Journal of Remote Sensing*, 19(8):1501–1518.

Lin, Y., Lee, Y. and Wahba, G. (2002) Support vector machines for classification in nonstandard situations. *Machine Learning*, 46(1–3):191–202.

Linde, Y., Buzo, A. and Gray, R.M. (1980) An algorithm for vector quantizer design. *IEEE Transactions on Communications*, 28(1):84–95.

Lindgren, B.W. (1976) *Statistical Theory*, 3rd edn., Macmillan, New York.

Lindsay, B.G. and Basak, P. (1993) Multivariate normal mixtures: a fast consistent method of moments. *Journal of the American Statistical Association*, 88:468–476.

Ling, R.F. (1973) A probability theory for cluster analysis. *Journal of the American Statistical Association*, 68:159–164.

Lingoes, J.C., Roskam, E.E. and Borg, I., eds (1979) *Geometric Representations of Relational Data.* Mathesis Press, Ann Arbor, MI.

Little, R.J.A. (1992) Regression with missing X's: a review. *Journal of the American Statistical Association*, 87:1227–1237.

Little, R.J.A. and Rubin, D.B. (1987) *Statistical Analysis with Missing Data.* Wiley, New York.

Liu, C. and Rubin, D.B. (1994) The ECME algorithm: a simple extension of EM and ECM with faster monotone convergence. *Biometrika*, 81(4):633–648.

Liu, H.-L., Chen, N.-Y., Lu, W.-C. and Zhu, X.-W. (1994) Multi-target classification pattern recognition applied to computer-aided materials design. *Analytical Letters*, 27(11):2195–2203.

Liu, J.N.K., Li, B.N.L. and Dillon, T.S. (2001) An improved naïve Bayesian classifier technique coupled with a novel input solution method. *IEEE Transactions on Systems, Man, and Cybernetics – Part C: Applications and Reviews*, 31(2):249–256.

Liu, K., Cheng, Y.Q. and Yang, J.-Y. (1992) A generalized optimal set of discriminant vectors. *Pattern Recognition*, 25(7):731–739.

Liu, K., Cheng, Y.Q. and Yang, J.-Y. (1993) Algebraic feature extraction for image recognition based on an optimal discriminant criterion. *Pattern Recognition*, 26(6):903–911.

Liu, W.Z. and White, A.P. (1995) A comparison of nearest neighbour and tree-based methods of non-parametric discriminant analysis. *Journal of Statistical Computation and Simulation*, 53:41–50.

Liu, W.Z., White, A.P., Thompson, S.G. and Bramer, M.A. (1997) Techniques for dealing with missing values in classification. In X. Liu, P. Cohen and M. Berthold, eds, *Advances in Intelligent Data Analysis.* Springer-Verlag, Berlin.

Logar, A.M., Corwin, E.M. and Oldham, W.J.B. (1994) Performance comparisons of classification techniques for multi-font character recognition. *International Journal of Human–Computer Studies*, 40:403–423.

Loh, W.-L. (1995) On linear discriminant analysis with adaptive ridge classification rules. *Journal of Multivariate Analysis*, 53:264–278.

Loh, W.-Y. and Vanichsetakul, N. (1988) Tree-structured classification via generalized discriminant analysis (with discussion). *Journal of the American Statistical Association*, 83:715–727.

Lowe, D. (1993) Novel 'topographic' nonlinear feature extraction using radial basis functions for concentration coding on the 'artificial nose'. In *3rd IEE International Conference on Artificial Neural Networks*, pp. 95–99, IEE, London.

Lowe, D., ed. (1994) Special issue on 'Applications of Artificial Neural Networks'. *IEE Proceedings on Vision, Image and Signal Processing*, 141(4).

Lowe, D. (1995a) Radial basis function networks. In M.A. Arbib, ed., *The Handbook of Brain Theory and Neural Networks*, pp. 779–782. MIT Press, Cambridge, MA.

Lowe, D. (1995b) On the use of nonlocal and non positive definite basis functions in radial basis function networks. In *Proceedings of the 4th International Conference on Artificial Neural Networks*, Cambridge, pp. 206–211, IEE, London.

Lowe, D. and Tipping, M. (1996) Feed-forward neural networks and topographic mappings for exploratory data analysis. *Neural Computing and Applications*, 4:83–95.

Lowe, D. and Webb, A.R. (1990) Exploiting prior knowledge in network optimization: an illustration from medical prognosis. *Network*, 1:299–323.

Lowe, D. and Webb, A.R. (1991) Optimized feature extraction and the Bayes decision in feed-forward classifier networks. *IEEE Transactions on Pattern Analysis and Machine Intelligence*, 13(4):355–364.

Lugosi, G. (1992) Learning with an unreliable teacher. *Pattern Recognition*, 25(1):79–87.

Lunn, D.J., Thomas, A., Best, N. and Spiegelhalter, D.J. (2000) WinBugs – a Bayesian modelling framework: concepts, structure and extensibility. *Statistics and Computing*, 10(4):325–337.

Luttrell, S.P. (1989) Hierarchical vector quantisation. *IEE Proceedings Part I*, 136(6):405–413.

Luttrell, S.P. (1994) Partitioned mixture distribution: an adaptive Bayesian network for low-level image processing. *IEE Proceedings on Vision, Image and Signal Processing*, 141(4):251–260.

Luttrell, S.P. (1995) Using self-organising maps to classify radar range profiles. In *Proceedings of the 4th International Conference on Artificial Neural Networks*, Cambridge, pp. 335–340, IEE, London.

Luttrell, S.P. (1997) A theory of self-organising neural networks. In S.W. Ellacott, J.C. Mason and I.J. Anderson, eds, *Mathematics of Neural Networks: Models, Algorithms and Applications*, pp. 240–244. Kluwer Academic Publishers, Dordrecht.

Luttrell, S.P. (1999a) Self-organised modular neural networks for encoding data. In A.J.C. Sharkey, ed., *Combining Artificial Neural Nets: Ensemble and Modular Multi-net Systems*, pp. 235–263. Springer-Verlag, London.

Luttrell, S.P. (1999b) An adaptive network for encoding data using piecewise linear functions. In *Proceedings of the 9th International Conference on Artificial Neural Networks (ICANN99)*, pp. 198–203.

Luttrell, S.P. (2002). Using stochastic vectors quantizers to characterize signal and noise subspaces. In J.G. McWhirter and I.K. Proudler, eds, *Mathematics in Signal Processing*. Oxford University Press, Oxford.

MacKay, D.J.C. (1995) Probable networks and plausible predictions – a review of practical Bayesian methods for supervised neural networks. *Network: Computation in Neural Systems*, 6:469–505.

Magee, M., Weniger, R. and Wenzel, D. (1993) Multidimensional pattern classification of bottles using diffuse and specular illumination. *Pattern Recognition*, 26(11):1639–1654.

Makhoul, J., Roucos, S. and Gish, H. (1985) Vector quantization in speech coding. *Proceedings of the IEEE*, 73(11):1511–1588.

Manly, B.F.J. (1986) *Multivariate Statistical Methods, a Primer*. Chapman & Hall, London.

Mao, J. and Jain, A.K. (1995) Artificial neural networks for feature extraction and multivariate data projection. *IEEE Transactions on Neural Networks*, 6(2):296–317.

Marinaro, M. and Scarpetta, S. (2000) On-line learning in RBF neural networks: a stochastic approach. *Neural Networks*, 13:719–729.

Marron, J.S. (1988) Automatic smoothing parameter selection: a survey. *Empirical Economics*, 13:187–208.

Matsuoka, K. (1992) Noise injection into inputs in back-propagation learning. *IEEE Transactions on Systems, Man, and Cybernetics*, 22(3):436–440.

McIntyre, R.M. and Blashfield, R.K. (1980) A nearest-centroid technique for evaluating the minimum-variance clustering procedure. *Multivariate Behavioral Research*, 2:225–238.

McKenna, S.J., Gong, S. and Raja, Y. (1998) Modelling facial colour and identity with Gaussian mixtures. *Pattern Recognition*, 31(12):1883–1892.

McLachlan, G.J. (1987) Error rate estimation in discriminant analysis: recent advances. In A.K. Gupta, ed., *Advances in Multivariate Analysis*. D. Reidel, Dordrecht.

McLachlan, G.J. (1992a) *Discriminant Analysis and Statistical Pattern Recognition*. Wiley, New York.

McLachlan, G.J. (1992b) Cluster analysis and related techniques in medical research. *Statistical Methods in Medical Research*, 1(1):27–48.

McLachlan, G.J. and Basford, K.E. (1988) *Mixture Models: Inference and Applications to Clustering*. Marcel Dekker, New York.

McLachlan, G.J. and Krishnan, T. (1996) *The EM Algorithm and Extensions*. Wiley, New York.

McLachlan, G.J. and Peel, D. (2000) *Finite Mixture Models*. Wiley, New York.

Meier, W., Weber, R. and Zimmermann, H.-J. (1994) Fuzzy data analysis – methods and industrial applications. *Fuzzy Sets and Systems*, 61:19–28.

Meinguet, J. Multivariate interpolation at arbitrary points made simple. *Zeitschrift Für Angewandte Mathematik and Physik*, 30:292–304.

Meng, X.-L. and Rubin, D.B. (1992) Recent extensions to the EM algorithm. In J.M. Bernado, J.O. Berger, A.P. Dawid and A.F.M. Smith, eds, *Bayesian Statistics 4*, pp. 307–320. Oxford University Press, Oxford.

Meng, X.-L. and Rubin, D.B. (1993) Maximum likelihood estimation via the ECM algorithm: a general framework. *Biometrika*, 80(2):267–27.

Meng, X.-L. and van Dyk, D. (1997) The EM algorithm – an old folk-song sung to a fast new tune (with discussion). *Journal of the Royal Statistical Society Series B*, 59(3):511–567.

Mengersen, K.L., Robert, C.P. and Guihenneuc-Jouyaux, C. (1999) MCMC convergence diagnostics: a review*ww*. In J.M. Bernardo, J.O. Berger, A.P. Dawid and A.F.M. Smith, eds, *Bayesian Statistics 6*, pp. 399–432. Oxford University Press, Oxford.

Merritt, D. and Tremblay, B. (1994) Nonparametric estimation of density profiles. *Astronomical Journal*, 108(2):514–537.

Merz, C.L. (1999) Using correspondence analysis to combine classifiers. *Machine Learning*, 36:33–58.

Micchelli, C.A. (1986) Interpolation of scattered data: distance matrices and conditionally positive definite matrices. *Constructive Approximation*, 2:11–22.

Michael, M. and Lin, W.-C. (1973) Experimental study of information measure and inter-intra class distance ratios on feature selection and orderings. *IEEE Transactions on Systems, Man, and Cybernetics*, 3(2):172–181.

Michalek, J.E. and Tripathi, R.C. (1980) The effect of errors in diagnosis and measurement on the estimation of the probability of an event. *Journal of the American Statistical Association*, 75:713–721.

Michalewicz, Z. (1994) Non-standard methods in evolutionary computation. *Statistics and Computing*, 4:141–155.

Michie, D., Spiegelhalter, D.J. and Taylor, C.C. (1994) *Machine Learning, Neural and Statistical Classification.* Ellis Horwood Limited, Hemel Hempstead.

Micó, M.L., Oncina, J. and Vidal, E. (1994) A new version of the nearest-neighbour approximating and eliminating search algorithm (AESA) with linear preprocessing time and memory requirements. *Pattern Recognition*, 15:9–17.

Miller, A.J. (1990) *Subset Selection in Regression*. Chapman & Hall, London.

Milligan, G.W. (1981) A Monte Carlo study of thirty internal measures for cluster analysis. *Psychometrika*, 46(2):187–199.

Milligan, G.W. and Cooper, M.C. (1985) An examination of procedures for determining the number of clusters in a data set. *Psychometrika*, 50(2):159–179.

Mingers, J. (1989) An empirical comparison of pruning methods for decision tree inductions. *Machine Learning*, 4:227–243.

Minsky, M.L. and Papert, S.A. (1988) *Perceptrons. An Introduction to Computational Geometry.* MIT Press, Cambridge, MA.

Mitchell, T.M. (1997) *Machine Learning*. McGraw-Hill, New York.

Mkhadri, A. (1995) Shrinkage parameter for the modified linear discriminant analysis. *Pattern Recognition Letters*, 16:267–275.

Mkhadri, A., Celeux, G. and Nasroallah, A. (1997) Regularization in discriminant analysis: an overview. *Computational Statistics and Data Analysis*, 23:403–423.

Mojena, R. (1977) Hierarchical grouping methods and stopping rules: an evaluation. *Computer Journal*, 20:359–363.

Mojirsheibani, M. (1999) Combining classifiers via discretization. *Journal of the American Statistical Association*, 94(446):600–609.

Mola, F. and Siciliano, R. (1997) A fast splitting procedure for classification trees. *Statistics and Computing*, 7:209–216.

Moran, M.A. and Murphy, B.J. (1979) A closer look at two alternative methods of statistical discrimination. *Applied Statistics*, 28(3):223–232.

Morgan, B.J.T. and Ray, A.P.G. (1995) Non-uniqueness and inversions in cluster analysis. *Applied Statistics*, 44(1):117–134.

Morgan, N. and Bourlard, H.A. (1995) Neural networks for the statistical recognition of continuous speech. *Proceedings of the IEEE*, 83(5):742–772.

Mukherjee, D.P., Banerjee, D.K., Uma Shankar, B. and Majumder, D.D. (1994) Coal petrography: a pattern recognition approach. *International Journal of Coal Geology*, 25:155–169.

Mukherjee, D.P., Pal, A., Sarma, S.E. and Majumder, D.D. (1995) Water quality analysis: a pattern recognition approach. *Pattern Recognition*, 28(2):269–281.

Munro, D.J., Ersoy, O.K., Bell, M.R. and Sadowsky, J.S. (1996) Neural network learning of low-probability events. *IEEE Transactions on Aerospace and Electronic Systems*, 32(3):898–910.

Murphy, P.M. and Aha, D.W. (1995) UCI repository of machine learning databases. Technical Report University of California, Irvine. `http://www.ics.uci.edu/\~mlearn/MLRepository.html`.

Murtagh, F. (1985) *Multidimensional Clustering Algorithms*. Physica-Verlag, Vienna.

Murtagh, F. (1992) Contiguity-constrained clustering for image analysis. *Pattern Recognition Letters*, 13:677–683.

Murtagh, F. (1994) Classification: astronomical and mathematical overview. In H.T. MacGillivray *et al.*, ed., *Astronomy from Wide-Field Imaging*, pp. 227–233, Kluwer Academic Publishers, Dordrecht.

Murtagh, F. (1995) Interpreting the Kohonen self-organizing feature map using contiguity-constrained clustering. *Pattern Recognition Letters*, 16:399–408.

Murtagh, F. and Hernández-Pajares, M. (1995) The Kohonen self-organizing map method: an assessment. *Journal of Classification*, 12(2):165–190.

Musavi, M.T., Ahmed, W., Chan, K.H., Faris, K.B. and Hummels, D.M. (1992) On the training of radial basis function classifiers. *Neural Networks*, 5:595–603.

Myles, J.P. and Hand, D.J. (1990) The multi-class metric problem in nearest neighbour discrimination rules. *Pattern Recognition*, 23(11):1291–1297.

Nadaraya, E.A. (1989) *Nonparametric Estimation of Probability Densities and Regression Curves*. Kluwer Academic Publishers, Dordrecht.

Narendra, P.M. and Fukunaga, K. (1977) A branch and bound algorithm for feature subset selection. *IEEE Transactions on Computers*, 26:917–922.

Neapolitan, R.E. (1990) *Probabilistic Reasoning in Expert Systems: Theory and Algorithms*. Wiley, New York.

Nie, N.H., Hull, C.H., Jenkins, J.G., Steinbrenner, K. and Brent, D.H. (1975) *SPSS: Statistical Package for the Social Sciences*, 2nd edn., McGraw-Hill, New York.

Niemann, H. and Weiss, J. (1979) A fast-converging algorithm for nonlinear mapping of high dimensional data to a plane. *IEEE Transactions on Computers*, 28(2):142–147.

Nilsson, N.J. (1965) *Learning Machines: Foundations of Trainable Pattern-Classifying Systems*. McGraw-Hill, New York.

Novovičová, J., Pudil, P. and Kittler, J. (1996) Divergence based feature selection for multimodal class densities. *IEEE Transactions on Pattern Analysis and Machine Intelligence*, 18(2):218–223.

O'Hagan, A. (1994) *Bayesian Inference*. Edward Arnold, London.

Okada, T. and Tomita, S. (1985) An optimal orthonormal system for discriminant analysis. *Pattern Recognition*, 18(2):139–144.

Oliver, J.J. and Hand, D.J. (1996) Averaging over decision trees. *Journal of Classification*, 13(2):281–297.

Olkin, I. and Tate, R.F. (1961) Multivariate correlation models with mixed discrete and continuous variables. *Annals of Mathematical Statistics*, 22:92–96.

O'Neill, T.J. (1992) Error rates of non-Bayes classification rules and the robustness of Fisher's linear discriminant function. *Biometrika*, 79(1):177–184.

Orr, M.J.L. (1995) Regularisation in the selection of radial basis function centers. *Neural Computation*, 7:606–623.

Osuna, E., Freund, R. and Girosi, F. (1997) Training support vector machines: an application to face detection. In *Proceedings of 1997 IEEE Computer Society Conference on Computer Vision and Pattern Recognition*, pp. 130–136, IEEE Computer Society Press, Los Alamitos, CA.

Overall, J.E. and Magee, K.N. (1992) Replication as a rule for determining the number of clusters in a hierarchical cluster analysis. *Applied Psychological Measurement*, 16:119–128.

Pal, N.R. and Bezdek, J.C. (1995) On cluster validity for the fuzzy *c*-means model. *IEEE Transactions on Fuzzy Systems*, 3(3):370–379.

Pao, Y.-H. (1989) *Adaptive Pattern Recognition and Neural Networks*. Addison-Wesley, Reading, MA.

Park, J. and Sandberg, I.W. (1993) Approximation and radial-basis-function networks. *Neural Computation*, 5:305–316.

Parzen, E. (1962) On estimation of a probability density function and mode. *Annals of Mathematical Statistics*, 33:1065–1076.

Pawlak, M. (1993) Kernel classification rules from missing data. *IEEE Transactions on Information Theory*, 39(3):979–988.

Pearl, J. (1988) *Probabilistic Reasoning in Intelligent Systems: Networks of Plausible Inference*. Morgan Kaufmann Publishers, San Mateo, CA.

Pearson, K. (1901) On lines and planes of closest fit to systems of points in space. *Philosophical Magazine*, 2:559–572.

Pedersen, F., Bergström, M., Bengtsson, E. and Langström, B. (1994) Principal component analysis of dynamic positron emission tomography images. *European Journal of Nuclear Medicine*, 21(12):1285–1292.

Peel, D. and McLachlan, G.J. (2000) Robust mixture modelling using the *t* distribution. *Statistics and Computing*, 10(4):339–348.

Pinkowski, B. (1997) Principal component analysis of speech spectrogram images. *Pattern Recognition*, 30(5):777–787.

Platt, J. (1998) Fast training of support vector machines using sequential minimal optimisation. In B. Schölkopf, C.J.C. Burges and A.J. Smola, eds, *Advances in Kernel Methods: Support Vector Learning*, pp. 185–208. MIT Press, Cambridge, MA.

Powell, G.E., Clark, E. and Bailey, S. (1979) Categories of aphasia: a cluster-analysis of Schuell test profiles. *British Journal of Disorders of Communication*, 14(2):111–122.

Powell, M.J.D. (1987) Radial basis functions for multivariable interpolation: a review. In J.C. Mason and M.G. Cox, eds, *Algorithms for Approximation*, pp. 143–167. Clarendon Press, Oxford.

Prabhakar, S. and Jain, A.K. (2002) Decision-level fusion in fingerprint verification. *Pattern Recognition*, 35:861–874.

Prakash, M. and Murty, M.N. (1995) A genetic approach for selection of (near-)optimal subsets of principal components for discrimination. *Pattern Recognition Letters*, 16:781–787.

Press, S.J. and Wilson, S. (1978) Choosing between logistic regression and discriminant analysis. *Journal of the American Statistical Association*, 73:699–705.

Press, W.H., Flannery, B.P., Teukolsky, S.A. and Vetterling, W.T. (1992) *Numerical Recipes. The Art of Scientific Computing*, 2nd edn., Cambridge University Press, Cambridge.

Provost, F. and Fawcett, T. (2001) Robust classification for imprecise environments. *Machine Learning*, 42:203–231.

Psaltis, D., Snapp, R.R. and Venkatesh, S.S. (1994) On the finite sample performance of the nearest neighbor classifier. *IEEE Transactions on Information Theory*, 40(3):820–837.

Pudil, P., Ferri, F.J., Novovičová, J. and Kittler, J. (1994a) Floating search methods for feature selection with nonmonotonic criterion functions. In *Proceedings of the International Conference on Pattern Recognition*, vol. 2, pp. 279–283, IEEE, Los Alamitos, CA.

Pudil, P., Novovičová, J. and Kittler, J. (1994b) Floating search methods in feature selection. *Pattern Recognition Letters*, 15:1119–1125.

Pudil, P., Novovičová, J. and Kittler, J. (1994c) Simultaneous learning of decision rules and important attributes for classification problems in image analysis. *Image and Vision Computing*, 12(3):193–198.

Pudil, P., Novovičová, J., Choakjarernwanit, N. and Kittler, J. (1995) Feature selection based on the approximation of class densities by finite mixtures of special type. *Pattern Recognition*, 28(9):1389–1398.

Quenouille, M.H. (1949) Approximate tests of correlation in time series. *Journal of the Royal Statistical Society Series B*, 11:68–84.

Quinlan, J.R. (1986) Induction of decision trees. *Machine Learning*, 1(1):81–106.

Quinlan, J.R. (1987) Simplifying decision trees. *International Journal of Man–Machine Studies*, 27:221–234.

Quinlan, J.R. and Rivest, R.L. (1989) Inferring decision trees using the minimum description length principle. *Information and Computation*, 80:227–248.

Rabiner, L.R., Juang, B.-H., Levinson, S.E. and Sondhi, M.M. (1985) Recognition of isolated digits using hidden Markov models with continuous mixture densities. *AT&T Technical Journal*, 64(4):1211–1234.

Raftery, A.E. and Lewis, S.M. (1996) Implementing MCMC. In W.R. Gilks, S. Richardson and D.J. Spiegelhalter, eds, *Markov Chain Monte Carlo in Practice*, pp. 115–130. Chapman & Hall, London.

Raju, S. and Sarma, V.V.S. (1991) Multisensor data fusion and decision support for airborne target identification. *IEEE Transactions on Systems, Man, and Cybernetics*, 21(5).

Ramasubramanian, V. and Paliwal, K.K. (2000) Fast nearest-neighbor search algorithms based on approximation-elimination search. *Pattern Recognition*, 33:1497–1510.

Ramaswamy, S., Tamayo, P., Rifkin, R., Mukherjee, S., Yeang, C.-H., Angelo, M., Ladd, C., Reich, M., Latulippe, E., Mesirov, J.P., Poggio, T., Gerald, W., Loda, M., Lander, E.S. and Golub, T.R. (2001) Multiclass cancer diagnosis using tumor gene expression signatures. *Proceedings of the National Academy of Sciences of the USA*, 98(26):15149–15154.

Ramsay, J.O. and Dalzell, C.J. (1991) Some tools for functional data analysis (with discussion). *Journal of the Royal Statistical Society Series B*, 53:539–572.

Ratcliffe, M.B., Gupta, K.B., Streicher, J.T., Savage, E.B., Bogen, D.K. and Edmunds, L.H. (1995) Use of sonomicrometry and multidimensional scaling to determine the three-dimensional coordinates of multiple cardiac locations: feasibility and initial implementation. *IEEE Transactions on Biomedical Engineering*, 42(6):587–598.

Raudys, S.J. (2000) Scaled rotation regularisation. *Pattern Recognition*, 33:1989–1998.

Rayens, W. and Greene, T. (1991) Covariance pooling and stabilization for classification. *Computational Statistics and Data Analysis*, 11:17–42.

Redner, R.A. and Walker, H.F. (1984) Mixture densities, maximum likelihood and the EM algorithm. *SIAM Review*, 26(2):195–239.

Reed, R. (1993) Pruning algorithms – a survey. *IEEE Transactions on Neural Networks*, 4(5):740–747.

Refenes, A.-P.N., Burgess, A.N. and Bentz, Y. (1997) Neural networks in financial engineering: a study in methodology. *IEEE Transactions on Neural Networks*, 8(6):1222–1267.

Remme, J., Habbema, J.D.F. and Hermans, J. (1980) A simulative comparison of linear, quadratic and kernel discrimination. *Journal of Statistical Computation and Simulation*, 11:87–106.

Revow, M., Williams, C.K.I. and Hinton, G.E. (1996) Using generative models for handwritten digit recognition. *IEEE Transactions on Pattern Analysis and Machine Intelligence*, 18(6):592–606.

Reyment, R.A., Blackith, R.E. and Campbell, N.A. (1984) *Multivariate Morphometrics*, 2nd edn., Academic Press, New York.

Rice, J.A. and Silverman, B.W. (1991) Estimating the mean and covariance structure nonparametrically when the data are curves. *Journal of the Royal Statistical Society Series B*, 53:233–243.

Rice, J.C. (1993) Forecasting abundance from habitat measures using nonparametric density estimation methods. *Canadian Journal of Fisheries and Aquatic Sciences*, 50:1690–1698.

Richardson, S. and Green, P.J. (1997) On Bayesian analysis of mixtures with an unknown number of components (with discussion). *Journal of the Royal Statistical Society Series B*, 59(4):731–792.

Ripley, B.D. (1987) *Stochastic Simulation.* Wiley, New York.

Ripley, B.D. (1994) Neural and related methods of classification. *Journal of the Royal Statistical Society Series B*, 56(3).

Ripley, B.D. (1996) *Pattern Recognition and Neural Networks.* Cambridge University Press, Cambridge.

Riskin, E.A. and Gray, R.M. (1991) A greedy tree growing algorithm for the design of variable rate vector quantizers. *IEEE Transactions on Signal Processing.*

Ritter, H., Martinetz, T. and Schulten, K. (1992) *Neural Computation and Self-Organizing Maps: An Introduction.* Addison-Wesley, Reading, MA.

Roberts, G.O. (1996) Markov chain concepts related to sampling algorithms. In W.R. Gilks, S. Richardson and D.J. Spiegelhalter, eds, *Markov Chain Monte Carlo in Practice*, pp. 45–57. Chapman & Hall, London.

Roberts, S. and Tarassenko, L. (1992) Analysis of the sleep EEG using a multilayer network with spatial organisation. *IEE Proceedings Part F*, 139(6):420–425.

Rocke, D.M. and Woodruff, D.L. (1997) Robust estimation of multivariate location and shape. *Journal of Statistical Planning and Inference*, 57:245–255.

Rogers, S.K., Colombi, J.M., Martin, C.E., Gainey, J.C., Fielding, K.H., Burns, T.J., Ruck, D.W., Kabrisky, M. and Oxley, M. (1995) Neural networks for automatic target recognition. *Neural Networks*, 8(7/8):1153–1184.

Rohlf, F.J. (1982) Single-link clustering algorithms. In P.R. Krishnaiah and L.N. Kanal, eds, *Handbook of Statistics*, vol. 2, pp. 267–284. North Holland, Amsterdam.

Rosenblatt, M. (1956) Remarks on some nonparametric estimates of a density function. *Annals of Mathematical Statistics*, 27:832–835.

Roth, M.W. (1990) Survey of neural network technology for automatic target recognition. *IEEE Transactions on Neural Networks*, 1(1):28–43.

Rousseeuw, P.J. (1985) Multivariate estimation with high breakdown point. In W. Grossmann, G. Pflug, I. Vincze and W. Wertz, eds, *Mathematical Statistics and Applications*, pp. 283–297. Reidel, Dordrecht.

Rumelhart, D.E., Hinton, G.E. and Williams, R.J. (1986) Learning internal representation by error propagation. In D.E. Rumelhart, J.L. McClelland and the PDP Research Group, eds, *Parallel Distributed Processing: Explorations in the Microstructure of Cognition*, vol. 1, pp. 318–362. MIT Press, Cambridge, MA.

Safavian, S.R. and Landgrebe, D.A. (1991) A survey of decision tree classifier methodology. *IEEE Transactions on Systems, Man, and Cybernetics*, 21(3):660–674.

Samal, A. and Iyengar, P.A. (1992) Automatic recognition and analysis of human faces and facial expressions: a survey. *Pattern Recognition*, 25:65–77.

Sammon, J.W. (1969) A nonlinear mapping for data structure analysis. *IEEE Transactions on Computers*, 18(5):401–409.

Sankar, A. and Mammone, R.J. (1991) Combining neural networks and decision trees. In S.K. Rogers, ed., *Applications of Neural Networks II*, vol. 1469, pp. 374–383. SPIE.

Saranli, A. and Demirekler, M. (2001) A statistical framework for rank-based multiple classifier decision fusion. *Pattern Recognition*, 34:865–884.

Schaffer, C. (1993) Selecting a classification method by cross-validation. *Machine Learning*, 13:135–143.

Schalkoff, R. (1992) *Pattern Recognition. Statistical Structural and Neural.* Wiley, New York.

Schapire, R.E. (1990) The strength of weak learnability. *Machine Learning*, 5(2):197–227.

Schapire, R.E. and Singer, Y. (1999) Improved boosting algorithms using confidence-rated predictions. *Machine Learning*, 37:297–336.

Schiffman, S.S., Reynolds, M.L. and Young, F.W. (1981) *An Introduction to Multidimensional Scaling*. Academic Press, New York.

Schölkopf, B. and Smola, A.J. (2001) *Learning with Kernels. Support Vector Machines, Regularization, Optimization and Beyond.* MIT Press, Cambridge, MA.

Schölkopf, B., Sung, K.-K., Burges, C.J.C., Girosi, F., Niyogi, P., Poggio, T. and Vapnik, V. (1997) Comparing support vector machines with Gaussian kernels to radial basis function classifiers. *IEEE Transactions on Signal Processing*, 45(11):2758–2765.

Schölkopf, B., Smola, A.J. and Müller, K. (1999) Kernel principal component analysis. In B. Schölkopf, C.J.C. Burges and A.J. Smola, eds, *Advances in Kernel Methods – Support Vector Learning.*, pp. 327–352. MIT Press, Cambridge, MA.

Schölkopf, B., Smola, A.J., Williamson, R.C. and Bartlett, P.L. (2000) New support vector algorithms. *Neural Computation*, 12:1207–1245.

Schott, J.R. (1993) Dimensionality reduction in quadratic discriminant analysis. *Computational Statistics and Data Analysis*, 16:161–174.

Schwenker, F., Kestler, H.A. and Palm, G. (2001) Three learning phases for radial-basis-function networks. *Neural Networks*, 14:439–458.

Sclove, S.L. (1987) Application of model selection criteria to some problems in multivariate analysis. *Psychometrika*, 52(3):333–343.

Scott, D.W. (1992) *Multivariate Density Estimation. Theory, Practice and Visualization.* Wiley, New York.

Scott, D.W., Gotto, A.M., Cole, J.S. and Gorry, G.A. (1978) Plasma lipids as collateral risk factors in coronary artery disease – a study of 371 males with chest pains. *Journal of Chronic Diseases*, 31:337–345.

Sebestyen, G. and Edie, J. (1966) An algorithm for non-parametric pattern recognition. *IEEE Transactions on Electronic Computers*, 15(6):908–915.

Selim, S.Z. and Al-Sultan, K.S. (1991) A simulated annealing algorithm for the clustering problem. *Pattern Recognition*, 24(10):1003–1008.

Selim, S.Z. and Ismail, M.A. (1984a) K-means-type algorithms: a generalized convergence theorem and characterization of local optimality. *IEEE Transactions on Pattern Analysis and Machine Intelligence*, 6(1):81–87.

Selim, S.Z. and Ismail, M.A. (1984b) Soft clustering of multidimensional data: a semi-fuzzy approach. *Pattern Recognition*, 17(5):559–568.

Selim, S.Z. and Ismail, M.A. (1986) On the local optimality of the fuzzy isodata clustering algorithm. *IEEE Transactions on Pattern Analysis and Machine Intelligence*, 8(2): 284–288.

Sephton, P.S. (1994) Cointegration tests on MARS. *Computational Economics*, 7:23–35.

Serpico, S.B., Bruzzone, L. and Roli, F. (1996) An experimental comparison of neural and statistical non-parametric algorithms for supervised classification of remote-sensing images. *Pattern Recognition Letters*, 17:1331–1341.

Sethi, I.K. and Yoo, J.H. (1994) Design of multicategory multifeature split decision trees using perceptron learning. *Pattern Recognition*, 27(7):939–947.

Sharkey, A.J.C. (1999) Multi-net systems. In A.J.C. Sharkey, ed., *Combining Artificial Neural Nets. Ensemble and Modular Multi-net Systems*, pp. 1–30. Springer-Verlag, Berlin.

Sharkey, A.J.C., Chandroth, G.O. and Sharkey, N.E. (2000) A multi-net system for the fault diagnosis of a diesel engine. *Neural Computing and Applications*, 9:152–160.

Shavlik, J.W., Mooney, R.J. and Towell, G.G. (1991) Symbolic and neural learning algorithms: an experimental comparison. *Machine Learning*, 6:111–143.

Sheather, S.J. and Jones, M.C. (1991) A reliable data-based bandwidth selection method for kernel density estimation. *Journal of the Royal Statistical Society Series B*, 53:683–690.

Sibson, R. (1973) Slink: an optimally efficient algorithm for the single-link cluster method. *Computer Journal*, 16(1):30–34.

Siedlecki, W., Siedlecka, K. and Sklansky, J. (1988) An overview of mapping techniques for exploratory pattern analysis. *Pattern Recognition*, 21(5):411–429.

Siedlecki, W. and Sklansky, J. (1988) On automatic feature selection. *International Journal of Pattern Recognition and Artificial Intelligence*, 2(2):197–220.

Silverman, B.W. (1982) Kernel density estimation using the fast Fourier transform. *Applied Statistics*, 31:93–99.

Silverman, B.W. (1986) *Density Estimation for Statistics and Data Analysis.* Chapman & Hall, London.

Silverman, B.W. (1995) Incorporating parametric effects into functional principal components analysis. *Journal of the Royal Statistical Society Series B*, 57(4):673–689.

Simpson, P., ed., (1992) Special issue on 'Neural Networks for Oceanic Engineering'. *IEEE Journal of Oceanic Engineering*.

Skurichina, M. (2001) *Stabilizing Weak Classifiers*. Technical University of Delft, Delft.

Smith, S.J., Bourgoin, M.O., Sims, K. and Voorhees, H.L. (1994) Handwritten character classification using nearest neighbour in large databases. *IEEE Transactions on Pattern Analysis and Machine Intelligence*, 16(9):915–919.

Smyth, P. and Wolpert (1999) Linearly combining density estimators via stacking. *Machine Learning*, 36:59–83.

Sneath, P.H.A. and Sokal, R.R. (1973) *Numerical Taxonomy*. Freeman, San Francisco.

Somol, P., Pudil, P., Novovičová, J. and Paclík (1999) Adaptive floating search methods in feature selection. *Pattern Recognition Letters*, 20:1157–1163.

Sorsa, T., Koivo, H.N. and Koivisto, H. (1991) Neural networks in process fault diagnosis. *IEEE Transactions on Systems, Man, and Cybernetics*, 21(4):815–825.

Späth, H. (1980) *Cluster Analysis Algorithms for Data Reduction and Classification of Objects.* Ellis Horwood Limited, Hemel Hempstead.

Spiegelhalter, D.J., Dawid, A.P., Hutchinson, T.A. and Cowell, R.G. (1991) Probabilistic expert systems and graphical modelling: a case study in drug safety. *Philosophical Transactions of the Royal Society of London*, 337:387–405.

Spragins, J. (1976) A note on the iterative applications of Bayes' rule. *IEEE Transactions on Information Theory*, 11:544–549.

Sridhar, D.V., Seagrave, R.C. and Bartlett, E.B. (1996) Process modeling using stacked neural networks. *Process Systems Engineering*, 42(9):387–405.

Sridhar, D.V., Bartlett, E.B. and Seagrave, R.C. (1999) An information theoretic approach for combining neural network process models. *Neural Networks*, 12:915–926.

Stäger, F. and Agarwal, M. (1997) Three methods to speed up the training of feedforward and feedback perceptrons. *Neural Networks*, 10(8):1435–1443.

Stassopoulou, A., Petrou, M. and Kittler, J. (1996) Bayesian and neural networks for geographic information processing. *Pattern Recognition Letters*, 17:1325–1330.

Stearns, S.D. (1976) On selecting features for pattern classifiers. In *Proceedings of the 3rd International Joint Conference on Pattern Recognition*, pp. 71–75, IEEE.

Stevenson, J. (1993) Multivariate statistics VI. The place of discriminant function analysis in psychiatric research. *Nordic Journal of Psychiatry*, 47(2):109–122.

Stewart, C., Lu, Y.-C. and Larson, V. (1994) A neural clustering approach for high resolution radar target classification. *Pattern Recognition*, 27(4):503–513.

Stewart, G.W. (1973) *Introduction to Matrix Computation.* Academic Press, Orlando, FL.

Stone, C., Hansen, M., Kooperberg, C. and Truong, Y. (1997) Polynomial splines and their tensor products (with discussion). *Annals of Statistics*, 25(4):1371–1470.

Stone, M. (1974) Cross-validatory choice and assessment of statistical predictions. *Journal of the Royal Statistical Society Series B*, 36:111–147.

Stuart, A. and Ord, J.K. (1991) *Kendall's Advanced Theory of Statistics*, vol. 2. Edward Arnold, London, fifth edition.

Sturt, E. (1981) An algorithm to construct a discriminant function in Fortran for categorical data. *Applied Statistics*, 30:313–325.

Sumpter, R.G., Getino, C. and Noid, D.W. (1994) Theory and applications of neural computing in chemical science. *Annual Reviews of Physical Chemistry*, 45:439–481.

Sutton, B.D. and Steck, G.J. (1994) Discrimination of Caribbean and Mediterranean fruit fly larvae (diptera: tephritidae) by cuticular hydrocarbon analysis. *Florida Entomologist*, 77(2):231–237.

Tarassenko, L. (1998) *A Guide to Neural Computing Applications*. Arnold, London.

Tax, D.M.J., van Breukelen, M., Duin, R.P.W. and Kittler, J. (2000) Combining multiple classifiers by averaging or multiplying? *Pattern Recognition*, 33:1475–1485.

Terrell, G.R. and Scott, D.W. (1992) Variable kernel density estimation. *Annals of Statistics*, 20(3):1236–1265.

Therrien, C.W. (1989) *Decision, Estimation and Classification. An Introduction to Pattern Recognition and Related Topics*. Wiley, New York.

Thisted, R.A. (1988) *Elements of Statistical Computing. Numerical Computation*. Chapman & Hall, New York.

Thodberg, H.H. (1996) A review of Bayesian neural networks with application to near infrared spectroscopy. *IEEE Transactions on Neural Networks*, 7(1):56–72.

Tian, Q., Fainman, Y. and Lee, S.H. (1988) Comparison of statistical pattern-recognition algorithms for hybrid processing. II. Eigenvector-based algorithm. *Journal of the Optical Society of America A*, 5(10):1670–1682.

Tibshirani, R.J. (1992) Principal curves revisited. *Statistics and Computing*, 2(4):183–190.

Tierney, L. (1994) Markov chains for exploring posterior distributions. *Annals of Statistics*, 22(4):1701–1762.

Titterington, D.M. (1980) A comparative study of kernel-based density estimates for categorical data. *Technometrics*, 22(2):259–268.

Titterington, D.M. and Mill, G.M. (1983) Kernel-based density estimates from incomplete data. *Journal of the Royal Statistical Society Series B*, 45(2):258–266.

Titterington, D.M., Murray, G.D., Murray, L.S., Spiegelhalter, D.J., Skene, A.M., Habbema, J.D.F. and Gelpke, G.J. (1981) Comparison of discrimination techniques applied to a complex data set of head injured patients (with discussion). *Journal of the Royal Statistical Society Series A*, 144(2):145–175.

Titterington, D.M., Smith, A.F.M. and Makov, U.E. (1985) *Statistical Analysis of Finite Mixture Distributions*. Wiley, New York.

Todeschini, R. (1989) *k*-nearest neighbour method: the influence of data transformations and metrics. *Chemometrics and Intelligent Laboratory Systems*, 6:213–220.

Todorov, V., Neykov, N. and Neytchev, P. (1994) Robust two-group discrimination by bounded influence regression. A Monte Carlo simulation. *Computational Statistics and Data Analysis*, 17:289–302.

Tou, J.T. and Gonzales, R.C. (1974) *Pattern Recognition Principles*. Addison-Wesley, New York.

Toussaint, G.T. (1974) Bibliography on estimation of misclassification. *IEEE Transactions on Information Theory*, 20(4):472–479.

Tukey, J.W. (1977) *Exploratory Data Analysis*. Addison-Wesley, Reading, MA.

Turkkan, N. and Pham-Gia, T. (1993) Computation of the highest posterior density interval in Bayesian analysis. *J. Statistical Computation and Simulation*, 44:243–250.

Unbehauen, R. and Luo, F.L., eds (1998) Special issue on 'Neural Networks'. *Signal Processing*, 64.

Valentin, D., Abdi, H., O'Toole, A.J. and Cottrell, G.W. (1994) Connectionist models of face processing: a survey. *Pattern Recognition*, 27(9):1209–1230.

Valiveti, R.S. and Oommen, B.J. (1992) On using the chi-squared metric for determining stochastic dependence. *Pattern Recognition*, 25(11):1389–1400.

Valiveti, R.S. and Oommen, B.J. (1993) Determining stochastic dependence for normally distributed vectors using the chi-squared metric. *Pattern Recognition*, 26(6):975–987.

van der Heiden, R. and Groen, F.C.A. (1997) The Box–Cox metric for nearest neighbour classification improvement. *Pattern Recognition*, 30(2):273–279.

van der Smagt, P.P. (1994) Minimisation methods for training feedforward networks. *Neural Networks*, 7(1):1–11.

van Gestel, T., Suykens, J.A.K., Baestaens, D.-E., Lambrechts, A., Lanckriet, G., Vandaele, B.V., De Moor, B. and Vandewalle, J. (2001) Financial time series prediction using least squares support vector machines within the evidence framework. *IEEE Transactions on Neural Networks*, 12(4):809–821.

Vapnik, V.N. (1998) *Statistical Learning Theory.* Wiley, New York.

Varshney, P.K. (1997) *Distributed Detection and Data Fusion.* Springer-Verlag, New York.

Venkateswarlu, N.B. and Raju, P.S.V.S.K. (1992) Fast isodata clustering algorithms. *Pattern Recognition*, 25(3):335–345.

Vidal, E. (1986) An algorithm for finding nearest neighbours in (approximately) constant average time. *Pattern Recognition Letters*, 4(3):145–157.

Vidal, E. (1994) New formulation and improvements of the nearest-neighbour approximating and eliminating search algorithm (AESA). *Pattern Recognition Letters*, 15:1–7.

Vio, R., Fasano, G., Lazzarin, M. and Lessi, O. (1994) Probability density estimation in astronomy. *Astronomy and Astrophysics*, 289:640–648.

Viswanathan, R. and Varshney, P.K. (1997) Distributed detection with multiple sensors: Part 1 – fundamentals. *Proceedings of the IEEE*, 85(1):54–63.

Vivarelli, F. and Williams, C.K.I. (2001) Comparing Bayesian neural network algorithms for classifying segmented outdoor images. *Neural Networks*, 14:427–437.

von Stein, J.H. and Ziegler, W. (1984) The prognosis and surveillance of risks from commercial credit borrowers. *Journal of Banking and Finance*, 8:249–268.

Wahl, P.W. and Kronmal, R.A. (1977) Discriminant functions when covariances are unequal and sample sizes are moderate. *Biometrics*, 33:479–484.

Waltz, E. and Llinas, J. (1990) *Multisensor Data Fusion.* Artech House, Boston.

Wand, M.P. and Jones, M.C. (1994) Multivariate plug-in bandwidth selection. *Computational Statistics*, 9:97–116.

Wand, M.P. and Jones, M.C. (1995) *Kernel Smoothing.* Chapman & Hall, London.

Ward, J.H. (1963) Hierarchical grouping to optimise an objective function. *Journal of the American Statistical Association*, 58:236–244.

Watanabe, S. (1985) *Pattern Recognition: Human and Mechanical.* Wiley, New York.

Webb, A.R. (1994) Functional approximation in feed-forward networks: A least-squares approach to generalisation. *IEEE Transactions on Neural Networks*, 5(3):363–371.

Webb, A.R. (1995) Multidimensional scaling by iterative majorisation using radial basis functions. *Pattern Recognition*, 28(5):753–759.

Webb, A.R. (1996) An approach to nonlinear principal components analysis using radially-symmetric kernel functions. *Statistics and Computing*, 6:159–168.

Webb, A.R. (2000) Gamma mixture models for target recognition. *Pattern Recognition*, 33:2045–2054.

Webb, A.R. and Garner, P.N. (1999) A basis function approach to position estimation using microwave arrays. *Applied Statistics*, 48(2):197–209.

Webb, A.R. and Lowe, D. (1988) A hybrid optimisation strategy for feed-forward adaptive layered networks. DRA memo 4193, DERA, St Andrews Road, Malvern, Worcs, WR14 3PS.

Webb, A.R., Lowe, D. and Bedworth, M.D. (1988) A comparison of nonlinear optimisation strategies for feed-forward adaptive layered networks. DRA Memo 4157, DERA, St Andrews Road, Malvern, Worcs, WR14 3PS.

Wee, W.G. (1968) Generalized inverse approach to adaptive multiclass pattern recognition. *IEEE Transactions on Computers*, 17(12):1157–1164.

West, M. (1992) Modelling with mixtures. In J.M. Bernardo, J.O. Berger, A.P. Dawid and A.F.M. Smith, eds, *Bayesian Statistics 4*, pp. 503–524. Oxford University Press, Oxford.

Weymaere, N. and Martens, J.-P. (1994) On the initialization and optimization of multilayer perceptrons. *IEEE Transactions on Neural Networks*, 5(5):738–751.

Whitney, A.W. (1971) A direct method of nonparametric measurement selection. *IEEE Transactions on Computers*, 20:1100–1103.

Wilkinson, L. (1992) Graphical displays. *Statistical Methods in Medical Research*, 1(1):3–25.

Williams, C.K.I. and Feng, X. (1998). Combining neural networks and belief networks for image segmentation. In T. Constantinides, S.-Y. Kung, M. Niranjan and E. Wilson, eds, *Neural Networks for Signal Processing VIII*, IEEE, New York.

Williams, W.T., Lance, G.N., Dale, M.B. and Clifford, H.T. (1971) Controversy concerning the criteria for taxonomic strategies. *Computer Journal*, 14:162–165.

Wilson, D. (1972) Asymptotic properties of NN rules using edited data. *IEEE Transactions on Systems, Man, and Cybernetics*, 2(3):408–421.

Wolfe, J.H. (1971) A Monte Carlo study of the sampling distribution of the likelihood ratio for mixtures of multinormal distributions. Technical Bulletin STB 72–2, Naval Personnel and Training Research Laboratory, San Diego, CA.

Wolpert, D.H. (1992) Stacked generalization. *Neural Networks*, 5(2):241–260.

Wong, S.K.M. and Poon, F.C.S. (1989) Comments on 'Approximating discrete probability distributions with dependence trees'. *IEEE Transactions on Pattern Analysis and Machine Intelligence*, 11(3):333–335.

Woods, K., Kegelmeyer, W.P. and Bowyer, K. (1997) Combination of multiple classifiers using local accuracy estimates. *IEEE Transactions on Pattern Analysis and Machine Intelligence*, 19(4):405–410.

Wray, J. and Green, G.G.R. (1995) Neural networks, approximation theory, and finite precision computation. *Neural Networks*, 8(1):31–37.

Wu, X. and Zhang, K. (1991) A better tree-structured vector quantizer. In J.A. Storer and J.H. Reif, eds, *Proceedings Data Compression Conference*, pp. 392–401. IEEE Computer Society Press, Los Alamitos, CA.

Wylie, C.R. and Barrett, L.C. (1995) *Advanced Engineering Mathematics*, 6th edn., McGraw-Hill, New York.

Yan, H. (1994) Handwritten digit recognition using an optimised nearest neighbor classifier. *Pattern Recognition Letters*, 15:207–211.

Yang, M.-S. (1993) A survey of fuzzy clustering. *Mathematical and Computer Modelling*, 18(11):1–16.

Yasdi, R., ed., (2000) Special issue on 'Neural Computing in Human–Computer Interaction'. *Neural Computing and Applications*, 9(4).

Young, T.Y. and Calvert, T.W. (1974) *Classification, Estimation and Pattern Recognition*. Elselvier, New York.

Zentgraf, R. (1975) A note on Lancaster's definition of higher-order interactions. *Biometrika*, 62(2):375–378.

Zhang, G.P. (2000) Neural networks for classification: a survey. *IEEE Transactions on Systems, Man, and Cybernetics – Part C: Applications and Reviews*, 30(4):451–462.

Zhang, P. (1993) Model selection via multifold cross validation. *Annals of Statistics*, 21(1):299–313.

Zhang, Y., de Silva, C.J.S., Togneri, R., Alder, M. and Attikiouzel, Y. (1994) Speaker-independent isolated word recognition using multiple hidden Markov models. *IEEE Proceedings on Vision, Image and Signal Processing*, 141(3):197–202.

Zhao, Q., Principe, J.C., Brennan, V.L., Xu, D. and Wang, Z. (2000) Synthetic aperture radar automatic target recognition with three strategies of learning and representation. *Optical Engineering*, 39(5):1230–1244.

Zhao, Y. and Atkeson, C.G. (1996) Implementing projection pursuit learning. *IEEE Transactions on Neural Networks*, 7(2):362–373.

Zois, E.N. and Anastassopoulos, V. (2001) Fusion of correlated decisions for writer verification. *Pattern Recognition*, 34:47–61.

Zongker, D. and Jain, A.K. (1996) Algorithms for feature selection: an evaluation. In *Proceedings of the International Conference on Pattern Recognition*, pp. 18–22, Vienna, IEEE Computer Society Press, Los Alamitos, CA.

Zupan, J. (1982) *Clustering of Large Data Sets*. Research Studies Press, Letchworth.

Index